SEISMIC BEHAVIOUR OF GROUND AND GEOTECHNICAL STRUCTURES

PROCEEDINGS OF DISCUSSION SPECIAL TECHNICAL SESSION ON EARTHQUAKE GEOTECHNICAL ENGINEERING DURING FOURTEENTH INTERNATIONAL CONFERENCE ON SOIL MECHANICS AND FOUNDATION ENGINEERING HAMBURG/GERMANY/6-12 SEPTEMBER/1997

Seismic Behaviour of Ground and Geotechnical Structures

Edited by
Pedro S. Sêco e Pinto
Portuguese Society for Geotechnique, Lisbon, Portugal

A.A. BALKEMA / ROTTERDAM / BROOKFIELD / 1997

Cover: Some consequences of the catastrophe of 1755, in an architectural context obviously alien to Portugal in a print of unknown origin (from The Lisbon earthquake of 1755; British accounts).

The texts of the various papers in this volume were set individually by typists under the supervision of each of the authors concerned.

Published by
A.A. Balkema, P.O. Box 1675, 3000 BR Rotterdam, Netherlands (Fax: +31.10.4135947)
A.A. Balkema Publishers, Old Post Road, Brookfield, VT 05036-9704, USA (Fax: +1.802.2763837)

ISBN 90 5410 887 8
© 1997 A.A. Balkema, Rotterdam
Printed in the Netherlands

Seismic Behaviour of Ground and Geotechnical Structures, Sêco e Pinto (ed.) © 1997 Balkema, Rotterdam, ISBN 90 5410 887 8

Table of contents

2 Seismic behaviour of geotechnical structures

3 *Miscellaneous*

Seismic Behaviour of Ground and Geotechnical Structures, Sêco e Pinto (ed.)© 1997 Balkema, Rotterdam, ISBN 90 5410 887 8

Preface

The Technical Committee n°4 (TC4) Earthquake Geotechnical Engineering under the auspices of ISSMFE began its activity in 1985 and during the last two tenues 1985-1989 and 1989-1993 has developed its activities under the leadership of Professor Kenji Ishihara.

The very successful activity of TC4 during this period includes the publication of the following special volumes and manual:
- Earthquake Geotechnical Engineering (1989);
- Performance of Ground and Soil Structures during Earthquakes (1993);
- Manual for Zonation of Seismic Geotechnical Hazards (1993).

For the period 1994-1997 the activities of TC4, under the leadership of Professor Pedro Sêco e Pinto, have begun on October 1994 with contacts with core members and members appointed by National Geotechnical Societies.

The final list was ready on December 1994 and includes 36 members from the following countries: Belgium, Bulgaria, Canada, Chile, Colombia, Costa Rica, Denmark, Ecuador, France, Greece, India, Iran, Israel, Italy, Japan, Norway, Pakistan, Peru, Portugal, Russia, Spain, Sweden, Taiwan, Turkey, United Kingdom, USA and Venezuela.

The following topics have been discussed: 1. Applications of the Manual for Zonation, 2. Comments related with Eurocode 8 – Part 5: Foundations, retaining structures and geotechnical aspects.

The following TC4 meetings took place on April 4, 1995 in St. Louis, USA, on May 31, 1995 in Copenhagen, Denmark, on November 15, 1995 in Tokyo, Japan, on June 15, 1996, in Acapulco, Mexico.

Active participation to divulge the Manual for Zonation in Seismic Geotechnical Hazards has been done in the following events:
- 3rd National Earthquake Engineering Conference in Turkey, 1995;
- Short Course on Geotechnical Aspects on Earthquake Engineering Pile Dynamics in Bangkok, 1995;
- 2nd Coloquio International sobre Microzonification Sísmica, in Venezuela, 1995;
- Introduction of the Manual of Zonation in Nice, October 1995, during the Fifth International Conference on Seismic Zonation.

Promotion of the following Conferences, Workshops and Special Sessions:
- IS Tokyo 95 'First International Conference on Earthquake Geotechnical Engineering', November 1995, Tokyo;
- Special Session on 'Lessons Learned from Northridge and Kobe Earthquake' during 11 WCEE in Mexico, June 1996;
- Workshop 'Earthquake Geotechnical Engineering', Roorkee, India, September 1996;

– Workshop 'Dynamic Analysis of Embankment Dams', Lima, October 1996.

Related with the above events the following documents were published:

– II Coloquio Internacional sobre Microzonificacion Sísmica y V Reunion De Cooperation Interamericana. Programa y Resumunes. Informe Funvisis, 1995;

– Earthquake Geotechnical Engineering. Proceedings of IS-Tokyo'95/The first International Conference on Earthquake Geotechnical Engineering. Edited by Kenji Ishihara. Published by A.A.Balkema, November 1995;

– Special Session on 'Lessons Learned from Northridge and Kobe Earthquakes'. A collection of 10 selected papers. Proceedings of 11 WCEE in Mexico, June 1996;

– Design Practices in Earthquake Geotechnical Engineering. Proceedings of the Workshop on Design Practices in Earthquake Geotechnical Engineering, Roorkee. Editors Swami Saran and R.Anbalagan, September 1996.

Joint activity with European Association for Earthquake Engineering and interaction with the International Association for Earthquake Engineering were established.

I would like to express my deep gratitude to the Society and members of TC4 (the names of the members are given below) who have collaborated so enthusiastically on the work of the Technical Committee.

To keep alive the momentum of TC4 this Special Volume was published with papers submitted by well known experts in these fields. This volume, that includes 40 papers from 17 countries of medium and high risk seismic, will be presented and discussed in a Special Technical Session on Earthquake Geotechnical Engineering, during the 14th International Conference on Soil Mechanics and Foundation Engineering in Hamburg, on September 10, 1997. This Session organized by TC4 will provide a forum for researchers and practitioners interested in geotechnical earthquake engineering from all over the world to exchange ideas and to share experiences.

I would like to thank the Portuguese Society for Geotechnique for supporting the activities of TC4 and for sponsoring the publication of this Special Volume.

The financial support given by 'Junta Nacional de Investigação Científica e Tecnológica' and 'Fundação Calouste Gulbenkian' for the publication of this Special Volume is greatly acknowledged.

Last but not least I would like to thank the authors for their cooperation and valuable contributions for this Special Volume.

Professor Pedro Simão Sêco e Pinto
President of the Portuguese Society for Geotechnique
Chairman of the Technical Committee on Earthquake Geotechnical Engineering (ISSMFE)

July 30, 1997

Seismic Behaviour of Ground and Geotechnical Structures, Sêco e Pinto (ed.)© 1997 Balkema, Rotterdam, ISBN 90 5410 887 8

Organization

Members of TC4 for Earthquake Geotechnical Engineering in ISSMFE:

Ansal, A. (Turkey)	Guzman, A.A. (Colombia)	Ortigosa, P. (Chile)
Bard, P.Y. (France)	Hurtado, J.A. (Peru)	Palacio, J. (Ecuador)
Bodare, A. (Sweden)	Idriss, I. (USA)	Pecker, A. (France)
Bouckovalas, B. (Greece)	Ilyichev, V.A. (Russia)	Pinto, P.S.S. (Portugal)
Cheng, H.C. (Taiwan)	Ishihara, K. (Japan)	Pires, J.A. (USA)
Diaz, R.M. (Ecuador)	Justo, J.L. (Spain)	Prakash, S. (USA)
Finn, W.D.L. (Canada)	Kokusho, T. (Japan)	Saeed, I. (Pakistan)
Foged, N. (Denmark)	Maugeri, M. (Italy)	Saran, S. (India)
Frydman, S. (Israel)	Mora, S. (Costa Rica)	Steedman, S. (UK)
Gatmiri, B. (Iran)	Murria, J. (Venezuela)	Van Impe, W.F. (Belgium)
Gazetas, G. (Greece)	Nadim, F. (Norway)	Yasuda, S. (Japan)
Germanov, T. (Bulgaria)	Nordal, S. (Norway)	Youd, T.L. (USA)

1 Microzonation and site effects

Seismic Behaviour of Ground and Geotechnical Structures, Sêco e Pinto (ed.) © 1997 Balkema, Rotterdam, ISBN 90 5410 887 8

A preliminary microzonation study for the town of Dinar

Atilla Ansal, Recep İyisan & Mustafa Özkan
Istanbul Technical University, Faculty of Civil Engineering, Maslak, Turkey

ABSTRACT: A preliminary microzonation study has been conducted with respect to ground shaking intensity for the town of Dinar using Grade-1 and Grade-2 methods recommended in the *Manual for Zonation on Seismic Geotechnical Hazards* based on the available information obtained from the damage survey and the geotechnical investigation. Insitu penetration tests, seismic wave velocity measurements and microtremor studies were performed to determine the variation of the soil profile as well as the characteristics of the soil layers within the town boundaries. The results obtained are compared with the damage distribution observed in Dinar Earthquake that clearly demonstrated the effect of local site conditions and soil amplification arising from the geological and geotechnical factors.

1. INTRODUCTION

An earthquake of magnitude M_s=6.1 took place on October 1, 1995 causing extensive damage in the town of Dinar. Approximately 40 percent of all the buildings have collapsed or were heavily damaged. Dinar is located partly on the hills and partly in a valley extending below the hills. The surficial geology of the hills to the east of the town consists of limestone, marl and schist. The flat zone is covered with alluvium deposit containing alternating layers of loose to medium dense silty sands and soft to medium stiff silty fat clays at some locations of organic nature.

The damage distribution in Dinar clearly indicates the effects of the differences in the geotechnical conditions. The buildings located on rocky hill slopes suffered relatively minor damage while heavy damage occurred in the valley. A detailed damage survey conducted by the General Directorate of Disaster Affairs showed large variations in damage ratios within different districts in the town of Dinar.

Following the earthquake relatively detailed soil investigation was carried out to evaluate the effects of soil conditions as well as to obtain the necessary soil characteristics for repair and reconstruction of partly damaged buildings. In addition a detailed study was conducted by recording microtremors at different districts within the town to evaluate soil amplification characteristics.

An attempt is made in this study to establish preliminary microzonation maps for the town of Dinar using Grade-1 and Grade-2 methods recommended in the *Manual for Zonation on Seismic Geotechnical Hazards* (1993) using available information from the damage surveys and the geotechnical investigation.

2. EARTHQUAKE CHARACTERISTICS

The earthquake sequence that affected Dinar was composed of small to medium size foreshocks, main shock and aftershocks. The foreshocks started on September 26, 1995 and the main shock took place on October 1,1995. Nine strong motion records for earthquakes with $M_L > 4$ were recorded at the Meteorological Station located in the valley on the alluvium deposit between September 26 and October 6, 1995 as listed in Table 1.

Table 1. List of major records of Dinar Earthquake

Date	Time	Epicenter	M_L	Ap.(gal)			R
dd/mm	(gmt)	coordinates		L	T	V	(km)
26/9	14:58	38.01N-30.18E	4.6	106	183	76	6
26/9	15:58	38.01N-30.18E	4.2	54	81	49	6
27/9	14:15	38.10N-30.07E	4.8	87	180	72	8
1/10	15:57	38.13N-30.07E	5.9	282	330	151	10
1/10	18:02	38.14N-29.97E	5	225	126	55	18
1/10	21:14	38.12N-29.95E	4.1	91	172	38	19
3/10	7:38	38.05N-30.00E	4.3	69	146	98	13
5/10	16:15	38.04N-30.15E	4.4	104	129	80	2
6/10	16:16	38.10N-30.10E	4.5	99	168	45	6

3

The epicentre distance, R, varied between 2-19 km while the recorded peak accelerations, Ap, varied between 54 gals to 330 gals in horizontal directions. Acceleration records from these major events are given in Figure 1.

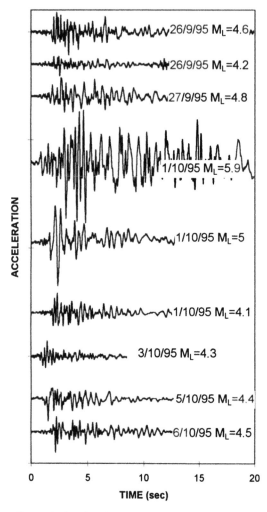

Figure 1. Acceleration records from major events

3. STRUCTURAL DAMAGE

The earthquake caused highly localised damage in Dinar. The building structures in the town centre range from one to five stories. Along the main streets, first stories of buildings are generally occupied for commercial purposes. Buildings with more than three stories are almost all reinforced concrete construction. Buildings with lower number of stories are partly reinforced concrete and mostly

brick masonry. Stone masonry and adobe buildings are very few. Almost all reinforced concrete buildings are of moment resisting frames with hollow brick and occasionally solid brick infill walls. In masonry buildings, load bearing wall are generally made of solid bricks. Prior to the Oct.1 event, the sequence of earthquakes, starting on Sept.26 (M_L=4.6) and continuing with Sept.27 (M_L=4.8) events, had caused light structural and mostly non-structural damage in some buildings in south-western part of the town. During the main shock of October 1, 1995, most of four and five storey reinforced concrete apartment buildings were either heavily damaged or totally collapsed. Some three storey buildings suffered similar damage (Erdik, et al.1995).

The damage ratios calculated based on the detailed damage survey conducted by General Directorate of Disaster Affairs are given in Table 2 for the eight districts in the town of Dinar. In this table D1-D8 are the different districts, T is the total number of damaged buildings, H, M and L are the number of heavily, medium and lightly damaged buildings. DR is the damage ratio calculated as the weighted average for all damaged buildings by assigning weights of 0.6, 0.4 and 0.2 for heavy, medium and light damage, respectively.

Table 2. Damage distribution in different districts

D	T	H	M	L	DR (%)
D1	482	282	106	94	48
D2	807	375	210	22	44
D3	221	32	49	140	30
D4	73	35	11	27	42
D5	221	157	26	38	51
D6	384	267	57	60	51
D7	285	56	174	55	40
D8	147	40	15	92	33

4. GEOTECHNICAL SITE CONDITIONS

In order to determine the variations in local geotechnical site conditions in the town of Dinar, a detailed geotechnical investigation composed of insitu penetration tests, seismic wave velocity measurements, microtremor measurements and laboratory tests was performed. Several borings to the depth of about 30 m were drilled in Dinar. In the most of these boreholes soil stratification consists of alluvium deposit composed of alternating layers of silty clay and clayey sand. The ground water table is almost at the ground surface. A typical soil profile located in flat zone determined based on subsurface borings is given in Figure 2. As can be seen from this figure, soil profile consists of a fill with thickness of about 0.5 m, underlain by a medium to stiff sandy brown coloured clay layer of SPT-N values about 17 with thickness of about 2.5 m, underlain by a fine gravely and silty sandy clay layers of low to medium plasticity with N values between 8 to 17 and with thickness of about 2 and 7 m, respectively. A fine gravely medium to dense sand layer of N value about 14 is located between the depths of 12 to 15 m. Then, sandy, silty clay and clayey silty sand layers are located alternatively, and N values in these layers are relatively high.

In every district several dynamic penetration tests were conducted and the obtained results were converted to equivalent SPT-N blow counts. Laboratory tests performed on soil samples obtained from the observation pits excavated near the footings of the collapsed buildings, indicate that plasticity index of soils encountered on the surface vary between 13% and 46%, and soil groups were generally silty lean clay (CL).

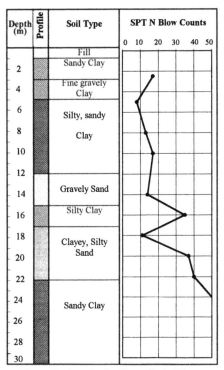

Figure 2. A typical soil profile in Dinar

5.MICROTREMOR MEASUREMENTS

After the Dinar Earthquake a detailed soil investigation was conducted based on insitu penetration tests and microtremor measurements. Microtremors are very low amplitude oscillations of the ground surface produced by natural sources such as wind, ocean waves, geothermal reactions and small magnitude earth tremors. Microtremor measurements are relatively easy and economically feasible method to estimate site response under earthquake excitations (Lermo, 1994; Gaull, 1995). Microtremors measurement were conducted at different locations within the town of Dinar to estimate soil amplification and predominant soil periods. The locations of measurement points are given in Figure 3.

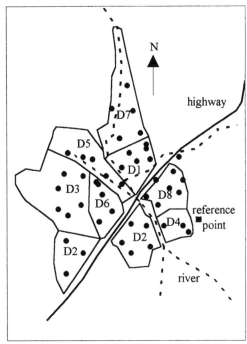

Figure 3. Locations of microtremor points

The microtremor measurements were conducted at points selected near to previously performed dynamic penetration tests to establish correlations with respect to soil characteristics. The microtremor measurements were recorded using two sets of three sensitive seismometers (two horizontal, one vertical), an amplifier and a digital recorder.

One set of seismometers were situated continuously at the selected reference station, located in the slopes of the hills on a firm ground away from the town centre. The other set of seismometers were moved to different locations within the town as shown in Figure 3. Microtremor measurements were taken at the same time simultaneously by the two sets of seismometers.

Two different methods; (a) spectral ratios relative to the reference station, (b) spectral ratios of horizontal to vertical component at each station (Nakamura, 1989, 1994) were used to estimate site effects from microtremor measurements.

After analysing the microtremor records using above mentioned two methods it was observed that amplification ratios computed using Nakamura's method varied in a relatively narrow range of 2.6 to 3.4. But the amplification ratios computed with respect to the reference station was observed to be scattered in a relatively wide range of 5.8 to 34. One possible reason for this difference between the two methods, may be different local source effects. The reference station was located in the slopes of the hills away from the town centre in a relatively quite area while the town was very noisy due to ongoing construction activities.

The amplification ratios are calculated for all the microtremor records taken within each district using Nakamura and reference point methods. In order to minimise the effects of local sources, an averaging procedure is adopted to obtain the representative amplification curve for each district as by these two methods as shown in Figure 4. The predominant periods and soil amplifications thus calculated by Nakamura and reference point methods are summarised in Table 3.

Table 3. The calculated average amplification factors and predominant periods based on Nakamura and reference point methods

District	Amplification, A_k		Pre. Period (sec)	
	Nakamura	Ref. Point	Nakamura	Ref. Point
D1	3.00	5.80	0.55	0.45
D2	2.60	11.00	1.00	0.20
D3	2.50	9.00	0.80	0.30
D4	3.40	12.00	0.55	0.70
D5	3.10	26.00	0.90	0.20
D6	3.20	16.00	0.80	0.30
D7	2.65	34.00	0.80	0.35
D8	2.70	11.00	0.35	0.10

6.COMPARISON OF STRONG MOTION AND MICROTREMOR RECORDS

During Dinar Earthquake sequence large number strong motion records were obtained at the Meteorological Station. The normalised elastic acceleration response spectra calculated for these records are shown in Figure 5. The site amplification was in the range of A=3-5 and predominant periods were in the range of T= 0.15-0.6 sec. In order to make a comparison, microtremor measurements were taken at the same place right by the strong motion instrument. The amplification ratio calculated for these microtremor records using Nakamura method is shown in Figure 6. As can be observed from this figure the amplification thus calculated is in the range of A=2.5-3 and predominant periods were in the range of T= 0.6-1.0 sec. The difference observed between microtremors and strong motion records is most likely due to the inelastic nonlinear behaviour of soil layers that becomes more dominant during earthquakes.

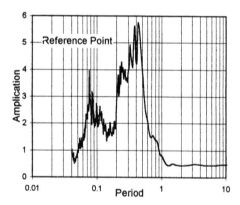

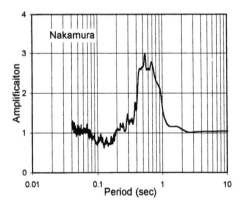

Figure 4. Average site amplification for D1 district calculated by reference point and Nakamura method

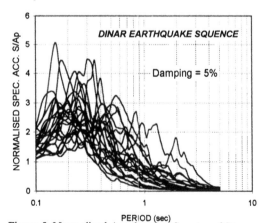

Figure 5. Normalised Acceleration Spectra of Strong motion records obtained at meteorological station

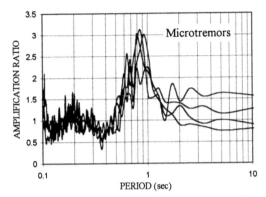

Figure 6. Amplification by Nakamura method of microtremor records at meteorological station

7. AMPLIFICATIONS FROM SHEAR WAVE VELOCITY

Shear wave velocity (V_s) of surface layers is very useful property for evaluating site amplification (Shima, 1978; Borcherdt, 1994). Equivalent shear wave velocities for the surface layers in the selected districts were determined from the calculated equivalent SPT-N blow counts (Ansal & Lav, 1995) using the correlation between V_s and N given by İyisan (1996) as:

$$V_s = 51.5 * N^{0.516} \quad (m/sec) . \qquad (1)$$

Shear wave velocities thus calculated as shown in Table 4, ranged between 175 and 217 m/sec.

The soil amplification ratios based on shear wave velocity can be determined using the relationship proposed by Midorikawa (1987).

$$A_k = 68 * V_s^{-0.6} \qquad (2)$$

Soil amplifications A_k values determined using Eq.(1) and Eq.(2) based on the calculated equivalent SPT-N blow counts are given in Table 4.

Table 4. Soil amplifications calculated from shear wave velocities

District	N	V_s (m/sec)	A_k
D1	11	175	3.07
D2	14	197	2.86
D3	11	215	2.71
D4	12	194	2.89
D5	13	189	2.93
D6	13	195	2.88
D7	16	216	2.70
D8	16	217	2.70

8. MICROZONATION BY GRADE-1 METHODS

Grade-1 Methods involve compilation and interpretation of information from historic documents, published reports and other available data. The most direct and simple way for determining local site effects is to compile data on the damage distribution during past earthquakes. In this study this approach was adopted and the damage distribution observed during October 1, 1995 earthquake used for microzonation of Dinar. The town centre was divided into eight zones with different hazard levels using the damage data of the 1995 Dinar earthquake. The damage ratios were calculated for every district as given in Table 2. A microzonation map of Dinar based on observed damage distribution was drawn as shown in Figure 7.

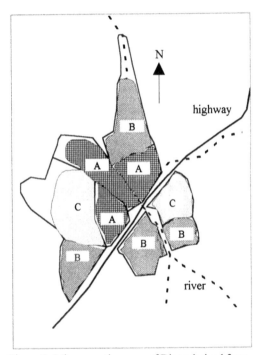

Figure 7. Microzonation map of Dinar derived from damage distribution.

As can be observed from Figure 7, it appears possible to delineate three different zones as A, B and C. Zone A is the high damage zone with damage ratio varying between 48%-51% , Zone B is the medium damage zone with damage ratio varying between 40%-44% and Zone C is the low damage zone with damage ratio varying between 30%-33%. Based on these observations, the earthquake design parameters need to be modified to increase the safety in Zone A and Zone B.

9.MICROZONATION BY GRADE-2 METHODS

Grade-2 Methods require considerably more soil investigations for evaluating the geotechnical properties at the studied site. These investigations may include geotechnical surveys such as insitu penetration test, insitu seismic tests for measuring shear wave velocity, microtremor measurements and soil sampling from boreholes for laboratory tests. A relatively detailed soil investigation was performed in Dinar for determining the geotechnical site conditions.

During this detailed post earthquake geotechnical investigation, insitu penetration tests were performed to determine the variation of the soil profile as well as the microtremor measurements were conducted within the town boundaries of Dinar. According to the ground classification proposed by Kanai (1961) using the microtremor measurements, the soil layers encountered in Dinar can be classified as Type IV according to the Japanese building code that is defined as soft, thick alluvium layers of loose sand and soft clays with high water table.

The damage distribution observed in Dinar Earthquake and soil amplifications calculated from microtremor measurements using Nakamura method and equivalent shear wave velocities calculated for different districts are given in Table 5.

Table 5. Damage distribution and soil amplification factors

District	Damage (%)	Amplification Factor, A_k	
		from V_s	from microt.
D1	48	3.07	3.00
D2	44	2.86	2.60
D3	30	2.71	2.50
D4	42	2.89	3.40
D5	51	2.93	3.10
D6	51	2.88	3.20
D7	40	2.70	2.65
D8	33	2.70	2.70

Although the structural properties of the buildings in Dinar were similar, different degrees of damage was observed in different districts. This indicates that the local soil conditions are one of the most important factor affecting in the damage distribution during earthquakes. Figure 8 shows the correlation obtained between soil amplification factors determined by microtremor measurements using Nakamura method and damage ratios. This linear correlation can be expressed as;

$$DR = 0.157 \, A_{kmic} - 0.293 \qquad (3)$$

where A_{kmic} represents the soil amplification factor and DR shows the damage ratio.

The basic purpose of the microzonation is to take into account the effects of local soil conditions and the related soil amplification characteristics as pointed out by Lu, et.al, (1992). However, as pointed out by Ansal and Siyahi (1995), the earthquake characteristics or in other terms, intensity of earthquake generated forces on structures located on the ground surface are also affected by the coupling between site conditions and earthquake source characteristics. Therefore in evaluating the damage distribution and in microzonation studies, it is not sufficient to base the analysis only on soil conditions. Thus, as can be observed in Figure 8, a significant scatter is observed when evaluating damage distribution only in term of soil conditions.

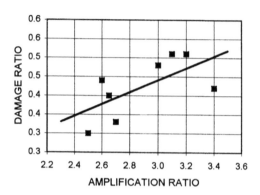

Figure 8. The relationship between damage ratio and soil amplification.

In every district a number of microtremor measurements were performed. The specific value of amplification factor for each district was determined by taking the average of all points in that region. Figure 9 shows the microzonation map of Dinar based on the amplification factors calculated from microtremor measurements using Nakamura method and amplification factors calculated based on the equivalent shear wave velocities.

In this case based, it appears possible to delineate two zones within the town of Dinar. In the Zone A, soil amplification factors calculated using Nakamura method based on microtremor measurements vary between 3.4-3.0 and amplification factors calculated from equivalent shear wave velocities range between by Midorikawa method vary between 3.07-2.88. In the Zone B, soil amplification factors from microtremor measurements vary between 2.5-2.7 and amplification factors from equivalent shear wave velocities vary between 2.86-2.7. These results may be more realistic in comparison to zonation map obtained from the damage survey as given in Figure 7 since structural characteristics of the buildings are excluded from the analysis. However as mentioned previously, since the amplification factors obtained from microtremor measurements and equivalent shear wave velocities only reflect the effects of soil conditions, in a real earthquake the observed damage distribution or the distribution of the intensity of earthquake generated forces in different regions could be very different due to coupling between soil conditions and earthquake source characteristics.

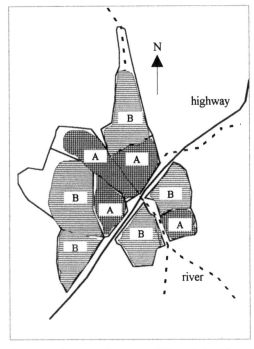

Figure 9. Microzonation map of Dinar derived from amplification factors

10. CONCLUSIONS

A preliminary microzonation study has been conducted with respect to ground shaking intensity for the town of Dinar using the available information obtained from the damage survey and the geotechnical investigation based on insitu penetration tests and microtremor measurements.

The soil amplification factors determined from microtremor measurements and from equivalent shear wave velocities are used to establish preliminary microzonation map for ground shaking intensity within the town of Dinar. The results obtained are compared with the damage distribution observed in Dinar Earthquake that clearly shows the applicability of the microtremor measurements and equivalent shear wave velocities for preliminary microzonation purposes.

REFERENCES

Ansal,A.M. & B.G.Siyahi 1995. Effects of Coupling Between Source and Site Characteristics During Earthquakes. *Proc. of Seced Con. on European Seismic Design Practice*. Chester. :83-89.

Ansal,A.M. & A.M.Lav 1995. Geotechnical Factors in 1992 Erzincan Earthquake. 5th Int. Conf. on Seismic Zonation. Nice. 1:667-674.

Borcherdt,R.D. 1994. Estimates of Site Dependent Response Spectra for Design (Methodology and Justification). *Earthquake Spectra*. 10:4:617-654.

Erdik, M., et al. 1995. *KOERI Reconnaissance Report 1st October 1995 Dinar (Turkey) Earthquake (Ms=6.1)*. Boğaziçi Uni. Kandilli Observatory and Earthquake Res. Inst. Istanbul.

Gaull,B.A., H.Kagami, Meeri & H.Taniguchi 1995. The Microzonation of Perth, Western Australia , Using Microtremor Spectral Ratios. *Earthquake Spectra*. 11:2:173-191.

Iyisan,R. 1996. Correlations Between Shear Wave Velocity an In-situ Penetration Test Results. *Technical Journal of Turkish Chamber of Civil Engineers*. 7:2:1187-1199.

Kanai,K. & T.Tanaka 1961. On Microtremors - VIII. *Bull. Earthquake Res. Inst.* 39:97-114.

Lermo,J. & J.Chavez-Garcia 1994. Are Microtremors Useful in Site Response Evaluation? *BSSA*, 84:5:1350-1364.

Lu,L., F.Yamazaki & T.Katayama 1992. Soil Amplification Based on Seismometer Array and Microtremor Observations in Chiba, Japan. *Earthquake Eng. and Struc. Dyn.* 21:95-108.

Manual for Zonation on Seismic Geotechnical Hazard 1993. The Japanese Society of Soil Mech. and Foundation Engineering. Tokyo. Japan.

Midorikawa,S. 1987. Prediction of Isoseismal Map in the Kanto Plain due to Hypothetical Earthquake. *Journal of Structural Eng.* 33B:43-48.

Nakamura,Y. 1989. A Method for Dynamic Characteristics Estimation of Subsurface Using Microtremor on the Ground Surface. *QR of RTRI.* 30:1:25-33.

Nakamura,Y. & J.Saita 1994. Characteristics of Ground Motion and Structures Around the Damaged Area of the Northridge Earthquake by Microtremor Measurement. *1st Preliminary Report, Railway Technical Research Institute.* Japan

Shima,E. 1978. Seismic Microzoning Map of Tokyo. *Proc. Second Int. Conference on Microzonation.* 1:433-443.

Seismic Behaviour of Ground and Geotechnical Structures, Sêco e Pinto (ed.)© 1997 Balkema, Rotterdam, ISBN 90 5410 887 8

Zonation of geotechnical seismic hazards in Tuscany, Italy

T. Crespellani, C. Madiai & G. Vannucchi
Dipartimento di Ingegneria Civile, Università di Firenze, Italy

A. Marcellini
Istituto di Ricerca sul Rischio Sismico, Milano, Italy

M. Maugeri
Istituto di Strade, Ferrovie e Aeroporti, Università di Catania, Italy

ABSTRACT :   With the aim of verifying the possibility of applying the criteria of the Manual for Zonation on Seismic Geotechnical Hazards (MZSGH) (TC4, 1993) to the Italian and Central-European seismic context, the northern region of Tuscany and surrounding areas included in Folio N° 96 of the Italian Geological Map at a scale of 1:100,000 were chosen as a test site.   The zone included two main seismic Italian areas, which have been repeatedly struck by destructive earthquakes up until quite recently.   The region is mainly mountainous and very rocky, and is crossed by different systems of active faults.   Many urban nuclei are situated on steep hillsides and along deep valleys.   As shown by the strongest earthquake of this century (that occurred on 7 September 1920, X MCS), landsliding is a severe problem during earthquakes.   The morphological and geological conditions are, in fact, nearly optimum for producing topographic amplification effects, rock falls and landsliding.   On the other hand, as in most parts of the Italian and European seismic zones, soil liquefaction does not represent a significant geotechnical seismic hazard.   For this reason,   analysis was limited to a quantification of the ground-shaking severity and landsliding.

A survey of the numerous and reliable seismic-risk studies published and the geological maps which cover the major part of the urban centres and surrounding areas at a scale of 1:10,000, as well as the scarcity of geotechnical data, have led to a limiting of the application of the Manual to the first grades of zonation.   Two areas of different sizes were taken into account: about 40 x 37 km^2 for Grade-1 and 6.5 x 11 km^2 for Grade-2. The mapping scales adopted are: 1 : 100,000 for Grade-1 and 1 : 10,000 for Grade-2.

1 INTRODUCTION

Despite the similarities of seismic and geographic conditions and of the historic tradition of many European countries, a comparison of the seismic regulations reveals considerable differences.   The instrument expected to achieve their harmonisation is the Eurocode 8 (EC8, 1993) which, in supplying the specifications for the design of buildings, establishes that seismic hazards are to be described in terms of a single parameter, i.e. the value a_g of the effective peak acceleration on rock or hard soil corresponding to a reference return period of 475 years.   However, the EC8 does not provide any indications on the procedures that are at the basis of the evaluation of

the hazard.   Therefore, in order to make possible comparative evaluations on seismic hazard and the levels of protection used by the different European countries, standardisation of the zonation procedures is also essential.   It is in this light that the possibility of extending to Europe the Manual for the Zonation of Seismic Geotechnical Hazards (1993), in the experimentation phase at a world level, was recently considered by the European representatives of several international technical committees (TC4 of the ISSMFE, ETC12 of the ESSMFE, and ESC&EAEE Joint Task Group). The present work is included within this framework. Indeed, the Manual for the Zonation of Seismic Geotechnical Hazards (MZSGH) represents an important attempt at taking into account the influence of geological and geotechnical factors and at making seismic zonation techniques more uniform. On the basis of the available information and of the investigation scale, three zonation levels and various procedures are proposed; in order to permit to choose the one that most conforms to the specific situation. The aim of the application performed was to point out possible gaps, incompleteness and difficulties in application of the procedures provided for by the Manual, so as to contribute to improving the final product but, above all, with a view to its application to the European reality.

2 APPLICATION OF THE MANUAL TO THE ZONATION OF SEVERAL SEISMIC AREAS IN TUSCANY

In this work, application has been made of the MZSGH to the zones of Lunigiana and Garfagnana and the surrounding areas included in Folio N° 96 of the geological map of Italy to a scale of 1:100,000 (Fig. 1).
There are three basic reasons that led to choosing this zone of Italy as the test site for the MZSGH experimentation.
The first reason is that pilot seismic studies have been underway in this zone of Italy for quite some time (Imbesi et al., 1986; Petrini, 1995) for a definition of seismic risk. These have included in-depth investigations of structural geology and historic seismicity, studies of the search for attenuation laws (Grandori et al., 1987; Crespellani et al., 1992), surveys of surface geology and proneness to landslides (Nardi, 1990; Nardi, 1992), surveys on the vulnerability of the residential patrimony (Petrini, 1995). In fact, the utility of having reliable data available is evident, both when these are necessary as input for the application of the various methodologies and also when they are used exclusively for the *a posteriori* control of the results obtained (for example, as was done in the specific case for an evaluation of the proneness to landslides). The second reason is that in 1920, the zone was struck by an earthquake that reached a maximum intensity of MCS X , the effects of which were also the subject of thorough investigations (Marcellini et al., 1986; Patacca et al., 1986) that thus make it possible in the specific case to control the reliability of the criteria suggested by the MZSGH. A survey map of the macroseismic intensities observed (Postpischl, 1985b) is available, and several slumps,

Figure 1 - Localisation of the regions examined for application of the methods for zoning ground motions and landslides of the Manual for Zonation on Seismic Geotechnical Hazards
G-1 ≡ *Grade-1 ;* **G-2** ≡ *Grade-2*

rock falls and landslides were localised (Patacca et al., 1986).
The third and perhaps most important reason is the fact that the zone is representative of the seismic regions of Western Central Europe. Indeed, with regard to the regions from which the empirical criteria and the predictive equations reported in the MZSGH have been inferred (which for the most part fall in areas of high and medium-to-high seismicity, and solely to which - in point of fact - the Manual is applicable), the zone under examination possesses three fundamental aspects which place it at the limits of applicability, namely:
- even if comparatively high for Tuscany, in reality the seismicity is moderate;
- the territory is mountainous: with the exception of a small coastal strip consisting of recent deposits, the zone is affected by typical Apennine foldings, with the widespread presence of Palaeozoic and Miocene rock formations, stratified and often weathered, characterised by steep slopes and intercalated by deep, narrow valleys, filled by alluvial deposits;
- the overall basic knowledge (general, geographic, historic, geological, topographical, hydrological, etc.) is high.

Moderate seismicity, the presence of upward-sloping reliefs of ancient origin, and a good basic documentation are the elements that exist in common with other seismic zones in Western Central Europe, where densely inhabited territories, with residential patrimonies that are often very old, are subjected to earthquakes of moderate severity which are sometimes frequent and the effects of which can be amplified by irregular, very steep morphologies. In these zones, the phenomena of topographical amplification of the seismic response and the landslide movements are geotechnical hazards to which the territory is mostly exposed, while instead the phenomena of liquefaction are inconsistent and not very probable. For this reason, in the recommendations of the EC8 (AFPS 90, 1990) and in the Guidelines for Seismic Microzonation Studies of the "Association Française du Génie Parasismique" (1995), considerable attention has been given to an evaluation of the topographic effects.

In many European countries, the level of knowledge of the territory is advanced. Also, ancient and recent seismic history is often well-documented, and for this reason in Europe the evaluation of seismic hazard profit notably from historical data. Even in the zone under examination, the basis of knowledge of the natural and built environment is extensive. Regarding a large part of the urban centres that fall in Folio N° 96, in addition to geological maps to a 1:10,000 scale drawn up by the National Earthquake Defence Group (Nardi, 1990; Nardi, 1992), there exist computerised documentation of the hydrologic and topographic data edited by the Tuscany Region. The seismic history, starting from the year 1000 A.D., is also well-documented. In addition, the maps of the intensities and accelerations expected and representative of the national hazard are also available, drawn up by the National Earthquake Defence Group for a return period of 475 years (Rebez A., 1996; Peruzza L., 1996). Contrary to the vast historic documentation and the high investigation level of the studies on seismic risk assessment, there exist instead only two strong-motion accelerometric stations of the ENEL-ENEA national seismic network, activated solely on the occasion of the earthquake of 29.4.1984. As they are equipped with a vertical trigger with a high threshold, they were not activated during any recent earthquakes of low energy content. A mobile network has recently been installed for microtremor measurements.

The information on the geotechnical properties of the various lithological units has been insufficient up until

now. For this reason, application of the MZSGH to the case under examination was restricted to the first two levels of zonation and, because of the scarce relevance of the liquefaction phenomena, was limited to the problem of zoning for ground motions and landsliding.

The work carried out consisted of a precise application of the MZSGH, choosing between the possible options the ones which best conformed to the nature of the territory and to the available data. Parallel to the description of the steps gradually taken, of the data and procedure utilised, the criteria adopted are subsequently illustrated and discussed, and the points are indicated which, in the opinion of the Authors, should be modified within the prospect of transferring the MZSGH to the European reality (at least to that of the zones of Central Western Europe).

3 ZONING FOR GROUND MOTIONS

3.1 *Grade-1 methods*

3.1.1 *General*

The scale adopted is 1:100,000, which is that of the Geological Map of Italy (used for reference in the national territory), based on surveys at the 1:25,000 scale. For the application, Folio N° 96, which contains a rectangular area of about 40 x 37 km^2 with the following values of latitude and longitude (with respect to Greenwich), was chosen:

- latitude: 44° - 44°20'
- longitude: 9°57'08",40 - 10°27'08",40

On this Folio are included a vast part of the Tuscan regions of Garfagnana and Lunigiana, the upper tip of Versilia, a small zone of Emilia-Romagna and of Liguria. The territory is mountainous and is furrowed by various rivers, mostly belonging to the hydrographic basin of the Magra and Serchio rivers, the most important in the zone. Of a torrential character, these rivers have been considered an important reference for the purposes of identifying the geographic and morphological configuration of the territory, and the main ones have in general been reported in the maps obtained through application of the MZSGH. The altitude varies between 0 and 2000 m. The rainfall is rather high, and the zone is at flood risk. Recently, imposing rainfall events have involved several inhabited centres, causing the collapse of buildings and landslide phenomena.

13

3.1.2 *Seismicity*

As has been said, the area considered is one of the most seismic zones of Tuscany. It constitutes the most westerly seismogenetic region of the Apennines, the seismotectonic events of which have been analysed in depth by Scandone et al. (1991). In brief, the high valley of the Serchio river is laid out in a band-shaped tectonic depression extended in a NW-SE direction, attributable to a structure of discarded blocks; it is limited by synthetic faults along the eastern border and by antithetical faults along the western border. It is crossed by a network of faults, some of which are seismically active (Fig. 2) and have recently also given rise to superficial earthquakes of high destructive potential. The main structures have a fairly regular development with a NW-SE direction, and are crossed by structures with anti-Apennine direction. The most important earthquakes in the area are in perfect agreement with the structures identified, with a lengthening of the isoseismals parallel with the structures.

The great amount of research on historical seismicity has led to the compilation of a catalogue, revised several times (Postpichl, 1985a; Stucchi et al., 1997). In Fig. 2 are shown the main earthquakes with an intensity greater than or equal to level V of the MCS scale which, according to the last version of this catalogue, have interested the zone since the year 1000.

Studies on the identification of the Italian seismic source zones and on the laws of attenuation conducted by the National Earthquake-Defence Group have recently led to the compiling of a map of the macroseismic intensities and of a map of the horizontal peak acceleration expected in Italy with a return period of 475 years (Peruzza et al., 1996; Rebez et al., 1996). These maps were assumed as the basis for application of the MZSGH. The return period corresponds to the one indicated by EC8, and to a probability of exceedance of 10% in 50 years. The intensities and the values of the peak acceleration expected in the zone under examination are shown in Fig. 3 and 4.

The maximum intensity expected is I = VIII MCS, while the peak ground acceleration on firm soil expected during the same number of years and with equal probability is 0.32 g.

a) Damage and intensity surveys

In-depth research on the effects produced by the 1920 earthquake has been carried out at different times by various researchers (Imbesi et al., 1986). Of particular usefulness was the news, obtained from documents, newspapers and eye-witnesses, on the structural damage and on the local effects of the earthquake (landslides, rock falls, etc.). The survey map of the observed intensities inferred on the basis of this information (Imbesi et al., 1986) has been given in Fig. 5. Instead, it has not been possible to reconstruct the isoseismals of the earthquake in a sufficiently reliable manner. Three different patterns, differing greatly one from the others, have been designated by Iaccarino (1968), Eva et al. (1978) and Postpischl (1985a); however, in this work, we have preferred to refer to the survey map, rather than to the latter.

b) Surface geology

The stratigraphic characteristics of the Northern Apennines have been studied in depth by many researchers, and are well known. On Folio N° 96 of the Geological Map of Italy emerge all the most important formations in the Northern Apennines, from the oldest to the most recent (Dalla, Nardi & Nardi, 1972). In fact, the tectonic breakdown into blocks with direct faults occurred mainly in the high Miocene age; subsequently, the uplifting blocks were

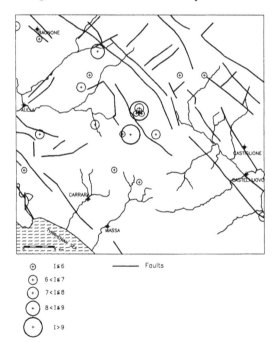

⊕	I ≤ 6
⊕	6 < I ≤ 7
⊕	7 < I ≤ 8
⊕	8 < I ≤ 9
⊕	I > 9

——— Faults

Figure 2 - Main faults and epicentres of historical earthquakes of an intensity greater than V MCS, that have occurred since 1000 A.D. in the region under consideration for application of Grade-1

14

subjected to more or less rapid erosion, while on the sinking blocks narrow lacustrine or fluvio-lacustrine basins, extended parallel with the Apennine axis, were initiated.

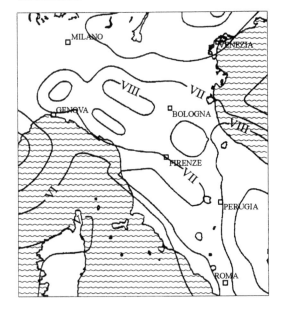

Figure 3 - Italian map of the MCS macroseismic intensities with a return period of 475 years (Peruzza et Al., 1996 ; Rebez et Al., 1996). The maximum intensity expected in the area under consideration is VIII MCS

The following basic groups of geological formations can be identified:
a) the metamorphic formations of the Apuan nucleus, consisting of Palaeozoic phyllites, marbles and Jurassic and Cretaceous limestones, and of Oligocene sandstones and schists;
b) the Tuscan-type formations, prevalently represented by an Oligocene and Miocene quartz-feldspathic sandstones known in Tuscany as 'macigno';
c) the conglomeratic quaternary formations with a prevalence of elements of 'macigno', and the terraced alluvial deposits;
d) recent alluvial deposits.

For the purposes of applying the MZSGH, in view of the large variety of geological units and of lithotypes present, the Authors thought to regroup them essentially on the basis of age. The following classification systems were chosen:

a) the proposal of Evernden and Thomson (1985) (Tab. 3.6 of the MZSGH), which considers 12 lithological classes, in order to evaluate the increments of intensity to be assigned to the lithological units ;
b) the correlation proposed by Midorikawa (1987) (Tab. 3.7 of the MZSGH), which distinguishes 5 classes, in order to evaluate the relative amplification factors.

There are essentially two reasons for this choice. Having available values of macroseismic intensity in the MCS scale, it was thought to reduce to a minimum the errors due to the changes in scale, choosing classification systems based on intensities measured on the MM scale or MSK. Also, as the territory consists mainly of Palaeozoic and Miocene geological formations, a classification of the lithological units on the basis of age, like those of Evernden and Thomson (1985) and of Midorikawa (1987), appeared in the absence of geotechnical information to be the least arbitrary. Moreover, this classification system was less linked to rather singular regional geological formations (as, for example, that of Borcherdt and Gibbs, 1976, appeared).

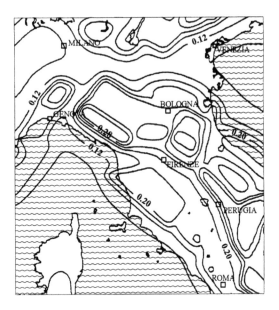

Figure 4 - Italian map of peak ground acceleration values with a return period of 475 years (Peruzza et Al., 1996 ; Rebez et Al., 1996). The maximum PGA expected in the area considered is 0.32 g

15

The results obtained are reported in Fig. 5 and 6. Fig. 5 indicates, with a closer traced line, the lithological formations in which the greatest increments of intensity are to be expected. In the same figure are reported the curves of isointensity of the reference map for the national hazard on firm soil (Peruzza et al., 1996; Rebez et al., 1996), for a return period T = 475 years. In the same figure are also shown the intensities observed during the 1920 earthquake which, as can be noted, harmonise well with the results obtained in the mountainous zone, while in the coastal band the experimental evidence does not agree with the proposal of Evernden and Thomson (1985). Figure 6 indicates the lithological units where a different seismic response is to be expected in terms of acceleration. Also reported are the curves of equal peak acceleration of the reference map for hazard (Peruzza et al., 1996 ; Rebez et al., 1996).

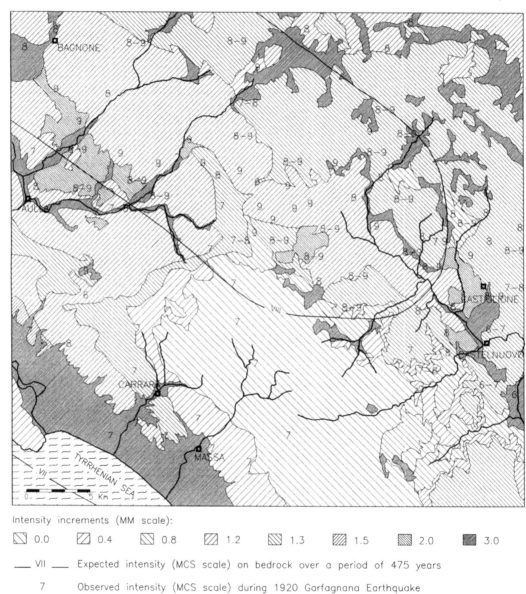

Intensity increments (MM scale):

| ▨ 0.0 | ▨ 0.4 | ▨ 0.8 | ▨ 1.2 | ▨ 1.3 | ▨ 1.5 | ▨ 2.0 | ▨ 3.0 |

—— VII —— Expected intensity (MCS scale) on bedrock over a period of 475 years

7   Observed intensity (MCS scale) during 1920 Garfagnana Earthquake

Figure 5 - Intensity increment map inferred from application of the proposal of Evernden and Thomson (1985) (Table 3.6 of the MZSGH). In the figure are shown the intensities expected according to the map of Figure 3, as well as the intensities observed during the 1920 earthquake

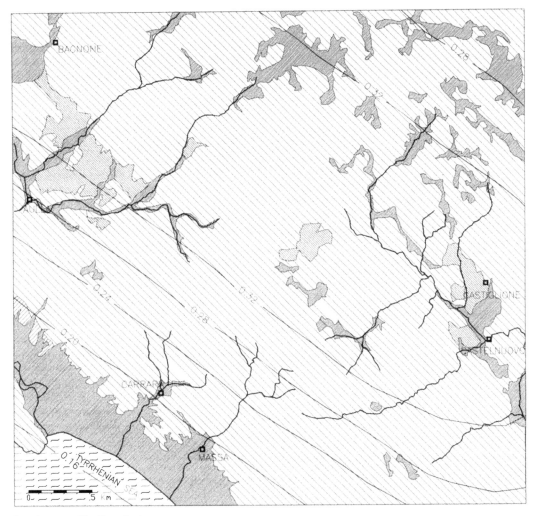

Relative amplification factor:

| | 3.0 | | 2.1 | | 1.5 | | 1.0 |

—— 0.16 —— Expected peak ground acceleration (g) on bedrock over a period of 475 years

Figure 6 - Map of the relative amplification factors inferred from application of the proposal of Midorikawa (1987) (Table 3.7 of the MZSGH). In the figure are also shown the peak ground acceleration values expected in accordance with the map of Figure 4

3.1.3 Comment

The problems encountered in the application of Grade-1 are essentially concerned with the following aspects:

1) the system of classification of the lithotypes ; there exists the risk that there is considerable influence of the subjective factors, since the categories provided for by the MZSGH, to which correspond very different increments of intensity, refer either to formations that are not very well-known in Europe - and whose characteristics it is therefore difficult to understand - or to categories the differentiating element of which is not clear ;

17

2) the fact that in Grade-1 the effects on ground motion of the slope of the reliefs are disregarded; from a seismic perspective, in the specific case the lithologies are in fact quite similar to each other (with the exception of the coastal and valley-bottom deposits that almost all fit into class B of the EC8), and the increments of intensity or of accelerations only on this basis are not justified ; the topographic effects are probably the most important differentiation element of seismic response in the area and disregarding them can lead to results that are not very reliable ; furthermore, it must noted that the greater increments of intensity and acceleration obtained with the application of the MZSGH in the coastal band do not agree with the data observed during historical earthquakes.

According to the Authors, in addition to a classification more specifically linked to the European geological formations and to a consideration of the topographical effects, to transfer the MZSGH to the European reality, explicit reference would have to be made to EC8 and to the return period of 475 years established in it for a definition of seismic hazard and generally used by European seismic codes.

3.2 Grade-2 methods

To apply the methods of Grade-2, a rectangular area has been considered on the inside of the one considered for Grade-1, having:

- latitude   44° 6' ÷ 44°, 12'
- longitude   10° 20' ÷ 10, 25'

The scale used is 1:10,000. For a classification of the lithological units into representative classes of the effects of site, since there is little geotechnical information in the area considered, the Authors resorted - as provided for in the MZSGH - to a classification system of the seismic regulation, and in the specific case it appeared logical to refer to EC8, which considers three subsoil classes:

Subsoil class A:
- Rock or other geological formation characterised by a shear wave velocity v_s of at least 800 m/s, including at most 5 m of weaker material at surface;

- Stiff deposits of sand, gravel or overconsolidated clay, up to several tens of m thick, characterised by a gradual increase in the mechanical properties with depth (and by v_s values of at least 400 m/s at a depth of 10 m);

Subsoil class B:
- Deep deposits of medium dense sand, gravel or medium stiff clays with thickness from several tens to many hundred of m, characterised by minimum values of v_s increasing from 200 m/s at a depth of 10 m, to 350 m/s at a depth of 50 m;

Subsoil class C:
- Loose cohesionless soil deposits with or without some cohesive layers, characterised by v_s values below 200 m/s in the uppermost 20 m;
- Deposits with predominantly soft to medium stiff cohesive soils, characterised by v_s values below 200 m/s in the uppermost 20 m.

In the area examined, stratigraphic conditions corresponding to subsoil classes A and B were identified, while no situations attributable to class C were recognised. In Fig. 7 are shown the lithological units which fall in classes A and B. Also reported are the intensities observed during the 1920 earthquake, which indicate that there is no evidence of correlation between lithology and the intensity observed.
For each of these classes, the elastic response spectra were obtained (Fig. 8) ; the acceleration values at period zero are the horizontal peak ground acceleration of the reference Italian map (Rebez et al., 1996; Peruzza et al., 1996) with an increased soil parameter S equal to 1.2 for subsoil class B (see EC8, Part 1, 4.2.2).

3.2.1 Comment

The application of Grade-2 essentially suggests a consideration, namely that also in this case, as for Grade-1, for application of the MZSGH to Europe, it is necessary to make explicit reference to EC8: in regard to the hazard specifications and return period, for a consideration of the topographic effects, and in regard to soil classes A and B provided for in it, for a determination of the elastic design spectra. Naturally, the empirical relations and the ways in which to introduce the topographical effects will have to be the result of appropriate experiments and observations during earthquakes.

18

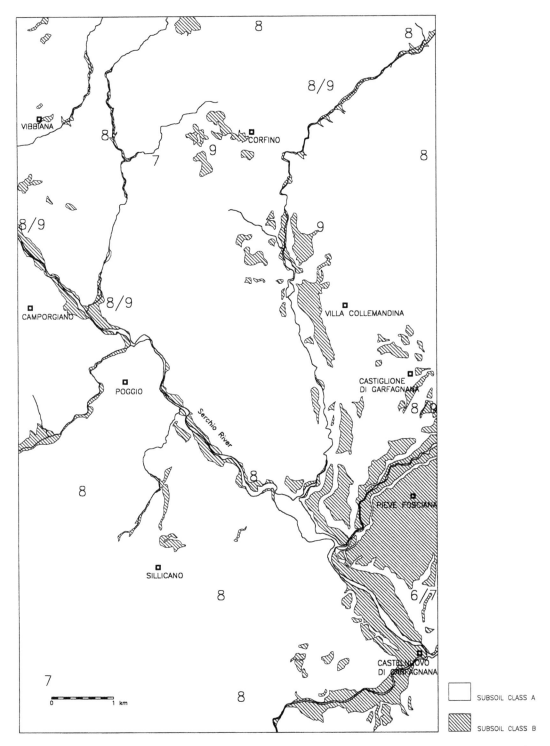

Figure 7 - Geological superficial materials grouped according to the subsoil classes of Eurocode 8. In the area subsoils of class C are not present at the scale used. The intensities observed during the 1920 Garfagnana earthquake are also shown

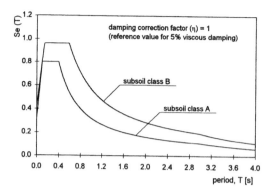

Figure 8 - Design spectra obtained for the reference return period of 475 years, for the two subsoil classes A and B of Eurocode 8

4 ZONING FOR SLOPE INSTABILITY

4.1 General

As the zone is prevalently mountainous, very steep, with lithologies that are often weathered and with extensive rainfall, the risk of movements on unstable slopes, of the reactivating of landslides and also of landslide phenomena on intact slopes during earthquakes, is high. During the 1920 earthquake, landslide phenomena of categories I and II, foreseen by Keefer and Wilson (1989), were identified.
For this reason, in the areas surrounding many urban centres, in-depth investigations have been made under the support of the National Earthquake-Defence Group, for the identification of active and potential landslides, as well as for an assessment of the lithological and morphological characteristics. In the zone, landslide maps to a scale of 1 : 10,000 are now available in all the most important urban nuclei (Nardi, 1990; Nardi, 1992). In applying the MZSGH, these maps together with the observations on landslide movements that occurred during the 1920 earthquake have been utilised only *a posteriori* for verifying the methods used.

4.2 Grade-1 methods

For the application of Grade-1, the reference scale was 1 : 100,000. With reference to Fig. 9, the following zonation criteria have been followed:

1. the zone inside the isoseismal of VIII MCS intensity, in which the most important

seismogenetic structures fall, has been circumscribed (with a dash-double dot line) and for the purposes of the hazard evaluation, this area has been defined as an 'epicentral area';
2. inside this zone, a value of magnitude, M_s, has been estimated, corresponding to the value of expected intensity I = 8, by using the equation proposed in the Catalogue of Italian Earthquakes (Postpitchl, 1985a):

$$I = 1.78 \, M_s - 1.93$$

   and obtaining $M_s \cong 6$;
3. by utilising the empirical equation of Fig. 4.7 of the MZSGH, the maximum epicentral distance of destructive landslide movements ($\Delta \cong 7$ km), evidenced with the dashed line, and the maximum epicentral distance of landslide movements ($\Delta \cong 40$ km), have been defined.

Since the line which defines the maximum epicentral distance of the landslide movements falls outside Folio N° 96, on the basis of the MZSGH criteria the entire region therein included thus results as being exposed to the risk of landslide.

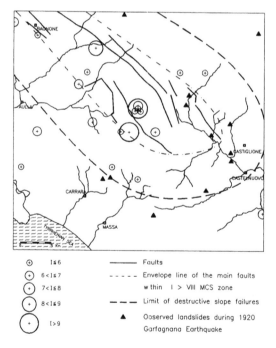

Figure 9 - Map showing the envelope line of the main faults and historical earthquake epicentres falling into the I> VIII MCS zone (dashed-dotted line), the limits of the zone of destructive slope failures (dashed line), and the landslides observed during the 1920 Garfagnana earthquake.

The reason for which this procedure was chosen is that in the region, the seismogenetic structures all have the same probability of being activated, and it appeared logical to consider their envelope as an epicentral area.

By way of verification of the reliability of the criteria followed, in Fig. 9 are reported the localities in which, during the 1920 earthquake, falls, block slides, slumps and earthflows were observed. It can be noted that, for the most part, these are included in the 'epicentral area' and in the surrounding zone.

4.2.1 *Comment*

On Folio N° 96 fall several flat areas, namely: areas of valley bottom that are not very extended (the borders of which, however, might be interested by the consequences of landslide movements) and a wide coastal strip of Versilia which, instead, cannot be exposed to landslide movements. By applying the methods of the MZSGH very strictly, it is not possible to identify these areas, since the slope of the relief is not introduced into the parameters considered (which are only the earthquake magnitude and the rainfall). It would therefore be important for specific reference to be made in the text to exclude the plain areas from the zonation.

A further improvement of the procedures for landslide zoning could come from a stricter link between zoning for ground motion and zoning for landsliding.

4.3 *Grade-2 methods*

On the basis of the available knowledge of the area under examination and of the morphological characteristics of the territory, the most appropriate method for the specific case among those included in the MZSGH has appeared to be that of Mora and Vahrson (1994). This method evaluates the landslide-hazard index on the basis of five parameters (relative relief, lithology, humidity, seismicity, rainfall precipitation intensity), which have been evaluated by considering a raster of 400x400 m², relative to an area of 6.5 x 11 km². The investigative scale considered is 1 : 10,000.

The digital model of the ground was processed starting from the 1 : 5000 cartography of the Tuscany Region, with level curves every 10 m. By utilising a GIS, the maps in Fig. 10 a, b and c were obtained: these report separately the different factors of susceptibility (relative relief, lithology and humidity). The triggering factors (seismicity and rainfall precipitation intensity) have not been represented because they are of a constant value throughout the area. Fig. 10d shows the values of the landslide hazard index.

Parameter R_r (see Table 4.4 of the MZSGH) has been redefined in order to take into account the dimension of the grid. Mora and Vahrson (1994) define this parameter as:

$$R_r = \frac{h_{max} - h_{min}}{km^2} \quad \left[m / km^2\right]$$

where

h_{max} = maximum elevation within one grid unit (m)

h_{min} = minimum elevation within one grid unit (m) when the size of the grid unit is 1 km² (side $L = 1$ km)

In the present work, having utilised square grids with side $l = 0.4$ km (for a corresponding area $A_p = 0.16$ km²), R_r has been defined as:

$$R_r = \frac{h_{max} - h_{min}}{\frac{L}{l} \cdot A_p} \quad \left[m / km^2\right]$$

For the evaluation of the index of influence of the natural humidity of the soil, reference was made to the average monthly rainfall data for the 1951-1970 period, and represented on a 1 : 500,000 cartography. The sum of the average monthly precipitations, according to the indications of Mora and Vahrson that are reported in Table 4.6 of the MZSGH, were included between 6 and 10 (see Table 4.7), distributed in four zones with increasing values from the far north-east to the far south-west of the zone under examination. A value for the index of influence for the natural humidity of the soil, S_h, resulted that was equal to 2 throughout the entire area, except for a small section in the far south-west in which S_h is equal to 3 (Fig. 10c). If the approximations of the available data linked to the reduced scale and to the limited observation period are considered, the choice of an S_h factor that is constant for the entire area might also be appropriate.

The value of the rainfall precipitation intensity T_p was determined starting from the data available for the rain gage stations that fall within the zone of interest.

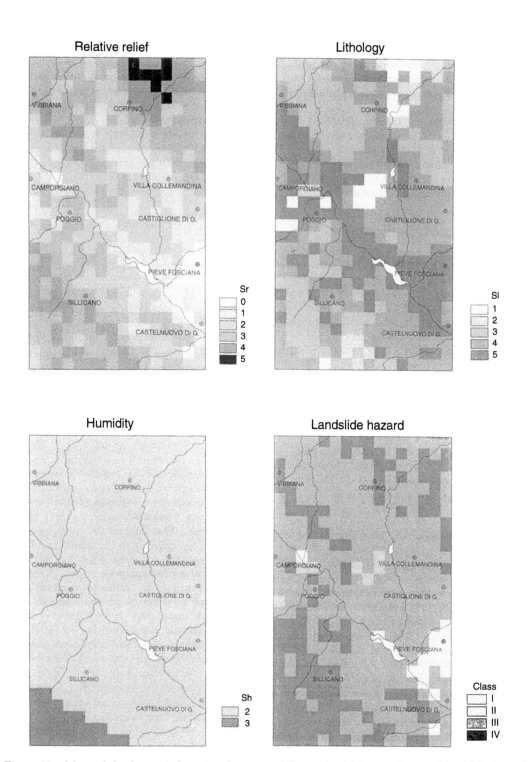

Figure 10 - Maps of the factors influencing the susceptibility to landslides and map of landslide hazard obtained by using the Mora and Vahrson (1994) method. Seismicity and rainfall precipitation intensity, are uniformly distributed in the area

There are 4 stations, and the number of years of activation varies from one station to another, from a minimum of 6 to a maximum of 44 years. In order to render the available information homogeneous, the mean values of the annual daily maximums were considered: these were equal to 100, 117, 127, 95 mm/d, showing a substantial uniformity of the data within the area. Therefore, the value of influence of rainfall precipitation intensity T_p was assumed to be constant and equal to 3.

Lastly, it is to be noted that the Mora and Vahrson method requires among the input parameters the seismic intensity expected during a return period of 100 years. In the specific case, the expected intensity is VI MCS; this value is uniformly distributed inside the area considered.

4.3.1 Comment

As a broad base of reliable information on the lithology, the morphology, the rainfall and the rainfall precipitation intensity was available in the specific case, and as the seismicity expected in 100 years is distributed in uniform manner, the Mora and Vahrson method should have led to a landslide map very similar to those of Nardi (1992), obtained with geological investigation 1 : 10,000 in the area being considered for application of Grade-2. Instead, the results agree only in part. In particular, the major discrepancies between the maps obtained with the calculation and those obtained with on-site investigations are noted in correspondence with slopes that, by slipping, have reached configurations characterised by gentler slopes. Probably in the specific case, the S_r factor should have a lesser weight than the unit.

According to the Authors, the Mora and Vahrson method is unquestionably a useful instrument for estimating proneness to landslides, but weights - to be evaluated locally on sample areas, with studies intentionally aimed at this - should be assigned to the various factors.

As noted for Grade-1, a further improvement in the procedures for landslide zoning could come from a stricter link between zoning for ground motion and zoning for landsliding.

5 CONCLUSIONS

Many research and retrofitting experiments have been promoted in recent years in the area considered by the Italian Government and the Tuscany Region.

At present, it is probably the most important laboratory of seismic analyses in Italy, and is also representative of some zones of medium seismicity in Central Western Europe. The extensive body of existing seismic and geological information that is available on these areas makes possible an effective control of the methodologies proposed by the MZSGH for the prediction of geotechnical hazards related to Grades 1 and 2 for zoning ground motions and landslides. Moreover, a programme of geotechnical investigations was recently established which will permit carrying out the zonation also of Grade-3 in the near future. For this reason, application of the MZSGH to this region appears to offer many opportunities for discussing the methods proposed and their application to the Italian and Central Western European context.

The main result of the study performed is that, in the region considered, the seismic evidence agrees quite well with the findings obtained by means of the MZSGH methods considered, resulting thus a sound tool for zoning seismic geological hazards. Several accurate surveys of geological and geotechnical conditions using field methods in some representative test areas of the region under consideration can further improve the predictive effectiveness of the procedures suggested.

In order to be able to extend the use of the MZSGH to other European countries, the Authors wish to stress the necessity of taking into account the guidelines of Eurocode 8, that have established the return period for evaluating seismic hazard, the classes of soils to be considered for assumption of the design spectra, and the consideration of the dynamic effects of earthquakes, such as topographic effects, strength softening and pore pressure increments, in slope stability evaluation.

ACKNOWLEDGEMENTS

The present work is the result of the decision of the Technical Committees TC4, ETC12, ESC&EAEE Joint Task Group, and of the Italian representatives, M. Maugeri, T. Crespellani, A. Marcellini, to extend in Europe the experimentation and discussion of the Manual for Zonation on Seismic Geotechnical Hazards.

The Authors wish to acknowledge the Tuscany Region and particularly Dott. Paolo Rosati and Dott. Claudia Di Passio, for having provided topographic and hydrological computerized data, and to thank Tiziana Pileggi for her help for the use of GIS.

REFERENCES

Association Française du Génie Parasismique 1990. Recommendation AFPS 90, Presses de l'École Nationale des Ponts et Chaussées, Paris.

Association Française du Génie Parasismique 1995. Guidelines for Seismic Microzonation Studies, Presses de l'École Nationale des Ponts et Chaussées, Paris.

Borcherdt R.D., Gibbs J.F. 1976. Effects of Local Geological Conditions in the S. Francisco Bay Region on Ground Motions and the intensities of the 1906 Earthquake, Bull. Seism. Soc. Am., Vol. 66, pp.467 - 500.

Crespellani T., Vannucchi G., Zeng X. 1992. Seismic hazard analysis in the Florence area. Eur. Earth. Eng., 3, pp. 33 - 42.

Dallan Nardi L., Nardi R. 1974. Schema stratigrafico e strutturale dell'Appennino Settentrionale. Mem. Acc. Lunig. Di Scienze G. Cappellini, Vol. XLII. 1972.

Eurocode 8 (1993): Earthquake Resistant Design of Structures. Second Draft CEN/TC250/SC8, PT1: General Rules.

Eva C., Giglia C., Graziano F., Merlanti F. 1978. Seismicity and its relation with surface structures in the North-Western Apennines.Boll. Geop. Teor. Appl., XX, 79, pp. 263-277.

Evernden J., Thomsom, J.M. 1985. Predicting Seismic Intensities, U.S. Geol. Survey Prof. Paper 1360, pp.151 -202.

Grandori G., Perotti F., Tagliani A. 1987. On the attenuationof macroseismic intensity with epicentral distance. Ground motion and Eng. Seism., A.S. Cakmak ed., Elsevier.

Iaccarino E. 1968. Attività sismica dal 1500 al 1965 in Garfagnana, Mugello e Forlivese, NEN, TR/GEO, 19.

Imbesi G., Marcellini A., Petrini V., Di Passio C., Ferrini M., (a cura di) 1986. Progetto Terremoto In Garfagnana e Lunigiana, CNR - Regione Toscana, La Mandragora Ed., Firenze.

Legge 28.10.1986, n. 730: Misure per l'attuazione degli interventi, diretti all'adeguamento sismico degli edifici pubblici nelle zone di Lucca e Massa Carrara.

Marcellini A., Tento A. 1986. Ricostruzione dello scuotimento. In: Progetto Terremoto in Garfagnana e Lunigiana, CNR - Regione Toscana, La Mandragora Ed., Firenze. pp.46-50.

Ministero dei LL.PP. Servizio Idrografico. Annali idrologici Anni 1940-1987.

Mora S., Vahrson W. 1994. Macrozonation methodology for Landslide Hazard Determination, Bull. Ass. Eng. Geol., XXXI (1), pp. 49-58.

Nardi R. (coord.) - 1985÷ 1992. Carta geologica e carta della franosità della Garfagnana e della media valle del Serchio (Lucca), scala 1 :10.000, CNR - Gruppo Nazionale per la Difesa dai Terremoti, S.E.L.C.A., Firenze.

Nardi R. (coord.) 1990. Carta geologica e carta della franosità della Lunigiana, scala 1 :10.000, CNR - Gruppo Nazionale per la Difesa dai Terremoti, S.E.L.C.A., Firenze.

Patacca E., Scandone P., Petrini V., Franchi F., Sargentini M., Vitali A. 1986. Revisione storica. In: Progetto Terremoto in Garfagnana e Lunigiana, CNR - Regione Toscana, La Mandragora Ed., Firenze, pp.40-46.

Peruzza L., Monachesi G., Rebez A., Slejko D and Zerga A. 1996. Specific macroseismic intensity attenuation of the seismogenic sources, and influences on hazard estimates. In: Thorkelsson B. (ed), Seismology in Europe, Icelandic Meteorological Office, Reykjavik, pp. 373 - 378.

Petrini V. (a cura di) 1995. Pericolosità sismica e prime valutazioni di rischio in Toscana. CNR-IRRS e Regione Toscana, Ed. Landini, Firenze.

Postpischl D. 1985a. Catalogo dei terremoti italiani dall'anno 1000 al 1980", CNR, Quaderni della Ricerca Scientifica, No 114, Vol. 2B, Bologna.

Postpischl D. 1985b. -Atlas of isoseismal maps of the Italian earthquakes. CNR, P.F.Geodinamica.

Rebez A., Peruzza L. and Slejko D. 1996. Characterization of the seismic input in the seismic hazard assessment of Italian territory. In: Thorkelsson B. (ed), Seismology in Europe, Icelandic Meteorological Office, Reykjavik, pp. 327 - 332.

Scandone P., Patacca E., Meletti C., Bellatall M., Perilli N., Santini U. 1991. Struttura geologica, evoluzione cinematica e schema sismotettonico della penisola italiana. In : Atti del Convegno 1990 Zonazione e riclassificazione sismica, Vol. 1 GNDT (ed), Tipografia Moderna, Bologna, pp.119-133.

Stucchi M. 1997. Catalogo dei Terremoti Italiani. Internet, http://emidius.itim.mi.cnr.it.

Technical Committee For Earthquake Geotechnical Engineering, TC4, ISSMFE, (1993) - Manual for Zonation of Seismic Geotechnical hazards, Japanese Society of Soil Mechanics and Foundation Engineering, Tokyo.

Università degli Studi di Firenze, Istituto di Idronomia. Atlante di Idrologia Agraria per la Toscana e l'Umbria. Incontro CNR-Regione Toscana, 24-25 maggio 1979, Firenze.

Seismic Behaviour of Ground and Geotechnical Structures, Sêco e Pinto (ed.) © 1997 Balkema, Rotterdam, ISBN 90 5410 887 8

Liquefaction potential map for Chimbote, Peru

J. E. Alva-Hurtado & D. Parra-Murrugarra
CISMID, National University of Engineering, Lima, Peru

ABSTRACT: The purpose of this paper is to present the assessment of soil liquefaction potential for the City of Chimbote, Perú, based on Grade-3 method proposed by TC-4 (ISSMFE, 1993). Results were compared with soil damages produced by the May 31, 1970 earthquake. Chimbote City is located on alluvial deposits of the Lacramarca River. Soil profiles are formed by clean and silty saturated sands with densities from loose to medium. A soil exploration program was undertaken with standard penetration and dutch cone testing throughout Chimbote. Besides, data from older borings were compiled in order to get a comprehensive set of soil data for the city. A map of the city of Chimbote showing the results of superficial damage caused by soil liquefaction is presented. A good comparison of liquefaction potential sites with the damage caused by the May 31, 1970 earthquake is obtained.

INTRODUCTION

One of the best documented cases of soil liquefaction in Peru is the one relevant to the May 31, 1970 earthquake in Chimbote. The city is located at a distance of about 400 kilometers north of Lima, the capital of Peru. (Alva-Hurtado, 1983).

On May 31, 1970 an earthquake of magnitude Ms=7.8 and focal depth of 45 kilometers occurred west of Chimbote. The epicenter was located 50 km off the coast. A strong motion record of the earthquake was obtained in Lima, with a maximum horizontal acceleration of 0.11 g. No record was obtained in Chimbote. Maximum intensity of IX in the Modified Mercalli Scale was determined.

Several soil borings were made in Chimbote after the May 31, 1970 earthquake (Morimoto et al, 1971; Carrillo, 1972; Parra, 1990), so it was possible to apply the Manual for Liquefaction Zonation of the Technical Committee for Earthquake Geotechnical Engineering of the ISSMFE. A design earthquake similar to the May 31, 1970 earthquake was used, resulting in good agreement between the liquefaction predicted and the soil effects produced by the quake.

SOIL LIQUEFACTION CAUSED BY 1970 EARTHQUAKE

A brief summary on liquefaction effects in Chimbote during the May 31, 1970 earthquake is presented.

Ericksen et al (1970) and Plafker et al (1971) indicated that in Casma, Puerto Casma and near the coast in Chimbote, lateral spreading of the ground caused by liquefaction of deltaic and beach deposits was produced. Cracks were observed on the ground that affected structures. Chimbote's central zone (Casco Urbano) was evidently an area of soil liquefaction and of differential compaction. In Chimbote, Casma and along the Panamerican Highway ground subsidence in the surface, because of liquefaction, was noticed.

Cluff (1971) reported ground failure in Chimbote because of saturated and loose beach deposits. Sand volcanoes and water ejection were observed in several areas where the water level was superficial. Berg and Husid (1973) verified the occurrence of soil liquefaction in the foundation of the Mundo Mejor school in Chimbote.

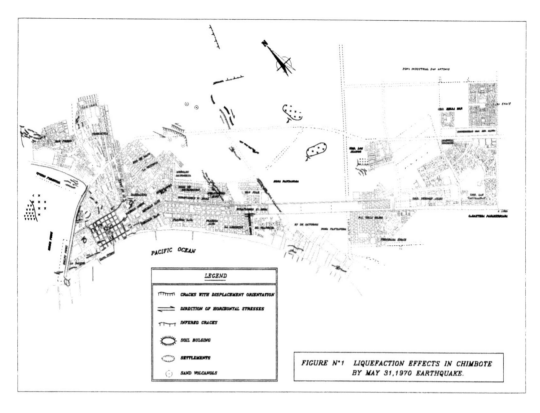

LEGEND

⟅⟆⟆⟆ CRACKS WITH DISPLACEMENT ORIENTATION

→→→ DIRECTION OF HORIZONTAL STRESSES

⟆⟆⟆ INFERRED CRACKS

SOIL BULGING

SETTLEMENTS

SAND VOLCANOS

PACIFIC OCEAN

FIGURE Nº1   LIQUEFACTION EFFECTS IN CHIMBOTE
BY MAY 31,1970 EARTHQUAKE.

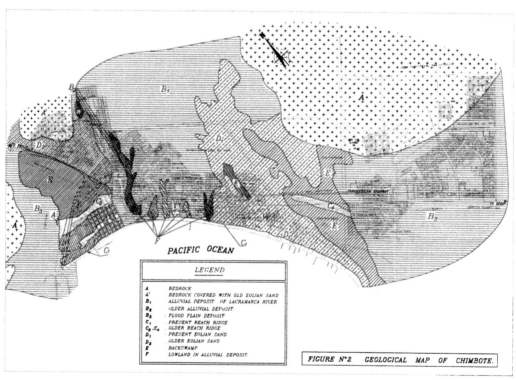

PACIFIC OCEAN

LEGEND

A   BEDROCK
A'   BEDROCK COVERED WITH OLD EOLIAN SAND
B_1 ALLUVIAL DEPOSIT OF LACRAMARCA RIVER
B_2 OLDER ALLUVIAL DEPOSIT.
B_3 FLOOD PLAIN DEPOSIT
C_1 PRESENT BEACH RIDGE
C_2, C_3 OLDER BEACH RIDGE
D_1 PRESENT EOLIAN SAND
D_2 OLDER EOLIAN SAND
E   BACKSWAMP
F   LOWLAND IN ALLUVIAL DEPOSIT.

FIGURE Nº2   GEOLOGICAL MAP OF CHIMBOTE.

26

Carrillo (1970) reported settlement of accesses to almost all of the bridges in the Panamerican Highway and subsidence of the Chimbote Port Terminal. He also presented evidence of saturated sand liquefaction at Elías Aguirre street in Chimbote.

Morimoto et al (1971) described soil liquefaction in Chimbote and presented a distribution map of ground cracks and sand volcanoes (Fig N°1). In the backswamps and lowlands in alluvial deposit, general liquefaction was developed with cracks due to differential compaction of soil deposits. In the alluvial deposit, subsurface liquefaction developed, generating cracks with sand volcanoes and damage to wells.

GEOTECHNICAL CHARACTERISTICS OF CHIMBOTE

Geology of the area is composed of bedrock and quaternary deposits. The rocks are Cretaceous andesitic volcanic with shale and sandstone and granitic intrusive rocks. Quaternary deposits are alluvial, beach ridges, eolian sands and backswamps. To the north and southeast of the city there are rocky mountains and hills which are covered partly with eolian sand. Chimbote is located on the alluvial plain of the Lacramarca River (Fig N°2).

The subsoil of Chimbote consists mainly of a thick sandy deposit with superficial water. The deposit is likely to suffer liquefaction and densification under seismic excitation. In most parts of the city the sand is medium to dense, with N values of 10 to 30, overlying very dense sands or gravel down to bedrock. However, in some areas, high water level and N values lower than 10 are found.

The northern part of the city includes the districts of San Pedro, Pensacola, Casco Urbano, La Caleta and Siderúrgica. In San Pedro the soil profile consists of loose to medium sand with the water level at 5.0 meters. In Siderúrgica there are medium to fine sand deposits with lenses of silt and gravel, covered by organic fill with the water level at 0.5 to 1.0 meters. Values of N range from 5 to 10 at the surface, increasing in value with depth. Casco Urbano is composed of fine to medium sand with caliche and gravel layers. The water level is at 1.50 meters. Value of N vary from 10 at the surface to over 50 at 5.0 meters in depth. La Caleta and El Puerto are characterized by loose silty sands with organic matter; the thickness goes from 1.5 to 4.0 meters, overlying medium compacted sands and gravels.

The central zone includes the areas known as Urb. 21 de Abril, Pueblo Libre, Villa María Baja, Miraflores, Miramar Alto and Miramar Bajo, Florida Alta and Florida Baja, La Libertad and Trapecio. In Miramar Bajo there is organic material at the surface, with fine to medium sand down to 10 meters, silty sand down to 20 meters and then gravel. Water level varies from 0.70 to 1.40 meters. In Miraflores values of N range from 8 to 12 at 2.0 meters, reaching 40 at 4.0 meters. The soil profile in Trapecio consists of fine and silty sands with seashells with a thickness of 4 to 6 meters. The sand overlies dense sand with clay and gravel layers.

In 27 de Octubre there is fine sand with the water level at 1.0 meters. In Villa María Baja the soil profile shows has fill at the surface overlying loose to medium sand with a high water level.

The southern part includes the districts of Villa María Alta, Buenos Aires, Nuevo Chimbote, Casuarinas and Canalones. The water level is below 16 meters in this area. The ground is composed of coarse to fine sand with gravel. In Buenos Aires the coarse to fine sand layer has a thickness of 20 meters whereas in Nuevo Chimbote it has a thickness of 4 meters, overlying fine to medium sand with clay down to 16 meters. Standard penetration values obtained were higher than 12 at the surface, increasing in value with depth.

CHIMBOTE SEISMIC MICROZONATION

Seismic microzonation of Chimbote was undertaken by Morimoto et al (1971) based upon geology, standard penetration testing, damage evaluation, aftershocks and microtremors, seismic refraction and soil amplification studies. Four different zones were established (Fig N°3).

Zone I. The ground consists of dense gravel or rock with the water level at least 10 meters below ground surface. The elevation of the ground is above 10 m.a.s.l. Subsidence of buildings or ground is not expected. Larger seismic forces are expected in this zone because of soil structure interaction.

Zone II. This area is covered by loose to medium sand several meters thick. Below there is either dense sand or cemented sand. Water level is about 5 meters below surface. No settlement is expected for two-story buildings, except those at the outer edge of sand dunes. It is recommended that buildings higher than

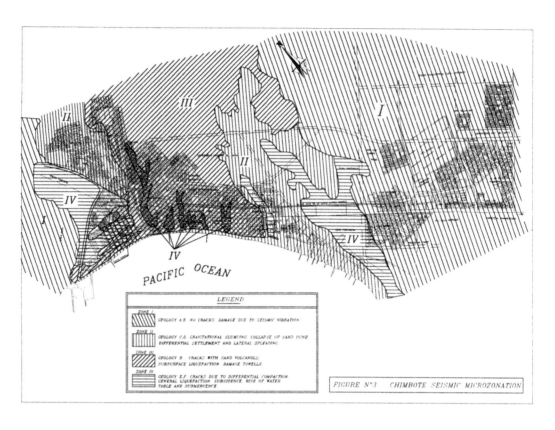

LEGEND

ZONE I
GEOLOGY A,B NO CRACKS DAMAGE DUE TO SEISMIC VIBRATION

ZONE II
GEOLOGY C,D GRAVITATIONAL SLUMPING, COLLAPSE OF SAND DUNE.
DIFFERENTIAL SETTLEMENT AND LATERAL SPREADING.
MUD

ZONE III
GEOLOGY B CRACKS WITH SAND VOLCANOLS.
SUBSURFACE LIQUEFACTION. DAMAGE TOWELLS.

ZONE IV
GEOLOGY E,F CRACKS DUE TO DIFFERENTIAL COMPACTION.
GENERAL LIQUEFACTION. SUBSIDENCE. RISE OF WATER
TABLE AND SUBMERGENCE.

FIGURE N°3   CHIMBOTE SEISMIC MICROZONATION

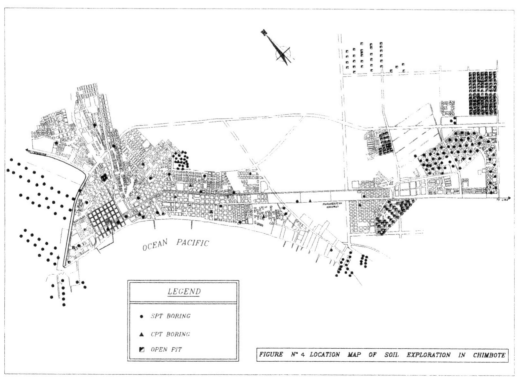

LEGEND

● SPT BORING

▲ CPT BORING

▨ OPEN PIT

FIGURE N° 4 LOCATION MAP OF SOIL EXPLORATION IN CHIMBOTE

two stories be supported by piles reaching the dense sand. Construction on sand dunes must be improved with vibroflotation.

Zone III. The soil profile is formed by sandy soil covered by thin agricultural soil. Gravel is found below 10 meters, with a shallow water level. Loose fine sand existing at different depths may liquefy during earthquakes. However no appreciable settlements will occur in building, except special cases.

Zone IV. This zone has a high water table. Most of the land is covered with water or swamps. Average elevation of the land is lower than 5 m.a.s.l. The soil profile consists mainly of sand covered with a very thin layer of organic silt. Damage to buildings is mainly by settlement and partly by seismic force. Liquefaction of the sand up to the surface will be produced by a severe earthquake. Pile foundations down to dense sand or improvement of the ground conditions, are recommended.

GEOTECHNICAL INVESTIGATION

Alva-Hurtado and Parra (1991) compiled data from standard penetration tests and open pits, carried out in Chimbote by public and private organizations after the May 31, 1970 earthquake. In addition a geotechnical investigation program was undertaken.

The research program consisted of 11 standard penetration and 9 dutch cone tests throughout Chimbote, to add information to the available soil data bank. SPT tests were made by wash boring and casing. Tests were made every meter, down to depths of 4 to 11 meters. Disturbed samples were obtained for soil classification. CPT tests were performed with a portable equipment with a 2-ton capacity. Cone penetration was recorded every 20 cm up to 6 meters in depth. Manual auger exploration was made near the CPT test sites to get soil samples for classification. Fig N°4 presents the area of the geotechnical study in Chimbote.

LIQUEFACTION POTENTIAL EVALUATION

The evaluation of liquefaction potential in Chimbote was undertaken by the application of the Grade-3 method proposed by TC-4.

A design earthquake with a magnitude of Ms=7.5 and a maximum acceleration of 0.3 g was used in the evaluation. Dutch cone results were converted into N values according to Robertson and Campanella (1983, 1985). Overburden correction was made with the formula given by Liao and Whitman (1986).

With the information from soil profiles and the design earthquake, the simplified method of liquefaction potential was used. The method proposed by Seed et al (1983) and Seed and De Alba (1986) was applied for SPT and CPT borings. Prof. Ishihara's (1985) criteria for damage in the presence of an unliquefiable surface layer was also included in the evaluation. Fig N°5 presents the potential liquefaction damage at ground surface for an earthquake with maximum acceleration equal to 0.30 g.

Liquefaction will occurr in the central part of Chimbote: Dos de Mayo, La Victoria, Alto Peru, Zona de Reubicación, Miraflores Alto, Miraflores 1st and 3rd zone and part of Florida Baja. These districts are located on the alluvial deposit of the Lacramarca River. The soil profile shows a superficial thin layer (silt and clay) over loose sand, that increases its compactness with depth, and has thin layers of clay. Water level is at the surface. Liquefaction will also occur in the districts of San Juan, Urb. Trapecio, Barrio Magisterial in the south and in part of Miraflores 2nd zone. These sites are of loose eolian sand at the surface, increasing in density with depth. Water level is at 1.5 to 2.0 meters. In the northern part, Miramar Bajo and La Caleta districts, the phenomena of liquefaction will occur. The soil profile shows fine and silty sands with a high water level. The backswamp zone that exists to the north and to the south of Chimbote will also liquefy.

Fig N°5 presents a map of Chimbote showing the results of the liquefaction potential evaluation. Comparison of liquefaction potential sites with damage caused by the May 31, 1970 earthquake, as presented in Fig N°1, shows good correlation.

CONCLUSIONS

The soil liquefaction phenomena that ocurred in the city of Chimbote during the May 31, 1970 earthquake was documented.

Boring logs performed after the earthquake by public and private organizations were compiled. In addition, a soil exploration program was undertaken by means of standard penetration and dutch cone testing. An important soil data bank was obtained for Chimbote to be used in its future development.

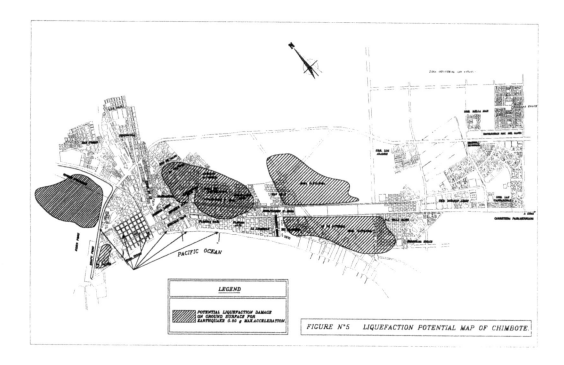

FIGURE N°5   LIQUEFACTION POTENTIAL MAP OF CHIMBOTE.

A case study for the application of the Manual for Zonation on Seismic Geotechnical Hazards proposed by TC-4 of ISSMFE was possible with the soil information and soil liquefaction damage available in Chimbote.

A good correlation was obtained between the predicted potential liquefaction areas in Chimbote and the actual soil damage produced by the May 31, 1970 earthquake.

REFERENCES

Alva Hurtado J.E. and Parra D. (1991), "Evaluación del Potencial de Licuación de Suelos en Chimbote, Perú". IX Congreso Panamericano de Mecánica de Suelos e Ingeniería de Fundaciones, Viña del Mar, Chile.

Alva Hurtado J.E. (1983), "Breve Historia del Fenóeno de Licuación de Suelos en el Perú", IV Congreso Nacional de Mecánica de Suelos e Ingeniería de Fundaciones, Lima, Perú.

Berg G.V. and Husid R. (1973), "Structural Behavior in the 1970 - Perú Earthquake", 5th World Conference in Earthquake Engineering, Rome, Italy.

Carrillo Gil A. (1970), "Algunas Apreciaciones del Comportamiento del Suelo en la Zona del Sismo de "Ancash", II Congreso Nacional de Mecánica de Suelos e Ingeniería de Fundaciones, Lima, Perú.

Carrillo Gil A. (1972), "Estudio de Suelos para Cimentación y Pavimentación", Plan Director de Desarrollo de Chimbote PNUD, Chimbote, Perú.

Cluff L.S. (1971), "Perú Earthquake of May 31, 1970; Engineering Geology Observations", Bulletin of the Seismological Society of America, Vol. 61, N°3, pp. 511-534.

Ericksen G.E., Plafker G. and Fernández-Concha J. (1970), "Preliminary Report on the Geological Events Associated with the May 31, 1970 Perú Earthquake", U.S. Geological Survey Circular 639.

Ishihara K. (1985), "Stability of Natural Deposits During Earthquakes", Proc. 11th Int. Conf. on Soil Mechanics and Foundation Engineering, San Francisco, Vol I, pp 321-376.

ISSMFE (1993), "Manual for Zonation on Seismic Geotechnical Hazards", Technical Committee for Earthquake Geotechnical Engineering, TC4, The Japanese Society of Soil Mechanics and Foundation Engineering.

Liao S. and Whitman R.V. (1986), "Overburden Correction Factors for SPT in Sand", Journal of Geotechnical Engineering ASCE, March, pp 373-377.

Morimoto R., Koizumi Y., Matsuda T., Hakuno M. and Yamaguchi I. (1971), "Seismic Microzoning of Chimbote Area, Perú", Overseas Technical Cooperation Agency, Government of Japan, March.

Parra D. (1990), "Evaluación del Potencial de Licuación de Suelos de la Ciudad de Chimbote", Tesis de Grado, Universidad Nacional de Ingeniería, Lima, Perú.

Plafker G. Ericksen G.E. and Fernández-Concha J. (1971), "Geological Aspects of the May 31, 1970, Perú Earthquake", Bulletin of the Seismological Society of America, Vol 61, N°33, pp 543-578.

Robertson P.K. and Campanella R.G. (1983), "Interpretation of Cone Penetration Test Part I: Sand" Canadian Geotechnical Journal, Vol 20, pp 718-733.

Robertson P.K. and Campanella R.G. (1985), "Liquefaction Potential of Sands Using the CPT", Journal of the Geotechnical Engineering Division, ASCE, Vol 111, N°3, pp 384-403.

Seed H.B., Idriss I.M. and Arango I. (1983). "Evaluation of Liquefaction Potential Using Field Performance Resistance of Sands", Use of Insitu Test in Geotechnical Engineering, ASCE, pp 281-302.

Seed H.B. and De Alba P. (1986), "Use of SPT and CPT Tests for Evaluating the Liquefaction Resistance of Sands", Use of Insitu Test in Geotechnical Engineering, ASCE, pp 281-302.

Seismic Behaviour of Ground and Geotechnical Structures, Sêco e Pinto (ed.) © 1997 Balkema, Rotterdam, ISBN 90 5410 887 8

Liquefaction potential assessment – Application to the Portuguese territory and to the town of Setúbal

C. Jorge & A. M. Vieira
Laboratório Nacional de Engenharia Civil (LNEC), Lisbon, Portugal

ABSTRACT: In this paper the methodology proposed by the Technical Committee for Earthquake Geotechnical Engineering, TC4, for the liquefaction potential assessment is applied to the Portuguese territory and to the town of Setúbal. An approach is made to three zoning levels proposed by this committee, which are associated to different scales of detail. In a first stage a zoning is carried out of the liquefaction potential on a territorial level (level 1 zoning); in a second stage a zoning of liquefaction potential is performed on a regional level (level 2 zoning). For the last case a numerical modelling is used, and a quantification of local effects is performed for the assessment of liquefaction potential on a site level (level 3 zoning).

1 - INTRODUCTION

Liquefaction is very often one of the most severe consequences of damage provoked by earthquakes. It produces high damages in water and power supplies, in industrial complexes, in the roadways and associated structures and waste disposal systems, in harbours and also in urban areas. The most frequent effects observed at ground surface are represented by sand boils and cracks and sometimes by flooding of a given area. The damages caused in several structures are related with failure of foundations, differential settlements, subsidence, uplift and flotation.

The hazard zoning map of liquefaction potential must incorporate both seismicity and soil conditions. This approach has been applied to generate a small scale zoning map of liquefaction hazard for the Portuguese territory and for the town of Setúbal (Jorge, 1993). This required the previous preparation of two basic zoning maps, following the methodology proposed by Youd and Perkins (1978): a Liquefaction Susceptibility Map based on geological data and a Liquefaction Opportunity Map based on seismicity data. The final map - the Liquefaction Potential Zoning Map - is obtained by superposition of the two preceding basic maps.

In the site study, the liquefaction potential is analysed for three stratigraphic profiles located in the town of Setúbal. The profiles belong to a ground surface formation composed by sandfills of the harbour zone, in accordance with the geotechnical map of Setúbal. The liquefaction potential was assessed by comparing the values of the average

shear stress induced by the seismic actions of the Portuguese Building Code with in situ strength of the soil based on Standard Penetration Test results.

2 - LIQUEFACTION SUSCEPTIBILITY MAP

2.1 - General approach

Liquefaction susceptibility is associated with the response of sedimentary deposits to undergo liquefaction induced by ground shaking during earthquakes. This response is a function of the mineralogical composition, texture, and of the geological history. Youd and Hoose (1997) have studied the influence of these geological conditions and have shown that liquefaction is related with specific geological formations. In fact, the Holocene fluvial, deltaic and eolian saturated deposits and poorly compacted sandfill have highest susceptibility to liquefaction. Thus, it has been possible to establish correlations between the liquefaction susceptibility and the geological and geomorphological units and to use the existing geological maps to obtain liquefaction susceptibility maps.

2.2 - Application to the Portuguese territory

The geomorphological and geological criteria for assessing liquefaction susceptibility have been applied to the Portuguese territory. The basic available information has been used for this purpose (Jorge 1993). This classification was assessed by the available data on liquefaction sites occurred in

Portugal during historical earthquakes. The classification of liquefaction susceptibility obtained for the Pleistocene and recent superficial deposits occurring in Portugal is presented in section 4. On basis of this classification, the Geological Map of Portugal 1:500 000 was modified into a Liquefaction Susceptibility Map. The character and meaning of the units of the existing geological map were modified to achieve this purpose.

The zoning map units adopted for Portugal are as follows:

- High to very high susceptibility
- Moderate susceptibility
- Low to very low susceptibility
- No susceptibility

3 - LIQUEFACTION OPPORTUNITY MAPPING

3.1 - General approach

The aim of the liquefaction opportunity map is to locate areas exposed to strong enough seismic ground shaking to cause liquefaction. For this purpose, the first step is to set up an appropriate empirical relation between magnitude and the maximum distance of liquefaction. Several researchers have provide correlations to estimate the maximum epicentral distance (R) or seismic source distance or fault distance (D) with the earthquake magnitude M (Figure 1).

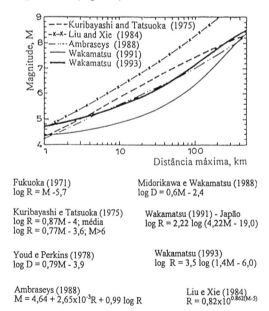

Fukuoka (1971)
log R = M -5,7

Kuribayashi e Tatsuoka (1975)
log R = 0,87M - 4; média
log R = 0,77M - 3,6; M>6

Youd e Perkins (1978)
log D = 0,79M - 3,9

Ambraseys (1988)
M = 4,64 + 2,65x10⁻³R + 0,99 log R

Midorikawa e Wakamatsu (1988)
log D = 0,6M - 2,4

Wakamatsu (1991) - Japão
log R = 2,22 log (4,22M - 19,0)

Wakamatsu (1993)
log R = 3,5 log (1,4M - 6,0)

Liu e Xie (1984)
R = 0,82x10^{0.862(M-5)}

Figure 1 - Epicentral distance to farthest liquefied sites versus magnitude (TC4, 1993).

In areas of moderate seismicity whenever such empirical relationship can be established from historical data, it is possible to use a probabilistic approach for the preparation of a Liquefaction Opportunity Map. Youd and Perkins (1978) have proposed an approach based on the classical method developed by Cornell (1968) for the probabilistic assessment of seismic hazard, modified by Algermissen and Perkins (1975).

That approach follows two basic steps:

- the definition of a seismotectonic model;
- the processing of seismic and liquefaction data by the modified Cornell method.

3.2 - Seismotectonic model

The strongest seismic activity in Portugal is located in the Gulf of Cadiz, offshore of the southern Portuguese coast, and can be associated with the Africa-Eurasian plates interface. The strongest earthquakes are not equally distributed and tend to be located in the area of the Horseshoe Abyssal Plain. The source of onshore earthquakes is more poorly understood. However, in order to create a seismotectonical model for the Portuguese territory one must cope with the problem related with definition of sources in a region where the seismic sources cannot be exclusively restricted to mapped fault zones. In practice, geological and seismological observations in a slow rate deforming area such as Portugal do not provide sufficient data to fully define the parameters of active crustal volumes and active fault systems for the elaboration of a synthesised seismotectonic model. Under these circumstance, the engineering requirements of a seismic hazard assessment may be fulfilled by a "seismicity source model" representing a simplification of the seismotectonic model designed to analyse the earthquake activity of a given site. The seismic source model generated for the assessment of liquefaction opportunity is based on area source zones representing geographic regions that are believed to contain at least one and perhaps a collection of faults that can generate earthquakes. Seismic parameters such as the frequency-magnitude relation and the maximum magnitude earthquake assigned to each zone are assumed to be homogeneous within the entire zone.

3.3 - Processing the seismic and liquefaction data by the modified Cornell method

The seismotectonic model made possible to apply the probabilistic method for the opportunity mapping following the procedure proposed by Youd and Perkins (1978).

The total accumulated liquefaction opportunity, resulting from the contribution of all seismic source zones is the annual cumulated frequency, f_c. For a period of time, T, the number of occurrence is given by $T.f_c$, which can be compared to the parameter m of the Poisson distribution.

$$P(n) = \frac{m^n}{n!} \cdot e^{-m} \qquad (1)$$

where P(n) is the probability of observing n occurrences and m is the mean of occurrences.

The determination of the annual cumulated frequency, f_c, for a large number of points uniformly distributed over the area to study (e.g. a grid) is one of the ways to construct a liquefaction opportunity map. If the return period is to be considered, the alternative is to use $1/f_c$ in the assessment.

3.4 - Application to the Portuguese territory

The study of historical records was a first and fundamental step for the preparation of the liquefaction opportunity map. The data collected made it possible to define a relationship between magnitude and maximum distance of liquefaction as well as to identify the sites that can undergo liquefaction in future incidents, since the phenomenon of liquefaction is a reincident process in the sedimentary formations.

For the Iberian Peninsula, the historical seismic record comprises almost two thousand years. The liquefaction phenomenon was identified in the literature survey for six earthquakes that developed intensive ground shaking in the Portuguese territory However, only the 1531, 1755, 1858 and 1909 earthquakes (Table 1) have provided more significant information concerning liquefaction. The first reference to liquefaction goes back to the XVI century, regarding the earthquake of 26[th.] January, 1531. The epicentre of this earthquake was located in the Lower Tagus Valley and developed liquefaction throughout the basins of the rivers Tagus and Sado.

Figure 2 shows the spatial distribution of liquefaction in Portugal for the major onshore and offshore historical earthquakes. The data concerning the description of these effects were classified as accurate, doubtful and very doubtful and also as reliable and non reliable.

Another interesting conclusion of this study is that the liquefaction effects of the 1755 earthquake have been reported to a distance over 400 km far from the presumed epicentral area. This fact can reflect a lower attenuation in this region comparing with other regions such as Japan or the West of the USA. Nevertheless, it is necessary to take into account the influence of different factors such as the

focal depth, the direction and the propagation of the seismic movement and the local effects of amplification. The seismic event of 1755 is considered to be the largest event ever felt.

3.4.1 - Definition of the relationship magnitude-maximum distance of liquefaction

The maximum distances for the four most important earthquakes that are related with liquefaction evidences is shown on Table 1.

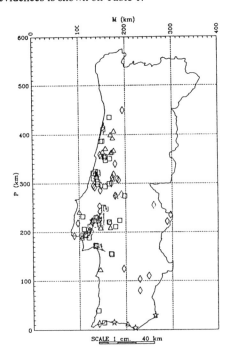

CLASSIFICATION OF HISTORICAL DATA

□ - Certain   ☆ - Credible
△ - Doubtful
◊ - Very doubtful

Figure 2 - Liquefaction sites associated with historical earthquakes (Jorge, 1993)

With these values it is possible to establish a relationship between the logarithm of the maximum distance of liquefaction and the earthquake magnitude. The following relation has been defined by means of a regression analysis:

$$\log R = 0,61M - 2,52 \qquad (2)$$

where R is the maximum distance of liquefaction and M is the Richter magnitude (M_s).

Table 1 - Maximum distance of liquefaction

Earthquakes	Magnitude	Maximum distance (km)
26.01.1531	7,1	80
01.11.1755	8,5	420
11.11.1858	7,2	66
23.04.1909	7,6	120

3.4.2 - Zoning for the Portuguese territory

By following the steps previously described the seismic source zones were defined based on the instrumental and historical seismological data (Sousa et al., 1992) and on the neotectonic features of the region (Cabral and Ribeiro, 1989). After an assessment of these data, seven seismic source zones were identified as capable of generating strong earthquakes ($M \geq 5$) to induce liquefaction in Portugal.

For the assessment of liquefaction opportunity a probabilistic approach was used. Primarily, it was necessary to calculate the annual frequency of earthquakes for each seismic source zone, per unit area. Each seismic source zone was divided into a certain number of unit square areas, considering the characteristics of each square unit as equal all over the area of the source-zone.

By taking the central value of each magnitude range, as the representative value for the determination of maximum distance of liquefaction by means of equation 2, the limit values beyond which liquefaction could not occur were then determined. These values are 7, 28, 114 and 462 km, respectively for the magnitude ranges of [5, 6[, [6, 7[, [7, 8[and ≥ 8.

The existing data and the relationship defined for the Portuguese territory made it possible to develop a probabilistic approach of the liquefaction opportunity for Portugal. After obtaining the maximum distances of liquefaction for each class of magnitude, the territory was covered by a grid with near 270 points and the conjugated contribution of all seismic source zones was determined for each point. The opportunities accumulated at a specific point due to the contribution of all seismic source zones produced a cumulative annual frequency of opportunity of liquefaction. The values obtained for each point of the grid made it possible to draw lines of equal values of the return period for the liquefaction opportunity, each interval of magnitude being taken into account. The overall picture is shown in Figure 3.

4 - LIQUEFACTION POTENTIAL MAP FOR PORTUGAL - LEVEL 1

By the superposition of the liquefaction opportunity and of the liquefaction susceptibility maps a liquefaction potential map for the Portuguese territory was obtained (Figure 4). This last map shows the locations where liquefaction susceptible sediments are more likely to occur and at the same time the return period for earthquakes strong enough to provoke liquefaction.

The units marked with different stripped patterns correspond to different probabilities of occurrence of liquefiable materials. Areas with a return period of liquefaction opportunity belonging to the same range of values are separated by isolines.

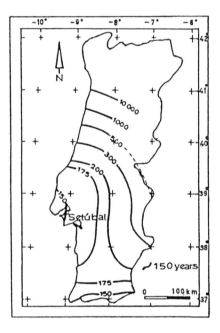

Figure 3 - Map of the return period of the liquefaction opportunity for the Portuguese territory

5 - ZONING OF THE LIQUEFACTION POTENTIAL FOR THE TOWN OF SETÚBAL - LEVEL 2

5.1 - General approach

The town of Setúbal was selected to illustrate the zoning of the liquefaction potential correspondent to level 2. Setúbal is located in an alluvial plane and has major harbour infrastructures. This choice is also justified by the fact that Setúbal is one of the Portuguese towns where liquefaction has occurred in historical earthquakes. It is also one of the few towns

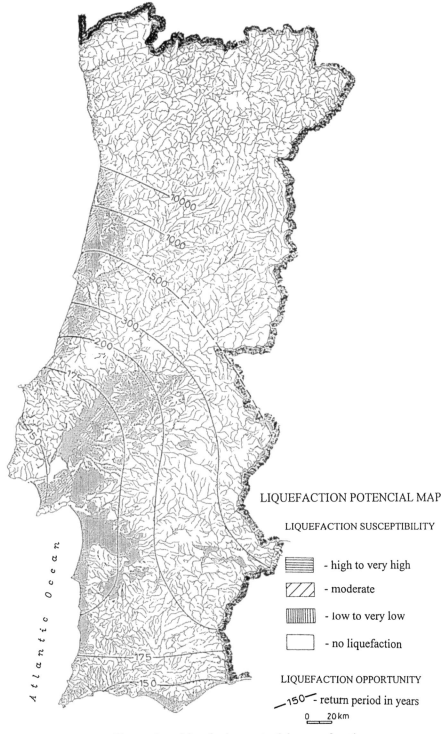

LIQUEFACTION POTENCIAL MAP

LIQUEFACTION SUSCEPTIBILITY

- high to very high

- moderate

- low to very low

- no liquefaction

LIQUEFACTION OPPORTUNITY

150 - return period in years

0   20 km

Figure 4 - Liquefaction potential map for the liquefaction opportunity for the Portuguese territory (Jorge, 1993) that apart from having a geological cartography it has also a geotechnical map (Gomes Coelho, 1980). For the zoning of the liquefaction potential the scale 1:25 000 was adopted.

For this zoning level the liquefaction opportunity map defined previously (Figure 3) was considered. According to this map the town of Setúbal is located in an area where the minimum return period of earthquakes capable of provoking liquefaction ranges from 150 to 175 years.

Thus, the study of the liquefaction potential was therefore confined to the preparation of a liquefaction susceptibility map (scale 1: 25 000) based on the following elements: geological maps, topographical maps, geotechnical maps of the region of Setúbal; aerial-photographic panchromatic coloured coverage; aerial-photographic infrared black and white coverage; inventory of 253 boreholes and in situ tests.

5.2 - Geological conditions of the area

The area can be divided into the following units as regards the set of ground surface Pleistocene and Holocene deposits (Gomes Coelho, 1980): afm - Fluvial-marine alluvia, with a significant thickness, composed by clayey, muddy and sandy deposits, unevenly stratified into oblique lenticular layers, forming a very heterogeneous group with several lateral and vertical variations of facies; af - fluvial clayey-sand alluvia with materials coming from the mountainous region; af' - fluvial sandy alluvia with reduced thickness, which may also reach 8m in the alluvial plain; Ca - coluvio-alluvial complex comprising the silty-sand heterogeneous materials with fragments of rock; S - Complex of ground surface sands resulting from desegregation of the Pliocene complex, with thickness ranging from 0 to 6 m, and with an average thickness of 2 to 3 m; at - earthfill consisting of coarse to medium sands, with frequent muddy passages, with thickness ranging from 12 to 15 m.

The low alluvial and colluvial zones with small gradient were defined (< 2%) based on the geotechnical map available and on the observation of aerial photographs. In these lower zones of small inclination, the phreatic level is at small depth (2-3m), and it varies according to the time of the year or with the tides. Table 2 shows the grain size distribution and the resistance to penetration for each unit defined.

5.3 - Assessment of the liquefaction potential

For the definition of the liquefaction susceptibility, an analysis was performed considering the data related with the composition of deposits, the thickness of the sand layers, their stiffness and the

hydrogeologic conditions obtained from the information of boreholes, from in situ tests and from the identification tests included in the inventory already mentioned. These data are summarised in Table 3 where the conditions favourable and/or unfavourable to the occurrence of liquefaction are defined.

The liquefaction susceptibility map presented in Figure 5 was established based on the classification defined in Table 3. This map is simultaneously a liquefaction potential map whenever the information of Figure 3 is considered.

Table 2 - Grain-size distribution and resistance to penetration

Units	Grain-size (Unified Classification)	SPT (N)
afm	sand complex SC, SM, SP	increase of the resistance to penetration with depth 0-20 - clay complex <8 - mud complex 10-30 - sand complex (below wl)
af	Clayey sands Sandy sands - well graded -	----------
af'	medium to fine sands SP; SP-SM	until 2m - 4 to 10 2 to 5m - 10 to 30
Ca	mixture of sands, pebbles and fines	----------
S	medium to coarse sands SP	----------
at	earthfill	9 a 60

6 - ZONING OF THE LIQUEFACTION POTENTIAL AT A LOCAL LEVEL - LEVEL 3

6.1 - Introduction

There are several methods for assessing the liquefaction potential. Most of these methods are empirical and are based in a comparison between the average cyclic shear stress induced by an earthquake (L) and the in situ shear resistance of the soil (R).

Table 3 - Calculation of the effect of the features of geological on the development of liquefaction

Geologic and geomorphologic mat. units	Sand Mat.	fine content (clay<2%)	high phreatic level	low compact-ness	gradient slope <2% bad-drain.	amplification of the seismic signal	favourable geologic structure	Susceptibility
fluvial marine alluvia (afm) sand complex	+	+	+	+/-	+	+	+/-	moderate to high
fluvial sand clay-alluvia (af)	-	-	-/+	----	+	+	+/-	≈ nil local. Moderate
sand fluvial alluvia (af')	+	+	-	+/-	-	+	+	low
coluvial-alluvial deposits (ca)	-	+	-	----	-/+	+	+/-	≈ nil
Ground surface sand layer (S)	+	+	-	----	+/-	+	+	very low
Compacted soil (at)	+	+	+	-	+	+	+/-	low to moderate

+ favourable conditions
- unfavourable conditions

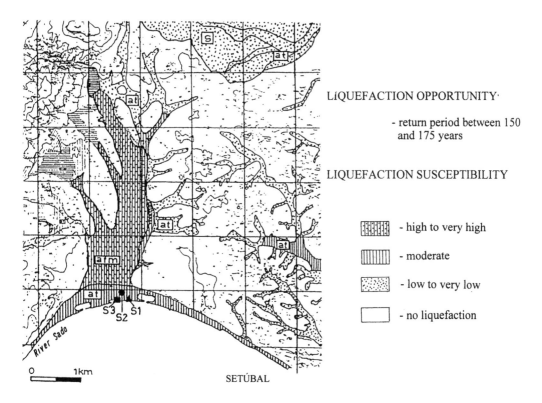

LIQUEFACTION OPPORTUNITY·

- return period between 150 and 175 years

LIQUEFACTION SUSCEPTIBILITY

- high to very high

- moderate

- low to very low

- no liquefaction

SETÚBAL

0   1km

River Sado

S3 S2 S1

Figure 5 - Zoning of the liquefaction potential for the town of Setúbal (Jorge, 1993)

39

$$F = \frac{R}{L} \qquad (3)$$

The shear stress induced by a seismic movement can be computed by numerical analyses or estimated using simplified empirical relations, such as the one proposed by Seed and Idriss (1971):

$$\frac{\tau_{av}}{\sigma'_0} = 0,65 \frac{a_{max}}{g} \times \frac{\sigma_0}{\sigma'_0} \times r_d \qquad (4)$$

where τ_{av} is the average shear stress induced by the earthquake; σ_0 is the overburden vertical total stress at the depth in question; σ'_0 is the overburden vertical effective stress at that depth; a_{max} is the maximum acceleration at the surface; g is the acceleration of the gravity and r_d is a stress reduction factor related with depth and the rigidity of the soil.

6.2 - Description of the performed analysis

The liquefaction potential was assessed through a comparison between the values of the average shear stress induced by the Portuguese standard seismic actions with the values of the resistant shear stress of the soil deposit calculated by means of SPT tests.

Three stratigraphic profiles (Figure 6) located in the lower zone of Setúbal town (see Figure 5) at the ground surface geological unit designated as at - earthfills (geotechnical map of Setúbal) were analysed. The sand materials and the muddy and clayey sands, with a percentage of fines of about 15%, were considered as susceptible to liquefy. Due to the proximity of the river an unfavourable scenario was assumed and the phreatic surface was

considered at the ground surface.

The methodology used for liquefaction potential assessment was based in Eurocode 8 (1994), which is generally similar to the one proposed by Seed et al. (1985).

In accordance with the methodology prescribed in EC8 (1994) the values of N are corrected for a reference effective stress of 100 kPa and for an impact energy of 60% of the free-fall theoretical energy. Regarding the effects of the overburden stress, the value of N is multiplied by the factor $(100/\sigma'_0)^{1/2}$. Following EC8 (1994) a 25% reduction of the N value for the initial three metres of ground surface was done.

For the determination of the shear stresses induced by the seismic movements defined in the Portuguese Code a stochastic analysis in the frequency domain was performed. This analysis considers a homogeneous and continuos soil deposit composed by horizontal layers with an infinite lateral dimension and it considers vertical propagation of shear waves (Vieira, 1994). The input seismic actions analysed are defined by means of acceleration power spectrums related to the seismic zone A of the Portuguese territory (the most severe) and concerning to type I soil (rock and hard soil) (RSA, 1983): type 1 standard seismic action (which corresponds to a moderate magnitude seismic event generated by a near seismic source) and type 2 standard seismic action (which corresponds to a high magnitude seismic event generated by a distant seismic source). A magnitude of 7 was associated to type 1 standard seismic action and a magnitude of 8 to type 2 standard seismic action (Campos Costa et al., 1991).

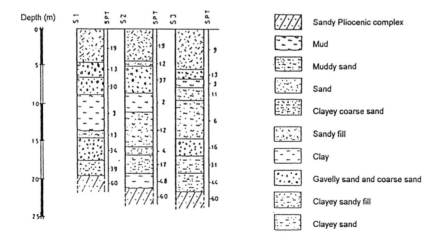

Figure 6 - Boreholes, lithological characterisation and SPT results

For the soil behaviour the equivalent linear method was used.

The values of G_{max} were obtained through correlations between SPT versus V_s (shear waves velocity). For the G/G_{max} versus γ (shear strain) and damping ratio versus γ (shear strain) variations, the relationships proposed by Vucetic and Dobry (1991) were used.

After the dynamic characterisation of the various layers the 50% probability shear stress peak response was calculated for the two standard actions.

Two series of calculations were carried out: in the first series the standard acceleration power spectrums were used directly (safety factor 1) and in the second series an increase of the unfavourable variable actions was taken into account by a safety factor of 1.5, as the values of the seismic action are characteristic values. This will correspond to a multiplication of the power spectral acceleration values by a 2.25 factor, since the peak response of a linear system for a given confidence level (probability of non-exceedence) is proportional to the square root of the area covered by this spectrum (Vieira e Correia, 1995).

The liquefaction graphs for clean sands and sands with 15% of fines proposed by the EC8 (1994) for magnitudes $7^{1/2}$ were used. These graphs are similar to the ones proposed by Seed et al. (1985). For other magnitude values correction factors were introduced.

6.3 - Results

The three average shear stress profiles (obtained by multiplying the median shear stress peak responses by 0.65) for the two seismic actions, for safety factors 1 and 1.5 are shown in Figure 7.

The liquefaction assessment for clean sands and silty sands, for type 1 and type 2 seismic actions, both for safety factors 1 and 1.5, is presented in Figures 8a) e 8b).

7 - CONCLUSIONS

The areas with a highest liquefaction potential are, accordingly to level 1 zoning, located in the southern coast of Algarve and in the region of Lisbon, Estuary of the river Sado and in the Lower Valley of the river Tagus. In the Northern part of the country the liquefaction potential decreases as a result of the fact that the liquefaction opportunity decreases significantly.

For the level 2 zoning, from the analysis performed, it can be concluded that the deposits with a highest liquefaction potential in Setúbal town are located in the downtown area, corresponding to the geological unit designated as afm. The sandfill of the harbour zone presents a moderate liquefaction potential, while for the other sedimentary units the liquefaction potential is low or very low.

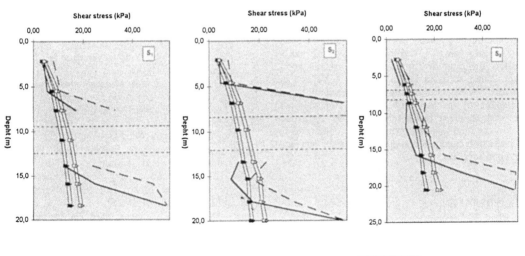

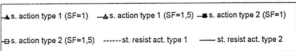

Figure 7 - Profiles of average induced shear stresses (S.F.=1 and S.F.=1.5) and resistant shear stresses for the two seismic actions.

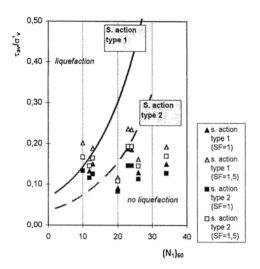

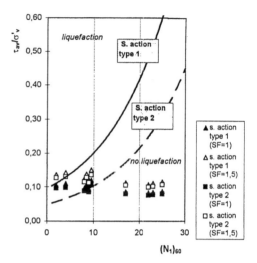

Figure 8a) - Relationship between τ_{av}/σ'_0 and $(N_1)_{60}$. Clean sands.

Figure 8b) - Relationship between τ_{av}/σ'_0 and $(N_1)_{60}$. Sands with 15% of fines.

In the site study, level 3 zonation, an analysis was made to the conditions existing in three soil profiles located in the earthfill of the riverine zone (harbour zone) of the town of Setúbal. The conclusions obtained for type 1 seismic action, was that the resistant shear stress exceeds always the induced shear stress for safety factor 1; regarding safety factor 1.5 it can be verified the occurrence of liquefaction in the site of borehole S3, in the sand (ground surface levels), where the conditions of the fill are less compact, and in the sites of S2 and S3 (below the mud level). Concerning seismic action type 2 the scenario is considerably different: the surface sand formations are susceptible to liquefaction in all profiles (for S2 and S3 considering safety factor 1.5) and for the formations below the muddy levels, there are also conditions for liquefaction occurrence in all sites both for increased and non-increased actions.

REFERENCES

Algermissen, S.T. & D.M. Perkins 1975. *A technique for seismic zoning: General considerations and parameters.* Proc. Int. Conf. on Microzonation. 865-874. Wash.:Seattle.

Cabral, J. & A. Ribeiro, A. 1989. *Neotectonic map of Portugal.* Descriptive note. Serviços Geológicos de Portugal. Lisboa.

Campos Costa, A., C.S. Oliveira & M.L Sousa 1992. *Seismic hazard consistent studies for Portugal.* Proceedings, 10th World Conference on Earthquake Engeneering. I:477-482.

Cornell, C.A. 1968. *Engineering seismic risk analysis.* Bull. Seism. Soc. Am.. 58(5):1583-1606.

EC8 - Eurocódigo 8 1994. *Design provisions for earthquake resistance of structures.* June.

Gomes Coelho, A. 1980. *Geotechnical cartography on the urban and regional planning. Application experience on the region of Setúbal.* Thesis for obtain the Degree of Specialist in LNEC. Lisboa.

Jorge, C. (1993) - *Zoning of liquefaction potential. Application attempt to the Portuguese territory.* Dissertation presented to the University Nova de Lisboa for obtain the Degree of Master of Sciences in Engineering Geology.

RSA (1983) - *Portuguese building code.* Decreto-Lei nº 235/83 de 31 de Maio.

Seed, H.B. & I.M. Idriss 1971. *Simplified procedure for evaluation of soil liquefaction potential.* JSMFD, ASCE. 79(SMD9).

Seed, H.B., K. Takimatsu, L.F. Harder & R.M. Chung 1985. *Influence of SPT procedures in soil liquefaction resistance evaluations.* JSMFD, ASCE. 79(SMD9).

Seed, H.B., I. Arango, C.K. Chan, A. Gomez-Masso & B. Grant Ascoli 1979. *Earthquake induced liquefaction near lake Amatitlan, Guatemala.* Report UCB/EERC-79/27, College of Engineering. University of California. Berkeley.

Seed, H.B., I.M Idriss & I. Arango 1983. *Evaluation of liquefaction potential using field performance data.* JGE, ASCE. 109(3):458-482.

Sousa, L.M., A. Martins & C. Sousa Oliveira 1992. *Seismic catalogues compilation for Iberian region.* Report 36/92-NDA, LNEC. Lisboa.

Vieira, A.M. 1994. *Seismic motions amplification by horizontal soil layers.* Thesis presented for obtain the degree of Master of Sciences in Soil Mechanics in the University Nova de Lisboa.

Vieira, A.M. & R. Correia 1995. *Liquefaction potencial assessment in EXPO'98 area.* Report 91/95-NEGE, Geotecnia, LNEC.

Vucetic, M. & R. Dobry 1991. *Effect of soil plasticity in ciclic response.* Journal of Geotechnical Enginneering, ASCE. 117(GT1):89-107.

TC4 1993. *Manual for zonation on seismic geotechnical hazards.* Japanese Society of Soil Mechanics and Foundation Engineering, ISSMFE. December.

Youd, T.L. & S.N. Hoose 1977. *Liquefaction susceptibility and geologic setting.* Proc. 6th World Conf. on Earthq. Eng.. 6: 37-42. India:New Delhi.

Youd, T.L. &. D.M. Perkins 1978. *Mapping liquefaction-induced ground failure potential.* JCED, ASCE. 104(GT4): 433-446.

Seismic Behaviour of Ground and Geotechnical Structures, Sêco e Pinto (ed.)© 1997 Balkema, Rotterdam, ISBN 90 5410 887 8

Site effects for long-period Rayleigh waves

Amir M. Kaynia & Farrokh Nadim
Norwegian Geotechnical Institute, Oslo, Norway

ABSTRACT: Surface waves are scattered by alluvial valleys in their propagation path. This article presents parametric studies of amplification of Rayleigh waves in a sediment-filled valley. The results of the study show that sediment-filled valleys can significantly amplify Rayleigh waves that have a wave length shorter than twice the width of the valley. As the wavelength of the surface waves increases, the sediment-filled valley starts to 'ride' on the wave and the site amplification effects become less significant. For Rayleigh waves with a wavelength of the same order as the length of the valley, significant spatial variation of the motion takes place within the valley.

1 INTRODUCTION

The effects of local soil conditions and surface topography on modifying the characteristics of seismic waves have been recognized since the 1940s. Despite considerable progress in modelling various aspects of seismic site effects, the state-of-practice in geotechnical earthquake engineering is still the one-dimensional solution which is based on vertically propagating shear waves in horizontally layered sites of infinite lateral extent (e.g., Schnabel et al. 1972). The simplicity of this method combined with the possibility of performing an approximate non-linear analysis, through the equivalent-linear approach of Seed and Idriss 1969, have contributed to its popularity. This method has been successful in simulating the earthquake recordings atop sedimentary valleys in a large number of cases where the site effects have been practically one-dimensional. It has been also applied to cases where the site response were primarily due to surface waves. The one-dimensional model is clearly not suitable for such applications. This model is based on the shear mode shapes of the site which are different from the surface mode shapes. Therefore, it cannot reproduce the correct magnification of surface waves, which in certain cases may be underpredicted. Moreover, the one-dimensional model cannot simulate the phase effect and spatial variability of ground motions. In such cases, one has to resort to *two- or three-dimensional* numerical models. The objective of this paper is to present certain representative results of a general, two-dimensional model for Rayleigh waves.

2 BACKGROUND

The local site effect has been observed in well-documented earthquakes. For instance, it has been shown that during the Skopje earthquake of July 26, 1963, the localized damage distribution was related to the surface topography (Poceski 1969). A more thoroughly studied case has been the great Michoacan earthquake of September 19, 1985 which resulted in great casualties and economic loss in Mexico City. The studies conducted on the geotechnical aspects of this earthquake have concluded that the existence of a thin layer of a soft clay and the lateral heterogeneity of the sedimentary basin have been the main culprits for the significant amplification of the seismic excitation (Sanchez-Sesma et al. 1988, Campillo et al. 1988, Kawase and Aki 1989, and Chavez-Garcia and Bard 1994).

The mathematical complexity of the boundary value problem related to wave scattering by canyons and sedimentary valleys has limited analytical solutions to a few cases only. Prominent contributions are due to Gilbert and Knopoff 1960, Trifunac 1971 & 1973, Wong and Trifunac 1974, Lee 1982 and Sanchez-Sesma 1985, among others. As for numerical solutions, one of the earliest solutions is due to Aki and Larner 1970 who studied scattering of SH waves by irregular interfaces. The Aki-Larner method was extended by Bouchon 1973 and Bard and Bouchon 1980 & 1985 to the numerical solution of two-dimensional wave scattering problems. Using this technique, Bard and Bouchon pointed out the significant role of sediment-induced surface waves in the response of the valley and the resonant characteristics of two-dimensional valleys.

Other, more general numerical techniques, namely the finite difference method, the finite element method and the boundary element technique, have been more extensively applied to the solution of wave scattering by surface topography or sedimentary valleys. Most of these solutions, however, have been related to body waves. (See Aki 1988 and de Barros and Luco 1995 for a comprehensive survey of the reported solutions.)

Dealing with surface waves, Ohtsuki et al. 1984 developed a time domain finite element-finite difference model which incorporated Dirichlet and Neumann conditions as well as viscous damping to act as absorbing boundaries. They studied the effect of topography and local soil conditions on the surface motions due to Rayleigh waves. Ohtsuki et al. 1984 showed that a Rayleigh wave propagating through hard ground is magnified as it enters a two-layered ground with a soft surface layer. They observed that for the case $\lambda/h = 4$ (λ and h being respectively the wave length and the thickness of the surficial layer) the vertical motion at the surface of the two-layered ground was about 1.6 times the vertical amplitude of the incident wave. This ratio increases to about 3.0 in the case of $\lambda/h = 2$.

The finite element method was applied to the problem of wave scattering by a rectangular sediment-filled valley by Tassoulas and Roesset 1991. They used transmitting boundaries in their frequency domain analyses and observed a remarkable sensitivity of the response of the valley to the frequency of the incident waves. Depending on the characteristics of the incident Rayleigh wave, they obtained both amplification and deamplification of the waves.

Despite successful applications of the finite difference and finite element methods, most of the available numerical solutions of wave propagation in sedimentary valleys have been obtained by direct or indirect boundary element methods. This is due to the special features of such techniques in reducing the dimensionality of the problem, and hence solution time, as well as in handling the wave radiation in a rigorous manner. One of the first solutions by this method was presented by Wong 1982 who studied the scattering of Rayleigh waves and body waves by semi-circular and semi-elliptical canyons. Wong's formulation was modified and extended by Dravinski 1982 & 1983 and Dravinski and Mossessian 1987 for the solution of wave scattering by multiple and non-homogeneous valleys. These studies have shown significant interaction between the incident surface waves and the valley and the possibility of large amplification of the waves. Other boundary element methods have been developed by Bravo and Sanchez-Sesma 1990 and Sanchez-Sesma et al. 1993. For the same problem, discrete wave number

direct boundary element methods have been developed by Kawase and Aki 1989 and Papageorgiou and Kim 1993.

For the present numerical modelling, the computer code *SASSI* (Lysmer et al. 1981) was used. this code has the attractive feature of the finite element method in modelling heterogeneous media and has the capability of simulating the radiation condition in infinite media. A short description of the model is given in the next section.

3 PROBLEM DEFINITION AND PARAMETERS

Figure 1 shows schematically the problem under investigation. A Rayleigh wave propagating in a half space strikes a sedimentary valley. Part of this wave is transmitted to the valley and part of it is scattered. The objective of the present work was to obtain surface motions in the valley as a result of this wave scattering. It was assumed that both the basin (half space) and the valley were made of horizontal layers, and that during the seismic disturbance, the ground materials remained linearly elastic. The soils in various regions/layers were characterized by their mass density, ρ, Poisson's ratio, ν, shear wave velocity, V_s, and hysteretic damping ratio, β.

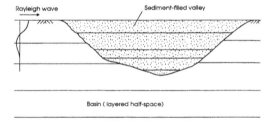

Figure 1. Sediment-filled valley in a layered half-space.

For the numerical simulations of the site response, a special finite-element-based formulation, *SASSI* (Lysmer et al. 1981), based on the *flexible volume Substructuring* was used. This is a numerical formulation most suitable for the soil-structure interaction analysis of embedded structures. According to this method, the complete system comprising the structure and the ground (i.e. the sedimentary valley and the basin for the wave scattering problem under investigation) is partitioned into two substructures: one, representing the basin in which the sediment is replaced by the basin-soil so as to form a uniform (undisturbed) ground, and the other representing the valley minus the basin-soil. Together, these substructures form the original site model. The second substructure is modelled by traditional finite elements and the first one is represented through the medium's

Green's functions at the same nodes. (For the mathematical derivations of *SASSI* see Lysmer et al. 1981.)

To validate the use of *SASSI* for the present site response analyses, the results obtained by this code were compared with those of Dravinski and Mossessian 1987 for a semi-circular valley. Figure 2 shows the finite element mesh used to model the valley. Four-noded quadrilateral elements were used for this purpose. For a reliable representation of the waves, the mesh size should be smaller than $\lambda/5$, where $\lambda = V_S T$ is the wave length of the shear waves and $T = 2\pi/\omega$ is the period of waves. The radius of the valley was R = 500 m, and the following values were used for the soil parameters in the valley:

shear wave velocity = 200 m/s
mass density = 1400 kg/m^3
Poisson's ratio = 1/3
damping ratio = 0.5%

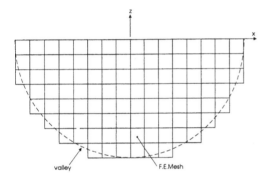

Figure 2. Finite element mesh for the semi-circular valley.

For the basin, the shear wave velocity and mass density were respectively 400 m/s and 2100 kg/m^3, while its Poisson's ratio and hysteretic damping ratio were the same as those of the valley. These values gave the same non-dimensional parameters used in the reference results. Comparisons are presented here for two values of frequency: $\omega = 1.257$ rad/s and $\omega = 1.885$ rad/s, which respectively correspond to periods of 5.0 s and 3.3 s.

Figure 3 (a) displays the transfer function of the horizontal displacement at various locations along the ground surface for $\omega = 1.257$ rad/s and Figure 3 (b) shows the corresponding transfer function for the vertical displacement. In both cases, the transfer function defines the ratio between the absolute value of the ground motion and the free-field horizontal motion. With this definition, the transfer functions then define the amplification of the incident wave. The results of the present study are plotted together with the rigorous results of Dravinski and Mossessian 1987. The figure shows good agreement between the two solutions.

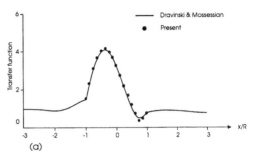

(a)

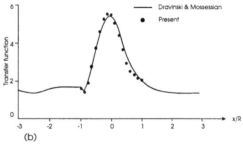

(b)

Figure 3. Absolute value of transfer functions for horizontal (a) and vertical (b) surface displacements in a semi-circular valley: T = 5.0 s.

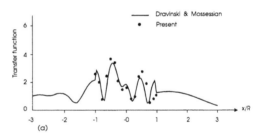

(a)

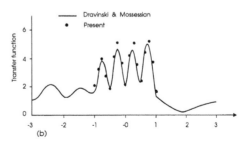

(b)

Figure 4. Absolute value of transfer functions for horizontal (a) and vertical (b) surface displacements in a semi-circular valley: T = 3.3 s.

The plots in Figure 3 clearly show that the valley significantly magnifies both horizontal and vertical surface displacements, especially in the central portion of the valley. The wave-length of Rayleigh wave in the basin is about 2000 m (i.e. twice that of the length of the valley). This explains the smooth variation of the transfer functions.

Figures 4 (a) and 4 (b) portray the transfer functions in the same medium for ω = 1.885 rad/s. Again good agreement with rigorous solution is observed. Because of smaller wave length in this case, the transfer functions exhibit stronger variations. The amplifications, however, are smaller than the previous case.

4 PARAMETRIC STUDIES

For the numerical simulations, a hypothetical rectangular valley of length L = 1000 m and depth h = 100 m was considered and the following values were used for the soil parameters in the valley and the surrounding rock basin.

• *Soil (valley) properties :*

shear wave velocity	V_s^{soil} =	200 m/s
mass density	ρ_s =	2000 kg/m³
pressure wave velocity	V_p^{soil} =	500 m/s

damping ratio	β =	0.5%

• *Rock (basin) properties :*

shear wave velocity	V_s^{rock} = 1000 m/s
mass density	ρ_r = 2200 kg/m3
pressure wave velocity	V_p^{rock} = 2000 m/s
damping ratio	β = 0.5%

The valley was modelled by 1000 four-noded quadrilateral elements with side dimension 10 m. Three cases corresponding to periods of 0.5, 1.0 and 2.0 seconds were considered in the simulations. In all cases, the wavelength in the basin was larger than the thickness of the valley (about 500m for the shortest period).

Figure 5 shows the real (top plot) and imaginary (bottom plot) parts of the displacement vectors in the valley for the case T = 0.5s. This case corresponds to a wavelength λ of about 500m in the basin. The figure shows a complex spatial site response in the valley which is a result of wave scattering and reflections at the three boundaries of the valley. If, for any of the two components of surface displacements, the ground motion amplification is defined as the ratio between that motion and the corresponding free-field motion, then the results show maximum amplifications of 7.4 for the horizontal motion and 4.5 for the vertical motion.

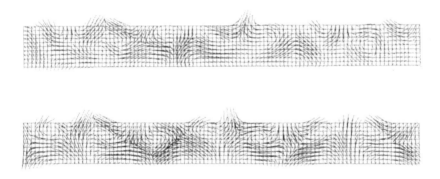

Figure 5. Real (top) and Imaginary (bottom) parts of displacement vectors for T = 0.5 s.

Figure 6 shows the same set of results for the case T = 1.0 s. As expected, due to the larger wavelength of this case (relative to the length of the valley), a smoother pattern of motions is displayed. This case, however, results in a stronger amplification of waves - 6.5 for the horizontal and 7.2 for the vertical cases. It is interesting to note that for the same model and assuming the 1-D shear model, one obtains almost no amplification in the horizontal motion.

Finally, Figure 7 shows similar results for the case T = 2.0s. The wavelength in this case is about 2000m in the basin and is twice the length of the valley. Therefore, it displays less spatial variation of ground motion. Whereas the horizontal motion is strongly amplified, the vertical motion seems to have approached that of the free-field. The amplification factors in the two directions in this case are 9.8 and 1.5, respectively. The reason for the large amplification of the horizontal motions is that the valley essentially responds in one-dimensional shear to the excitation in the rock which happens to have a period equal to the natural period of the layer, T = 4 h/V_s =

2.0 s. The fairly uniform horizontal motion in the central portion of the valley clearly shows this feature. A 1-D shear model, on the other hand, gives an amplification of 5.3. If the impedance contrast between the soil and the rock is increased then the amplification of the 1-D model increases, and the two models give more consistent amplifications. For instance, if the shear wave velocity of the rock is increased to 3000 m/s then the 1-D model gives an amplification of about 14.5, compared to 12.5 for the Rayleigh wave model.

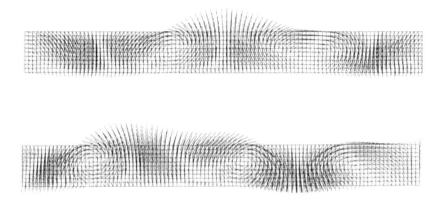

Figure 6. Real (top) and Imaginary (bottom) parts of displacement vectors for T = 1.0 s.

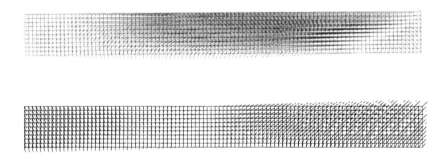

Figure 7. Real (top) and Imaginary (bottom) parts of displacement vectors for T = 2.0 s.

Simulations with larger periods (not presented here) showed reduced amplifications in both directions. For example, for the cases with T = 4.0 s and T = 6.0 s, the horizontal amplifications were respectively about 1.5 and 1.2. In fact, as the period increases, the sediment-filled valley 'rides' on the waves rather than interacting with it.

wave increases, the sediment-filled valley starts to 'ride' on the wave and the site amplification effects become less significant.

For Rayleigh waves with a wave length of the same order as the length of the valley (e.g. Fig. 5), significant spatial variation of the motion takes place within the valley.

5 CONCLUSIONS

The numerical simulations performed in this study show that sediment-filled valleys can significantly amplify Rayleigh waves that have a wave length shorter than twice the width of the valley. As the period, and hence the wave length, of the surface

ACKNOWLEDGEMENT

The work described in paper was supported by the Commission of European Communities, Contract No. EV5V-CT-94-0491: Long Period Earthquake Risk in Europe, and by Statoil. Their support is gratefully acknowledged. The opinions expressed in the paper are those of the authors and do not necessarily represent the standpoint of the sponsors.

REFERENCES

Aki, K. 1988. Local site effects on strong ground motion. *Geotechnical Special Publication No. 20*, ASCE, pp. 103-155.

Aki, K. and Larner, K. L. 1970. Surface motion of a layered medium having an irregular interface due to incident plane SH waves. *J. Geophys. Res.* 75: 933-953.

Bard, P. Y. and Bouchon, M. 1980. The seismic response of sediment-filled valleys, Part 1. The case of incident SH waves, and Part 2. The case of incident P and SV waves. *Bull. Seism. Soc. Am.* 70: 1263-1286 and 1921-1941.

Bard, P. Y. and Bouchon, M. 1985. The two-dimensional resonance of sediment-filled valleys. *Bull. Seism. Soc. Am.* 75: 519-554.

Bouchon, M. 1973. Effect of topography on surface motion. *Bull. Seism. Soc. Am.* 63: 615-632.

Bravo, M. A. and Sanchez-Sesma, F. J. 1990. Seismic response of alluvial valleys for incident P, SV and Rayleigh waves. *Soil Dyn. Earthquake Eng.* 9: 16-19.

Campillo, M., Bard, P. Y., Nicollin, F. and Sanchez-Sesma, F. J. 1988. The incident wave field in Mexico City during the great Michoacan Earthquake and its interaction with the deep basin. *Earthquake Spectra*, 4 (3): 591-608.

Chavez-Garcia, F. J. and Bard, P. Y. 1994. Site effects in Mexico City eight years after the September 1985 Michoacan earthquake. *Soil Dyn. Earthquake Eng.* 13: 229-247.

de Barros, F. C. P. and Luco, J. E. 1995. Amplification of obliquely incident waves by a cylindrical valley embedded in a layered half-space. *Soil Dyn. Earthquake Eng.* 14: 163-175.

Dravinski, M. 1982. Scattering of elastic waves by an alluvial valley. *J. Eng. Mech.* ASCE, 108: 19-31.

Dravinski, M. 1983. Amplification of P, SV and Rayleigh waves by two alluvial valleys. *Soil Dyn. Earthquake Eng.* 2: 66-77.

Dravinski, M. and Mossessian, T. K. 1987. Scattering of plane harmonic P, SV and Rayleigh waves by dipping layers of arbitrary shape. *Bull. Seism. Soc. Am.* 77: 212-235.

Gilbert, F. and Knopoff, L. 1960. Seismic scattering from topographic irregularities. *J. Geophys. Res.* 65: 3437-3444.

Kawase, H. and Aki, K. A. 1989. A study on the response of a soft soil basin for incident S, P and Rayleigh waves with special reference to the long duration observed in Mexico City. *Bull. Seism. Soc. Am.* 79: 1361-1382.

Lee, V. W. 1982. A note on the scattering of elastic plane waves by a hemispherical canyon. *Soil Dyn.*

Earthquake Eng. 1: 122-129.

Lysmer, J., Tabatabaie-Raissi, M., Tajirian, F., Vahdani, S. and Ostadan, F. 1981. SASSI - A system for analysis of soil-structure interaction. *Report UCB/GT 81-02*, Univ. of California, Berkeley.

Ohtsuki, A., Yamahara, H. and Harumi, K. 1984. Effect of topography and surface inhomogeneity on seismic Rayleigh waves. *Earthquake Eng. Struc. Dyn.* 12: 37-58.

Papageorgiou, A., S. and Kim, J. 1993. Propagation and amplification of seismic waves in 2-D valleys excited by obliquely incident P- and SV-waves. *Earthquake Eng. Struct. Dyn.* 22: 167-182.

Poceski, A. 1969. The ground effect of the Skopje July 26, 1963 earthquake. *Bull. Seism. Soc. Am.* 59: 1-29.

Sanchez-Sesma, F. J. 1985. Diffraction of elastic SH waves by wedges. *Bull. Seism. Soc. Am.* 75: 1435-1446.

Sanchez-Sesma, F., Chavez-Perez, S., Suarez, M., Bravo, M. A. and Perez-Rocha, L. E. 1988. On the seismic response of the valley of Mexico. *Earthquake Spectra*, 4: 569-589.

Sanchez-Sesma, F. J., Ramos-Martinez, J. and Campillo, M. 1993. An indirect boundary element method applied to simulate the seismic response of alluvial valleys for incident P, S and Rayleigh waves. *Earthquake Eng. Struc. Dyn.* 22: 279-295.

Schnabel, P. B., Lysmer, J. and Seed, H. B. 1972. SHAKE : A computer program for earthquake response analysis of horizontally layered sites. *Report EERC 72-12*, Univ. of California, Berkeley.

Seed, H. B. and Idriss, I. M. 1969. The influence of soil conditions on ground motions during earthquakes. *J. Soil Mech. Found. Eng.* ASCE, 94 (SM1) : 93-137.

Tassoulas, J. L. and Roesset, J. M. 1991. Wave propagation in a rectangular valley. *Structural Safety*, 10: 15-26.

Trifunac, M. D. 1971. Surface motion of a semi-cylindrical alluvial valley for incident plane SH waves. *Bull. Seism. Soc. Am.* 61: 1755-1770.

Trifunac, M. D. 1973. Scattering of plane SH waves by a semi- cylindrical canyon. *Earthquake Eng. Struc. Dyn.* 1: 267-281.

Wong, H. L. 1982. Effect of surface topography on the diffraction of P, SV and Rayleigh waves. *Bull. Seism. Soc. Am.* 72: 1167-1183.

Wong, H. L. and Trifunac, M. D. 1974. Surface motion of a semi-elliptical alluvial valley for incident plane SH wave. *Bull. Seism. Soc. Am.* 64: 1389-1408.

Seismic Behaviour of Ground and Geotechnical Structures, Sêco e Pinto (ed.) © 1997 Balkema, Rotterdam, ISBN 90 5410 887 8

Seismic behavior of the alluvial clay layer at Kobe Port Island during the 1995 Hyogoken-Nanbu Earthquake

M. Kazama, E. Yanagisawa & A. Yamaguchi
Tohoku University, Department of Civil Engineering, Sendai, Japan

ABSTRACT: The authors studied the seismic behavior of the ground at Kobe Port Island during the 1995 Hyogo-ken Nanbu Earthquake. In this paper we focused on the seismic behavior of the alluvial clay layer and conducted cyclic simple shear tests using the stress wave inferred from array records. The specimen used in this study was the clay sampled in Rokko Island near Kobe Port Island. The clay was from the same clay layer that was at Kobe Port Island, but it was not as consolidated as the clay at Kobe Port Island. After consolidation with several confining pressure levels, we conducted a dynamic simple shear test under the same stress rates as the actual earthquake. We compared the shear strain time histories inferred from array records with those from simple shear test. The strain time history obtained from simple shear test gave good comparison to that obtained from array records.

1 INTRODUCTION

On January 17, 1995, the Hyogo-ken Nanbu Earthquake, also called "The Great Hanshin-Awaji Earthquake," attacked the Kansai district of Japan. Port Island (P.I.) and Rokko Island (R.I.), which are man-made islands in Kobe city, had subsidence of several tens of cm, and the harbor facilities at the islands suffered severe damage. While studying the damage to these man-made islands, it is very important to know how much force acted on the ground and how the ground behaved during the earthquake. Fortunately, the strong motion observation system installed at Port Island successfully recorded the main shock and a series of aftershocks.

As it is well known that Port Island and Rokko Island were covered with sand boils due to liquefaction, there is no doubt that the reclaimed ground liquefied during the earthquake. The reclaimed soil was mainly granite-origin sandy soil called Masado. From the facts that Masado consists of much gravel and fines, many researchers are interested in how the liquefaction developed and how much strength against liquefaction Masado had. Ishihara et al. (1995) concluded, from a simple analysis based on liquefaction strength curve, that the shear stress ratio acting on the reclaimed ground was 0.4 to 0.7 and that the first or second pulse caused liquefaction. The authors also studied the liquefaction process at Kobe Port Island by using array records (Kazama et al., 1996b). From the study it was found that the maximum stress ratio be-

ing acted on the reclaimed ground was about 0.4 and that after the five to seven seconds the ground reached complete liquefaction.

On the other hand, we should not overlook that an underlying alluvial clay layer was very soft. The soft clay layer must have affected the seismic response of the man-made islands. According to the results of a pore pressure observation conducted in Rokko Island and Port Island second stage (after Kobe City report 1995), excess pore water pressure was developed after the earthquake. Especially for Rokko Island the excess pore water pressure due to earthquake reached about 20kPa, it corresponds to almost the same as the overburden pressure. This excess pore water pressure dissipated gradually in two weeks. This fact indicates that earthquake loading causes the effects like dynamic consolidation.

In this paper we focused on the seismic behavior of an underlying alluvial clay layer at array observation site in Kobe Port Island. First we have estimated the stress time history being acted on alluvial clay layer by using array records. Second by using the stress time history we have conducted dynamic simple shear test of the clay sampled from Rokko Island.

2 THE GROUND OF TWO MAN-MADE ISLANDS

Kobe City is located at the northwest side of Osaka Bay, and its population is about 1.5 million. The flat land area is only a two to four km wide alluvial plain

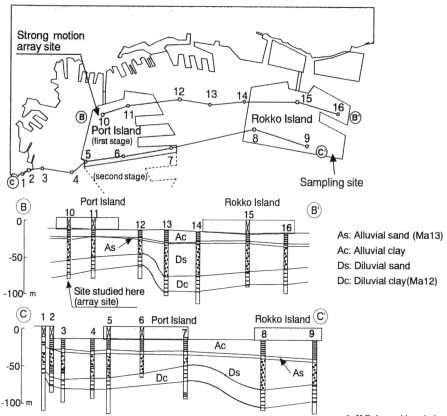

As: Alluvial sand (Ma13)
Ac: Alluvial clay
Ds: Diluvial sand
Dc: Diluvial clay(Ma12)

*: K.P. is an abbreviation for Kobe Pile,
which is mean sea level at Kobe Port.

Figure 1 Soil structure under the Kobe man-made islands

between Rokko Mountain and the coast line. Kobe City has been developed with Kobe Port opening in 1868. Today Kobe port plays an important role as an international trading port. Since the city area of Kobe City was a narrow area as written above, new development had to go towards the mountain or coastal area. To enlarge the city area, the construction of Port Island started in 1966, and the construction of Rokko Island started in 1972 (Akedo, J., 1993).

2.1 Geologic profile along the islands

Figure 1 shows the location of two man-made islands and the geologic profile along the islands (after Kobe City report, 1995). As shown in this figure, an alluvial clay layer (Ma13) underlying the man-made island was the old sea floor ground. Its thickness ranges from 10 to 20 meters. This alluvial layer is continuous between two islands.

The layer underlying the alluvial clay, represented by Ds, has alternately accumulated a thin layered structure with clayey and sandy soil. There is a

diluvial clay (Ma12) layer widely spread in starting from K.P.[*] -40m to -70m. The difference of soil structure between two islands is the depth of this layer. In this figure the site studied in Port Island and the sampling site in Rokko island are shown.

2.2 History of reclamation

For reclamation, several areas of Rokko mountain were cut, and the soil was transported to the construction area as fill. Figure 2 shows the reclamation history of the two man-made islands. Fill in Port Island was placed from 1968 to 1980. The reclamation of the site shown here had been completed in 1968-69. On the other hand, the fill in Rokko island was place between 1976 and 1992. The reclamation of the sampling site was completed in 1992. Comparing the time difference between the two sites, there is a 23-24 years time lag. Therefore the clay sampled from Rokko Island is unconsolidated younger clay compared to the clay at the array site in Port Island.

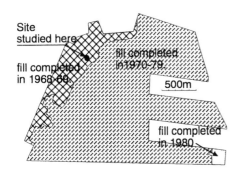
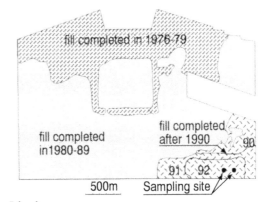

Figure 2 Reclamation history of Port Island and Rokko Island

Table 1 Soil properties of alluvial clay at Kobe man-made islands.

| | Port Island | Rokko island | | |
| | Array site | Center | Sampling site | |
	studied here		Sample-1	Sample-2
reclamation	1968-69	1981	1992	1992
sampling	1995.5	1989.11	1996.1	1996.1
G.L. (m)	17.7-28.0	35.4-45.7	23.0-23.8	25.0-26.8
N-value	4-5	---	1	1
V_S (m/s)	180	---	---	---
ρ_d (g/cm^3)	1.0-1.1	---	0.817	0.755
ρ_t (g/cm^3)	1.6-1.7	1.5-1.6	1.515	1.479
ρ_s (g/cm^3)	2.7	2.6	2.707	2.698
w_p (%)	25-30	30-45	41	42-43
w_n (%)	50-60	55-70	81	95
w_l (%)	85-90	100-115	116	112-115
I_p	60	70	75	70-72
e	1.45-1.55	1.45-1.90	2.313	2.575
q_u (kPa)	130-200	120-24	---	42
p_c (kPa)	250-300	250-500	44	60
C_c	0.7-0.9	0.7-1.0	0.55	1.02

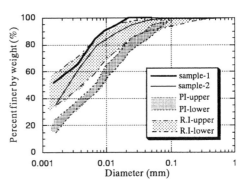

Figure 3 Grain-size distribution curve of the alluvial clay at Kobe man-made islands

2.3 Soil properties of alluvial clay

Table 1 shows the soil properties of alluvial clay at the Kobe man-made islands. This table shows the soil properties of array site in Port Island, the center in Rokko Island and the sampling site in Rokko Island. Two kinds of samples were used for simple shear testing, these two have somewhat different soil properties. As written previously, the property of the soil changes with the consolidation time. We have to pay close attention to the difference in plastic index (I_p) between the two islands.

Figure 3 shows the grain size distribution of the alluvial clay in the Kobe man-made island area. From this curve it is found that the sample used here had much finer content than the clay at the array site. It

seems that the clay sampled from Rokko Island is much more clayey than that at the array site.

3 REVIEW OF THE STRONG MOTION ARRAY AT KOBE PORT ISLAND

As is well known, four seismographs were installed in depth direction from the ground surface K.P.+4.0m to a depth of K.P.-79m in Kobe Port Island. The first research interest in the field of soil dynamics is to examine the nonlinear behavior of the ground subjected to significant strong motion. A number of earthquake response analyses using the strong motion records from Port Island have been already performed by many researchers. The dynamic properties of ground were also determined by Sato et al. (1996) and Yoshida et al. (1995).

Figure 4 shows the generalized soil profile, as well as the wave velocity structure obtained from PS logging, at the observation site in Kobe Port Island. From the surface to the depth of about K.P.-14m there is reclaimed Masado soil. The Underlying alluvial

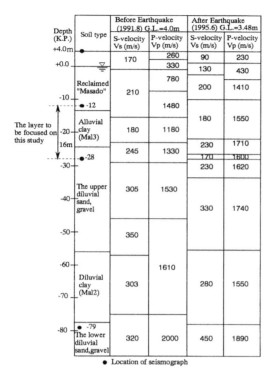

Depth (K.P.)	Soil type	Before Earthquake (1991.8) G.L.=4.0m		After Earthquake (1995.6) G.L.=3.48m	
		S-velocity Vs (m/s)	P-velocity Vp (m/s)	S-velocity Vs (m/s)	P-velocity Vp (m/s)
+4.0m					
+0.0	Reclaimed "Masado"	170	260	90	230
			330	130	430
-10		210	780	200	1410
-12			1480		
-20	Alluvial clay (Mal3)	180	1180	180	1550
16m		245	1330	230	1710
-28				170	1600
-30				230	1620
-40	The upper diluvial sand, gravel	305	1530		
				330	1740
-50		350			
-60	Diluvial clay (Mal2)	303	1610	280	1550
-70					
-79 -80	The lower diluvial sand, gravel	320	2000	450	1890

The layer to be focused on this study

● Location of seismograph

Figure 4 Soil profile of the array site at Kobe Port Island (K.P. is an abbreviation for Kobe Pile, which is mean sea level at Kobe Port.)

The accuracy of the technique with regard to wave velocity, seismograph installation spacing, and the frequency content has also been examined by the authors (1996a).

The first and second authors applied this technique to Kobe Port Island array records and have examined stress-strain relationship of the ground. The hysteretic deformation properties have been already analyzed from the following point of view: (1) Softening of the ground stiffness during the earthquakes, (2) Nonlinear property changes during the main shock and the aftershocks and (3) Liquefaction process of the reclaimed ground during the main shock. Our next interest is how are consistent the results obtained from array records with the results of the traditional elementary test, such as cyclic triaxial test and cyclic simple shear test. In this chapter we explain the outline of shear stress-strain relationship of the alluvial layer obtained from array records. Details of the method are available (Kazama et al., 1996b).

4.1 Evaluation of the shear stress

If the wave propagation can be assumed purely one dimensional in the depth z direction, shear wave equation can be written by equations (1):

$$\frac{\partial \tau_{xz}}{\partial z} = \rho \frac{\partial^2 u_x}{\partial t^2}, \qquad \frac{\partial \tau_{yz}}{\partial z} = \rho \frac{\partial^2 u_y}{\partial t^2} \tag{1}$$

Integrating the equations (1) from surface to depth z with the stress free surface boundary condition, the stress at level z can be expressed by equations (2).

$$\tau_{NS}(z,t) = \int_0^z \rho(z)\alpha_{NS}(z,t)dz$$
$$\tau_{EW}(z,t) = \int_0^z \rho(z)\alpha_{EW}(z,t)dz \tag{2}$$

Where subscripts NS and EW represent the direction of the acceleration component. Using the distribution of the acceleration $\alpha(z,t)$ and mass density $\rho(z)$, we can calculate the stress at any depth z. In this study, the distribution of acceleration can be approximately determined from array records using linear interpolation.

4.2 Evaluation of average shear strain

The average shear strain in the second layer can be calculated from the relative displacement between -12m and -28m divided by the thickness of the layer. We have obtained the displacements by double integration of the acceleration records.

Since we assumed uniform deformation in the layer, we can not extract the stress and strain due to the high frequency motion, in which the assumption of the lin-

clay layer extends from about K.P.-14m to -24m. Its thickness is about 10 meters. The layer to be focused on in this study is this alluvial layer. We have to pay attention to that the second layer sandwiched between -12m and -28m including the upper reclaimed layer (2m) and the lower diluvial layer (4m).

The PS logging data shown in the figure were the results of both before the earthquake (1991.8) and after the earthquake (1995.6). From the data, the shear wave velocity of the reclaimed ground decreased after the earthquake, but the shear wave velocity of the alluvial clay layer had no change.

To estimate the stress-strain relationship, we have used the main portion of the main shock data from 10s to 28s.

4 EVALUATION STRESS AND STRAIN HISTORIES INFERRED FROM ARRAY RECORDS

Few attempts have been made so far to evaluate the stress-strain relationship directly under actual earthquake loading. In recent years, however, a new technique to evaluate the stress-strain relationship directly from actual vibration data was developed (Koga et al., 1990, Zeghal et al., 1994,1995, Kazama et al., 1995).

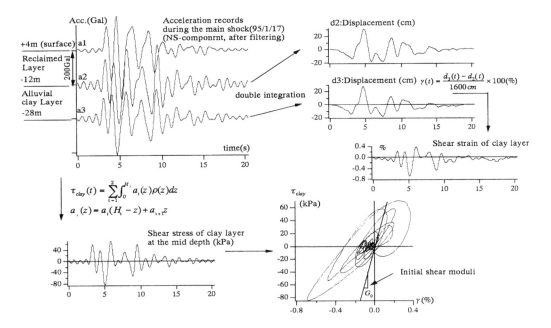

Figure 5 Estimation of average shear stress and strain of the alluvial clay layer from array records.

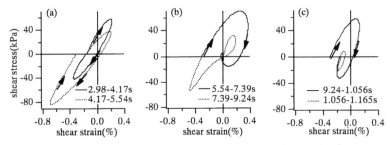

Figure 6 Change of the stress strain relationship of the clay layer with time

ear interpolation is not valid. According to the accuracy of the technique examined by the first author, if 10% approximation error is allowable, the highest frequency we can evaluate is 1.2Hz for the main shock horizontal components. These values are obtained from the wave velocity identified by Sato et al. (1996).

4.3 Stress-strain relationship of the alluvial clay layer obtained

Figure 5 shows the schematic diagram of the method to evaluate stress and strain. Since the locus analysis of array records in NS-EW plane showed that the principal axis of the motion was N20W at Port Island (Sato et al., 1996), we showed the results for NS component. In the following section, we mainly explain

the data for NS component. The maximum shear strain generated was about 0.7% for the alluvial clay layer (-12m to -28m). The maximum shear stress at K.P.-20m, which is the value at the mid depth of the second layer, was about 80kPa.

Using the stress and the strain time histories evaluated, we can plot stress-strain loops during main shock as shown in Figure 5. Calculating the initial shear moduli of the alluvial clay from shear wave velocity $V_s = 180 m/s$ and its unit weight $\gamma_t = 1.6 tf/m^3$, we can obtain a value of 53MPa. The evaluation of shear moduli from V_s and γ_t was made through the following relationship.

$$G_0 = (\gamma_t/g)V_s^2 \qquad (3)$$

where g is acceleration due to gravity. In Figure 5 a solid line with a gradient of an initial shear moduli

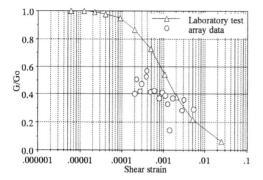

Figure 7 Shear modulus versus shear strain during the main shock

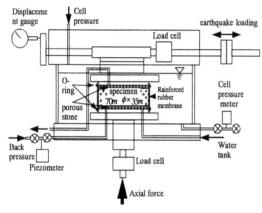

Figure 8 NGI type simple shear testing apparatus

Table 2 Test condition

Test No.	1	2	3	4	5
Sample No.	sample-1			sample-2	
Sampling depth (G.L.)	(23-23.8m)			(25-26.8m)	
Pre-loading pressure (kPa)	245	343	441	245	245
Pre-loading time (day)	3	3	3	3	30
OCR	1	1.4	1.8	1	1

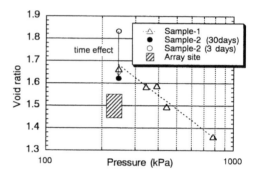

Figure 9 Relation between the pressure-yoid ratio curve of the sample used here and (e, σ_v) of the P.I. alluvial clay.

G_0 is drawn. It is found that significant nonlinear behavior occurred during main shock compared to the initial shear moduli.

Figure 6 shows the change of the stress-strain behavior during the earthquake. From the gradient of an each loop we can estimate the change of the shear modulus with strain amplitude during the main shock. Figure 7 shows the shear modulus reduction pattern with strain amplitude. In the figure we evaluated a shear moduli of each loop by the following method.

$$G = (\tau_{max} - \tau_{min}) / (\gamma_{max} - \gamma_{min}) \qquad (4)$$

The line plotted in Figure 7 is the G-γ curve obtained from laboratory test (after Kobe City report, 1995). It is found that the shear moduli reduced to about 25% of the initial shear moduli. It is also found that shear modulus reduction occurred roughly along the solid line obtained from laboratory test.

Concerning the damping ratio, there is difficulty in evaluating a damping ratio under random loading. In particular, few obvious loops were found in the stress-strain plots. Incidentally, the damping ratio evaluated from loops for Figure 6 (a), which is relatively obvious loop compared with others, were from 14% to 24%.

5 SIMULATION OF SEISMIC BEHAVIOR OF ALLUVIAL CLAY LAYER USING SIMPLE SHEAR TEST

For studying the dynamic behavior of soils, laboratory element tests such as the cyclic triaxial test and the torsional cyclic test have often been used to obtain stress-strain relationships. It is important to know how consistent laboratory tests are with results obtained from actual earthquake records. The authors have conducted simple shear test using the time history of shear stress obtained from previous study.

The test apparatus used here was an NGI-type simple shear apparatus as shown in Figure 8. The dimensions of a specimen are 70mm diameter and 35mm height. Pore pressure was measured through the porous stone in the center of the lower pedestal.

Table 2 shows the test case condition. Pre-loading pressure and duration time for preconsolidation were testing parameters. The 245kPa corresponds to the effective overburden pressure at K.P.-20m, it is the mid depth of alluvial clay layer studied here.

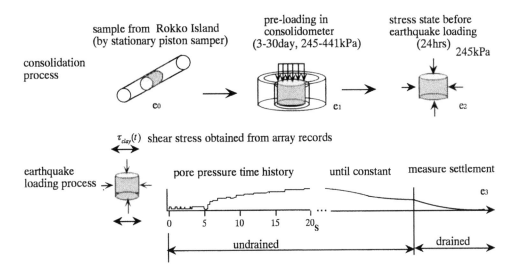

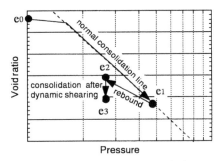

Figure 10   Simple shear testing procedure

5.1 Sample used in simple shear test

Since the average stress-strain relationship was obtained at Port Island K.P. -20m, it is the best choice to use the sample taken from the same point and depth as the array observation site. However, we could not sample an alluvial clay from Port Island, but we got undisturbed alluvial clay sample from Rokko island. The undisturbed sample was taken with a stationary piston sampler from a depth of 23.0-26.8m below the ground surface. As we mentioned before, the alluvial clay layer under the two man-made islands is continuous and there is only small differences between two samples except for the degree of consolidation. Figure 9 shows the e-logP relationship obtained from the clay sample used this study. In this figure a hatched region covering the extent of the void ratio of a 1.45-1.55 and a pressure of a 206-264kPa corresponds to the array site data. We can read 0.15-0.25 difference for void ratio between normal consolidation line and array site data at 245kPa. It is considered to be a time effect for 23-24 year difference in age.

5.2 Testing procedure

The testing procedure is shown in Figure 10. First, the specimens with initial dimensions of 74mm in diameter and 50mm in height were trimmed from the sampling core. Next specimens were preconsolidated in consolidometer with a vertical pressure of 245-441kPa for several days as prescribed. Then termination of consolidation was confirmed by $\sqrt{t}$ method. After consolidation, the specimens were set up in the cell and consolidated isotropically to 245kPa. To ob-

Figure 11 Pressure-void ratio change during the test.

tain a high degree of saturation, a back pressure of 40kPa was applied for 24 hours. B-values of more than 95% were observed in all specimens used in the tests. At this time in case of OCR>1 the sample generates elastic rebound due to unloading. Figure 11 shows schematic diagram of pressure and void ratio relationship of a OCR>1 case. where e_0 is an initial void ratio before the consolidation, e_1 is the void ratio after the preconsolidation, e_2 is the void ratio just before the earthquake loading and e_3 is the void ratio after a dissipation of pore pressure due to earthquake loading. We obtained e_3 from a volumetric change during a drain process and the water content of specimen after the earthquake loading.

The dynamic shear stress, which was inferred from array data, was applied at an actual rate up to 1.2Hz under the undrained condition. Then we measured pore pressure and displacement response during the dynamic loading. After the earthquake loading, keeping the undrained condition, we waited until an excess pore pressure would be stable. After that, the pore pressure generated was drained through both ends.

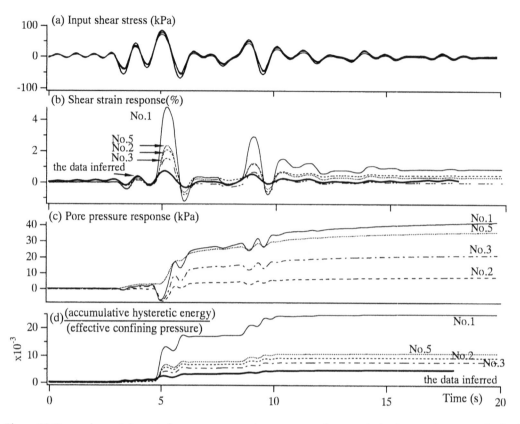

Figure 12 Comparison of shear strain, excess pore water pressure and accumulative hysteretic energy obtained from simple shear testing with the data inferred from array records

Variation of the pore pressure and vertical displacement were also measured statically after the dynamic loading.

5.3 Test results and discussion

Figure 12 shows the time histories obtained from the simple shear test. Thick solid line represents the data inferred from vertical array records. Here we will focus on dynamic shear strain response and the settlement due to earthquake loading.

It is very important to know how much strain developed during the earthquake in relation to an earthquake response analysis. It is found from Figure 12 (b) that the maximum response of dynamic shear strain is far different from the data inferred from array records. However, its change pattern is consistent between each other. Figure 13 shows the relation between e_2, void ratio just before the earthquake loading, and the maximum dynamic shear strain response. Extrapolating the results obtained from the test result seems to be harmonious with the data in-

ferred from the array records. Furthermore, looking at the shear strain time histories carefully, we also found that, until 4.8 seconds before the maximum shear strain response appeared, all cases show the same magnitude of response. This fact suggests that the shear strength of a clay affects the magnitude of dynamic shear strain response.

The next interesting point is how much settlement of the clay layer developed due to earthquake. Since there is no transducer for dynamic pore pressure change at array site in Kobe Port Island, it is impossible to discuss how the excess pore water pressure developed. However we can calculate an accumulative hysteretic energy consumed as a plastic strain energy during the earthquake using stress-strain relationship. It is known that this energy corresponds to excess pore water pressure response (Towhata, et al, 1985, Sugano, 1994). Therefore if an accumulative hysteretic energy agree with that of the data inferred, it is considered that the excess pore water pressure will agrees with also. For instance, the volumetric strain obtained from test No.3 was 0.07%, then the total settlement of clay layer in 10m thickness will be

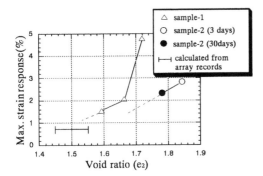

Figure 13 Void ratio versus maximum shear strain response

Table 3 Test results

Test No.	1	2	3	4	5
e1	1.663	1.580	1.494	1.84	1.620
e2	1.718	1.662	1.592	1.844	1.781
e3	1.655	1.643	1.589	1.753	1.769
$\Delta H / H$ (%)	1.4	0.44	0.07	2.01	0.27
γ_{max} (%)	4.770	2.088	1.550	2.860	2.352
$\gamma_{residual}$ (%)	0.829	0.482	0.005	0.285	0.350

about 0.7cm. This value is not so large value compared with the settlement due to liquefaction. However, for Rokko Island with unconsolidated soft clay, there is possibility that settlement of alluvial clay layer due to earthquake is much larger than array site in Port Island. In addition, this difference might cause a different seismic response between two man-made islands. Matsuda (1996) studied the method to evaluate the settlement of a clay layer due to earthquake. Because of the difficulty in distinguishing the settlement due to earthquake from an ordinary consolidation it is not clear how much settlement occurred in alluvial clay layer due to earthquake.

6 CONCLUSION

In this study the seismic behavior of the alluvial clay layer at the array site in the Kobe Port Island during the 1995 earthquake has been studied. The conclusions obtained from this study summarized as follows:

1. Using the vertical array records, we inferred the stress-strain response during the earthquake. The maximum shear strain developed in the clay layer was estimated to be about 0.7% for NS component. Shear moduli decreased during the main shock to about 25% of initial shear moduli.

2. Using the shear stress time history inferred from array records, we have conducted cyclic simple shear testing. When we consider the change depended on the void ratio of specimen, the magnitude of dynamic shear strain was harmonious with the data inferred from array records.

3. Settlement of the alluvial clay layer at the array site due to the earthquake is not large. However, the settlement of unconsolidated soft clay in Rokko Island may be much larger than that at array site.

Because of the limitation of the clay sample used, we could not consider the time effects on the dynamic response. Further study will be required for the time effects due to consolidation and cementation on dynamic response of a clay layer during a significant large earthquake.

REFERENCES

Akedo, J. 1993. Construction of Kobe Rokko Island, *Special issue of man-made island of JSCE*, Vol.78-12 : 107-109. (in Japanese)

Ishihara, K., Yasuda, S. and Nagase, H. 1996. Soil characteristics and ground damage, *Special issue of SOILS AND FOUNDATIONS*: 109-118.

Kazama, M., E. Yanagisawa, H. Toyota and I. Towhata 1995. Stress strain relationship of sandy soils directly obtained from 1-D centrifuge shaking table tests. *Proc. of IS-Tokyo '95/1st. Intern. Conf. on Earthquake Geotechnical Engrg.*, Vol.2 : 711-716.

Kazama, M., Toyota, H., Towhata I. and Yanagisawa, E. 1996a. Stress strain relationship of sandy soils obtained from centrifuge shaking table tests. *Proc.of JSCE*, No.535/III-34 :73-82.(in Japanese).

Kazama, M., Yanagisawa, E., Inatomi T., Sugano, T. and Inagaki, H. 1996b. Stress strain relationship in the ground at Kobe Port Island during 1995 Hyogo-ken Nanbu earthquake inferred from strong motion array records. *Proc. of JSCE*, No.547/III-36 :171-182.(in Japanese)

Kobe City report 1995. Investigation of ground deformation of reclaimed ground due to Hyogo-ken Nanbu earthquake (Port Island, Rokko Island) : 1-119. (in Japanese)

Koga, Y. and Matsuo O. 1990. Shaking table tests of embankments resting on liquefiable sandy ground. *SOILS AND FOUNDATIONS*, Vol.30, No.4 : 162-174.

Matsuda, H. 1996. Eathquake-induced settlement-time relation of clay layer, *Proc.of the 31th Japan national conference on geotechnical engineering.* : 1061-1062. (in Japanese)

Sato, K., Kokusho, T., Matsumoto, M. and Yamada, E. 1996. Nonlinear seismic response and soil property during strong motion, *Special issue of SOILS AND FOUNDATIONS*: 41-52.

Sugano, T. 1994. Experimental study on the effects of principal stress direction and rotation on the deformation behavior of sand. *Doctor Thesis submitted to Tohoku university.* (in Japanese)

Towhata, I. and Ishihara, K. 1985. Shear work and pore water pressure in undrained shears. *SOILS AND FOUNDATIONS, Vol.25, No.3.*: 73-84.

Yoshida, I. and Kurita, T. 1995. Back analysis of soil properties of Port Island with the observation data. *Tsuchi-to-Kiso,* Vol.43, No.9 :44-48. (in Japanese)

Zeghal, M.and A.-W. Elgamal 1994. Analysis of site liquefaction using earthquake records. *J. of Geotech. Engrg. Div., ASCE*, Vol.120, No.6: 996-1017.

Zeghal, M., A.-W. Elgamal., H.T. Tang and J.C. Stepp 1995. Lotung downhole array. II: Evaluation of soil nonlinear properties. *J. of Geotech. Engrg. Div., ASCE*, Vol.121, No.4 : 363-377.

Seismic Behaviour of Ground and Geotechnical Structures, Sêco e Pinto (ed.)© 1997 Balkema, Rotterdam, ISBN 90 5410 887 8

Nonlinear site response during the Hyogoken-Nanbu earthquake recorded by vertical arrays in view of seismic zonation methodology

Takeji Kokusho
Chuo University, Tokyo, Japan

Masaki Matsumoto
Kansai Electric Power Company, Osaka, Japan

ABSTRACT: During the 1995 Hyogoken-Nanbu earthquake, vertical array records were obtained at four sites with much different epicentral distances around the earthquake fault zone. These acceleration records demonstrated conspicuous nonlinearity effect in seismic response due to different input accelerations. These vertical array records are analyzed in this research by means of the inversion technique to back-calculate dynamic soil properties exhibited during the strong earthquake motion. Based on these vertical array records for the main shock and the aftershocks, relationships between S-wave velocity ratio and acceleration amplification ratio are examined in the light of proposed methodologies. A good correlation can be found between these parameters for linear site responses while lower amplification is evident for nonlinear response for strong seismic motions.

1 INTRODUCTION

Local site amplification is one of the most important factors in seismic zonation study. The site amplification is correlated to properties of soil layers such as soil densities, wave velocities and material dampings. At the same time it is expected to be highly dependent on the nonlinearity of soil properties in soft soil sites in particular.

Nonlinear seismic response of soft ground due to nonlinear soil properties was numerically evaluated either by equivalent linear analyses (e.g. Schnabel 1972) or by step-by step nonlinear analyses (e.g. Constantopoulos et al. 1973) for the past two decades. In model tests, Kokusho et al. (1979) performed a shake table tests of a model ground in a laminar shear box of about one meter in depth consisting of fine sand and demonstrated a very clear reduction in dynamic amplification due to increasing input acceleration level. The same authors also compared the test results with the equivalent linear analysis based on the soil properties of the model ground under a very low confining pressure to find a fair agreement between them. More recently several centrifuge shake table tests have been conducted for sand layers in laminar shear boxes to find clear amplification reduction with increasing acceleration again. Thus, the nonlinearity of site amplification

due to strong input motions is obviously shown in numerical analyses and model tests. However, due to absence of vertical array records during strong earthquakes, not many seem to have believed the nonlinear seismic response to actually occur in the local site amplification.

2 SITE CONDITION AND SEISMIC AMPLIFI-CATION

Vertical arrays which could record the main-shock of the 1995 Hyogoken-Nanbu earthquake (M_J=7.2) were located in four sites in the coastal zone around the Osaka-Bay area as shown in Fig.1. The same figure also indicates the fault zone including the epicenters of the main-shock as well as aftershocks. The four sites were very properly distributed by chance in terms of distances from the fault zone which can be estimated from the aftershock epicenters plotted in Fig.1. PI (the Port Island) array belonging to the Kobe Municipal Office was located just next to the fault zone, while other three arrays SGK, TKS and KNK belonging to the Kansai Electric Power Company were approximately 15km, 35km and 65km far from the fault zone respectively. The soil profiles and the depth of three dimensional down-hole seismographs are shown for the four sites in Fig.2

together with P and S-wave velocities measured by the down-hole logging method and SPT N-values along the depth. The deepest seismographs at the base layers of the four sites were located GL-83m in PI, GL-97m in SGK, and GL-100m in TKS and KNK respectively, and the geological condition there were Pleistocene dense gravelly soils except for KNK (a hard rock). Upper soil conditions at the four sites are rather similar as shown in Fig.2, consisting of sandy fill at the surface in most sites underlain by Holocene clay and/or sand and further underlain by Pleistocene soils. The S-wave velocity, Vs, at the base layer of Pleistocene gravelly soil in PI, SGK and TKS is 380-480 m/s while Vs at the base rock in KNK is as high as 1630 m/s.

In Fig.3, down-hole distribution of maximum acceleration in two horizontal (NS,EW) and one vertical directions (UD) at the four sites are shown. The maximum accelerations at the deepest level are different in a wide range (from 26gal in KNK to 679gal in PI) in the four sites, leading to obvious difference in amplification in upper layers. In the horizontal direction the amplification between the surface and the deepest level was 4 to 5 in KNK for

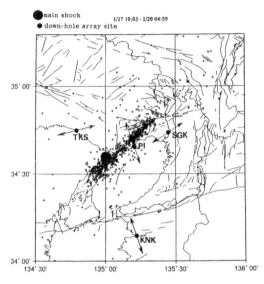

Fig.1   Location of vertical array sites around Osaka Bay and epicenters of main-shock and aftershocks

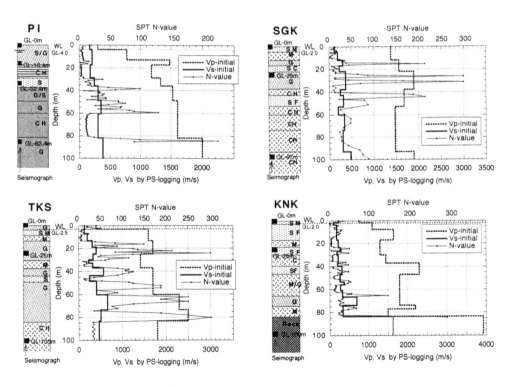

Fig.2 Soil profiles , wave logging test results and SPT N-values at four down-hole array sites

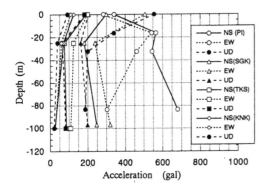

Fig.3 Distribution of maximum acceleration along depth at vertical array sites

the base acceleration of about 20gal, while in TKS it was about 2 for the acceleration of 100gal. It was between 1 to 2 for the acceleration of about 300gal and about one half in PI for the base accelerations of 300 to 680gal. In PI, the surface sandy fill layer experienced very extensive liquefaction during the main-shock which obviously led to the deamplification in the maximum acceleration in the upper sandy fill layer.

3 SEISMIC RECORDS AND INVERSION ANALYSES

The main-shock records obtained in the vertical arrays in the four sites were first examined to know directional drift of the buried seismographs in the horizontal plane (Sato et al. 1995). This examination revealed the following directional drifts or errors; a) 15 degrees clockwise rotation at GL-83.4m in PI, b) reverse in the NS-component at GL0m and 34 degrees anti-clockwise rotation at GL-97m in SGK, c) 30 degrees anti-clockwise rotation at GL-25m in TKS and d) reverse in the EW-component and 60 degrees clockwise rotation at GL0m in KNK. All data were accordingly corrected for later analyses except for the PI record in which 15 degrees of rotation was judged to have negligible effects on subsequent analyses.

A computer code used for the inversion analyses was originally developed by Ohta[1975], in which multi-reflection of vertically propagating SH-wave in a horizontally layering system is assumed. In the analysis transfer functions (Fourier spectrum ratio) between measured seismic motions at different levels were developed. Next, a soil layer model was made

with given thickness and density of each layer based on bore-hole data. Then starting from initial guess, the S-wave velocity, V_s, and the damping ratio, h were back-calculated so that the transfer functions of recorded motions be best reproduced by the soil layer model. Only the absolute value of the transfer function was used for the back-calculation. The initial guess for the S-wave velocity was made by reducing the measured V_s indicated in Fig.2 by some amount according to estimated strain-dependent velocity reduction. For the damping ratio laboratory cyclic loading test data were referred for the initial guess.

Since the details of the inversion analysis and their interpretation from the viewpoint of strain-dependent soil properties are available in other literature (Kokusho et al. 1996), only some principal results are addressed here. In Figs.4 to 7 time histories of the main-shock recorded at the four vertical array sites are shown with thick curves. The thin curves in the same figures are computed results based on the optimized S-wave velocity, V_s, and damping ratio, h, evaluated from the inversion analyses.

In Fig.8, S-wave velocities, V_s, in PI back-calculated for the major portion of the time history in the main-shock (T=20 seconds) are plotted together with initial values corresponding to the wave logging test against the depth for NS and EW directions. In NS direction, optimized V_s is approximately 20% and 40% smaller than the initial value for the Pleistocene and Holocene layer respectively. For the fill layer, 80% and 50% reduction of V_s can be identified for saturated and unsaturated layer respectively presumably because of the extensive liquefaction in the fill.

In the same diagram, V_s distribution optimized only for a partial time history (T=5.12s, from 4.10s to 9.22s in Fig.4) in which SH-wave is dominant in the measured time history is also shown. No meaningful difference can be seen between the results for the major portion time history and the partial time history, implying that reliable soil properties can be back-calculated from the major portion time history although different types of waves such as P and S-wave and surface waves are all together involved in it [Sato et al. 1995]. Based on this finding, major portions of time histories (T=20s for PI and SGK, T=40 for TKS and KNK) were chosen in all records for the back-calculation of properties hereafter.

Another interesting results can be obtained from the analysis for an aftershock with maximum surface acceleration of about 0.01G which occurred only about two minutes after the main-shock. As shown in the graph, V_s for Pleistocene and Holocene layers

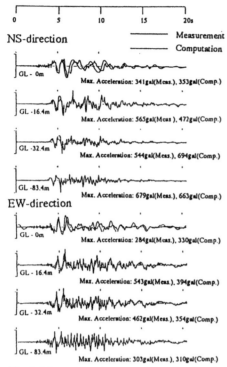

Fig.4 Measured and computed acceleration time histories at PI

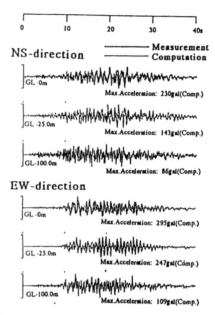

Fig.6 Measured and computed acceleration time histories at TKS

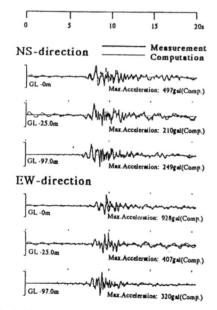

Fig.5 Measured and computed acceleration time histories at SGK

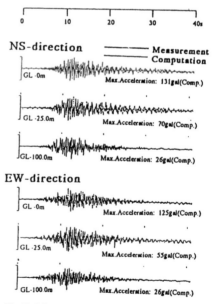

Fig.7 Measured and computed acceleration time histories at KNK

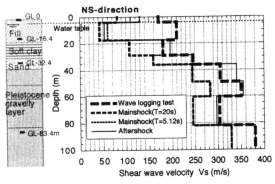
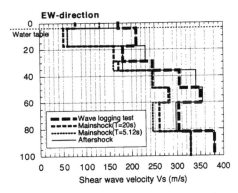

Fig.8 Back-calculated S-wave velocity for NS and EW-motions compared with wave logging test at PI

can be interpreted to have returned to the initial value because of small magnitude of the induced strain during the after-shock while those in the fill layer take almost the same reduced value as the main-shock, indicating the liquefaction in the fill was still sustained. With regard to the main-shock in the EW direction in Fig.8, the reduction rate of Vs is a little smaller than that in the NS direction because of the smaller input acceleration, while for the aftershock the result is exactly the same as that in the NS direction.

Fig.9 indicates the damping ratio, h, for horizontal motions in NS and EW directions, respectively. For the main-shock, h was optimized as 6% for the Pleistocene soils and the Holocene sand layer, while it was as high as 33 to 50% for the fill layer, indicating that equivalent damping ratio for the liquefied layer can be evaluated very high. The optimized h for the aftershock despite induced small strain is evidently larger, 5% for the Pleistocene soils and 10 to 12% for Holocene and fill layers, than the estimated value based on laboratory test data as shown on the same diagram.

In SGK site no trace of liquefaction such as sand eruptions, ground fissures, soil subsidence, etc. was found after the earthquake. The S-wave velocity, Vs, and the damping ratio, h, back-calculated in two directions are taken against the depth in Fig.10. In upper Holocene and Pleistocene layers of silts or sands, reduction rate in Vs compared with Vs corresponding to those in PS-logging is evaluated as 30 to 50% while in lower Pleistocene layers it is less than 20%. The damping ratio amounts to 2 to 6% in the Pleistocene soils in contrast to mostly more than 10% in Holocene soils.

In TKS site, soil eruptions and fissures were witnessed after the earthquake indicating the occurrence of liquefaction near the surface. S-wave velocities and damping ratios back-calculated for NS

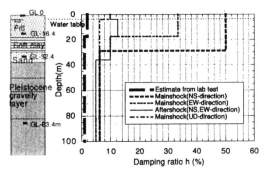

Fig.9 Back-calculated damping ratio for horizontal and vertical motions compared with lab test value at PI

and EW directions are plotted against the depth in Fig.11. Vs is drastically reduced from initial Vs corresponding to wave logging test more than 80% and 50% in silt and silty sand between GL-3m and GL-13m, implying the occurrence of liquefaction in these layers, whereas in the layers below the reduction rate is less than 25% and mostly less than 10%. The damping ratio is evaluated as 3 to 5% in the lower layers and even in the liquefied upper layers it is back-calculated as 6 to 10% in contrast to 33 to 50% in the liquefied soil in PI.

S-wave velocities and damping ratios back-calculated for NS and EW directions in KNK are plotted against the depth in Fig.12. Vs is reduced from the initial values by maximum 20% in the upper layers while in the lower layers the reduction rate is less than 5%. h takes 1% in the layers deeper than 40m and about 2% in the upper layers except in the very shallow part where h is evaluated as about 4%. With epicentral distance of 65km, the max. acceleration at the base rock in this site was about 0.025G, resulting in almost linear response in this site.

65

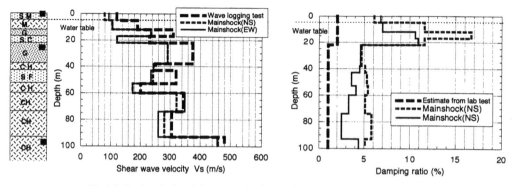

Fig.10 Back-calculated S-wave velocity and damping ratio at SGK

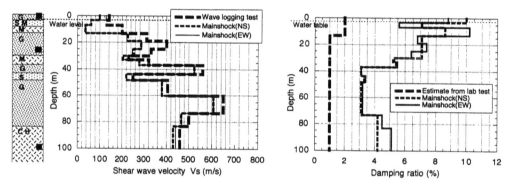

Fig.11 Back-calculated S-wave velocity and damping ratio at TKS

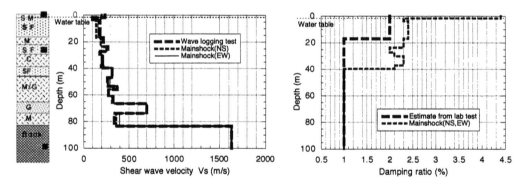

Fig.12 Back-calculated S-wave velocity and damping ratio at KNK

A normal multi-reflection analysis of the main-shock has been carried out based on the optimized properties to compare the results with the measurement. In Fig.4, time histories computed from the lowest level are indicated for the four vertical array sites with thin lines. In Fig.13 distribution of maximum accelerations along the depth computed by the analyses are compared with the measured values for the four sites. The agreement seems good in some sites quantitatively while in other sites at least qualitatively.

In Fig.14 computed max. shear strains, γ_{max}, are shown against the depth for two horizontal directions for the four sites. This indicates that the shear strain evaluated by the equivalent linear analysis exceeds 1 to 2% in the liquefied layers in PI and TKS and it also exceeds 0.1% in all other deeper layers in PI and SGK. Contrary to that, shear strain in KNK stays less than 0.05% indicating that almost linear seismic response took place in this site.

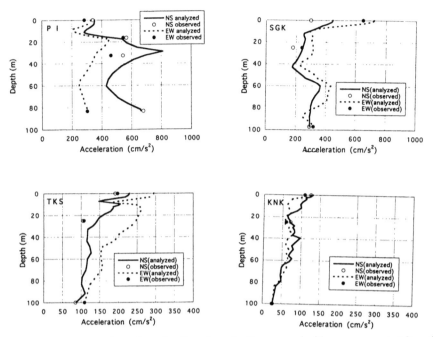

Fig.13 Computed max. horizontal acceleration along depth compared with measurement at four sites

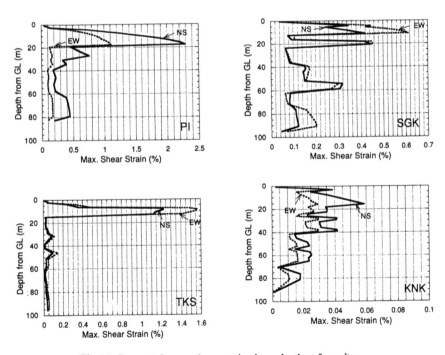

Fig.14 Computed max. shear strain along depth at four sites

4 AMPLIFICATION OF MAXIMUM ACCELE-RATION

Based on the findings on the degree of soil nonlinearlity exhibited during the earthquake, amplification ratios at the four vertical array sites have been studied to know the applicability of the current methodologies delineated in the Manual with emphasis on the nonlinear response effect.

In Fig.15 the ratios of maximum horizontal accelerations between surface and base are plotted against the ratio of S-wave velocities between the two layers. In the figure, open marks represent the mostly linear response for the aftershocks(AS). It should also be noted that the response for the main-shock in KNK was almost linear as indicated in the inversion analyses. The dashed line in the same figure indicates the line proposed by Shima(1978) for the relationship between the peak amplification values of transfer functions between surface and base and the S-wave velocity ratio. Althoug the open marks approximating linear seismic response are widely scattered, they are about 0.4 times of the Shima's line. Thus it is clearly indicated that the acceleration amplification for linear response is primarily dependent on the Vs-ratio between the base and the top and is about two fifths of the amplification under the sinusoidal motion in average. On the other hand, the solid marks representing the nonlinear response for the main-shock(MS) in PI, SGK and TKS are located obviously lower than the line approximating the linear response.

In some previous researches Vs averaged in upper 30m of a surface layer is taken instead of Vs in the top surface layer (e.g. Midorikawa 1987). In Fig.16 Vs-ratio is taken between the average in top 30m layers and the base. Although the data concentrates around 2 in the Vs-ratio, the linear response may be approximated by the same straight line as that in Fig.15, while nonlinear acceleration ratios are located mostly below.

In Fig.17 the maximum horizontal acceleration ratio between the surface and the base are plotted against the maximum base acceleration at the base. the solid marks are for hard rock base in KNK with the Vs-ratio of 7, which are clearly located higher than other sites, indicating a significant influence of Vs-ratio on the amplification. The open marks in the same figure corresponding to the sites of the Pleistocene soil base with the Vs-ratio of 2 to 4 appear to show nonlinear effect with increasing base accelerations as indicated by a pair of dashed thin curves. Despite rather large scatters the trend may be approximated by the regression equation;

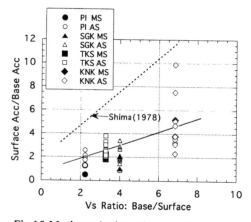

Fig.15 Maximum horizontal acceleration ratio plotted against Vs-ratio between surface and base for four vertical array sites

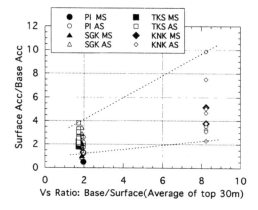

Fig.16 Maximum horizontal acceleration ratio plotted against Vs-ratio between top 30m average and base for four vertical array sites

$$Acc_{surface}/Acc_{base} = 2.0 \exp(-1.7 \, Acc/980)$$

where Acc stands for the maximum acceleration in cm/s^2. It is noted that the acceleration ratio between the surface and the Pleistocene soil base lowers 1.0 at about $Acc_{base}=400cm/s^2$, which is coincidental with a similar research result by Idriss(1990) although the base was rock in his research.

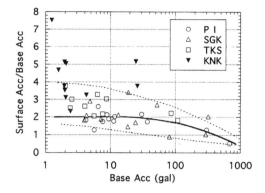

Fig.17 Maximum horizontal acceleration ratio plotted against maximum base acceleration

5 CONCLUSIONS

Strong motion records obtained during the Hyogoken-Nanbu earthquake in four vertical array sites have been incorporated to evaluate nonlinearity in site amplifications exerted during the main-shock and aftershocks. Variations of wave velocity and damping ratio from small-strain value due to strong motion have been quantified by means of inversion analyses, which demonstrates strong nonlinear effect during the main-shock in three sites nearer to the focal zone.

Base on these results, linear and nonlinear acceleration amplification between ground surface and base is studied with respect to wave velocity ratio between the same two levels. Strong correlation between the linear amplification and the wave velocity ratio can be seen very clearly as indicated in the Zonation Manual, and nonlinear amplification is evidently lower than that. Further study is obviously needed, however, to examine quantitative applicability of the amplification ratio based on methodologies previously proposed and to modify them for better estimation especially for nonlinear soil response during strong earthquakes.

REFERENCES

Constantopoulos,I.V., Roesset,J.M. and Christian, J.T. " A comparison of linear and exact nonlinear analyses of soil amplification, Proc. 5th International Conference on SMFE, Rome, pp.1806-1815, 1973

Idriss,I.M."Response of soft soil sites during earthquakes" Proc. H.Bolton Seed Memorial Symposium, pp.273-290, 1990

Kokusho,T. and Iwatate,K. "Scaled model tests and numerical analyses on nonlinear dynamic response of soft grounds" Journal of Japan Society for Civil Engineers (in Japanese), No.285, pp.57-67, 1979

Kokusho,T., Tohma,J., Yajima,Y., Tanaka,Y., Kanatani,M. and Yasuda,N. "Seismic response of soil layer and its dynamic properties" Proc. 10WCEE, Madrid, pp6671-6680, 1992

Kokusho,T. Sato,K. and Matsumoto,M."Nonlinear dynamic soil properties back-calculated from strong motions during Hyogoken-Nanbu Earthquake" Proc. 11th WCEE, Acapulco, 1996

Midorikawa, "Prediction of isoseismal map in the Kanto plain due to hypothetical earthquake, Journal of Structural Engineering, Vol.33B, pp.43-48 (in Japanese), 1987

Ohta,H. "Application of optimization method to earthquake engineering (Part I) -Estimation of underground structure of SMAC observation site in Hachinohe Harbor-" Journal of Japan Society of Architectural Engineering (in Japanese), 1975

Sato,K., Kokusho,T., Matsumoto,M. and Yamada,E. " Nonlinear seismic response and soil property during 1995 Hyogoken Nanbu earthquake" Soils and Foundations Special Issue for the Hyogoken Nanbu earthquake, January, pp41-52, 1996

Schnabel, P.B., Lysmer,J. and Seed, H.B. "SHAKE, A computer program for earthquake response analysis of horizontally layered sites" Report EERC 72-12, University of California Berkeley, 1972

Shima, E. "Seismic microzoning map of Tokyo" Proc. 2nd International Conference on Microzonation, Vol.1, pp.433-443, 1978

Technical Committee for Earthquake Geotechnical Engineering, TC4, ISSMFE "Manual for Zonation on Seismic Geotechnical Hazards", 1993

Seismic Behaviour of Ground and Geotechnical Structures, Sêco e Pinto (ed.) © 1997 Balkema, Rotterdam, ISBN 90 5410 887 8

Seismic zoning on ground motions in Taiwan area

Chin-Hsiung Loh & Wen-Yu Jean
Department of Civil Engineering, National Taiwan University, Taipei, Taiwan

ABSTRACT: For the purpose of establishing seismic design criteria for engineering structures, it is desirable to assess the expected intensity of seismic strong ground motion at a site for various probability levels (i.e., seismic hazard analysis). In order to consider the seismic hazard at a site, the assessment of regional seismicity and the attenuation of ground motion intensity must be made. In this paper, seismic zoning on ground motions in Taiwan area was made as the application of the Manual of Seismic Zonation and the results was expressed as the zonation map of Taiwan area.

1 INTRODUCTION

Seismic zoning for earthquake ground motions is one of the most fundamental aspects of seismic hazard assessment. The Japanese Society of Soil Mechanics and Foundation Engineering (JSSMFE), under the authorization of the International Society for Soil Mechanics and Foundation Engineering (ISSMFE), published a Manual for Zonation on Seismic Geotechnical Hazards [Angelier, 1986]. It is hoped that this manual will provide a useful document for those who are entrusted in preparing zonation maps in earthquake-prone areas of the world.

Zonation on earthquake ground motions are affected by several factors such as source, path and site effects. An assessment of ground motion therefore depends on the followings:

a) Regional seismicity;
b) Attenuation of ground motion intensity;
c) Local site effects on ground motion.

Key parameters underlying these effects must be examined. In this paper, methods for conducting a seismic hazard analysis of Taiwan area are presented for the purpose of establishing seismic design criteria for engineering structures, and served as one of the applications of the Manual of Seismic Zonation. Approaches to seismicity and attenuation do not depend to the same degree on the zonation level and detail description on the seismic hazard analysis must be discussed from region to region.

2 SEISMICITY

Taiwan is located at a complex juncture between the Eurasian and Philippine Sea Plate (Fig. 1). North and east of Taiwan, the Philippine Sea Plate subducts beneath the Eurasian Plate to the north, along the Pyukyu trench; while south of the island the Eurasian Plate underthrusts the Philippine Sea Plate to the east, along the Manila trench. However, the nature of the boundary near the southern tip of Taiwan is not very clear at present. It is known that the frequent seismic activity in Taiwan region is associated with the complicated interact in between the Eurasian Plate and the Philippine Sea Plate near Taiwan. Basically, the Taiwan area can be devided seismologically into three zones according to the plate tectonic setting. Earthquakes in the northward seismic zone are associated with the northward subduction of the Philippine Sea Plate. Earthquakes in the eastern seismic zone are associated with the

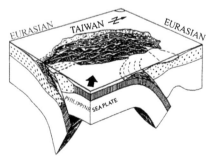

Fig. 1 Lithospheric plate structure in and around Taiwan (Angelier, 1986).

oblique collision between the Philippine Sea Plate and the Eurasian Plate along the longitudinal valley in eastern Taiwan.

The history of earthquake activity in the Taiwan region dated back to the sevententh century. Prior to 1897, only disastrous earthquakes had been described. Ninety-five earthquakes were documented in the period 1644-1895. In 1897, seismographs were first installed in Taiwan by the Taiwan Weather Bureau. These early instrumentations were few in number and had a low magnification, only those earthquakes that cause substantial damage (with magnitude exceeding 5.5) were recorded. Prior to 1935 the record for small earthquakes (i.e., ≤ 5.5) is incomplete. From 1936 to 1979 a catalog of earthquakes was compiled with magnitude ≥ 4.0. However, the seismographic instruments used during the period were not good enough to record all the smaller earthquakes. Therefore, the record is believed to be complete only for earthquakes with magnitude ≥ 5.0. After 1976, the record is complete for earthquakes with magnitude greater than 4.0. Figure 2a shows the epicenters for earthquakes between 1900 and 1996 with focal depth less than 35 km, and Fig. 2b shows the epicenters for earthquakes between 1900 and 1996 for focal depth greater than 35 km.

3 ANALYSIS OF HAZARD PARAMETERS

If an earthquake occurs in a region, the probability that the intensity, Y, will exceed some value y at the site is given as

$$P[y > y] = \sum_{i=1}^{n} P[Y > y|E_i]\, P[E_i] \quad (1)$$

where E_i = occurrence of the earthquake in source i, and n = the number of potential earthquake sources in the region. Assuming the average occurrence rate in source i relative to that over the region remains constant with time, the probability of occurrence of the event E_i may be expressed as $P(E_i) = \nu_i/\nu$, where ν is the average occurrence rate of earthquakes with magnitude m greater than m_0 in source i. The probability of the intensity Y exceeding y at the site in one year is give by:

$$P[Y > y]_{1\,year}$$
$$= \sum_{i=1}^{n} \left[\int_{m_0}^{m_u} P[Y > y|E_i(m)]\, f_M(m)\, dm \right] \nu_i \Big/ \nu \quad (2)$$

where m_u and m_0 are the upper and lower bound magnitudes, and $f_M(m)$ is the probability density function of earthquake magnitude. The upper bound magnitude, m_u, and function, $f_M(m)$, can be different for each earthquake source. As expected, much of the effort in a seismic hazard analysis requires the determination of the parameters and their physical relationships. The estimation of the hazard parameters is described as follow for Taiwan region:

Seismogenic zones: Based on the geological structure, subduction model and seismicity, a detail seismogenic zoning scheme was selected as shown in

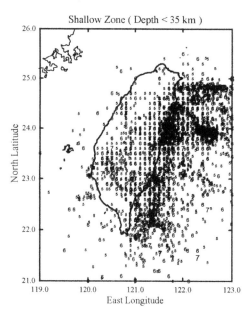

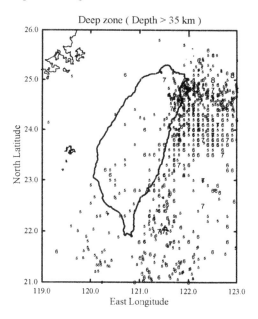

Fig. 2   Epicenters of earthquakes with $M_L \geq 4.5$ from 1900 through 1996: (a) shallow; (b) deep

Fig. 3. The occurrence rate of earthquakes in each subzone is distributed evenly within the respective subzones. Based on geophysical data and fault trend, type I source model and type III source model [Der Kiureghian & Ang, 1977] are used.

Source type model: The fault rupture is taken as the seismic hazard model, two possible source models are used [Der Kiureghian & Ang, 1977] depending on the information available for the fault system in the region. Known active faults are modeled as Type I source (well-defined faults). All other potential earthquake sources are modeled as Type III sources (unknown faults); that is, there is no information on the position and direction of the faults in the area. Type III sources are assumed to have an equal probability of rupturing in any direction. The identified active faults have been considered in seismic hazard analysis as also shown in Fig. 3(a).

Magnitude recurrence relationships: The Gutenberg-Richter magnitude recurrence relationship is given by: $\log N = a - b\,M$, where N is the number of earthquakes with magnitude greater or equal to M. For the purpose of evaluating the parameters a and b, two different sets or subsets of the available magnitude data for Taiwan may be considered appropriate. These are as follows:

(1) Data set A: Earthquakes which occurred between January 1973 through June 1996, with magnitude $M_L \geq 4.5$.

(2) Data set B: Earthquakes which occurred between January 1936 through June 1996, with magnitude $M_L \geq 4.5$.

Prior to the installation of the telemetered seismographic network in Taiwan in 1972, the record of earthquakes of significant magnitudes may not be totally complete. For this reason, the b value (slope of the magnitude recurrence relation) may be more reliably obtained using the data set A. Two different methods can be used to estimate the b-value: the least square method and the maximum likelihood method.

However, the $24\frac{1}{2}$ yr period of observation (1973 through June 1996) is too short to determine the value of parameter a, representing the mean occurrence rate of all earthquakes with significant magnitude. For this purpose, a hybrid procedure was devised to make use of the larger set of earthquake data, and at the same time make use of the b value determined on the basis of data set A. On the basis of data set B, the occurrence rate at a specified local magnitude (e.g., $M_L \geq 6.0$) is estimated. Then, with the estimated b value (determined from data set A), the magnitude occurrence line is passed through the data point just established for $M_L \geq 6.0$. On this basis, the annual magnitude occurrence line can be obtained, this is, the annual occurrence rates for earthquakes of different magnitudes can be estimated. The procedure mentioned above is illustrated in Fig. 4. Table 1 lists the estimate b-value in each subzone.

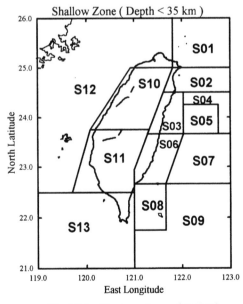

Fig. 3(a) Zoning scheme (shallow)

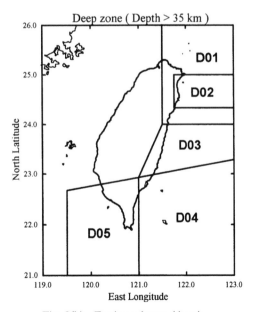

Fig. 3(b) Zoning scheme (deep)

Table 1   Parameters for seismic hazard analysis model

| Zone | b-value for G-R relationship | | Upper-bound magnitude | |
	maximum likelihood method	least square method	estimated value	recorded value
S01	0.1395	0.4052	7.76	7.70
S02	1.1759	1.2980	7.65	7.50
S03	1.2984	1.1606	8.10	8.10
S04	0.9486	0.9670	6.71	6.50
S05	1.1913	1.1013	7.16	6.80
S06	1.1657	1.2641	7.41	7.00
S07	0.8257	0.9146	7.11	7.09
S08	0.9017	0.9771	7.38	7.20
S09	0.8902	0.8017	7.15	7.00
S10	1.3371	1.1805	7.31	7.10
S11	1.6277	1.3778	7.42	7.30
S12	0.9050	1.0984	6.59	6.50
S13	0.9524	0.8508	6.84	6.80
D01	0.9349	0.8170	8.12	8.10
D02	1.2101	1.2014	7.48	7.30
D03	0.8707	1.0465	7.24	6.90
D04	0.9227	0.8580	7.53	7.30
D05	1.3581	1.1627	6.26	5.90

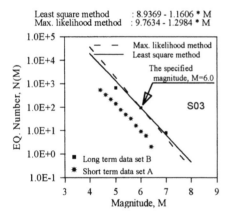

Fig. 4   Determination of b-value, magnitude recurrence line

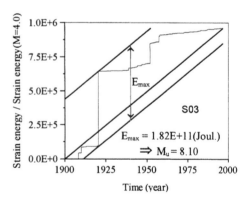

Fig. 5   Explanatory diagram for graphical representation of the energy release method of calculating upper bound energy (E_{max})

Minimum and maximum magnitudes: The minimum magnitude, m_0, is difficult to determine because there appears to be no reliable record for magnitudes less than 4.5. Moreover, a sensitivity study must be carried out to calculate site-specific hazards in order to identify the appropriate minimum magnitude. As for the maximum magnitude, the energy release method will be used to calculate the maximum energy (E_{max}) and mean annual energy (E) release. Thus, the corresponding maximum magnitude (M_{Lu}) of such an earthquake can be estimated as illustrated in Fig. 5. The energy-magnitude formula used in this study is:

$$\log E = 12.66 + 1.40\,M_L \qquad \text{[Iida, 1976]} \qquad (3)$$

The estimated upper bound magnitude in each sub-zone is also shown in Table 1.

Focal depth: Earthquakes in the Taiwan area may originate in the shallow zone (to a depth of 35 km), or in the deep subduction zone (as shown in Fig. 3). In either zone, the hypocenters of past earthquakes appear to vary widely in depth. Furthermore, for many of the recorded earthquakes, the focal depths are not well known. In the light of this it would be appropriate to describe the focal depths of future earthquakes as a random variable. In addition, as there is no evidence of preferred depths, a uniform distribution over a pertinent range of possible depths may be specified. This means that for the shallow zones the focal depths can be expected to be distributed uniformly from 5 to 35 km, whereas for the deep subduction zone the focal depths can be expected to be distributed uniformly through the 50 km thick Benioff zone dipping at an angle of 45°. A type III source model is used for shallow zones, while for deep zones the point source model is used.

Rupture length-magnitude relationship: The actual rupture lengths for past recorded earthquakes in Taiwan are unknown. To estimate these rupture lengths, isoseismal contours may be used. Based on this method, the following mean rupture length-magnitude relationship has been established [Tsai et al., 1987].

$$L = \exp(1.006\,M_L - 3.232) \qquad (4)$$

where L = the rupture length (in kilometer); M_L = the local magnitude.

The standard deviation of the logarithm of rupture length, $\sigma_{\ln L}$, is equal to 0.422.

Attenuation relationship: To describe the attenuation of peak ground acceleration (PGA) with magnitude and distance, the Joyner and Boore [1981] form is selected to characterize the ground motion attenuation in the Taiwan area. Coefficients for the median attenuation equation are determined through nonlinear regression using this attenuation model. The resulting equation is

$$\log_{10} \text{PGA} = 0.0278\,\exp\left[1.20\,M_L\right]$$
$$\left[R + 0.1413\,\exp(0.6918\,M_L)\right]^{-1.7347} \quad (5)$$

in which PGA(g) represents the median peak ground acceleration, M_L is the local magnitude, and R (km) is the shortest distance. The standard deviation of the logarithm of PGA is 0.539. This value is used to determine the upper bound acceleration (at a nonexceedance probability of 0.977 which corresponded to median plus 2σ level) and to truncate the seismic hazard curves. Figure 6 shows the PGA attenuation form for M_L = 7.0 and 6.5. Comparison with the regression form of Northridge earthquake data and

Kobe earthquake data is also shown in this figure. The site condition for Eq. (5) includes all types of site conditions from soft soil to hard site.

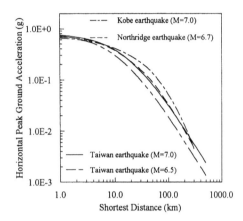

Fig. 6 Comparison on the PGA attenuation form of the Northridge earthquake data, Kobe earthquake data and the Taiwan soft site earthquake data

4 RESULTS OF SEISMIC HAZARD ANALYSIS

In calculating the hazard curves at a particular site, the estimated dispersion of PGA must be considered in the analysis. We define the actual intensity $Y_a = NY$ and N is the correction necessary on the predicted intensity Y. The required probability, including the effects of the dispersion, becomes

$$P\left[Y_a > y\right] = \int_0^\infty P\left[Y > y \,\middle|\, N = n\right] f_N(n)\, dn \quad (6)$$

where $f_N(n)$ is the probability density function of N. It is assumed that N is a random variable and follows the modified log-normal distribution [Loh, et al., 1994]. Finally, the iso-acceleration map of Taiwan area can be constructed. The area under consideration in developing seismic hazard map was divided into 74 quadrilaterals with side length of one-fourth degree of latitude and longitude, respectively. The seismic hazard analysis was calculated for every center point of this mesh. Using fault-rupture model (for shallow zone) and point source model (for deep zone), the PGA contours related to a certain risk level can be plotted. Figures 7a and 7b show the iso-acceleration map of Taiwan for return period of 475 years with one and half standard deviation (1.5 σ) and two standard deviation (2 σ) PGA dispersion correction, respectively.

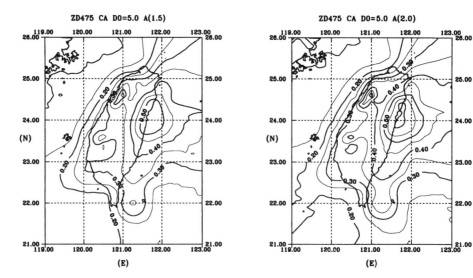

Fig. 7　The iso-acceleration map of Taiwan for return period of 475 years: (a) 1.5 σ; (b) 2.0 σ

5　UNIFORM HAZARD RESPONSE SPECTRUM

In the development of response spectra for use in designing seismic resistant structure, it has been a common practice to establish first sets of response spectra normalized to 1.0 g and then scale down to a specified PGA level in its design application. For Taiwan area the design spectra were constructed for two different site conditions. Figure 8 shows the design spectrum for soil site and hard site.

It should be recognized that the resulting spectra do not represent the same probability of exceedance over the full frequency range of interest (or say, not an uniform hazard). To generate site-specific uniform-hazard response spectra an alternative approach is presented. The uniform hazard response spectra is defined as the response spectra with equal probability of exceedance in all structural period. In the design application, the PGA levels are taken directly from a site-specific seismic hazard curve with a given specified annual probability of exceedance. Use these PGA levels to scale-down the normalized response spectra value. Let A denote the random PGA value. Its annual probability of exceedance function, $Q_A(a)$, is given directly by the seismic hazard curve for PGA. Further, let S denote the random normalized acceleration response spectral value $S_{an}(T,\xi)$ for a specific period T and damping ratio ξ. The desired probability distribution function for random variable $S_a(T,\xi) = A \times S$ representing the non-normalized acceleration response spectral value for period T and damping ratio ξ can be expressed

as

$$P(c) = P\big[S_a > c\big] = \int_{s=0}^{\infty} \int_{a=0}^{c/s} p_{A,S}(a,s)\, da\, ds \quad (7)$$

Generally it is reasonable to assume that random variables A and S are statistically independent, in which case their joint probability density function $p_{A,S}(a,s)$ is of the simple form $p_{A,S}(a,s) = p(a) \cdot p(s)$. Having obtained the probability distribution function $P(S_a)$ for the full range of discrete values of T and for specified discrete values of ξ, uniform hazard acceleration response spectrum curves can be obtained directly therefrom. For system subjected to horizontal ground excitation, statistical analysis on response spectrum had been studied [Loh, et al., 1994].

Probability distribution function of S_a: Let S denote the random value of $S_{na}(T,\xi)$. From a set of normalized acceleration response spectrum curves, let S denote the random value of $S_{na}(T,\xi)$. The mean value of the acceleration response spectra, $\mu(T,\xi)$, and the coefficient of variation $\big(\sigma(T,\xi)\big/\mu(T,\xi)\big)$ of $S_{na}(T,0.05)$ can also be evaluated as a function of period T. The coefficient of variation (C.O.V.) of $S_{na}(T,0.05)$ is shown in Fig. 9. The probability density function for random variable S (spectral acceleration, S_{na}) can be expressed using the lognormal form [Kiremidjian, 1978; Loh, 1994]:

$$p(S) = \frac{1}{\sqrt{2\pi}\,\eta\,S} \exp\left[-\frac{1}{2}\left(\frac{\ln S - \lambda}{\eta}\right)^2\right]$$

$$0 < S < \infty \quad (8)$$

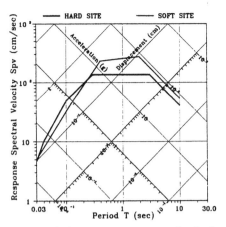

Fig. 8   The design response spectra of soft site and hard site for Taiwan

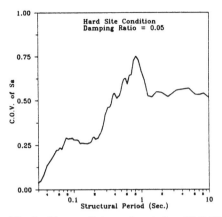

Fig. 9   The coefficient of variation (C.O.V.) of $S_{na}(T, 0.05)$

in which η and λ are shape and scale parameters, respectively. These two parameters can be estimated at each structural period by using maximum-likelihood method.

Seismic hazard curve for PGA: The seismic hazard analysis gives the annual probability of exceedance for a given specified ground intensity. As discussed before $Q_A(a)$ is given directly from the seismic hazard curve. Thus, the corresponding probability distribution and density functions of PGA can be obtained numerically using the relations

$$p_A(a) = 1 - Q_A(a) \quad \text{and} \quad p_A(a) = \frac{dP_A(a)}{da} \quad (9)$$

Substituting Eqs. (8) and (9) into Eq. (7) gives the probability distribution function for random variable $S_a(T, \xi) = A \times S$ representing the non-normalized acceleration response spectral value for period T and damping ratio ξ.

Uniform hazard response spectrum (UHRS):   The schematic diagram for the calculation of UHRS is shown in Fig. 10. Procedures for generating UHRS are listed as follows: (1) Generate the hazard curve from Eq. (7) which shows the annual probability of exceedance of acceleration spectral value for a given structural period, i.e., curve 3 in Fig. 10; (2) Given a specific value of return period, the acceleration response spectra is determined from step (1), i.e., curve 3; (3) Same calculation will be done by considering different structural period and generate curve 3 for each structural period. With the same annual probability of exceedance, spectral accelerations can be determined from different structural period. The uniform hazard response spectrum is constructed by plotting the spectral acceleration with respect to period. In Fig. 10 there are two other curves; curve

1 is the general seismic hazard curve. It is plotted the annual probability of exceedance with respect to PGA value. Curve 2 is obtained by scaling the curve 1 with the normalized acceleration response spectral value of the specified structural period, i.e., $C_d(T, \xi) = A \cdot S_{na}(T, \xi)$. For design purpose with a given annual probability of exceedance, the PGA value was selected from curve 1. This PGA value will be used to scale the normalized acceleration spectra $S_{na}(T, \xi)$ so as to obtain the design spectral

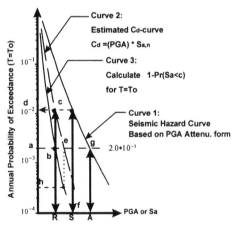

Fig. 10   Schematic diagram for generating UHRS and design iso-intensity map

77

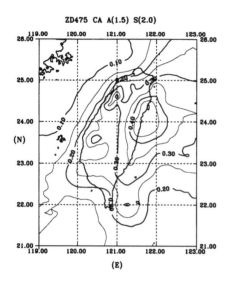

ZD475 CA A(1.5) S(2.0)

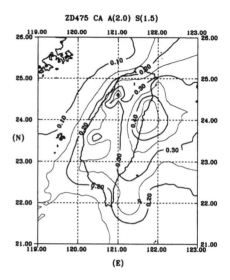

ZD475 CA A(2.0) S(1.5)

Fig. 11 The iso-intensity map generated from UHRS for return period = 475 years:
(a) PGA $= 1.5\,\sigma$, $S_a = 2.0\,\sigma$; (b) PGA $= 2.0\,\sigma$, $S_a = 1.5\,\sigma$

value. Curve 2 is obtained from this idea. Difference between curve 2 and curve 3 is obvious. To construct the uniform hazard response spectrum only curve 3 is needed.

Figures 11a and 11b show the iso-intensity map of Taiwan which corresponding to uniform hazard response spectra with return period of 475 year at $T = 1.0$ sec. In Fig. 11a, the dispersion consideration for PGA attenuation is $1.5\,\sigma$ and for S_a is $2.0\,\sigma$, and Fig. 11b with the dispersion consideration for PGA attenuation is $2.0\,\sigma$ and for S_a is $1.5\,\sigma$. With these information the final seismic zoning map of Taiwan area is generated, as shown in Fig. 12.

6 CONCLUSIONS

This paper presents the procedures for generating seismic zoning map of Taiwan. The regional seismicity and the attenuation of ground motion intensity were used as the key hazard parameters to perform the seismic hazard analysis. At the present study, the seismic zoning of Taiwan area is limited to Grade-1 method because of no detail soil information. It requires new surface investigation for evaluating the geotechnical properties at specific sites to perform the Grade-2 methods.

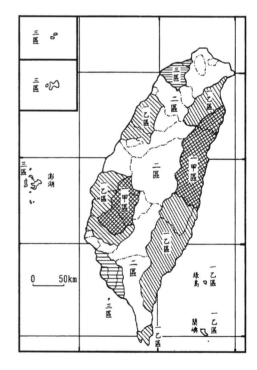

Fig. 12 Design seismic zoning map of Taiwan area
一甲區: $Z = 0.33$; 一乙區: $Z = 0.28$;
二區 : $Z = 0.23$; 三區 : $Z = 0.18$

REFERENCES

1. Angelier, J., 1986. Preface of Special Issue on Geodynamics of the Eurasia-Philippine Sea Plate Boundary. Tectonophysics, 125(1–3): IX–X.

2. Bonilla, M G., 1977. Summary of Quaternary Faulting and Elevation Changes in Taiwan. Mem. Geol. Soc. China, 2: 43–55.

3. Campbell, K .W., 1981. Near-Source Attenuation of Peak Horizontal Acceleration. Bull. Seismol. Soc. Am., 71(6): 2039–2070.

4. Cornell, C. A., 1968. Engineering Seismic Risk Analysis. Bull. Seismol. Soc. Am., 58(5): 1583–1606.

5. Der Kiureghian, A. and Ang, A. H-S., 1977. A Fault-Rupture Model for Seismic Risk Analysis. Bull./ Seismol. Soc. Am., 67(4): 1173–1194.

6. Ho, C. S., 1982. Tectonic Evolution of Taiwan, Explanatory Text of the Tectonic Map of Taiwan. Ministry of Economic Affairs, Taiwan, ROC.

7. Ho, C. S., 1986. A Synthesis of the Geologic Evolution of Taiwan. Tectonophysics, 125: 1–16.

8. Hsu, M. T., 1980. Earthquake Catalogues in Taiwan. Center Earthquake Eng. Res., National Taiwan University.

9. Hsu, T. L. and Chang, H. C., 1979. Quaternary Faulting in Taiwan. Mem. Geol. Soc. China, 3: 155–165.

10. IES, 1987. Seismic Hazard Analysis of Taiwan Power Company's Nuclear Power Plant No. 3 at Maanshan. Institute of Earth Sciences, Report ER8701, Taipei, ROC.

11. Iida, K., 1976. Earthquake Energy-Magnitude Relation, Strain Rate and Stress Drop in Earthquakes, Seismic Efficiency Factor, and Parameters of Faulting. Dept. Earth Sci., Nagoyo University.

12. Joyner, W. B. and Boore, D. M., 1981. Peak Horizontal Acceleration and Velocity from Strong-Motion Records Including Records form the 1979 Imperial Valley, California, Earthquake. Bull. Seism. Soc. Am., 71(6): 2011–2038.

13. Krinitzsky, E. L., Chang, F. K. and Nuttli, O. W., 1988. Magnitude-Related Earthquake Ground Motions. Bull. Assoc. Eng. Geol., 25(4): 399–423.

14. Lee, W. H. K., Wu, F. T. and Jacobsen, C., 1976. A Cataglo of Historical Earthquakes in China Compiled from Recent Chinese Publication. Bull. Seismol. Soc. Am., 66(6): 2003–2016.

15. Loh, C. H., Jean, W. Y. and Penzien, J., 1994. Uniform-Hazard Response Spectra − An Alternative Approach. Earthquake Engineering and Structural Dynamics, 23, 433–445.

16. McGuire, R. K., 1976. Fortran Computer Program for Seismic Hazard Analysis. USGS Open File Series 7667.

17. Tsai, Y. B., 1978. Review of Strong Earthquakes in Western Taiwan Before the 20th Century. Science Monthly, 9: 31–35.

18. Tsai, Y. B., 1985. A Study of Disastrous Earthquakes in Taiwan, 1683–1895. Bull. Inst. Earth Sci. Acad. Sin., 5:1–44.

19. Tsai, Y. B., 1986. Seismotectonics of Taiwan. Tectonophysics, 125: 17–37.

20. Tsai, Y. B., Teng, T. L., Chiu, J. M. and Liu, H. L., 1977. Tectonic Implications of the Seismicity in the Taiwan Region. Mem. Geol. Soc. China, 2: 13–41.

21. Tsai, C. C., Loh, C. H. and Yeh, Y. T., 1987. Analysis of Earthquake Risk in Taiwan Based on Seismotectonic Zones. Mem. Geol. Soc. China, 9: 413–446.

22. Wu, F. T., 1978. Recent Tectonics of Taiwan. J. Phys. Earth, 26 (suppl. S265–S299): 265–299.

Seismic Behaviour of Ground and Geotechnical Structures, Sêco e Pinto (ed.)© 1997 Balkema, Rotterdam, ISBN 90 5410 887 8

Microzonation for ground motion during the 1980 Irpinia earthquake at Calabritto, Italy

M. Maugeri
Facoltà di Ingegneria, Università di Catania, Italy

P. Carrubba
Facoltà di Ingegneria, Università di Padova, Italy

ABSTRACT: Earthquake hazard in the urban area of Calabritto was related either to the vulnerability of the existing old structures and to the site effects. Such evidence was well focused during the 23 November 1980 Irpinia earthquake which caused more than 2,000 victims in the whole seismic basin. A comparison between three different degree of zonation is presented; a first level takes into account soil instabilities and amplification phenomena only by geological survey. Referring to the site effects only, a second zonation criterion examines the one-dimensional response of soil deposits in the frequency domain. In this case geotechnical soil properties from in situ tests are employed. Using dynamic soil properties from laboratory tests, a third zonation level is achieved from the one-dimensional soil response in the time domain.

1 INTRODUCTION

Earthquake hazard in urban sites has two principal aspects: the first is related to the vulnerability of building to the seismic forces, while the second concerns the site behaviour under shaking such as amplification, landsliding, settlement and liquefaction. The latter aspects, which depend on geotechnical soil properties, can affect the building safety in different ways.

In order to mitigate the earthquake effects on structures, predictions of the dynamical site behaviour have been taken into consideration in many countries. Seismic risk is presented on a zoning map in which zones with different levels of potential hazard are identified. The Technical Committee for Earthquake Geotechnical Engineering of the International Society for Soil Mechanics and Foundation Engineering (T.C.4 1993) have indicated three levels of zonation. The first is based on geological survey and interpretation of the existing information from historic documents, including magnitudes and focal mechanisms. Existing correlations of ground motion attenuation with distance allow to draw preliminary maps of expected acceleration. A second level of zonation may be achieved at moderate cost by matching use of additional source of data such as microtremor measurements, field and laboratory collection of geotechnical data. Where a third level of detailed zonation is required, additional field and laboratory investigations will be needed for a specific site. Computer analyses of seismic ground response allow to appreciate amplification effect, slope instability, settlement and liquefaction potential hazard.

In this paper a comparison between three different zonation levels is presented for the urban site of Calabritto, which was strongly damaged during the 23 November 1980 Irpinia earthquake (Italy). A first level of zonation takes into account soil instabilities and amplification phenomena only by geological survey. Referring to the more relevant site effects, a second zonation criterion examines the one-dimensional response of soil deposits in the frequency domain. In this case geotechnical soil parameters have been obtained from in situ tests. Using dynamic soil properties from laboratory tests, a third zonation level is achieved from the one-dimensional soil response in the time domain.

2 THE FIRST MICROZONATION LEVEL

2.1 *Foreword*

Calabritto is located in the upper Sele valley, just in

the epicentral area of 1980 Irpinia earthquake which affected a large portion of the southern Italy. The effects of the earthquake were disastrous for the small irpinian town which population, of almost 3200 people, suffered a loss of nearly 100 lives and the damage of about the 95% of its 600 buildings. Just after the event, Italian authorities carried out some urgent studies with the aim of collecting the existing information on earthquake effects, regional seismicity and geology.

The project developed by the Consiglio Nazionale delle Ricerche (C.N.R. 1983) produced the zonation maps for some urban areas which suffered extensive damages. The C.N.R. microzonation map of Calabritto can be considered a first level map mainly based on historic research and geologic survey.

2.2 Historic seismicity

The Irpinia region affected by the 23 November 1980 earthquake has a severe historic seismicity with more than 50 earthquakes of maximum intensity I_{MM} (Modified Mercalli) between IX-XI, occurred in the last 2,000 years. Analysis performed on a sample of 48 earthquakes later than the 14th century has indicated a nearly gaussian distribution of maximum intensities at the epicentre, with a mean value I_{MM}=IX and standard deviation of almost I_{MM}=± 1.

Detailed studies on historic seismicity of Irpinia have been carried out by the C.N.R. (1983) who compiled the isoseismal maps of the most well documented historical seismic events. Remarkable earthquakes which occurred close to Calabritto are those of the 1694, 1732, 1853, 1910 and 1930 with maximum intensity I_{MM} between IX-X

2.3 The 23 November 1980 Irpinia earthquake.

The earthquake had epicentre about 90 km east of Naples, and occurred at 18h 34' 53'' GMT (local time one hour later) with the following ipocentral data: latitude 40° 46' N, longitude 15° 18' E and depth of 18 km. The estimated local magnitude was 6.5 and the maximum intensity was X MSK. (Mercalli, Sponheuer, Karnik). The isoseismals map of the earthquake is shown in fig.1; the isoseismals are elongated in the Apenninic direction (NW-SE), with an area of nearly 3,000 km^2 of intensity greater than VIII MSK. The earthquake lasted for about 50 s with the maximum horizontal acceleration of 0.33g recorded by ENEL accelerometer at Sturno, about

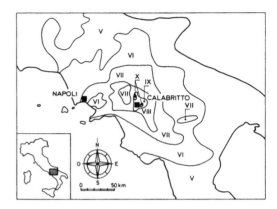

Fig.1 The 23 November 1980 Irpinia earthquake: isoseismal lines according the MSK scale (C.N.R. 1983)

35 km from the epicentre. The main shock was followed by numerous aftershocks, about six hundred, the maximum of which had magnitude 4.8 recorded on 14 February 1981. Analyses regarding the focal mechanism indicate that the earthquake was associated with a normal fault striking approximately NW-SE. This hypothesis was supported by the differences in land elevations observed after the earthquake in the epicentral area.

Fig.2 shows a map of damages occurred in Calabritto during the seismic event. The old centre was founded on a limestone rock formation; in this area the old masonry buildings were affected by total or partial collapse (zone I). The areas of new expansion had concrete building founded on stiff loose soil; slight damage occurred in zone II, whereas remarkable damage occurred in zone III.

2.4 Geological setting of the Calabritto area

Limestone and dolomitic limestone of the Campano-Lucana platform constitute the rises and the hill in which is located the old centre of Calabritto. The whole formation is several hundred meters thick and may be ascribed to the Triassic-Jurassic period (Mesozoic). The rock is characterised by intense fracturing and no karst phenomena are observed. The valley is formed by loose soils which show some overconsolidation or cementation processes. Varicolori clays of the Upper Oligocene are overconsolidated marly clays with embedded small calcareous and arenaceous inclusions. The unit has remarkable thickness of the order of 30-40 m. The Quaternary deposits of continental origin have a

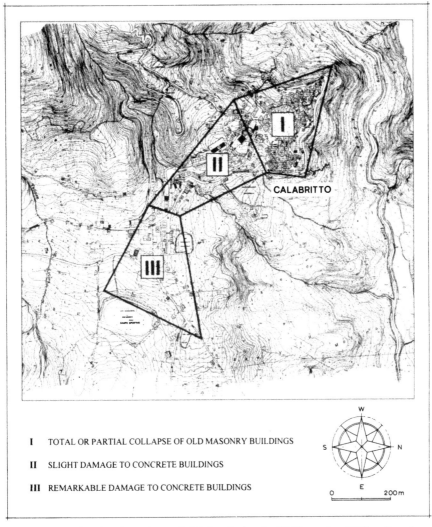

I   TOTAL OR PARTIAL COLLAPSE OF OLD MASONRY BUILDINGS

II SLIGHT DAMAGE TO CONCRETE BUILDINGS

III REMARKABLE DAMAGE TO CONCRETE BUILDINGS

Fig.2 Map showing damaging at Calabritto during the 23 November 1980 earthquake

sandy-clay matrix containing calcareous, arenaceous and marly inclusions of different size. These soils cover the varicolori clays and are widespread around the old centre.

2.5 *Microzonation of first level*

Fig.3 shows the microzonation map compiled by C.N.R. (1983) on the base of in situ recognition just after the event. Being the centre located in the epicentral area, it was not necessary to suppose an attenuation law for the maximum horizontal acceleration; the latter was compatible with an earthquake of intensity IX-X I_{MM}, with values abouth

$a_{max}=(0.1-0.3)g$. Basing on structural damaging and soil instability phenomena occurred at Calabritto, several zones have been classified for aseismic design purposes. The only zone where soil amplification is expected is **c**, where loose soil deposits are found. For the zone **a** and **b**, on rock formations, amplification is not expected. In excessively sloping sites with potential or effective instabilities, **d** and **e**, building is discouraged. However this microzonation approach was not supported by geotechnical ground investigations, but only by surface geology. The next two microzonation approaches concern the analysis of ground motion, to evaluate local soil amplification effects. Liquefaction did not occur during the 1980

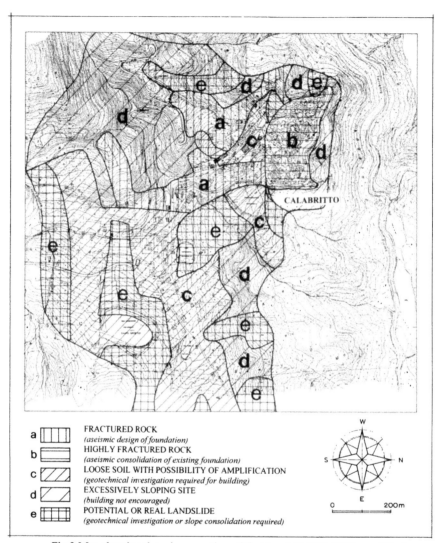

Fig.3 Map showing the microzonation of first level (C.N.R. 1983, modified)

earthquake and no potential liquefeable soil was noted in the area; instability associated to slope sliding is outside of the aim of this paper.

3 THE SECOND MICROZONATION LEVEL

3.1 *Foreword*

This microzonation was carried out just after the seismic event (Maugeri 1981, Maugeri and Carrubba 1985) by performing one-dimensional analyses in the frequency domain. Only the results of in situ Standard Penetration Test and seismic explorations were employed for geotechnical characterisation.

Full dynamic soil properties were not known at this stage; soil non-linearity as well as damping ratio were established from literature.

3.2 *Geotechnical characterisation by in situ tests*

Site investigations were carried out on a area of about 120 ha in both the old and the new expansion area. The investigation consisted in 29 boreholes driven at various depth between 25 and 100 m. Geoelectric and seismic in situ tests were also performed, including seismic refraction, cross-hole and down-hole. To characterise the in situ shear strength of the overconsolidated clay soils, Standard

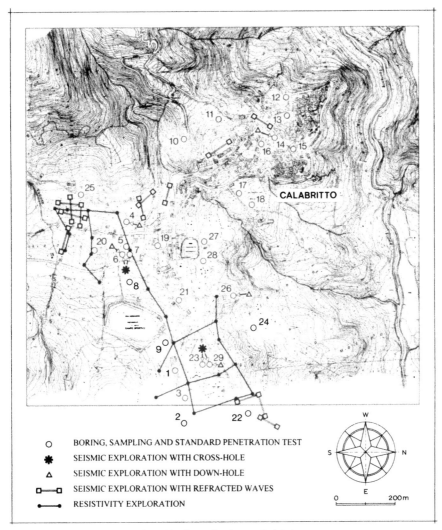

Fig.4 Site investigation at Calabritto

Penetration Tests were employed (Fig.4). The tests were carried inside the boreholes, at intervals between 2 and 3 m, while the maximum number of blows N_{SPT} was raised up to 150 for a penetration of 300 mm of the sampler. This procedure allowed a more wide data base to be taken into account in the analyses.

The aim of in situ tests was to correlate N_{SPT} with initial shear modulus G_0 of soil (Ohta and Goto, 1978).

Using the standard penetration test data, correlations between N_{SPT} with depth were at first evaluated. Fig.5 shows such correlations in the case in which the data were analysed according the A.S.T.M. criterion (Fig.5a) and when a proportional criterion

was employed for N_{SPT} data exceeding 100 blows per foot (Fig.5b) (Maugeri and Carrubba 1983).

Likewise using the results of in situ seismic tests, the correlation between shear wave velocities V_S with depth were also evaluated (Fig.5c). Then correlations between N_{SPT} and V_S have been obtained (Fig.6). The mean law $V_s = 48 N_{SPT}^{0.55}$ (m/s) has been selected for the overconsolidated Calabritto clay.

By using the law of elasticity $G_o = \rho V_s^2$, the expression of the in situ shear modulus was obtained:

$$G_o = 4.6 N_{SPT}^{1.1} \quad (Mpa) \qquad (1)$$

having assumed the unit mass $\rho \cong 2,000$ kg/m^3.

85

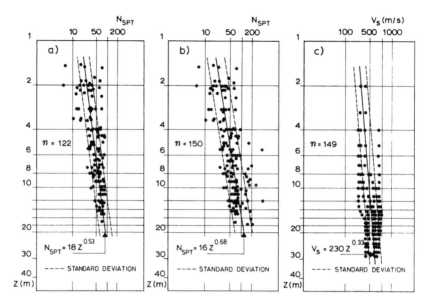

Fig.5 Correlations between in situ test results and depth: a) N_{SPT} evaluated according the A.S.T.M. criterion, b) N_{SPT} evaluated according a proportional criterion, c) V_S evaluated by cross-hole and down-hole tests

3.3 Dynamic model in the frequency domain

The one-dimensional model used in this approach considers a soil column of unit cross section vibrating in free field, without taking into account any boundary effects. The soil column behaves as an

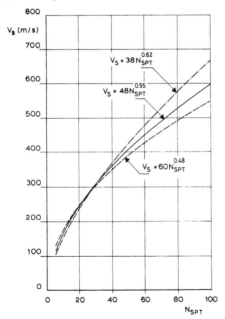

Fig.6 Correlations between N_{SPT} and V_S for the overconsolidated Calabritto clay

generalised oscillator with distributed shear stiffness and mass along the depth. The seismic input is applied at the base of the model in terms of response spectrum. Bedrock is assumed horizontal and stiff enough to maintain unchanged the patterns of the seismic motion. For this purpose the bedrock was assumed to be either the same of in situ rock formation or a layer in which shear wave velocity approaches $V_S \cong 750\text{-}770$ m/s (International Conference of Building Officials, 1979), corresponding to an initial shear modulus $G_0 \cong 1,000$ MPa and a value of $N_{SPT}\cong150$ blows per foot. The base of the model was excited with the E-W component of the response spectrum of seismic motion recorded at Bagnoli Irpino (Fig.7b), the closest to Calabritto accelerometric station founded on calcareous rock. The dynamic soil response has been evaluated in terms of maximum horizontal acceleration, velocity and displacement at the free-field, by using the Rayleigh method. The equivalence between maximum kinetic and potential energy, mobilised during a generic vibration mode by the soil column, allows an approximate undamped natural frequency of vibration to be evaluated according to the following expression:

$$\omega_n^2 = \frac{\int_0^H G(\gamma,z)[\phi'(z)]^2\,dz}{\int_0^H \rho(z)[\phi(z)]^2\,dz} \qquad (2)$$

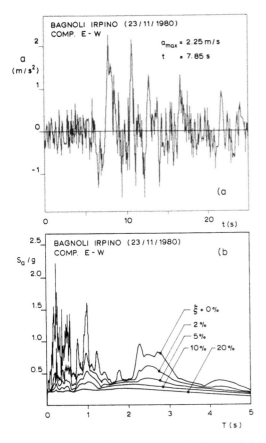

$$\phi_{i+1} = \int_0^H \frac{\int_0^H \rho(z)[\phi_i(z)]dz}{G(\gamma,z)}dz + C \qquad (3)$$

in which

$\phi_i(z) =$ the improved shape function at the (i)th iteration

$\phi_{i+1}(z) =$ the improved shape function at the (i+1)th iteration

C = the integration constant depending on the boundary conditions.

The shape function of first iteration was assumed to be:

$$\phi_0(z) = 1 - \frac{z}{H} \qquad (4)$$

Replacing the improved shape functions (3) in equation (2), a quick convergence is reached in a few iterations, which are stopped as soon as two successive frequency values, ω_i and ω_{i+1}, differ by less than 1%.

The variation of initial shear modulus G_0 with depth has been related to the N_{SPT} data by means of equation (1), while the dependence from shear strain level was not known at that time because dynamic laboratory tests on undisturbed samples where still not available. However such a dependence was established with the equation (Yokota et al. 1981):

$$G(\gamma) = G_0 \frac{1}{1 + \alpha\gamma^\beta} \qquad (5)$$

in which

$G(\gamma)=$ strain dependent shear modulus

$\gamma =$ shear strain

$\alpha, \beta =$ soil constants

The values of $\alpha= 150$ and $\beta= 0.80$ were selected for Calabritto clay.

Once the fundamental vibration frequency of soil layers has been obtained, the dynamic response is evaluated in terms of maximum horizontal displacement X_{max}, velocity $\dot{X}_{max}$ and acceleration $\ddot{X}_{max}$ at the free-field, according to the equations:

$$X_{max} = \phi_n(z = 0)\psi \frac{S_a(\xi,\omega_n)}{\omega_n^2} \qquad (6)$$

$$\ddot{X}_{max} = \phi_n(z = 0)\psi \frac{S_a(\xi,\omega_n)}{\omega_n} \qquad (7)$$

$$\ddot{X}_{max} = \phi_n(z = 0)\psi S_a(\xi,\omega_n) \qquad (8)$$

Fig.7 The 1980 Irpinia earthquake (Italy) recorded at Bagnoli Irpino: a) time history of horizontal acceleration, b) response spectra for different damping ratio ξ

in which

H = thickness of soil stratum

$G(\gamma,z)$ = shear modulus variable with strain and depth

$\rho(z)$ distributed soil mass with depth

$\phi(z)=$ estimated shape function for the vibration mode

In this analysis only the first vibration mode was considered so that the fundamental frequency of vibration of a shear beam soil column was evaluated with equation (2).

Accuracy of equation (2) can be raised by improving the shape function in an iterative manner. Using local equilibrium of elastic and inertia forces some improved shape functions, closest to the real main mode of vibration, are evaluated at each iteration The following expression has been employed:

in which
$\phi_n(z = 0)$ = value of the final shape function at zero depth
$S_a(\xi, \omega_n)$ = applied spectral acceleration to the base model
ξ = damping ratio
ψ = participation factor for the first vibration mode
$\omega_n = 2\pi / T_n$ = value of the undamped natural frequency of vibration
Dependence of damping ratio from shear strain was neglected in this analysis and a constant value $\xi = 10\%$ was assumed.
The participation factor can be evaluated according to the equation:

$$\psi = \frac{\int_0^H \rho(z)\phi_n(z)dz}{\int_0^H \rho(z)[\phi_n(z)]^2 dz} \qquad (9)$$

The simplified model employed in the analyses is based on the dynamical behaviour of a generalised oscillator with distributed parameters, vibrating with the fundamental mode. However the effect of the higher mode of vibration can be easily taken into consideration by evaluating the sequence of higher undamped frequencies and than superimposing the effects by means of a modal superposition. Analyses performed by Maugeri et al. (1988a,b) have employed the Holzer method to evaluate the eigenvalues of a soil column; modal superposition have indicated that only the first 3 to 5 mode of vibration may be significant in the dynamic response of a shear beam soil column model.

3.4 Dynamic soil response

The second microzonation level could be based only on static soil properties; however dynamic in situ test results were also taken into consideration.
The free field response has been evaluated for soil deposits related to boreholes, by means of equations (6), (7) and (8).
The selected spectrum to be applied at bedrock had damping ratio $\xi = 10\%$ (Fig.7b), compatible with soil damping.
The surface elastic response spectra were evaluated according to the Newmark's empirical method (Newmark and Hall 1982) for a damping ratio $\xi = 5\%$, compatible with the typical damping of concrete structures.

According to this procedure spectral displacement approaches to the field maximum displacement X_{max} in the frequency range lower than $f \cong 0.03$ Hz ($T \cong 33$ s), while in the frequency range higher than $f = 33$ Hz ($T \cong 0.03$ s) spectral acceleration approaches to the field maximum acceleration $\ddot{X}_{max}$. In the frequency range between 0.1 Hz $\leq f \leq 8$ Hz (0.125 s $\leq T \leq 10$ s), amplification takes place, with spectral ordinates increasing to maximum values. In this range spectral ordinates may be evaluated by statistical coefficients of the maximum kinematic field parameters. For a linearly elastic oscillator with damping ratio $\xi = 5\%$, such a coefficients are 2.01 for spectral displacement, 2.30 for spectral velocity and 2.71 for spectral acceleration. This values refer to a cumulative probability level of 84%.
Results of analyses have been summarised in terms of homogeneous spectra which show almost the same dynamic response of the sites.
Three elastic response spectra have been selected in fig.8; the spectrum A refers to zones where a fundamental period of vibration of subsoil $T_n \geq 0.30$s is expected. The spectrum B and C refer to zones where fundamental vibration periods $T_n \cong 0.25$s and $T_n \leq 0.15$s are respectively expected.

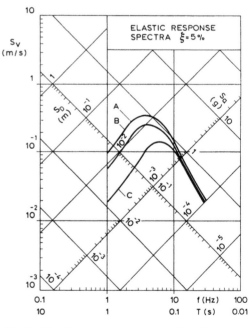

Fig.8 Elastic response spectra for the second microzonation level

3.5 *Microzonation of second level*

Italian seismic code has classified the whole territory into three zones known as the first (I), the second (II) and the third (III) category, in which response spectra are given for aseismic design purposes (Ministero dei Lavori Pubblici 1986). Depending on the seismicity of the site, the seismic category can be selected together to the appropriate response spectrum. However such a zonation criteria does not consider the role played by the foundation subsoil, being the rules mainly based on seismological considerations. The anelastic response spectra considered in the Italian code are reported in fig. 9a, while the relevant elastic response spectra have been obtained by increasing the anelastic spectra ordinates with a ductility factor $\lambda \cong 8$ for old concrete structures (fig.9b).

In order to compare the elastic response spectra obtained for Calabritto with those proposed by Italian authorities, the effective peak acceleration (EPA) and the effective peak velocity (EPV) approach was employed (Applied Technology Council 1978). Following this approach EPA and EPV may be considered as normalisation factors for comparing amplitude and frequency content of

different seismic events given in terms of elastic response spectrum. The EPA parameter is proportional to the mean spectral acceleration value in the range of $0.1\ \text{s} \leq T \leq 0.5\ \text{s}$, while EPV is proportional to spectral velocity evaluated at a period of 1 s. For a damping ratio $\xi = 5\%$ the constant of proportionality is assumed 2.5 for both parameters. The EPA and EPV values for both the Calabritto and the Italian code elastic response spectra are reported in table I

Tab.I Second zonation level: effective peak acceleration and velocity related to the Calabritto and the Italian code spectra

Response spectrum		EPA (g)	EPV (m/s)
Italian Code	I Category	0.320	0.431
	II Category	0.224	0.300
	III Category	0.128	0.168
Calabritto	A	0.400	0.040
	B	0.320	0.020
	C	0.230	0.006

The ratio between the EPA values gives the ε_1 coefficient which allow the modification of the reference spectrum in the range of lower vibration periods, while the ratio between the EPV values gives the ε_2 coefficient which allow the modification of the reference spectrum in the range of higher vibration periods. The coefficients ε_1 and ε_2, evaluated respect to the Italian II category response spectrum, are reported in table II. For higher vibration periods all the three spectra A, B and C are well covered by the Italian II category spectrum, being always $\varepsilon_1 < 1.0$. At lower vibration periods the spectrum C is almost equivalent to the Italian II category spectrum, being $\varepsilon_1 \cong 1.0$, while the spectra A and B show an amplification effect with $\varepsilon_1 = 1.4$ for zone B and $\varepsilon_1 = 1.8$ for zone A.

Tab.II Second zonation level: amplification coefficients for the II category Italian response spectrum

Response spectrum at Calabritto	ε_1	ε_2
A	1.80	0.13
B	1.40	0.07
C	1.00	0.02

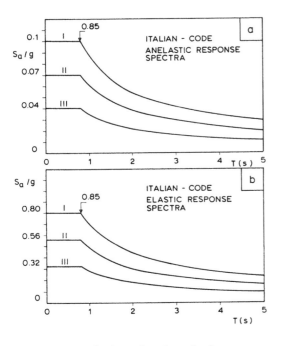

Fig.9 Italian seismic code: a) anelastic response spectra for the three seismic categories, b) elastic response spectra

Then for a seismic event similar to that of 23 November 1980, higher horizontal accelerations may

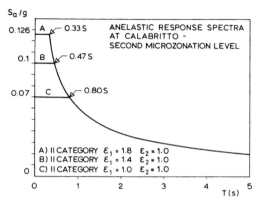

Fig.10 Modified Italian code spectra for the second microzonation level of Calabritto

be expected in zone **A** and **B**, respect those provided by the II category Italian code response spectrum. These results are more important for structures with low vibration periods.

The modified Italian code response spectra for the Calabritto area are shown in fig 10; moreover the proposed map for the second microzonation level is reported in Fig.11 in terms of homogeneous zone in which the spectra of fig.10 can be employed. Comparison of results with the damage map, confirm the validity of model adopted here, being the larger damages associated with higher expected spectral accelerations (zone **A**).

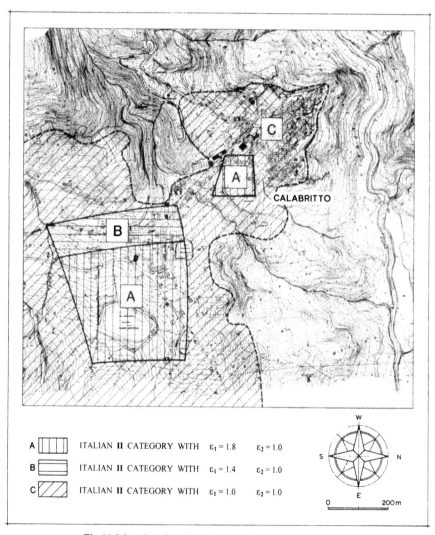

Fig.11 Map showing the microzonation of second level

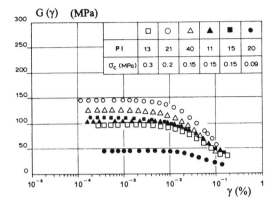

Fig.12 Degradation of Calabritto clay shear modulus by resonant column test results

4. THE THIRD MICROZONATION LEVEL

4.1 *Foreword*

This microzonation was carried out on the base of the results of dynamic laboratory tests performed on undisturbed samples of cohesive soil. In addition the geological model of subsoil was available.

The analyses were performed in the time domain by means of a one-dimensional model of shear beam with variable shear modulus with either depth and strain, and variable damping ratio with strain.

4.2 *Geotechnical characterisation by laboratory tests*

A dynamic geotechnical characterisation of the overconsolidated Calabritto clay was performed in laboratory by means of resonant column tests. The tests allowed the evaluation of shear modulus and damping ratio with respect to stress and strain level, stress history and plasticity properties.

Fig 12 shows the result of resonant column test in terms of degradation of shear modulus with strain for different levels of confining isotropic stress σ_c.

The same results have been represented also in the normalised form of fig.13 by dividing the shear modulus $G(\gamma)$ for the initial value G_0 at very low strain.

The latter representation shows that normalised shear modulus has a little dependence from stress history, plasticity index and stress level, while it is mainly dependent only on the shear strain level.

The following normalised law was established for the Calabritto Clay:

$$\frac{G(\gamma)}{G_0} = \frac{1}{1 + 20\gamma\,(\%)^{1.28}} \qquad (10)$$

The relationship between initial shear modulus G_0 and effective confining stress is shown in fig.14. A certain dependence of G_0 from the initial void index was noted in this case, while the gradient of growth of G_0 with stress level was almost uniform for all the tests. The following law was proposed:

$$G_0 = 70\exp(1.80\sigma_c) \qquad (11)$$

Expression (11) was able to describe the complete

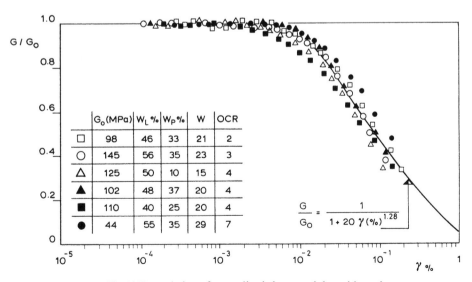

Fig.13 Degradation of normalised shear modulus with strain

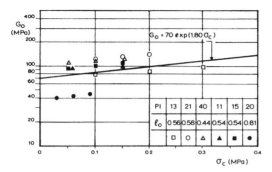

Fig.14 Relationship between initial shear modulus and effective confining stress for Calabritto clay

variation of initial shear modulus with depth, being the confining stress related to depth by effective unit weight of soil, while expression (10) allows the complete shear modulus degradation to be considered with strain level.

The evaluation of damping ratio was carried out using two different procedures: following the steady-state method, the damping ratio was obtained during the resonance condition of the samples (fig.15a); following the amplitude decay method (fig.15b) it was obtained during the decrement of free vibration. In both cases the damping ratios were comparable.

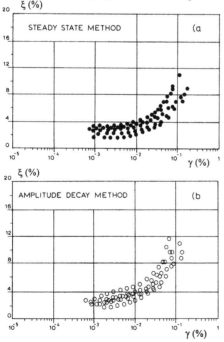

Fig.15 Damping ratio measurements versus strain: a) steady-state method, b) amplitude-decay method

The influence of effective confining stress, plastic index and stress history on damping ratio was not significant for this clay soil; otherwise the shear strain was the more influent parameter, causing damping to increase with strain level.

As suggested by Yokota et al. (1981), the inverse variation of damping ratios with respect to normalised shear moduli had an exponential form such that reported in fig.16 for the Calabritto clay:

$$\xi(\gamma)\% = 25\exp\left\{-2.2\left[G(\gamma)/G_0\right]\right\} \qquad (12)$$

with maximum value $\xi_{max} = 25\%$ for $G(\gamma)/G_0 = 0$.
and minimum value $\xi_{min} = 3\%$ for $G(\gamma)/G_0 = 1$.

So the complete dependence of damping ratio on shear strain is given by the existing relationship between normalised damping ratio and normalised shear modulus, according to the expression:

$$\xi(\gamma)/\xi_{max} = \exp\left\{-2.2\left[G(\gamma)/G_0\right]\right\} \qquad (13)$$

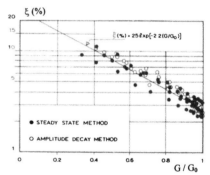

Fig.16 Relationship between damping ratio and normalised shear modulus for Calabritto clay

4.3 Dynamic model in the time domain

A one-dimensional non-linear model has been employed in the dynamical analysis of soil. The model (Maugeri and Frenna 1987) considers the soil deposit as a shear beam varying in mass, stiffness and damping with depth. In addition shear modulus and damping can vary with the mobilised strain. Seismic input is applied at the base of the model in terms of time history of the horizontal acceleration. Dynamic soil response is evaluated at any depth in terms of time history of displacement, velocity and acceleration.

The shear beam model is formed by discrete masses connected together with non linear springs and

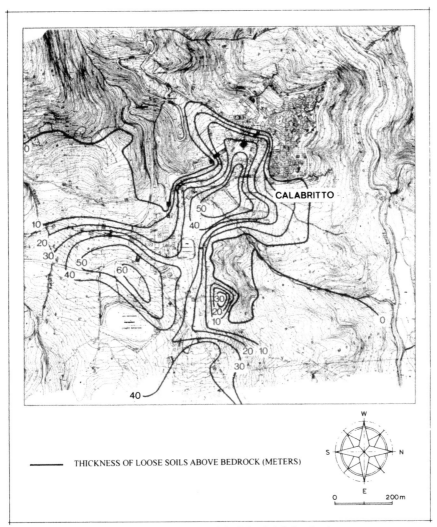

Fig.17 Geological model of the Calabritto basin

dampers. The hysteretic behaviour of soil during loading and unloading is treated as linear equivalent, (Idriss and Seed 1968) so that strain dependent shear modulus and damping can be employed according to the equations (10) and (13).

The differential equations of motion are expressed in the form:

$$|M|\left|\ddot{X}(t)\right| + |C(\gamma)|\left|\dot{X}(t)\right| + |K(\gamma)||X(t)| = -|M|\left|\ddot{X}_g(t)\right|$$

(14)

in which $|M|$ is the diagonal matrix of masses, $|C(\gamma)|$ and $|K(\gamma)|$ are respectively the damping and the stiffness strain dependent matrix. The terms $|X(t)|$, $\left|\dot{X}(t)\right|$ and $\left|\ddot{X}(t)\right|$ are respectively displacement, velocity and acceleration vectors related to the relative motion of masses respect to the base model, while the term $\left|\ddot{X}_g(t)\right|$ represents a vector with the same components $\ddot{X}_g(t)$ of the imposed horizontal acceleration at the bedrock. In this case the time history shown in fig.7a was employed in the analysis. The viscous strain dependent damping coefficients $C(\gamma)$ were derived from the viscous damping ratio $\xi(\gamma)$ by means of the following expression:

$$C(\gamma) = \frac{2\xi(\gamma)\sqrt{G(\gamma)\rho(z)}}{\phi(z)^2}$$

(15)

93

in which $\rho(z)$ is the unit mass of soil layer forming lumped mass and $\phi(z)$ an approximate shape function of the first mode of vibration.

Equations (14) are solved numerically by using the Newmark's integration operator with the linear acceleration hypothesis. In addition to make quicker the analysis, the Newmark's method is carried out by evaluating the relative equilibrium of lumped masses rather than the absolute one.

4.4 Dynamic soil response

Based on the results of in situ exploration, a geological model for the Calabritto basin was prepared. For dynamical analyses purposes the formations were subdivided into two principal units: the basal rock and the covering soils. The equal-thickness lines of loose soil above bedrock are reported in fig.17. Dynamic analyses were carried out for layer thickness ranging from H = 10 m up to H = 60 m. The analyses have produced the time histories of horizontal acceleration, velocity and displacement at the free field, from which the response spectra of horizontal acceleration have been evaluated for concrete structures.

According to the selected layer thickness, fig.18 summarise such a response spectra for a damping ratio ξ=5%. For the rock formation without covering (H=0), the response spectrum is the same of fig.7b with ξ = 5%.

Following the hypothesis of this microzonation approach, response spectra from 2 to 6 represent surface spectra, meaningful of the dynamical response of the damped simple oscillator varying in period and resting on shallow rock or on soil deposits.

The comparison between these spectra gives an idea of the amplification and the attenuation effects connected with the layer thickness.

4.5 Microzonation of third level

The EPA and EPV parameters for the third level elastic response spectra at Calabritto are reported in table III. The seismic coefficients which allow the II Italian code response spectra to be modified are reported in table IV.

Tab.III Third zonation level: effective peak acceleration and velocity related to the Calabritto elastic response spectra

Response spectrum		EPA (g)	EPV (m/s)
Calabritto	H=0m	0.139	0.337
	H=10m	0.224	0.332
	H=20m	0.292	0.424
	H=40m	0.204	0.856
	H=60m	0.192	0.544

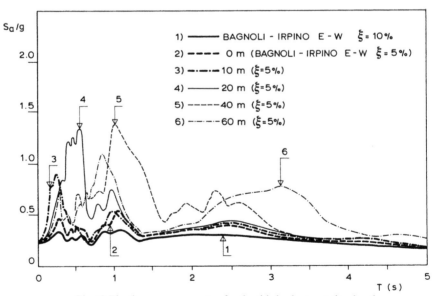

Fig.18 Elastic response spectra for the third microzonation level

94

Tab.IV Third zonation level: amplification coefficients for the II category Italian response spectrum

Response spectrum at Calabritto	ε_1	ε_2
H=0m	0.62	1.12
H=10m	1.00	1.11
H=20m	1.30	1.41
H=40m	0.91	2.85
H=60m	0.86	1.81

On the base of this results three homogeneous zone with specified response spectra have been selected for microzonation purposes (fig.19).

The spectrum **A** (fig.20) refers to the zones where the thickness of soil deposits ranges from 30m to 60m; in this case the II category Italian response spectrum can be used with seismic coefficients about $\varepsilon_1 = 1.0$ and $\varepsilon_2 = 2.3$

The spectrum **B** refer to zones where the thickness of soil deposits ranges from 10m to 30m; in this case the II category Italian response spectrum can be used with seismic coefficients about $\varepsilon_1 = 1.3$ and $\varepsilon_2 = 1.4$.

The spectrum **C** can be used for zones were shallow rock or soil deposits with thickness lower than 10m are present; the latter is the same of the II category Italian response spectrum, being seismic coefficients about $\varepsilon_1 = 1.0$ and $\varepsilon_2 = 1.0$.

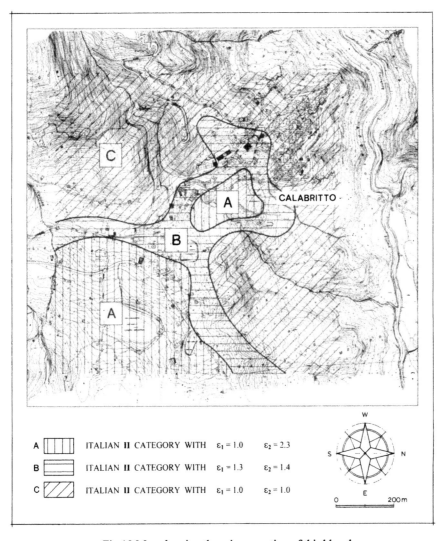

A		ITALIAN II CATEGORY WITH	$\varepsilon_1 = 1.0$	$\varepsilon_2 = 2.3$
B		ITALIAN II CATEGORY WITH	$\varepsilon_1 = 1.3$	$\varepsilon_2 = 1.4$
C		ITALIAN II CATEGORY WITH	$\varepsilon_1 = 1.0$	$\varepsilon_2 = 1.0$

Fig.19 Map showing the microzonation of third level

95

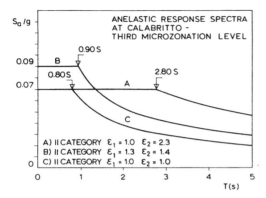

Fig.20 Modified Italian code spectra for the third microzonation level

Comparisons between the second and the third microzonation levels was suitable because the same three homogeneous zones **A**, **B**, **C** have been obtained in both cases. In addition the related design response spectra are well comparable; the response spectrum of zone **C** was the same for both the microzonation approaches, with a maximum value of horizontal spectral acceleration $S_a \cong 0.07g$. Also the spectrum of the zone **B** was almost the same, with a maximum value about $S_a \cong 0.1g$. The zone **A** has been revealed the most critical for the second microzonation level, where horizontal spectral acceleration can achieve values of almost $S_a \cong 0.126g$; however with the third microzonation approach the zone **A** was characterised by a lower horizontal spectral acceleration ($S_a \cong 0.07g$), but acting in a more relevant range of frequency, up to T=3s.

5 CONCLUSION

A comparison between three different degree of seismic zonation at Calabritto is presented in this paper. A first level takes into account soil instabilities and amplification phenomena only by geological survey (CNR 1983). Ground motion zonation is achieved in the other two levels by analysing the one-dimensional response of soil layers in the frequency and the time domain. The 1980 Irpinia earthquake has been used as seismic input. The second microzonation level was mainly based on static and dynamic in situ test results, while the third was founded on the dynamic laboratory test results and on the geological model of the Calabritto basin. The last two microzonation levels were in a well agreement each other and congruent with damaging occurred in Calabritto.

REFERENCES

Applied Technology Council 1978. *Tentative provisions for the development of seismic regulations for buildings.* ATC Publ. ATC3-06, Spec. Publ. 510, U.S. Govern. Print. Off., Wash.
Clough, R.W. & J. Penzien 1982. *Dynamics of structures.* McGraw-Hill Intern. Book Comp.
C.N.R. 1983. *Progetto finalizzato geodinamica.* Consiglio Nazionale delle Ricerche.
Idriss, I.M., & H.B. Seed 1968. Seismic response of horizontal soil layers. *Journ. Soil Mech. Found. Div., ASCE.* 94: 1003-1031.
International Conference of Building Officials 1979. *Uniform building code.* The Conf. Whittier, California.
Maugeri, M. & P. Carrubba 1983. Correlazione in situ della velocità delle onde di taglio con i valori N_{SPT}. XV Conv. Naz. Geotec., Spoleto, Italy, 4-6 Maggio 1983, pp.203-208.
Maugeri, M. & P. Carrubba 1985. Microzoning using SPT data. Proc.12th Intern. Conf. Soil Mech. Found. Eng., San Francisco, 12-16 August 1985, pp. 1831-1836.
Maugeri, M., A. Carrubba & P. Carrubba 1988a. Caratterizzazione dinamica e risposta del terreno nella zona industriale di Catania. *Ingegneria Sismica.* 2: 9-18.
Maugeri, M., P. Carrubba & S.M. Frenna 1988b. Frequenze e modi di vibrazione di terreni eterogenei. *Rivista Italiana di Geotecnica.* 3: 163-171.
Maugeri, M., & S.M. Frenna 1987. Modello isteretico semplificato per la determinazione della risposta dei terreni in campo non lineare. 3° Conv. Naz. "L'Ingegneria Sismica in Italia", Roma, 30 Settembre-2 Ottobre 1987, pp.269-288.
Ministero dei Lavori Pubblici 1996. *Norme tecniche relative alle costruzioni sismiche.* D.M. 16 Gennaio 1996, Gazz. Uff. Repub. Ital., N. 29.
Newmark, N.M., & W.J. Hall 1982. *Earthquake spectra and design* .Earthq. Eng. Res. Inst., Berkeley, California.
Ohta, Y. & N. Goto 1978. Empirical shear wave velocity equations in terms of characteristic soil indexes. *Earthq. Eng. Struc. Dyn.* 6:167-187.
T.C.4 1993. *Manual for zonation on seismic geotechnical hazard.* Tech. Com. Earthq. Geotech. Eng., Inter. Soc. Soil Mech. Found. Eng., publ. by Jap. Soc. Soil Mech. Found. Eng.
Yokota, K., T. Imai & M. Konno 1981. Dynamic deformation characteristics of soils determined by laboratory tests. *OYO Tec. Rep.* 3: 13-37.

Seismic Behaviour of Ground and Geotechnical Structures, Sêco e Pinto (ed.) © 1997 Balkema, Rotterdam, ISBN 90 5410 887 8

Estimation of ground motion intensity in Kobe during the 1995 Hyogoken-Nanbu, Japan earthquake by a simplified zonation method

Saburoh Midorikawa
Tokyo Institute of Technology, Yokohama, Japan

Masashi Matsuoka
Remote Sensing Technology Center of Japan, Tokyo, Japan

ABSTRACT: Ground motion intensity in Kobe during the 1995 Hyogoken–Nanbu, Japan earthquake (M_J 7.2) is estimated by a simplified method in the zonation manual by the ISSMFE TC4 Committee (1993). The peak ground velocity on rock is calculated by the empirical attenuation relation. The site amplification factor at each site is estimated from the surface geology. The peak ground velocity at each site is computed as a product of the peak velocity on rock and the site amplification factor. The compute peak ground velocities show good agreement with the observed records and isoseismal map. The results suggest that the simplified method is effective for preliminary assessment of ground motion during a destructive scenario earthquake.

1 INTRODUCTION

A shallow earthquake with M_J 7.2 struck the southern part of Hyogo prefecture, Japan on January 17, 1995. The damage extended to Kobe city and the western part of Osaka metropolis as well as Awaji Island. The numbers of fatalities and injuries are about 6,400 and 44,000, respectively. About 200,000 houses were collapsed or severely damaged. The direct monetary loss is estimated over 100 billion dollars.

One of major reasons for the enormous damage is very intense ground motion in the affected area. In the earthquake, the seismic intensity VII on the Japan Meteorological Agency (J.M.A.) scale, which corresponds to XI or XII on the M.M. scale, was observed. Since many strong motion data were observed in the affected area, the earthquake provides an opportunity to check applicability of existing ground motion zoning methods at high intensity level. This paper describes estimation of ground motion intensity in Kobe by a simplified ground motion zoning method shown in the zonation manual by the ISSMFE TC4 Committee (1993).

2 GEOLOGICAL AND SEISMOLOGICAL ASPECTS OF THE EARTHQUAKE

Kobe is surrounded by the Rokko Mountains on the north, Osaka plain on the east and Awaji Island on the southwest, as shown in Fig. 1 (Kaji et al., 1995).

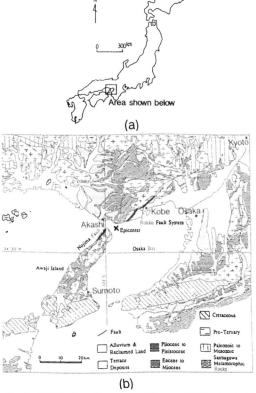

(a)

(b)

Fig. 1. (a) Site location map of Japan. (b) Geological map of affected area (after Kaji et al., 1995).

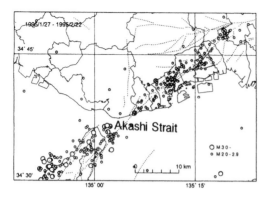

Fig. 2 Aftershock distribution (After Hirata, 1995).

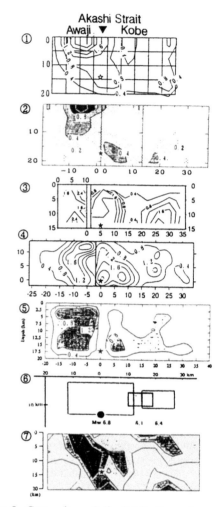

Fig. 3. Comparison of slip distributions of proposed fault models (After Takemura, 1996).

Many active faults are found in the surrounding area. The epicenter of the earthquake was located about 15 km southwest of Kobe, and the focal depth was about 15 km. As shown in Fig. 2 (Hirata, 1995), the aftershocks are distributed approximately on a line along the NE–SW direction, suggesting that the Rokko faults at Kobe and the Nojima fault at Awaji Island caused the earthquake. A maximum offset of 1.7 m was observed on the Nojima fault, but no significant offset on the Rokko faults.

Seismological analyses indicate that the earthquake is due to right–lateral faulting and the seismic moment is 2 to 3 x 10^{26} dyne·cm, corresponding to M_w 6.9. The rupture length is estimated to be 30 to 50 km. Figure 3 shows the comparison of the proposed fault models compiled by Takemura (1996). The earthquake consists of three subevents; the first one is at Awaji Island, the second and third ones are at Kobe. Although the first one is located at very shallow depth, the second and third ones are at depth of about 10 km.

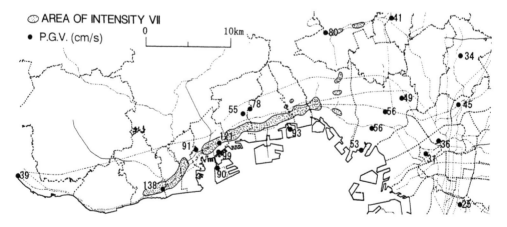

Fig. 4 Location of disastrous belt zone and distribution of observed peak ground velocities.

The J.M.A. assigned the area of intensity VII based on the field investigation, as shown in Fig. 4. This area corresponds to the disastrous belt zone where the ratio of collapsed houses is 30 % or more. The disastrous belt zone is located 1 to 2 km south of the estimated fault with a width of 1 to 2 km. In the figure, the distribution of observed peak horizontal velocities is also shown. Although few records were obtained in the disastrous zone, the observed peak ground velocities in or at the edge of the zone are 120 to 140 cm/s. Around the disastrous zone, many strong records with peak ground velocity of several tens cm/s were obtained.

3 ESTIMATION OF PEAK GROUND VELOCITY

The ISSMFE TC-4 Committee (1993) published the zonation manual in which the zonation methods on seismic geotechnical hazards with different levels are compiled. It would be of interest to check applicability of the methods through the case study of a damaging earthquake. Among the Grade 1 (simplest) methods of ground motion zoning in the manual, combination of the empirical attenuation relations of peak ground velocity and the relative amplification factor given for each geology is selected and applied to the Hyogoken–Nanbu earthquake.

The attenuation relation by Joyner and Boore (1981) is used to estimate the peak ground velocity on rock, v (cm/s):

$$\log v = 0.49\,M_w - \log r - 0.00256\,r - 0.67 \quad (1)$$
$$r = \sqrt{D^2 + 4^2}$$

where D is closest distance to surface projection of fault rupture in km. This equation was derived mainly from the observations during very shallow intraplate earthquakes in California. Therefore, the equation may be applicable to the Hyogoken–nanbu earthquake which is also an intraplate earthquake. Locations of the subevents at Kobe, however, are not very shallow but at depth of about 10 km. Considering the depth of the subevents as has been shown in Fig. 3, Equation (1) is modified as follows:

$$\log v = 0.49\,M_w - \log r' - 0.00256\,r' - 0.67 \quad (1)'$$
$$r' = \sqrt{D^2 + 8^2}$$

This modification means that the assumed source depth is changed 4 km to 8 km. The attenuation curves from Eqs. (1) and (1)' are compared in Fig. 5. The peak velocity at very close distance is reduced by the modification.

The relative site amplification factor (AF) at each site is estimated from the surface geology by using the empirical relation by Midorikawa (1987):

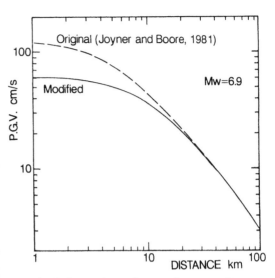

Fig. 5 Comparison of attenuation curves from Equations (1) and (1)'.

$$AF = \begin{array}{ll} 2.0 & \text{for Holocene} \\ 1.4 & \text{for Pleistocene} \\ 1.0 & \text{for Miocene} \\ 0.7 & \text{for Pre–Tertiary} \end{array} \quad (2)$$

The peak ground velocity (P.G.V.) at each site is computed as a product of the peak velocity on rock and the site amplification factor. The geology map in and around Kobe used for estimation of the site amplification is shown in Fig. 6. Pre–Tertiary rock dominates on the north of the Rokko faults. The area on the south, however, is covered by Holocene or Pleistocene sediments with higher amplification. This geological discontinuity is formed by the continuous activity on the faults.

The compute peak ground velocities is shown in Fig. 7. In the figure, the surface projection of the assumed fault rupture is shown by solid line. The higher P.G.V. zone (larger than 120 cm/s) is not found just on the fault but on the south of the fault, whose location is consistent with the observed disastrous belt zone. The outline of the disastrous belt zone is approximately reproduced by the simplified method used here. This suggests that one of the major causes for the disastrous belt zone is higher amplification of Holocene sediments.

However, it should be noted that the cause of the disastrous belt zone is also discussed as the effect of irregularity of the underground structure (Kawase, 1996; Pitarka et al., 1996; Motosaka and Nagano, 1997). The continuous activity on the Rokko faults forms a basin with a sharp edge on the south of the faults. This type of the structure (the basin–edge structure) can produce coupling of the direct S–wave and Rayleigh waves diffracted from the edge, resulting the stronger ground motion in a narrow zone near the basin edge. As the basin–edge effect have

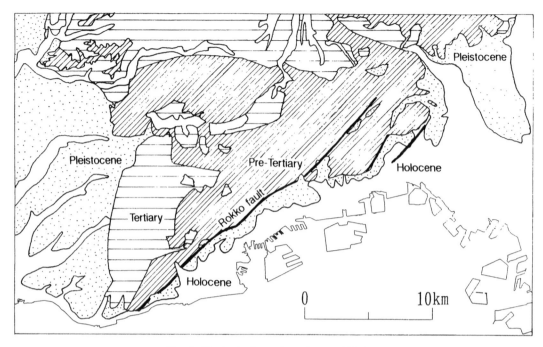

Fig. 6. Geological map in and around Kobe.

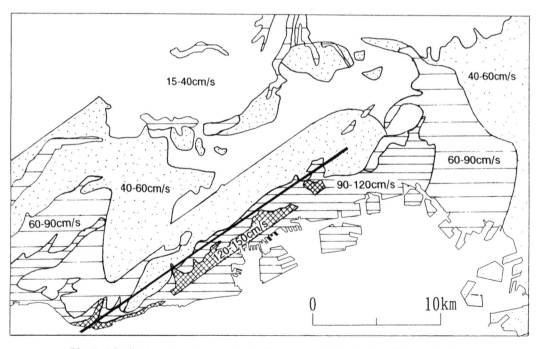

Fig. 7. Distribution of peak ground velocities computed by the simplified method.

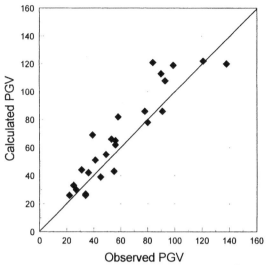

Fig. 8 Comparison of computed and observed
peak ground velocities.

an additional amplification of about 1.5 times
(Kawase, 1996), the P.G.V. in the disastrous belt
zone might reach 200 cm/s if the basin-edge effect
is added to the computed results shown in Fig. 7.

Figure 8 shows the correlation of the computed
values with the observations. The computed peak
ground velocities show good agreement with the
observed records. Although this agreement may be
attributed to scarcity of strong motion records in the
disastrous belt zone where the basin-edge effect is
expected, validity of the computed values is confirmed at least around the disastrous zone. The
results suggest that the simplified method used here
is effective for preliminary assessment of ground
motion of a destructive scenario earthquake.

4 CONCLUDING REMARKS

As the case study by a simple method shown in the
zonation manual (ISSMFE TC4 Committee, 1993),
ground motion intensity in Kobe during the 1995
Hyogoken-Nanbu, Japan earthquake is estimated.
The peak ground velocity on rock is calculated by
the empirical attenuation relation by Joyner and
Boore (1981) with modification. The site amplification factor at each site is estimated from the surface
geology by using the empirical relation by Midorikawa (1987). The peak ground velocity at each site
is computed as a product of the peak velocity on
rock and the site amplification factor. The compute
peak ground velocities show good agreement with
the observed strong motion records and isoseismal
map. The results suggest that the simplified ground
motion zonation method is effective for preliminary
assessment of ground motion of a destructive scenario earthquake.

REFERENCES

Hirata, N. 1995. Aftershock activity of the 1995
Hyogoken-nanbu earthquake, *Chikyu Monthly*,
13, 63–70 (in Japanese).
ISSMFE TC4 Committee. 1993. *Manual for Zona-
tion on Seismic Geotechnical Hazards*, The Japan
Society of Soil Mechanics and Foundation
Engineering.
Joyner, W. and D.M. Boore. 1981. Peak horizontal
acceleration and velocity from strong-motion
records including records from the 1979 Imperial
Valley, California, earthquake, *Bull. Seism. Soc.
Am.*, 71, 2011–2038.
Kaji, H. et al. 1995. *A Call to Arms: Report of the
17 January 1995 Great Hanshin Earthquake*,
UNCRD Discussion Paper No.95-2, United
Nations Center for Regional Development.
Kawase, H. 1996. The cause of the damage belt in
Kobe: "The basin-edge effect," constructive
interference of the direct S-wave with the basin-
induced diffracted/Rayleigh waves, *Seismologi-
cal Research Letters*, 67-5, 25–34.
Midorikawa, S. 1987. Prediction of isoseismal map
in the Kanto plain due to hypothetical earthquake,
Journal of Structural Engineering, 33–B, 43–48
(in Japanese with English abstract).
Motosaka, M. and M. Nagano. 1997. Analysis of
amplification characteristics of ground motions in
the heavily damaged belt zone during the 1995
Hyogo-ken Nanbu earthquake taking into account the deep irregular underground structure in
Kobe, Japan, *Earthquake Engineering and Structural Dynamics*, 26 (in print).
Pitarka, A., K. Irikura, T. Iwata and T. Kagawa.
1996. Basin structure effects in the Kobe area
inferred from the modeling of ground motion
from two aftershocks of the January 17, 1995,
Hyogo-ken Nanbu earthquake, *Journal of Physics of the Earth*, 44 (in print).
Takemura, M. 1996. Review of source process
studies for the 1995 Hyogo-ken-Nanbu earthquake; Part 1 Results from waveform inversion,
*Programme and Abstracts, The Seismological
Society of Japan*, 1996–2, A49 (in Japanese).

Seismic Behaviour of Ground and Geotechnical Structures, Sêco e Pinto (ed.) © 1997 Balkema, Rotterdam, ISBN 90 5410 887 8

Earthquake hazard zonation

R. D. Sharma, Sushil Gupta, Sanjeev Kumar & B. K. Bansal
Nuclear Power Corporation of India, Mumbai, India

ABSTRACT: This paper discusses Earthquake Hazard Zonation in view of the existing concerns related to earthquake hazard mitigation. Identification of the issues influencing the preparation of earthquake hazard zonation maps within the present state of the art is taken up. The need of making the basic information the basis of zonation maps is emphasized. It is recognized that a seismic hazard zonation map is only an aide in the assessment of earthquake hazard, rather than being an instrument to measure it.

1 INTRODUCTION

The term zonation is used for demarcation of geographically contiguous areas on the Earth's surface for which certain specified values of a selected set of geological and/or physical and/or socioeconomic parameters may be applied. Zonation to identify the exposure to earthquake hazard, so that a specified hazard mitigation strategy can be applied within each zone, is the primary objective of the earthquake hazard zonation exercise. Depending on the detailing of the information (geological, geotechnical and seismological data on the one hand and information on cultural and economic development on the other) the nomenclature (macro, micro or site specific) and utility of the zonation maps will vary. On a global scale the plate boundaries provide a macrozonation of the Earth's surface. Detailed information on a regional basis forms the basis of microzonation. The concerns of a particular site location are covered under site specific investigations.

Microzonation has become an important subject in disaster mitigation - minimizing the disastrous effects of earthquakes, cyclones, floods, landslides and avalanches. The first international conference on this subject was held in the year 1972, and several such conferences have been held in different parts of the world since then. In addition to this the issue has been discussed in relation to regional and local earthquake concerns. Microzonation finds an important place in the programmes of the International Decade of Natural Disaster Reduction (IDNDR). In India a workshop was held at the Indian Institute of Technology (IIT), New Delhi in 1991 to discuss microzonation of the Delhi area in this context.

Techniques for use in preparation of microzonation maps appear to have progressed substantially, and it is believed that microzonation maps could become part of the aseismic design code implemented through legislation, if the only apparent needs of developing a "suitable" methodology and collecting "adequate" data are met.

A formal definition of microzonation was given by Sherif (1980) as follows:

"Microzonation is a process for identifying relevant geological, seismological, hydrological and geotechnical site characteristics in a specific region and incorporating them into land use planning and the design of safe structures in order to reduce damage to human life and property resulting from earthquakes."

Detailing the information contained in this definition requires that the geological and seismotectonic regimes are understood sufficiently well to forecast occurrences of earthquakes in future (not necessarily their exact times and dates, as required in earthquake prediction), and to predict their effects in the area under microzonation. This definition of microzonation, thus, encompasses the domains of macro and micro zonation as well as site specific investigations. The word zonation will be used, henceforth, in this paper.

The term zonation has been in use in the context of earthquake hazard mitigation much before microzonation came in. It meant different things in different situations, though with a common objective. Zones of different earthquake intensities were demarcated in the map of India in the 1960's for depicting probable intensities from future earthquakes. These demarcations formed the basis of the aseismic design criteria for engineering structures in different parts of the country (IS-1893). The entire area of India was subdivided into five seismic zones, and each seismic zone was assigned a seismic coefficient based on the horizontal acceleration value. The zoning of this code was revised after the occurrence of the Magnitude 6.5 Koyna earthquake of December 10, 1967, and is again likely to change on account of the September 30, 1993 Latur earthquake. For a meaningful assessment of earthquake hazard site specific information is required on the factors affecting the ground motion parameters (amplitudes of ground acceleration, velocity and displacement), e.g. information on locations, magnitudes and source mechanisms of earthquakes, local geology and topography as well as those determining the potential of ground failure due to surface faulting, liquefaction, subsidence and collapse. Zonation to be considered as a subdivision of an area into zones depicting variation of earthquake hazard must incorporate details on all these aspects.

2 CHARACTERISTICS OF ZONATION

A zonation map is expected to provide up-to-date reliable information for assessment and mitigation of the impending earthquake hazard at a specified location. The following characteristics of zonation may be easily recognized:

(i) Zonation is a multidisciplinary exercise covering geology, geophysics, seismology, engineering and planning etc.

(ii) Zonation is knowledge based (both information and state of the art) making it a dynamic process requiring revision and upgradation as the knowledge base advances.

(iii) The interpretations and deductions, which enter the zoning maps, should be unambiguous, and traceable.

(iv) Reliability of the information should be high.

The multidisciplinary aspect of zonation requires a large quantity of information to be used, and presented. It may not be possible to present the entire information on, or integrate it into, one or two maps. No.(ii) above requires presentation of the basic (raw-unmutilated) information, e.g. geological faults,

locations and magnitudes of past earthquakes, observed intensities, geology etc. in such a manner that new information can be added and suspect information can be identified and deleted, as and when found necessary. In view of No.(iii) care is required in placing any derived information (e.g. return periods and probabilities of events, values of ground motion parameters etc.) on the zonation maps. The question of reliability - No.(iv) - arises from the possibility of such maps being based on incomplete or inaccurate information, suspect methodologies and assumptions or non uniqueness of interpretations of the available information. Absence of information on vital issues (e.g. potential of ground failure phenomena) tends to minimize risk. Reliable zonation maps may be possible only when sufficient information to identify locations of future earthquakes has been collected and geological mapping of the area has been carried out and well studied. Though it has been well recognized that earthquakes occur along faults and faults can be discovered through planned field investigation programs, faults are discovered, often, only after occurrence of an earthquake. This is, often, attributed to lack of investigations in the area before earthquake occurrence. Zonation should, therefore, be attempted on a regional or local basis. This paper deals with geological and seismological issues only.

3 GEOTECTONIC COMPONENTS OF ZONATION - REQUIRED DATA BASE

In almost every zonation exercise the available information (data base) is found inadequate, data on earthquakes in particular. Ignoring the gaps in the data, or filling these on questionable assumptions makes the zonation exercise unreliable. Assessment of earthquake hazard at a place requires the following geological and seismological information:

(i) Locations and magnitudes (or intensities) of past earthquakes.

(ii) Locations, dimensions, types and seismotectonic status (movement history) of geological faults in the area.

(iii) Occurrence rates of earthquakes of different magnitudes in space and time.

(iv) Magnitude the maximum credible earthquake (MCE) associated with each fault.

(v) Probabilities of ground motion of different severity.

(vi) Geological conditions in the site area (soil conditions, topography etc.)

Apart from the seismological, geological and

tectonic factors, earthquake hazard at a place also depends a great deal on the state of development in the area.

Application of the procedures of zonation is agog with severe constraints. On the one hand it is believed that different stages between earthquake occurrence and the resulting damaging effects can be predicted and, on the other, inability to do so is attributed to lack of data. The fact is that zonation is relevant only in areas where sufficient information on earthquakes and earthquake effects is not available. In areas where such information exists, experience is likely to take precedence over the analytical jargon. The uncertainties regarding recurrence rates of strong earthquakes (assuming that well defined recurrence intervals exist), ground motion attenuation, earthquake source mechanism and the seismic signal modification characteristics of the Earth's crust etc. will remain, always. Under these conditions it would be more appropriate to treat the problem of earthquake ground motion statistically - associating values of ground motion parameters with probabilities - allowing uncertainties. Figure-1 shows one such example. Here, the variation in the peak ground acceleration for a specified period is estimated for three levels of seismicity, each characterised by a magnitude frequency relationship (Sharma and Achuthankutty, 1984). The first and foremost requirement of zonation is, therefore, the recognition of what is adequate information in a given environment, and whether it can be generated. Information generation must, then, be an integral part of the zonation exercise. Very naturally, the information directly required for fixing the location(s) and magnitude(s) of the earthquake(s), against which the structures are to be protected, must be generated first.

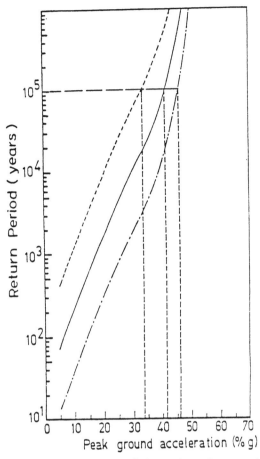

Figure 1. Return periods and peak ground acceleration values for three levels (High, Average and Low) of seismicity at a site

4 END USE OF ZONING MAPS

Zonation may be used for different purposes with varying degrees of effectiveness. For example these can find applications in:

(a) Predicting earthquake hazard at a specified site so that it can be classified into low, moderate or high risk area.

(b) Identifying those issues in the area, which should be investigated further before executing a development plan.

(c) Fixing aseismic design parameters for new structures (like those in a building code).

(d) Retrofitting of old structures to ensure safety.

(e) Land use planning.

(f) Predicting earthquake hazard in existing centres of activity, population centres and industrial areas for better emergency preparedness.

(g) Insurance purposes.

For each of the above applications the zonation has to be carried out keeping in view the particular requirement.

5 THE INFORMATION BASE

Availability of information on historical earthquakes varies a great deal from one region to another. Earthquake data in some parts of the world are available for several thousand years. In India a period of less than 200 years is covered. Except for the north eastern region, these data are also very sketchy

(Krishna, 1992). It has now become evident that absence of records of earthquakes during such a short interval is, by no means, an indication of seismic stability. Any prediction on the basis of such scanty records is not, at all, reliable. Though, geological field techniques for searching evidence of major earthquakes of the Holocene period have developed, these techniques have not yet been extensively applied, except in a certain geologically favourable environment (Seigh, 1996). In areas, where earthquake observations are sparse, earthquake potential of the area can be based only on the tectonic set up. Depending on the information on the geological faults (dimensions and evidence of fault movement) and the existing data bases (magnitude vs fault dimensions or rupture length) for different fault types a maximum magnitude is to be assigned to each fault to represent its earthquake potential. For a probabilistic analysis this is to be followed by associating a magnitude frequency relationship with fault. Operating a microearthquake network in the area to collect accurate data on locations and magnitudes of earthquakes down to magnitude, say 2, can prove useful in achieving this objective, provided earthquakes are recorded and accurately located during this period (Lee and Stewart, 1981).

5.1 Seismotectonic maps

In order to infer future earthquake occurrences the earthquake sources need to be identified, and earthquake potential of each source is to be assessed. A seismotectonic map is required to achieve this objective. Preparation of the seismotectonic map is an important step towards zonation. This map may be prepared through the following steps:

(a) Demarcate the area, say 100 km x 100 km, for zonation. Call it A1. Also, mark a bigger area A2 of, say 600 km x 600 km, around the area A1.

(b) Obtain the latest tectonic map (1:1000,000 scale) of area A2 showing all the major lineaments in the area. Mark the confirmed and suspected faults (by separate symbols) and the fault types, if known.

(c) Carry out independent checks of the maps, for the area A1, using remote sensing data (satellite imageries and aerial photographs) and other geological information to locate any fault, which might not be included in the original map.

(d) Superimpose the epicentres of the historical earthquakes on the tectonic maps along with the magnitude information.

(e) Plot the epicentres of the instrumentally recorded earthquakes (along with the magnitude information) on a copy of the map on the same scale.

For the earthquake data to be adequate for the purpose the instrumental data on earthquakes should be added from the continuously operating microearthquake network until a reasonably good picture of seismicity of the area emerges, i.e. data for a longer time is not likely to change the information content significantly. The seismotectonic maps can, then, be utilized for zonation of the area A1.

The seismotectonic map contains a number of lineaments of different lengths and orientations. All the lineaments on the map do not represent faults. All the faults are not seismogenic, and all the seismogenic faults may not be producing earthquakes (a fault may be undergoing creep movement without producing earthquakes or passing through a seismic gap). Rates of movement along different faults would be different. Thus, all the faults are not equally hazardous. Assessment of earthquake hazard using the information on the map is based on one's ability to delineate the faults from the lineaments, and categorize each fault as active, potentially active or inactive. Only then seismic potential (or its absence) can be decided on. Accept for some thoroughly investigated areas in seismically active belts (e.g. the faults in California), such information can be expected to become available only after detailed regional studies are completed under an industrial project (e.g. a nuclear power plant). Under these circumstances it would be more appropriate to list out the seismotectonic features (faults, epicenteres and earthquake magnitudes) along with the sources of such information and an attribute of quality and completeness of each information (quality factor). Only if the information can be considered of very high quality, demarcation of zones can be proceeded with.

5.2 Geological data

Published geological maps, normally, follow fairly stringent practices, and are of reasonably high quality. Zonation based on geological consideration is of two types, namely: (i) to show the areas, for which maps of different scales are available, and (ii) to show different geological provinces demarcated on the basis of the available information. Preparation of a geological map requires considerable effort in information collection, ground truth verification and preparation, checking, release and final publication of maps. If geological mapping for the area under consideration has been completed (say on

1:25,000/1:50,000) scales, then the geological provinces can be demarcated. More detailed mapping (on, say, 1:5000 scale) can be taken up during the site investigation stage along with investigations of soil characteristics.

Geological data are needed to enable the geologist and the engineer to apply the knowledge acquired elsewhere on the basis of geological similarity. From the available information it should be possible to infer the suitability of the site and requirements of foundation on the one hand and to analytically study the seismic signal modification properties of the site on the other. The former requires static and dynamic characteristics of the soil cover while the latter needs the seismic velocity profile of the area. A farily long list of geological data exists for geotechnical applications at sites of important engineering structures (IAEA, 1979). Information for evaluating the liquefaction potential includes grain size and distribution, density, shear wave velocity, shear history, age of sediments, level of the water table, penetration resistance. Similarly, properties leading to slope instability, subsidence and collapse in the site area, which could pose serious risk to the stability of the foundations must be evaluated. Apart from geological data this includes information on topography, natural and man made drainage, meteorological data, tidal waves, tsunamis, population density etc. All this information is not required for microzonation. It is, therefore, desirable to distinguish between the two categories of such information, namely:

(i) the information required for zonation purposes, and

(ii) the information needed for geotechnical applications at a site.

The geologist will list the information in the first category whereas that in the second category is to be listed by the geotechnical engineer as a requirement, not much to do with zonation as such. Identification of faults capable of giving rise to damaging earthquakes (or undergoing aseismic creep), and assigning the maximum earthquake (or creep) potential to each fault is a very important and difficult task in evaluating earthquake hazard. Any inaccuracy/uncertainty in this area can render the zonation exercise, practically, useless.

6 QUALITY OF DATA

In engineering applications the adequacy of the seismological and geological data base is, often, questioned. Though, geological and geophysical field investigations can be undertaken to collect data to fill some gaps, inadequacy of these data is the inherent nature of some regions characterized by infrequent occurrences of earthquakes and inaccessible geology. While decisions have to be taken in the absence of certain data, salient features of the available data should be reflected in the decisions. A few cases are discussed below:

(1) In a specified area a moderate earthquake, say magnitude 6.0, occurred about every one hundred years. Then a magnitude frequency relationship of the type Log (N) = a - b.M may be postulated by adopting a global b-value (1.0), and the seismic risk calculations can be proceeded with. If additional data on earthquake occurrences are available to estimate the recurrence intervals of earthquakes, of magnitude 5.0 or 7.0, say) the b-value can be estimated slightly more accurately. If the `a' and `b' values can be accurately estimated, and seismicity can be apportioned between different earthquake sources, seismic risk calculations will be more accurate. Unless additional data lead to an improvement over this situation, the increase in the bulk of data is of little value.

(2) Reservoir induced seismicity (RIS) is a very exciting topic in earthquake science. Almost all codes and guides on the subject caution about taking into account the RIS potential while estimating seismic risk. A reservoir (or any other earthquake inducing agent) is effective only in those areas where sufficient strain energy has accumulated to give rise to an earthquake (and the background seismicity is not high making the RIS irrelevant). RIS is, thus, possible only when a fault having a slip potential existed in the vicinity of the earthquake inducing agent (a reservoir in the present case). The only difference the RIS phenomenon could make is to alter the estimated probabilities in an unknown manner. No elaborate theory involving seismic hazard calculations is likely to add to the degree of confidence in the calculations for near future. The only issue, then, is whether there is a fault capable of accumulating strain energy in the neighborhood of the reservoir. This is not to suggest that research on RIS should be discontinued.

(3) Zonation maps based on readily available data on seismicity, geology, tectonics and ground motion parameters etc. and with severe constraints of theory must be distinguished from those, which were carefully prepared to meet contractual requirements.

(4) There may be records of a good number of historical earthquakes in a region, and some data on instrumentally recorded earthquakes may also exist. However, unless the sources of the earthquakes can

be identified this information will only allow a qualitative attribute, e.g. the area is moderately seismic.

(5) Consideration of the site conditions is of paramount importance in determining earthquake effects, and aseismic design parameters (peak values of acceleration, velocity and displacement and response spectra). In the latest practices of aseismic design use of site specific response spectra is recommended. Reasonably consistent results have been obtained by classifying sites into rock and soil sites for the purpose of fixing ground motion parameters (Seed et al., 1976; Mohraz, 1976). Mean + σ response spectral shapes for rock and soil sites from a selected set of accelerograms, including some from India, are shown in Figure-2 (Kumar et al., 1993). Unless the applicability of more detailed site specific information is established beyond doubt, use of such information tends to reduce the reliability of zonation based on such information.

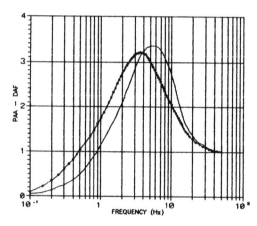

Figure 2. 5% Mean+sigma response spectral shapes for rock (-) and soil (*) sites. (Kumar et al., 1993)

(6) Often, the earthquake data base consists of information on small earthquakes. This is true for frequencies of earthquake occurrences as well as ground motion parameters. Extrapolations in the data without considering the nonlinearity in observed correlations may lead to incorrect values of the derived parameters (e.g. probabilities, return periods, peak values of ground motion parameters and response curves) based on such data.

(7) In a certain area discovery of new faults may alter the predicted values of a derived ground motion parameter considerably. Figure-3 shows a typical seismotectonic environment around a construction site. Contours of peak ground acceleration (PGA) values in Figure-4 show how the PGA values are

modified after the newly discovered fault in the neighborhood of the site is taken into account, and when the `b'- value (which determines the slope of the magnitude frequency curve) is altered.

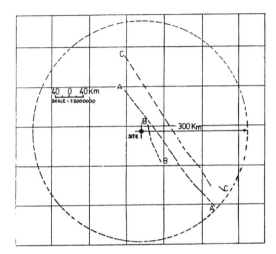

Figure 3. A typical seismotectonic environment in the area of a construction site. Seismicity along the faults is assumed to follow a magnitude frequency relationship of the type Log N = a - b.M.

(8) Historical records of earthquakes do not exist for some areas, and such regions are often considered aseismic. The occurrences of the December 10, 1967 Koyna earthquake on the west coast of India and the September 30, 1993 Latur earthquake and the seismic activity in penninsular India, which followed these earthquakes have drastically changed the seismotectonic status of this region, which was considered practically aseismic. The effect of these earthquakes on any zoning exercise can be easily seen (Figure-5).

It is, therefore, desirable that the character of the available data is reflected in the estimates of seismic risk, distinction is made between the situations, which can and which cannot be corrected by collecting additional data. Whenever required, additional data should be collected, rather than basing the zonation exercise on the available data only. Further, any methodology adopted for preparation of zonation maps should be sufficiently flexible to accommodate surprises of the future. Occurrence of an isolated event in an area should not render the zoning maps useless. Zonation should be undertaken only when assurance of availability of adequate data exists.

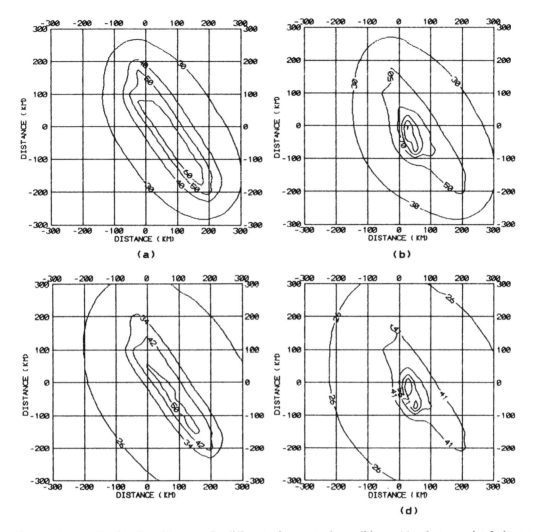

Figure 4. Iso-acceleration (in gals) curves for different seismotectonic conditions: (a) only two major faults are known and, (b) after a smaller fault closer to the site is discovered . (c) and (d) are same as (a) and (b), respectively, with b-value of 1.0 against 0.54 used in (a) and (b).

7 ESTIMATING GROUND MOTION PARAMETERS

The philosophy behind zonation is based on the assumption that future earthquakes will occur at places where they occurred earlier, with the same intensity and recurrence intervals. Experience has, however, shown that such an assumption is only partially valid. This situation is corrected, to a certain extent, by adopting a seismotectonic approach, in which past earthquakes are associated with tectonic structures (faults or tectonic provinces), and it is assumed that future ones could occur anywhere along the tectonic structure in question. Rather than using fixed values of earthquake parameters (magnitude, hypocentral location and recurrence rates) in a deterministic approach, the possible range of each of these parameters is, then, used to evaluate the seismic hazard in a probabilistic analysis. Values of ground motion parameters being exceeded with specified probabilities during a specified period (useful life of engineering structures at a site, for example) are estimated (Cornell, 1968; Algermissen and Perkins, 1972; McGuire, 1976; Basu and Nigam,

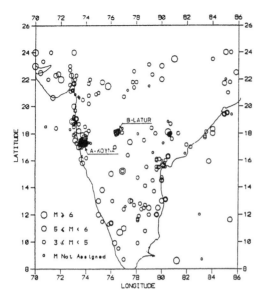

Figure 5. Earthquakes of peninsular India. A- and B-show the locations of Dec.10, 1967 Koyna and Sept.30, 1993 Latur earthquakes, respectively.

1978). In this analysis the probability of exceeding a certain specified value of a ground motion parameter (peak ground acceleration - PGA -`a' is often used) is estimated by evaluating the expression:

$$P(a) = \int_{M=Mo}^{M=Mmax} \int_{S=a}^{S=Smax} P1(M) \cdot P2(M,S) \, dM \, dS \quad (1)$$

Here, P1(M) is the probability of an earthquake of magnitude M occurring in the area under consideration and P2(M,S) is the probability that given an earthquake of magnitude M the value `a' of the PGA is exceeded during the useful life (L years) of a certain utility located at the site. The return period, T, of such a causative earthquake is given by (Lomnitz, 1976):

$$T = -L / \ln[1.0 - P(a)] \quad (2)$$

Thus for estimating the ground motion parameters for a specified exceedance probability the following information is required:

(a) Distribution of earthquake magnitudes in space and time,

(b) Values of the ground motion parameter as a function of the earthquake magnitude, and

(c) Attenuation characteristics (frequency dependent) for the ground motion parameter in question as a function of distance.

It is apparent that a large uncertainty (or inaccuracy) in any of these areas will render any refinements in another area ineffective. In order to make the analysis of the effects of the local geological/soil characteristics meaningful it is necessary that:

(i) The values of the ground motion parameter (as a function of frequency) are known accurately at the base rock level.

(ii) The soil conditions at the site are modelled accurately in terms of the known soil parameters (thickness, shear velocities etc.)

(iii) Uncertainties arising from limitation of applicability of the model and in the various soil parameters can be translated into the confidence limits of different levels of ground motion.

The above description on estimating the ground motion brings out reasonably well that zonation in terms of values of ground motion parameters (or any other derived parameter) is unlikely to be very rewarding from the point of view of design. It will be more useful if the available seismological and geological information is compiled, and additional information is generated by using geological and geophysical methods (geological field mapping and microearthquake recording, for example) for the chosen area. The experiences on earthquake occurrences in shield areas have demonstrated that the priorities of taking up the data collection task is to be fixed keeping in view the development efforts in the region, rather than taking up the so called "seismically active" areas.

8 GROUND FAILURE PHENOMENA

Surface faulting, liquefaction, subsidence and collapse can become major causes of damage during a strong earthquake. In certain cases it may be possible to mitigate the effects of some of these, but it is not always possible to demonstrate the efficacy of the engineering solutions. Sufficient geological information can be collected to identify the potential of ground failure at a site by either of these causes (IAEA, 1979). It is, therefore, desirable that the zonation exercise fully accounts for these.

9 CONCLUSIONS

The issues related with earthquake hazard zonation have been discussed. The importance of making the basic information on geology and seismotectonics, and geotechnical parameters the basis of zonation, rather than using any doubtful derived information has been emphasised. Differentiating between the information required for zonation and that required for the zones (but is not to be necessarily available for zonation) is necessary for making any progress in the zonation exercise. Lack of data in certain areas, if these cannot be obtained by undertaking a program of systematic investigations, is to be recognised as characteristics of those areas, and zonation is to be proceeded with. Zonation has greater relevance for areas where earthquake data are sparse. Knowledge of the limitations of the information used in zonation is important.

10 ACKNOWLEDGEMENTS

This paper, originally, resulted from an invitation to the Workshop on Design Practices in Earthquake Geotechnical Engineering held at Roorkee, India on September 26-27, 1996. The authors are grateful to the organizers of this workshop, from where it found its way to the ICSMFE. The authors are also grateful to Mr. Y.S.R.Prasad, Managing Director, NPC and Mr. A. Sanatkumar, Director (Engineering), NPC for encouragement and support.

11 REFERENCES

Algermissen, S.T. & D.M. Perkins. 1972. A technique for seismic zoning: General considerations and parameters. Proc. Int. Conf. on Microzonation, Seattle, pp. 865-878.

Basu, S. & N.C. Nigam. 1978. Seismic risk analysis of Indian peninsula. Proc. World Conf. Earthquake Engg., 6th, New-Delhi, 2:425-431.

Cluff, L.S. 1978. Geologic considerations for seismic microzonation. Proc. of the Second Int. Conf. on Microzonation for safer Construction - Research and Application, San Francisco, California, U.S.A., Vol. 1, pp. 135-152.

Cornell, C.A. 1968. Engineering seismic risk analysis. Bull. Seism. Soc. Am., 58:1583-1606.

IAEA 1979. Earthquakes and associated topics in relation to nuclear power plant siting - Safety Guide 50-SG-S1, International Atomic Energy Agency, Vienna.

I.S.: 1893. 1984. Indian standard criteria for earthquake resistant design of structures. Bureau of Indian Standards, New Delhi, INDIA.

Krishna, J. 1992. Seismic zoning maps of India. Current Science, (special Issue) Vol. 62, Nos. 1&2, Indian Academy of Sciences.

Kumar, S., S. Gupta & R.D. Sharma. 1993. Site-dependent response spectra for rock and soil sites. Proc. 2nd International Conference on Vibration Problems. A.C. College, Jalpaiguri, W.B., INDIA, 109-114.

Lee, W.H.K. & S.W. Stewart. 1981. Principles and applications of microearthquake networks. Advances in Geophys., Supp. 2, Academic Press, New York and London.

Lomnitz, C. 1976. Global tectonics and earthquake risk. Development in Geotectonics, 5, Elsevier Scientific Publishing Company.

McGuire, R.K. 1976. EQRISK - Evaluation of earthquake risk to site. U. S. Geological Survey, Open-File Report 76-67.

Mohraz, B. 1976. A study of earthquake response spectrum for different geological conditions. Bull. Seis. Soc. Am., 66, 915-935.

Seigh, K. 1996. The repetition of large earthquake ruptures. Proc. Natl. Acad. Sci., U.S.A., 93, 3764-3771.

Seed, H.B., C. Ugas & J. Lysmer. 1978. Site-dependent spectra for earthquake resistant design. Bull. Seis. Soc. Am., 66, 221-243.

Sharma, R.D. & I. Achuthankutty. 1984. On uncertainty in design basis ground motion parameters having inadequate data. Bull. Indian society of Earthquake Technology, 21, 1-14.

Sherif. 1980. Definition of microzonation. Newsletter, Earthquake Research Institute, V.14, No.4, p. 68.

Seismic Behaviour of Ground and Geotechnical Structures, Sêco e Pinto (ed.) © 1997 Balkema, Rotterdam, ISBN 90 5410 887 8

Evaluation of liquefaction potential of a site located in the South of Portugal

P. Sêco e Pinto, J. Correia & A. Vieira
Laboratório Nacional de Engenharia Civil, Lisbon, Portugal

ABSTRACT: This paper presents the evaluation of liquefaction potencial of a site located in the South of Portugal using the methodology recommended in the Manual for Zonation on Seismic Geotechnical Hazards. The geotechnical characteristics of soil profile were obtained by geophysical techniques, borings with SPT tests, CPT tests, vane shear and pressurometer tests as well as static and dynamic laboratory tests.
The assessment of settlements in sands deposits is performed by two methodologies.

1 INTRODUCTION

A microzonation study was conducted with respect to liquefaction potencial for a site located in the South of Portugal, within zone A of seismic Risk Map of Portugal, using Grade - 1, Grade - 2 and Grade - 3 methods recommended in the Manual for Zonation on Seismic Geotechnical Hazards (TC4, 1993).

The general geological and seismic characterists of the site are presented.

The geotechnical survey to assess the static and dynamic properties of the materials was based on field tests and laboratory tests.

For the evaluation of settlements in sand deposits two methodologies were used.

Some final conclusions are presented.

2 GENERAL GEOLOGICAL CHARACTERISTICS

The area exhibits a stratigrafic sequence with units from Holocenic to Triassic. The sandy materials are predominant and the rivers are filled with alluvia material composed by sands and muds.

Three complexes were identified: (i) complex 2A composed by mud materials with some incrustations of sand material; (ii) complex 2B predominantly composed by sandy material and

(iii) complex 2C composed by silty - sandy materials with gravels.

The bedrock is composed by shale and grawackes formations.

3 EARTHQUAKE CHARACTERISTICS

The site is located within the maximum intensity zone of Portugal seismic risk map. It is near the transform Azores - Gibraltar fault the source of 1755 and 1969 earthquakes.

The 1755 Lisbon Earthquake with a Ritcher magnitude 8.5 and a fault length 1 500 km has provoked liquefaction in places located 420 km of the fault.

For the 1969 Earthquake with a Ritcher magnitude 7.1 and a fault length of 100 km, there was no evidences of liquefaction.

The 1909 Benavente Earthquake with a continental fault and a magnitude 7.5 has provoked liquefaction in places located 100 km of the fault.

It is also important to point the Messejana fault with an orientation NE-SW that has provoked in the last 1950 years 15 earthquakes with a magnitude higher than 5 but only 3 earthquakes with magnitudes 6 to 7. It can only provoke liquefaction for sites located less than 100 km of the fault.

4 GEOTECHNICAL SURVEY

To evaluate the static and dynamic properties of the materials field tests as well as laboratory tests were performed.

Geophysical tests using cross-hole and down-hole techniques, 25 borings with SPT tests 2.0 m apart, CPT tests, vane shear tests and pressurometer tests were performed.

Laboratory tests such as classification tests, static triaxial tests, oedometer tests and dynamic triaxial tests were also done.

The results of SPT tests and geophysical tests are summarized in Table 1.

The results of CPT tests, pressurometer tests and vane shear tests are described in Table 2.

The results of classification tests and oedometer tests are shown in Table 3.

The results of static triaxial tests and dynamic triaxial tests are shown in Table 4.

Table 1. Summary of SPT and geophysical tests

Type	SPT (N)		P Wave Velocity (m/s)		S Wave Velocity (m/s)		Maximum Shear Modulus (MPa)	
	Range	Average	Range	Average	Range	Average	Range	Average
Complex 2A	1 - 5	2	1 300 - 1 600	1 400	60 -230	110	5 - 79	23
Complex 2B	3 - 20	10 - 15	1 000 - 1 500	1 300	105 -360	170	17 -199	53
Complex 2C	6 - 60	20 - 40	1 500 - 1 800	1 600	115 - 400	260	22 - 288	127

Table 2. Summary of CPT, pressurometer and vane shear tests

Type	CPT (MPa)		Pressurometer				Vane Shear	
	Range	Average	Horizontal Range	Stress (kPa) Average	G Modulus Range	(MPa) Average	Undrained Range	Strength (kPa) Average
Complex 2A	0.3 -1.0	0.6	126 - 275	214	13.2 - 27.8	20.5	15 -25	20
Complex 2B	1 - 5	2.5						
Complex 2C	2 - 20	10						

Table 3 Summary of classification and oedometer tests

Type	Classification Tests		Oedometer Tests		Consolidation Coefficient (m²/s)
	Liquid Limit (%)	Plastic Limit (%)	Initial Void Ratio	Compressibility Index	
Complex 2A and 2B	40.6 - 47.2	15.6 - 20.0	0.99 - 1.03	0.19 - 0.29	4.8×10^{-7} - 6.1×10^{-7}

Table 4 Summary of Static and Dynamic Triaxial Tests

Type	Static Triaxial Test			Dynamic Triaxial Test	
	Cohesion (kPa)	Friction Angle (°)	Elasticity Modulus (MPa)	Shear Modulus (MPa)	Damping Ratio (%)
Complex 2A	30	10	10	20 - 40	6 - 10
Complex 2B	0	33	10	10 - 40	7 - 17

5 ZONING FOR LIQUEFACTION

Grade - 1 method

Based on historical earthquakes and particularly on 1755 Lisbon earthquake Jorge (1994) proposed the following equation:

$$Log\ R = 0.61\ M - 2.52 \qquad (1)$$

where R is the epicentral distance, in km, and M is the earthquake magnitude defined by Richter.

This relationship can be used to assess the liquefaction opportunity for earthquakes similar to the 1755 Lisbon earthquake and is similar to other relationships proposed by other authors (Kuribayashi and Tatsuoka, 1975; Wakamatsu, 1993).

The liquefaction susceptibility was inferred by geological and geomorphological criteria.

The analysis was based on the Portugal Geological Map and a scale 1:200 000 was used.

This map refers to the existence of alluvial deposits composed by sands and mud materials.

The location of the different materials is shown in Fig. 1.

For the liquefaction assessment the criteria proposed by Youd and Perkins (1978) was used and has given the following results: (1) alluvial soils with high susceptibility; (2) sandy materials with moderate susceptibility; (3) quaternaly material with low susceptibility; and (4) Triassic material with nil susceptibility.

Following the proposal of Johnson and Kissenpfenning (1977) the isointensity map for Lisbon 1755 earthquake is presented in Fig. 2 and the liquefaction susceptibility for the different zones is shown.

Jorge (1994) has developed for Portugal a map of liquefaction opportunity (Fig. 3) based on 7 seismic zones, which shows that for the area the return period for liquefaction opportunity is 150 years. Using the Poisson distribution this value can be related with an annual excedence probability of 6.57×10^{-3}.

Grade - 2 method

To improve the quality of the Grade - 1 zonation map an attempt was performed. Existent historical information was poor and with additional laboratory test data, using the results of sieve analyses, for grade 2 a corographic map with a scale 1:50 000 is presented in Fig. 4.

Grade - 3 method

A simplified method to assess liquefaction potencial based on field tests such as SPT tests, CPT tests and seismic tests following the procedure proposed by Seed et al (1983) was used.

To express the ability of a soil element to resist liquefaction, a liquefaction resistant factor F_L is defined as F_L: R/L, where R is the in situ resistance of the soil element to an earthquake loading, and L is the earthquake load in the soil element induced by a seismic motion.

Following Portuguese Code (RSA, 1983) the seismic actions should be defined for near sources and far sources. The near sources are related with continental faults and magnitude 6.9 to 7.5 and a peak acceleration 170 gal. The far sources are related with offshore faults and magnitude 8.5 - 9.0 and a peak acceleration 100 gal.

Normalisation with respect to the overburden effects were performed by multiplying the measured SPT value or penetration resistance q_C value by the factor $(100/\sigma'_{vo})^{1/2}$ where σ'_{vo} (kPa) is the effective over burden pressure acting at the depth where the SPT or CPT measurement has been made.

The SPT values where corrected following the energy ratios for SPT procedures proposed by Skempton (1986).

Based on geological conditions and geometry conditions two profiles were selected.

The value of L is estimated by the relation developed by Seed and Idriss (1971):

$$\frac{\tau_{av}}{\sigma'_0} = 0.65\ \frac{a_{max}}{g}\ \frac{\sigma_0}{\sigma'_0}\ r_d \qquad (2)$$

where τ_{av} is the average cyclic shear stress during a particular time history, σ'_0 is the effective overburden stress at the depth in question, σ_0 is the total overburden stress at the same depth, g is the acceleration of gravity and r_d is a stress reduction factor.

Following the procedure proposed by EC8 (1994) the liquefaction assessment is performed for safety factors of 1.0 and 1.25.

Typical grain size distribution curves for the sandy materials are shown in Fig. 5.

Also this liquefaction assessment is presented in Figs. 6 to 9 for SPT tests and CPT tests.

The seismic response was also obtained by a finite element computer program developed at LNEC (Vieira,1995).

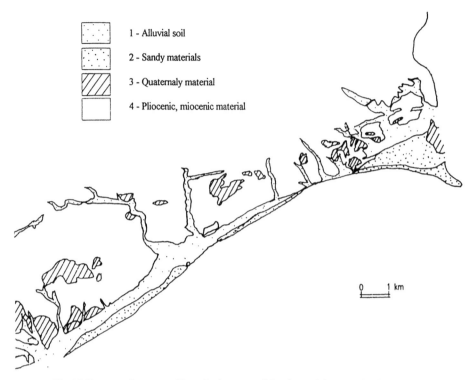

Fig. 1 Microzonation map of liquefaction potential estimated from geological conditions

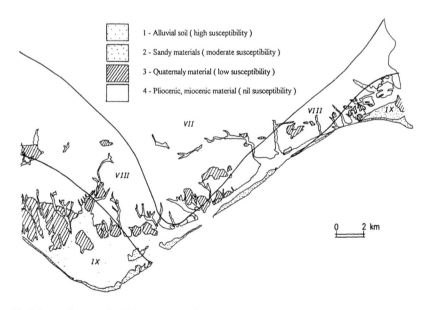

Fig 2 Isotensity map for Lisbon 1755 earthquake. Assessment of liquefaction susceptibility

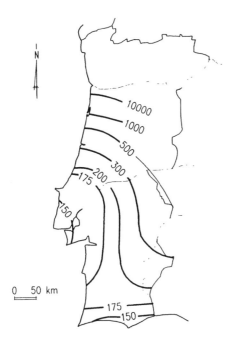

Fig 3 Liquefaction opportunity map of return periods for Portugal (Jorge,1994)

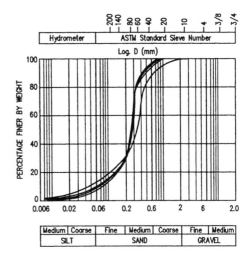

Fig 5. Grain size distribution curves

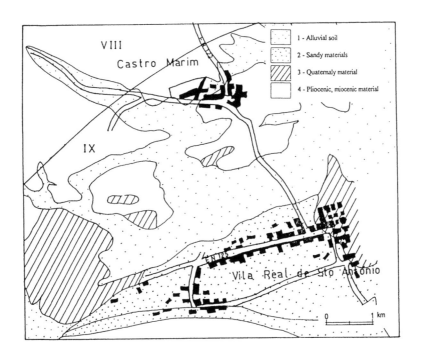

Fig 4. Site corographic map

117

PROFILE 1

PROFILE 2

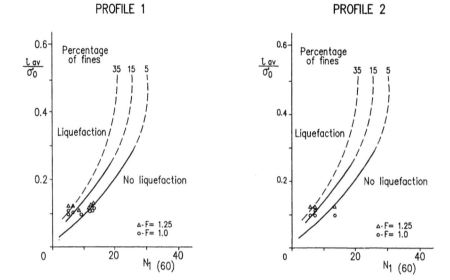

Fig .6 Liquefaction assessment based on SPT tests. Near source earthquake

PROFILE 1

PROFILE 2

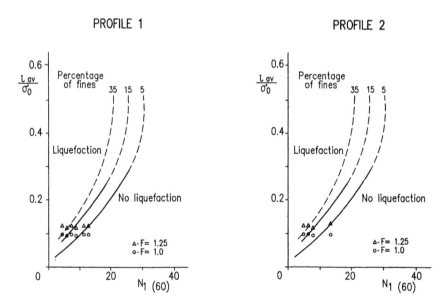

Fig .7 Liquefaction assessment based on SPT tests. Far source earthquake

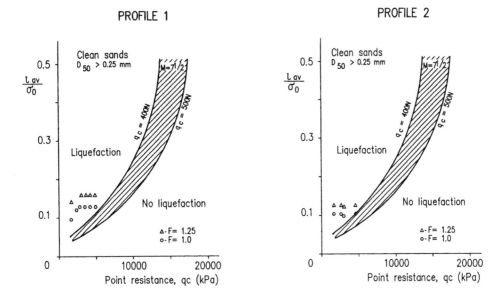

Fig .8 Liquefaction assessment based on CPT tests. Near source earthquake

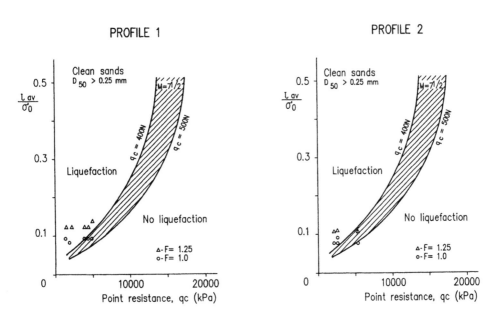

Fig .9 Liquefaction assessment based on CPT tests. Far source earthquake

Input motions are incorporated by base horizontal and vertical acceleration power spectra. These can be obtained by direct records of seismic motions by response spectra or by trilogaritme diagram.

This analysis is based on the solution of the equation of motion considering a homogenous and continuous soil deposit composed by horizontal soil layers and assuming a vertical propagation of shear waves.

For the soil behaviour the equivalent linear

The shear stresses distribution and the accelerations distribution are presented in Figs. 10 and 11.

The Table 7 summarizes for near source and far source the transfer functions of acceleration between the bedrock and the ground level (TraFA), the fundamental period of the layer (T_F), the maximum ground acceleration (Max A) and the amplification ratio (AR). It can be noticed that the amplification effects are higher for the far source.

Table 5 - Summary of the shear modulus and damping ratio values for profile P1

Complex	Depths	Thickness	Shear Modulus (MPa)			Damping Ratio (%)		
	(m)	(m)	MIN.	AVER.	MÁX.	MIN.	AVER.	MÁX.
Fill	0.00-1.00	1.0	25	25	25			
C2A	1.00-3.00	2.0	25	25	25			
C2B	3.00-20.8	17.8	47	65	76	8	12	15
C2A	20.8-38.30	17.5	41	54	68	9	10	11
C3-Pliocen. clay	38.30-43.10	4.8	68	73	78	8	9	9
C4-Grawackes	43.10-46.50	3.4	500	500	500	4	6	8
Bedrock	46.50	-	800	800	800	4	4	4

Table 6 - Summary of the shear modulus and damping ratio values for profile P2

Complex	Depths	Thickness	Shear Modulus (MPa)			Damping Ratio (%)		
	(m)	(m)	MIN.	AVER.	MÁX.	MIN.	AVER.	MÁX.
Fill	0.00-0.90	0.9	25	25	25			
C2A	0.90-2.70	1.8	25	25	25			
C2B	2.70-14.70	12.0	35	46	60	9	12	16
C2A	14.70-22.10	7.4	35	36	37	9	10	10
C2C	22.10-37.00	14.9	270	480	530	3	4	5
C3-Pliocen. clay	37.00-44.90	7.9	75	81	83	8	9	9
Bedrock	44.90	-	800	800	800	4	4	4

Table 7 - Summary of the seismic analyses results

	Near Source		Far Source	
Profile	P1	P2	P1	P2
TraFA	3.670	3.790	3.634	3.777
T_F (s)	1.005	0.537	1.026	0.851
MaxA (m/s^2)	1.800	1.671	1.525	1.638
A (m/s^2)	1.521	1.600	0.956	0.950
AR	1.18	1.04	1.46	1.72

method is used and the shear modulus and damping ratio are adjusted in each iteration until the convergence has occurred.

The two mentioned profiles for the seismic actions near and far source were analysed.

The values for the shear modulus and damping ratio assumed for profiles P1 and P2 are summarized in Tables 5 and 6.

The variations of shear modulus and damping ratios, for the two profiles and for the seismic actions are presented in Figs 12 to 13.

A comparison between the soil resistance, the cyclic stresses computed by Seed method and 65% of the shear stresses computed by the code for the two seismic actions is shown in Figs. 14 and 15. It can be noticed that for the initial depths the computed stresses

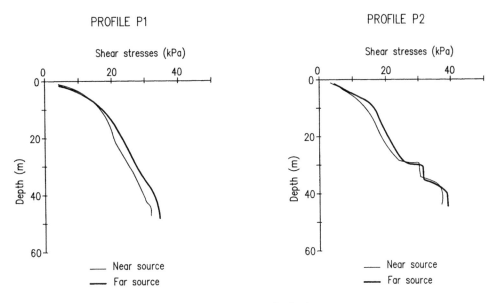

Fig 10 Shear stresses distribution

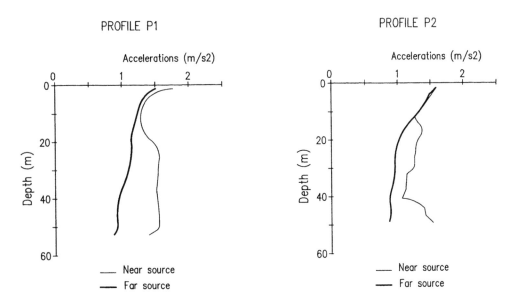

Fig 11 Accelerations distribution

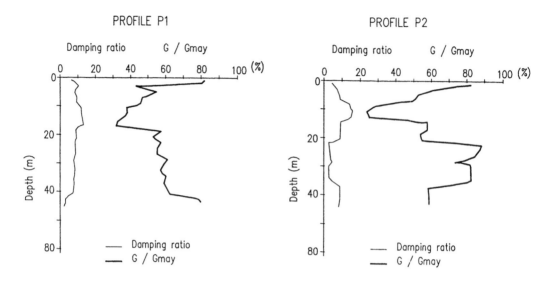

Fig 12 Variation of shear modulus and damping ratio. Near source earthquake

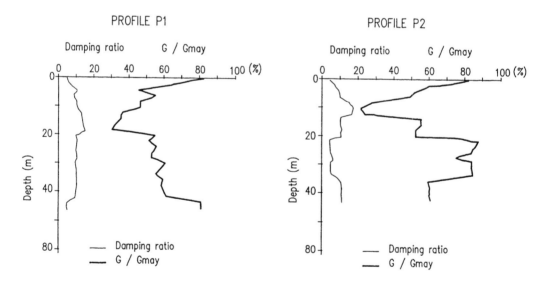

Fig 13 Variation of shear modulus and damping ratio. Far source earthquake

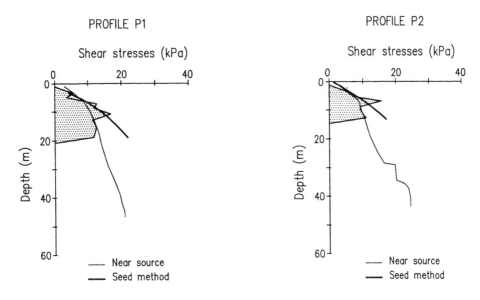

Fig 14 Liquefaction assessment for near source earthquake

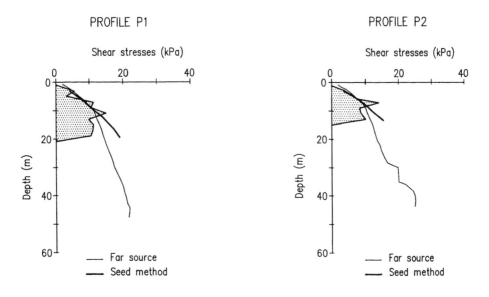

Fig 15 Liquefaction assessment for far source earthquake

by the two approaches are similar, but for depths higher than 10 m the values computed by Seed method are higher than the shear stresses obtained by the computer code.

6 ASSESSMENT OF SETTLEMENTS IN SAND DEPOSITS

Saturated sands subjected to earthquake shocks generate pore pressures leading to liquefaction or loss of strength. The dissipation of the pore water pressure is accompanied by volume change of sand deposits which tends to settle.

It is important to predict the earthquake induced settlements of sands as they have an important effect on the performance of the structures.

Tokimatsu and Seed(1984) have presented relationships between the shear strain developed in situ during earthquakes by a combination of cyclic shear stress ratio and normalized SPT N- values.

For the complex 2B considering an average value N_1= 12 a stress ratio value of $\tau/\sigma' = 0.17$ obtained from the dynamic analysis, the post-liquefaction volumetric strain estimated is 2.5% following the Tokimatsu and Seed (1984) procedure. Assuming for the complex C2B a thickness of 17.8 m (profile P1) the estimated settlement is 0.45 m.

The post-liquefaction volumetric strain, following the procedure proposed by Ishihara (1996), is estimated to be about 2.8%. Thus the surface settlement is 0.50 m.

For these settlement values extensive damages will occur on ground surface with large cracks, spouting of sands, large offsets and lateral movements.

To mitigate the detrimental liquefaction effects for the site the following countermeasures can be used: densification, drainage, solidification and inclusions.

7 CONCLUSIONS

A liquefaction study for a site located in south of Portugal was performed following the methodology recommended in the Manual for Zonation on Seismic Geotechnical Hazards.

For grade 3 the cyclic shear stresses computed by the Seed simplified method were much higher than the shear values obtained by the computer code considering an equivalent linear method for the soil behavior.

For the prediction of earthquake induced settlements of sands two procedures were used. The estimated values were of the same order of magnitude.

To mitigate the detrimental liquefaction effects the following countermeasures were reccommended: densification, drainage, solidification and inclusions.

REFERENCES

Ishihara, K. 1996. Soil behaviour in earthquake geotechniques. Oxford Engineering Science Series.

Eurocode 8. 1994 Design provisions for earthquake resistance of structures - Part 5 Foundations, Retaining Structures and Geotechnical Aspects.

Johnson, W. J. and Kissenpfenning, J. F. 1977 Vibratory ground motion from a distance large magnitude earthquake. A discussion of the 1755 Lisbon earthquake. Proc. VI World Conference Earthquake Engineering, India.

Jorge, C. 1994 Zonation of liquefaction potencial - Application to Portugal. Master thesis (in Portuguese). University New of Lisbon.

Kuribayashi, E. and Tatsuoka, F. 1975. Brief review of soil liquefaction during earthquakes in Japan, Soils and Foundations, Vol. 15, N° 4, pp. 81-92.

RSA, 1983 Regulamento de segurança e acções para estruturas de edificios e pontes.

Seed, H. B. and Idriss, I.M. 1971. Simplified procedure for evaluation soil liquefaction potencial, J. SMFD, ASCE, Vol. 97, N° 9, pp. 1249 - 1273.

Seed, H. B., Idriss, I. M. and Arango, I. 1983 Evaluation of liquefaction potential using field performance data, JGE, ASCE, Vol. 109, N° 3, pp. 458 - 482.

Skempton, A.W. 1986 Standard penetration test procedures and the effects in sands of overburden pressure, relative density, particle size, aging and overconsolidation, Geotechnique, Vol. 36, N° 3, pp. 425 - 447.

Technical Committee for Earthquake Geotechnical Engineering, TC4, ISSMFE, 1993 - Manual for zonation of seismic geotechnical hazards, Published by the Japanese Society of Soil Mechanics and Foundation Engineering, Tokyo.

Tokimatsu, K. and Seed, H.B, 1984 Simplified procedure for the evaluation of settlements in clean sands. Report n^o UCB/EERC 84/16. University of California.

Vieira, A. 1995. Amplification of seismic movements for horizontal soil layers. Master thesis (in Portuguese). University New of Lisbon.

Wakamatsu, K. 1993. History of soil liquefaction in Japan and assessment of liquefaction potential based on geomorphology. A Thesis in the Department of Architecture presented in partial fulfillment of the requirements for the degree of Doctor of Engineering, Waseda University of Tokyo, Japan, 245 pp.

Youd, T. L. and Perkins, D. M. 1978. Mapping of liquefaction induced ground failure potencial, J. GED, ASCE, Vol. 104, N° 4, pp. 433 - 446.

Seismic Behaviour of Ground and Geotechnical Structures, Sêco e Pinto (ed.)© 1997 Balkema, Rotterdam, ISBN 90 5410 887 8

Site-dependent seismic response including recent strong motion data

R. B. Seed & J. D. Bray
University of California at Berkeley, Calif., USA

S. W. Chang
Washington State University, Pullman, Wash., USA

S. E. Dickenson
Oregon State University, Oreg., USA

ABSTRACT: This paper presents a brief summary of recently completed studies of site-dependent seismic site response incorporating the wealth of strong motion data provided by recent earthquakes. The empirical data, results of back analyses of various strong motions recording sites, and analyses of the response of sites to various design levels of shaking are combined to develop recommendations for site classification, prediction of site-dependent amplification, and site-dependent design spectra. The adequacy of current U.S. building codes and provisions in addressing site-dependent site response is assessed in light of the strong motion data from these recent earthquakes.

1 INTRODUCTION

In recent years, the importance of site effects on seismic site response has been repeatedly demonstrated during earthquakes such as Mexico City (1985), Armenia (1988), Loma Prieta (1989), the Philippines (1990), Northridge (1994), and Hyogo-ken Nanbu (1995). This paper presents a brief overview of recently completed studies on the seismic response of (a) soft cohesive sites (b) deep, stiff cohesive sites, and (c) deep, stiff cohesionless soil sites which incorporate the wealth of empirical data and analytical results, principally from the Loma Prieta and Northridge Earthquakes (Chang, 1996 and Dickenson, 1994).

The results of these studies were used to develop recommendations for site classifications and site-dependent design spectra for code-based design. The resulting recommendations are then compared with the design levels recommended by the 1994 Uniform Building Code (UBC) and the 1994 National Earthquake Hazards Reduction Program (NEHRP) Provisions.

Maps of the areas affected by the Loma Prieta and Northridge earthquakes are presented in Figure 1 and Figure 2, respectively. The locations of strong motion stations are shown, along with a simplified overview of the regional geology. In Figure 1, soft and deep cohesive soil sites are primarily located along the San Francisco Bay margins; deep stiff soil sites of interest are generally located in the East Bay (Oakland) area. The soil sites in Figure 2 are predominantly deep stiff soil sites.

2 SOFT AND DEEP COHESIVE SOILS

Strong motion records were obtained at ten soft and/or deep cohesive soil sites throughout the San Francisco Bay region during the Loma Prieta earthquake for moderate levels of shaking (A_{max} = 0.14g to 0.33g). Dickenson (1994) back-analyzed these sites and developed one-dimensional site response models using both equivalent linear (SHAKE90) and fully nonlinear (MARDESRA) analysis methods. SHAKE90 is a slightly modified version of the original SHAKE (Schnabel et al., 1972), and MARDESRA is similar to DESRA-2 (Lee and Finn, 1978) except that the dynamic properties of the soil are represented by the Martin-Davidenkov (Martin, 1975) model. The predictive capabilities of these methods can be excellent, as illustrated in the following analysis of Treasure Island, one of the ten soft and/or deep cohesive soil sites of interest.

The generalized soil profile for Treasure Island (TI), shown in Figure 3, indicates that the site consists of loose sandy fill and loose silty sand

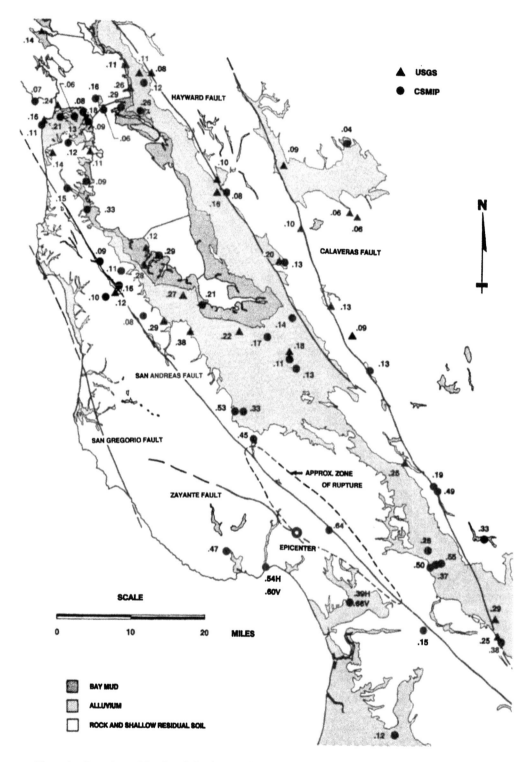

Figure 1: Overview of Regional Geology and Recorded Peak Horizontal Ground Acceleration During the 1989 Loma Prieta Earthquake (Seed et al., 1990).

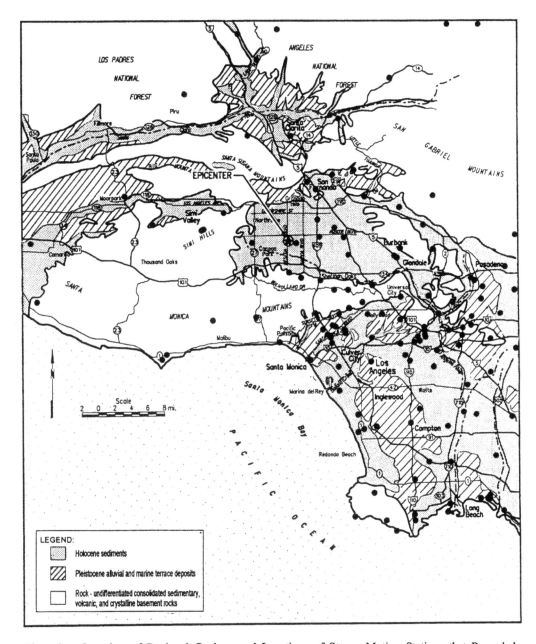

Figure 2: Overview of Regional Geology and Locations of Strong Motion Stations that Recorded Motions from the 1994 Northridge Earthquake.

underlain by a significant thickness of soft Young San Francisco Bay Mud, a Holocene marine clay. Below the Bay Mud, the site is underlain to a depth of about 90 meters by older materials, such as dense sands and stiff to hard silty clays. The rock motion recorded at Yerba Buena Island (YBI), located approximately 2 km from the TI instrument, was used as the input motion. Figure 4 presents the shear wave velocity profile at TI developed from downhole shear wave velocity measurements and regional correlation studies (Dickenson, 1994).

Figure 3 also shows the acceleration time histories recorded at TI and YBI, along with the calculated response spectra (5% damping) of the

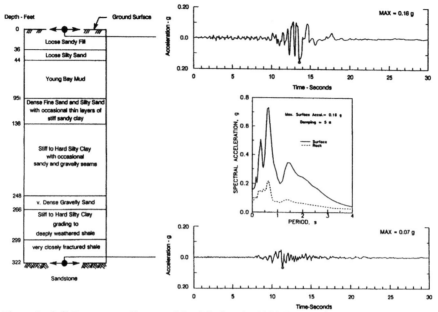

Figure 3: Soil Response at Treasure Island during the 1989 Loma Prieta Earthquake (Seed and Dickenson, 1992).

recorded motions. By comparing the response spectra, it can be observed that A_{max} was amplified by a factor of two, and that spectral values were amplified by factors of up to four to five. The

results of the SHAKE and MARDESRA analyses of the TI site are presented in Figure 5. Both methods provided excellent agreement with the actual strong motion recording.

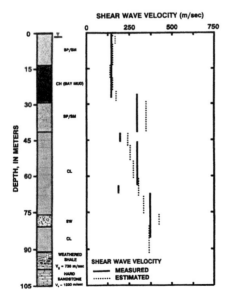

Figure 4: Soil Profile and Shear Wave Velocity Profile at Treasure Island (Dickenson, 1994)

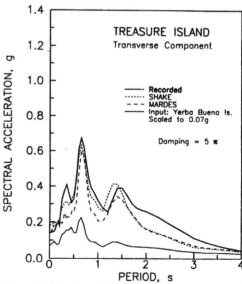

Figure 5: Results of Equivalent Linear and Fully Nonlinear Analyses for Treasure Island (Seed and Dickenson, 1992).

3 GENERAL SITE CONDITIONS

Seed and Dickenson (1992) performed additional studies to evaluate the general seismic site response characteristics of various idealized sites. The idealized soil profiles for "soft clay" sites, shown in Figure 6, were subjected to varying levels of shaking (varying A_{max}, magnitude, and frequency content). SHAKE90 analyses (more than 100) were performed for motions with $A_{max,rock}$ between 0.10g and 0.30g. A lesser number of fully nonlinear

MARDESRA analyses were performed for $A_{max,rock}$ between 0.30g and 0.50g. The resulting nonlinear variation of A_{max} on soft or deep cohesive soil versus $A_{max,rock}$ is presented in Figure 7. Both the empirical data available and the results of the analyses are included in this figure.

The mean and mean ± one standard deviation normalized response spectra calculated for the idealized sites for (a) low to moderate and (b) high levels of rock input motion are presented in Figure 8, illustrating the nonlinearity of response to

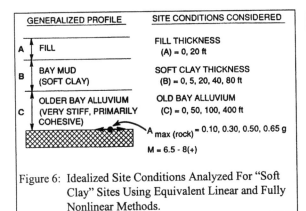

Figure 6: Idealized Site Conditions Analyzed For "Soft Clay" Sites Using Equivalent Linear and Fully Nonlinear Methods.

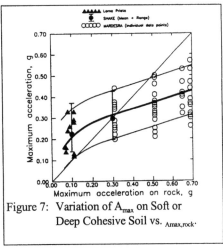

Figure 7: Variation of A_{max} on Soft or Deep Cohesive Soil vs. $A_{max,rock}$.

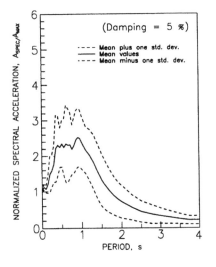

(a) Input Motions: $a_{max,rock}$ = 0.1 - 0.3 g

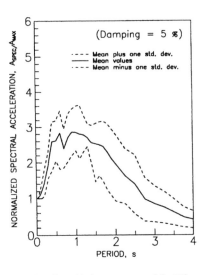

(b) Input Motions: $a_{max,rock}$ = 0.5 - 0.65 g

Figure 8: Mean and Mean ± 1 Standard Deviation Response Spectra for San Francisco Bay Mud Sites at (a) Low to Moderate and (b) High Levels of Input Rock Motion.

differing levels of excitation. These curves were used to develop proposed normalized response spectra for soft and deep cohesive soil sites similar to those presented in Figure 10 (Dickenson, 1994).

Chang (1996) performed similar studies of deep, stiff soil sites using the same general approach. Both equivalent linear (SHAKE91) and fully nonlinear analyses (D-MOD; Matasovic, 1993) were performed for strong motion recording sites from the Loma Prieta and Northridge Earthquakes to better calibrate these analytical tools. Additional analyses were then performed for hypothetical sites subjected to differing levels of excitation. These analyses, along with the available database of strong motion records, were then used to develop recommendations for evaluation of site-dependent amplification and spectral response. This, in turn, resulted in modification of the global site-dependent response recommendations of Seed and Dickenson (1992), with the principal modifications applying to deep, stiff cohesionless and deep, stiff cohesive site conditions.

4   RECOMMENDED DESIGN VALUES

The results of these calibrated analytical studies, as well as the empirical strong motion data, have next been further extended to consider a broader suite of site conditions. The resulting recommendations for evaluation of site-dependent seismic response for all site conditions are presented in Table 1 and Figures 9 and 10. These are based on the available strong-motion database (not just from the Loma Prieta and Northridge Earthquakes), on both equivalent linear and fully nonlinear analyses of response to varying levels and characteristics of excitation, and judgment.

The site-dependent amplification relationships presented in Figure 9 are set at "mean" levels, and they do not incorporate any explicit allowance for variability.   As illustrated in two examples in Figures 7 and 11, such variability can be pronounced. Similarly, as illustrated in one example in Figure 12, variability of site-dependent response spectra can be very pronounced. Taken together, these two sources of variability lead to a significant likelihood that spectral response at any given single site will strongly exceed the "mean" spectral levels predicted based on averaging numerous sites (and responses) within a class of site characteristics. It is the authors' experience that some additional

conservatism is often applied when developing design recommendations based on response analyses for a particular site, and it is our recommendation that such conservatism is warranted. Accordingly, the response spectral values of Figure 10 are set at approximately a mean plus one-half standard deviation level. Approximate mean levels can be developed by dividing all spectral response levels (at T>0 seconds) by about 1.15, and mean plus one full standard deviation levels can be developed by multiplying all spectral response values (at T>0 seconds) by about 1.15.

Evaluation of design level response for a given site proceeds in three steps as follow. First, the peak horizontal ground acceleration ($A_{max,rock}$) that would occur if the site had been a competent rock site (site class A) is evaluated, based on regional seismicity. This is then modified to estimate the mean estimate site-specific peak acceleration (A_{max}) for the actual site conditions using Figure 9. The appropriate normalized spectrum from Figure 10 is then scaled to this value of A_{max} (at T = 0 seconds).

5   DISCUSSION

An illustration of the adequacy of the "fit" provided by these recommended site classification amplification relationships and spectra, as well as the variability and degree of conservatism, is presented in Figures 11 and 12. Taking "deep, stiff" soil sites as a class, Figure 11 presents a plot of both recorded and calculated amplification factors. The recorded data are situated by plotting recorded peak ground acceleration (for deep, stiff soil sites) vs. $A_{max,rock}$ based on regional attenuation patterns developed for the earthquakes in question (Chang, 1996). The calculated values were developed using both equivalent linear and fully nonlinear analyses methods previously calibrated and verified by means of back analyses of recordings made at soil sites with good subsurface characterization data (Chang, 1996). As shown in Figure 11, there is considerable scatter to the data, but the recommended amplification curve from Figure 9 (for site class C3) provides good, representative median values.

Figures 12(a) and (b) present the 5% damped elastic response spectra for these same sites and records, along with recommended spectral shapes from the 1994 UBC provisions (site classes S_2 and S_3) and the 1994 NEHRP provisions (site class D), along with the earlier recommendations of Seed and

Table1: Proposed Site Classification System

Site Class	Site Condition	General Description	Site Characteristics
(A_0)	A_0	Very hard rock	V_S (avg.) > 5,000 ft/s in top 50 ft.
A	A_1	Competent rock with little or no soil and/or weathered rock veneer.	2,500 ft/s $\leq V_S$ (rock) $\leq$ 5,000 ft/s, and $H_{soil+weathered rock}$ < 40 ft with V_S > 800 ft/s (in all but the top few feet[3]).
AB	AB_1	Soft, fractured and/or weathered rock.	For both AB_1 and AB_2: 40 ft $\leq H_{soil+weathered rock} \leq$ 150 ft, and $V_S \geq$ 800 ft/s (in all but the top few feet[3]).
AB	AB_2	Stiff, very shallow soil over rock and/or weathered rock.	
B	B_1	Deep, primarily cohesionless [4] soils. ($H_{soil} \leq$ 300 ft.)	No "soft clay" (see Note 5), and $H_{cohesive soil}$ < 0.2 $H_{cohesionless soil}$
B	B_2	Medium depth, stiff cohesive soils and/or mix of cohesionless with stiff cohesive soils; no "soft clay".	$H_{all soils} \leq$ 200 ft, and V_S (cohesive soils) > 500 ft/s (see Note 5).
C	C_1	Medium depth, stiff cohesive soils and/or mix of cohesionless with stiff cohesive soils; thin layer(s) of soft clay.	Same as B_2 above, except 0 ft < $H_{soft clay} \leq$ 10 ft (see Note 5).
C	C_2	Very deep, primarily cohesionless soils.	Same as B_1 above, except H_{soil} > 300 ft.
C	C_3	Deep, stiff cohesive soils and/or mix of cohesionless with stiff cohesive soils; no "soft clay".	H_{soil} > 200 ft, and V_S (cohesive soils) > 500 ft/s
C	C_4	Soft, cohesive soil at small to moderate levels of shaking.	10 ft $\leq H_{soft clay} \leq$ 100 ft, and $A_{max,rock} \leq$ 0.25 g
D	D_1	Soft, cohesive soil at medium to strong levels of shaking.	10 ft $\leq H_{soft clay} \leq$ 100 ft, and 0.25 g < $A_{max,rock} \leq$ 0.45 g, or (0.25 g < $A_{max,rock} \leq$ 0.55 g and M $\leq$ 7-1/4)
(E)[6]	E_1	Very deep, soft cohesive soil.	$H_{soft clay}$ > 100 ft (see Note 5).
(E)[6]	E_2	Soft, cohesive soil and very strong shaking.	$H_{soft clay}$ > 10 ft and either: $A_{max,rock}$ > 0.55 g, or $A_{max,rock}$ > 0.45 g and M > 7-1/4
(E)[6]	E_3	Very high plasticity clays.	H_{clay} > 30 ft with PI > 75% and V_S < 800 ft/s
(F)[7]	F_1	Highly organic and/or peaty soils.	H > 20 ft of peat and/or highly, organic soils
(F)[7]	F_2	Sites likely to suffer ground failure due either to significant soil liquefaction or other potential modes of ground instability.	Liquefaction and/or other types of ground failure analysis required.

1. H= total (vertical) depth of soils of the type or types referred to.
2. V_s=seismic shear wave velocity (ft/sec) at small shear strains (shear strain ~ 10^{-4}%)
3. If surface soils are cohesionless, V_s may be less than 800 ft/sec in top 10 feet.
4. "Cohesionless soils" = soils with less than 30% "fines" by dry weight. "Cohesive soils" = soils with more than 30% "fines" by dry weight, and 15% $\leq$ PI (fines) $\leq$ 90%. Soils with more than 30% fines, and PI (fines) < 15% are considered "silty" soils herein, and these should be (conservatively) treated as "cohesive" soils for site classification purposes in this Table.
5. "Soft Clay" is defined as cohesive soil with: (a) Fines content $\geq$30%, (b) PI(fines) $\geq$20%, and (c) V_s $\leq$500 ft/s.
6. Site-specific geotechnical investigations and dynamic site response analyses are strongly recommended for these conditions. Response characteristics within this Class (E) of sites tends to be more highly variable than for Classes A_0 through D, and the response projections herein should be applied conservatively in the absence of (strongly recommended) site-specific studies.
7. Site-specific geotechnical investigations and dynamic site response analyses are *required* for these conditions. Potentially significant ground failure must be mitigated, and/or it must be demonstrated that the proposed structure/facility can be engineered to satisfactorily withstand such ground failure.

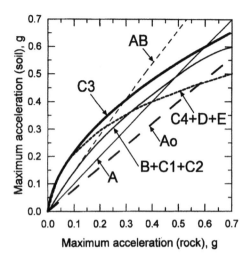

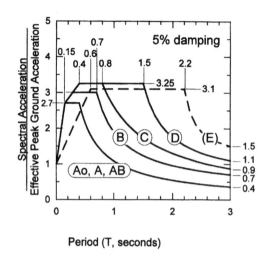

Figure 9: Proposed Site-Dependent Relationship Between Mean A_{max} and A_{max} for "Competent" Rock Sites.

Figure 10: Proposed Site-Dependent Response Spectra (with 5% damping).

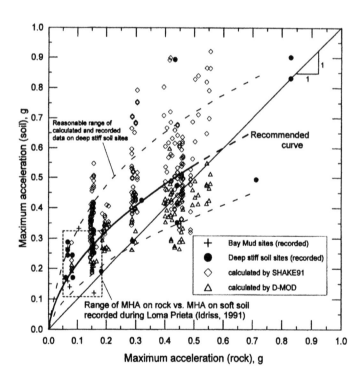

Figure 11: Relationship for A_{max}, vs. $A_{max, rock}$ for deep stiff soil sites based on available empirical data from the Loma Prieta and Northridge Earthquakes and calculations using both equivalent linear and fully nonlinear site response methods (modified from Chang, 1996).

132

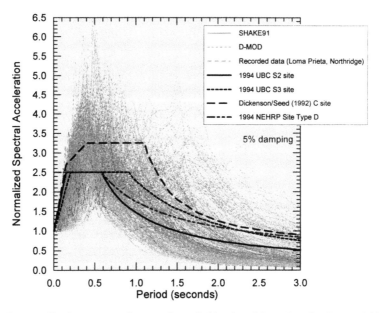

Figure 12(a): Calculated Normalized Response Spectra from Oakland and Los Angeles Deep, Stiff Sites Compared to Current Design Spectra (Chang, 1996).

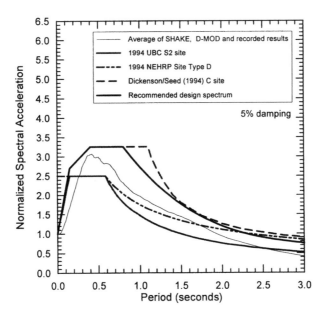

Figure 12(b): Average of Normalized Response Spectra from Oakland and Los Angeles Sites; Recommended Design Spectrum for Deep Stiff Soil Sites (Chang, 1996).

Dickenson (1992). As shown in these figures, scatter or variability is high, and the UBC and NEHRP provisions may be significantly exceeded for any given individual site and motion. The original spectral shape proposed by Seed and Dickenson (1992) was judged to be slightly overconservative at T ~ 0.8 to 1.2 seconds, and so was reduced in this range to produce the spectral shape now recommended in Figure 10.

6 CONCLUSIONS

Recent earthquakes have provided a large number of strong motion recordings that allow us to update the state-of-practice in estimating site-dependent design response spectra. The empirical data from these earthquakes and back-analyses of the response of numerous sites have been combined to develop recommendations for classification of sites based on response characteristics, and for site-specific evaluation of (a) amplification and (b) elastic response spectra, for use in simplified code-based seismic design. Early versions of this work were incorporated in the recently revised 1994 NEHRP Provisions treatment of site effects on response, but the now completed results presented herein lead to recommendations of stronger response levels for some site classes than those currently embodied in the NEHRP or UBC code provisions.

REFERENCES

Building Seismic Safety Council 1994. *NEHRP Recommended Provisions for Seismic Regulations for New Buildings*.

Chang, S.W. 1996. *Seismic Response of Deep Stiff Soil Deposits*, dissertation submitted in partial satisfaction of the requirements for the degree of Doctor of Philosophy in Civil Engineering of the University of California at Berkeley.

Dickenson, S.E. 1994. *Dynamic Response of Soft and Deep Cohesive Soils During the Loma Prieta Earthquake of October 17, 1989*, dissertation submitted in partial satisfaction of the requirements for the degree of Doctor of Philosophy in Civil Engineering of the University of California at Berkeley.

International Conference of Building Officials 1994. *Uniform Building Code*, Vol. 2, Structural Engineering Design Provisions.

Lee, M.K.W. & Finn, W.D.L. 1978. *DESRA-2, Dynamic Effective Stress Response Analysis of Soil Deposits with Energy Transmitting Boundary Including Assessment of Liquefaction Potential*, Soil Mechanics Series No. 36, Dept. of Civil Engineering, University of British Columbia, Vancouver, Canada, 60 p.

Martin, P.P. 1975. *Non-Linear Methods for Dynamic Analysis of Ground Response*, thesis presented to the University of California at Berkeley, in partial fulfillment of requirements for the degree of Ph.D. in Engineering.

Matasovic, N. 1993. *Seismic Response of Composite Horizontally-Layered Soil Deposits*, Ph.D. Dissertation, Civil and Environmental Engineering Department, University of California, Los Angeles, 483 p.

Schnabel, P.B., Lysmer, J. and Seed, H.B. 1972. *SHAKE: A Computer Program for Earthquake Response Analysis of Horizontally Layered Sites*, Report No. EERC/72-12, University of California at Berkeley, December.

Seed, R.B. & Dickenson, S.E. 1992. *Site-Dependent Seismic Site Response*, Proceedings , Second Annual CALTRANS Research Workshop, Sacramento, California, March.

Seed, R.B., Dickenson, S.E., Riemer, M.F., Bray, J.D., Sitar, N., Mitchell, J.K., Idriss, I.M., Kayen, R.E., Kropp, A., Harder, L.F., and Power, M.S. 1990. *Preliminary Report on the Principal Aspects of the October 17, 1989 Loma Prieta Earthquake*: Report No. UCB/EERC-90/05, University of California at Berkeley, 124 p.

Stewart, J.P., Bray, J.D., Seed, R.B., and Sitar, N. 1994. *Preliminary Report on the Principal Geotechnical Aspects of the January 17, 1994 Northridge Earthquake*, Report No. UCB/EERC 94-08, University of California at Berkeley.

Seismic Behaviour of Ground and Geotechnical Structures, Sêco e Pinto (ed.) © 1997 Balkema, Rotterdam, ISBN 90 5410 887 8

Application of the TC4 Manual for soil liquefaction assessment to the 1995 Hyogoken-Nanbu (Kobe) earthquake

Kazue Wakamatsu
Institute of Industrial Science, University of Tokyo, Japan

ABSTRACT: Applicability of the TC4 Manual for soil liquefaction assessment to the 1995 Hyogoken-nunbu earthquake is presented. The liquefaction potential maps compiled based on the manual was compared with the distribution of liquefaction occurrences generated by the 1995 earthquake. As the results of the comparison, the assessment was generally consistent with the actual performance of the ground.

1 INTRODUCTION

The Hyogoken-nanbu earthquake directly struck the densely populated Kobe and its neighboring cities at 5:46 a.m. local time, on January 17, 1995 and caused approximately a hundred billion dollars of direct damage to public infrastructures and private properties. It is reported that 6308 deaths, 2 missing, and as many as 43,177 injures were sustained, with more than 100,000 buildings and houses are completely destroyed (National Land Agency, 1996).

The earthquake caused soil liquefaction in the zones along Osaka Bay, which resulted in serious damage to buried lifeline facilities and foundations of bridges and buildings.

In this paper, Grade-1 and -2 methods for soil liquefaction assessment presented in the manual entitled "Manual for Zonation on Seismic Geotechnical Hazards" prepared by the TC4 of the International Society for Soil Mechanics and Foundation Engineering (1993) (simply called the TC4 Manual below) were applied for the 1995 Hyogoken-nanbu earthquake. The applicability of the TC4 Manual is discussed.

2 THE EARTHQUAKE AND INTENSITY DISTRIBUTION

The 1995 earthquake was assigned a JMA magnitude of 7.2 by the Japan Meteorological Agency (JMA) and the surface wave magnitude of 6.9 by the USGS. The epicenter of the earthquake was located about 20 km southwest of downtown Kobe between the northern tip of Awaji Island and main land, which was placed at 34.61°N and 135.04°E by the JMA (Fig.1). The focal depth was given as 14.3 km by the JMA.

Based on teleseismic body wave modeling, the main shock consists of three sub-events where total rupture time was 11 seconds with seismic moment of 2.5×10^{19} N.m, yielding a moment magnitude of 6.9 (Kikuchi, 1995). The three earthquake sub-faults are shown in Fig.1.

Seismic intensity of VII, the maximum on the JMA scale was assigned to several wards of Kobe and the neighbouring cities of Ashiya, Nishinomiya, Takarazuka and the towns of Hokudancho and Ichinomiya on Awaji Island as shown in Fig. 2. VII on JMA corresponds a level of X-XII on the MM and MSK scales. This was the first time for the intensity scale classification of VII to be issued in Japan.

Strong motion data were recorded at more than 350 sites through the southwestern half of Japan. The highest recorded horizontal acceleration was 818 cm/s^2 at JMA Kaiyo station in Kobe, 15.5 km northeast of the epicenter (Architectural Institute of Japan, 1996). Several examples of the peak values of the recorded accelerations are shown in Fig.2.

3 ESTIMATION OF THE MAXIMUM EXTENT OF A LIQUEFACTION SUSCEPTIBLE AREA BASED ON GRADE-1 METHOD

The earthquake induced extensive liquefaction in low-lying areas along Osaka Bay. Figure 1 shows the location of liquefied sites found by many reconnaissance teams (e.g. Shibata et al., 1996) as well as by the author. Sand boils occurred in substantial areas along Osaka Bay that stretched from Takasago, west of Kobe, to Kishiwada, south of Osaka. Most of the liquefied sites in the figure lie on recently reclaimed land areas from the sea, but some of them located at natural ground and fill on the

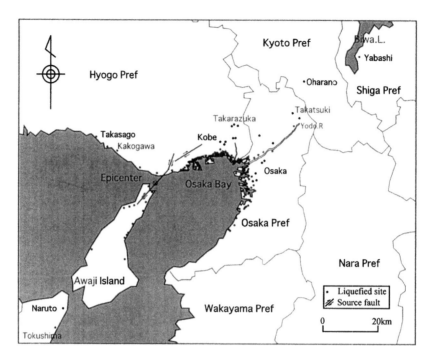

Figure 1 Map of Awaji-Kobe region showing distribution of liquefied sites (Hamada and Wakamatsu, 1996) and the earthquake faults (Kikuchi, 1995)

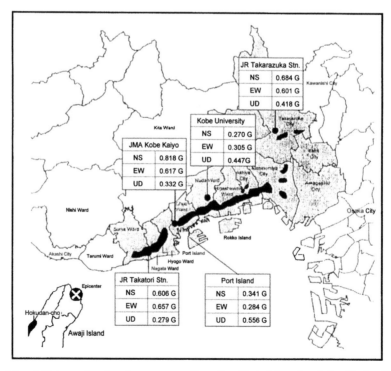

Figure 2 Map of Awaji-Kobe region showing areas most heavily affected (gray) and classified as JMA Intensity VII (black) and PGA values for stations in the areas (Reproduced from JMA, 1995; AIJ, 1996).

former pond. Soil liquefaction was also observed coastal areas of Awaji Island and north eastern part of Tokushima Prefecture, Shikoku Island as shown in the figure.

In order to examine the validity of Grade-1 method for liquefaction, magnitude-maximum distance criteria were applied to the 1995 event. The maximum distance of liquefaction from the epicenter, identified up to date, is 90 km in Yabashi, Kusatsu city located at the shore of Biwa Lake, where minor effects of liquefaction such as sand boils were observed (Shibata et al., 1996). Similarly, the maximum distance of significant effects of liquefaction is 70 km at Rakusai New Town in Oharano, about 8 km southwest of downtown Kyoto, where school buildings and a gymnasium were collapsed due to liquefaction-induced ground settlement and lateral spreading (Uemurta, 1996).

The Ms versus R for these sites are plotted in Fig. 3. The latter distance agree with the Wakamatsu (1993)'s bound for significant effects of liquefaction. The former one lies within the bound developed by Wakamatsu (1991) whereas it lies beyond the limiting bounds proposed by Liu and Xie (1984), Kuribayashi and Tatsuoka (1975), and Ambraseys(1988), which were developed for all effects of liquefaction. These bounds were originally developed for different magnitude scales which are represented as follows:

$$\log R = 0.77\, M_J - 3.6 \qquad (1)$$

$$R = 0.82 \cdot 100.8^{62(M_L - 5)} \qquad (2)$$

$$M_W = 4.64 + 2.65 \times 10^{-3}\,R + 0.99 \log R \qquad (3)$$

$$\log R = 2.22 \log (4.22\, M_J - 19.0) \qquad (4)$$

where M_J = JMA magnitude; M_L = Rihiter's magnitude; and M_W = moment magnitude.

Substituting M_J = 7.2 and M_w = 6.9 in each equation, the following values are obtained; R = 88 km from eq. (1) proposed by Kuribayashi and Tatsuoka (1975); R = 102 km from eq. (3) proposed by Ambraseys (1988); and R = 221 km from eq. (4) proposed by Wakamatsu (1991). The values estimated from eqs. (1) and (3) also lie within the actual distance, however, the order of the underestimate is within acceptable range of the assessment. Whereas the value predicted from eq. (4) is too conservative. This is because that the equation was developed for predicting the maximum range of liquefaction effects including even minor sign of liquefaction for a particular magnitude of earthquake, given the presence of potentially liquefiable Holocene sediments, as have been described in the TC4 Manual.

In calculating the R by the Liu and Xie's eq. (2), the value of M_L should be known. However, it is not known because a special kind of seismograph is required to estimate it. At any rate, some discrepancy is inevitable between the predicted distance and the actual one because the equation (2) was originally developed for not a maximum bound but an average one based on Chinese liquefaction data.

Although the epicentral distance, R, can generally be estimated easily but the distance from the seismic energy source is more suitable for the present earthquake with defined source faults shown in Fig.1. The furthest sites of minor and significant liquefaction from the faults also correspond to Yabashi and Oharano, respectively. The farthest distances from these sites, 68 km and 45 km, respectively, are plotted in Fig. 4, indicating that the predicted values underestimate the actual ones; the former distance is about 20 km beyond the bound, for all effects of liquefaction, proposed by Ambraseys (1988) and the latter one is 10 km beyond bound, for significant effect of liquefaction, by Youd and Perkins (1978).

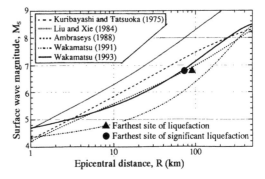

Figure 3 Comparison between results obtained from the 1995 earthquake and relationships proposed for Grade-1 zonation work (added to TC4, ISSMFE, 1993)

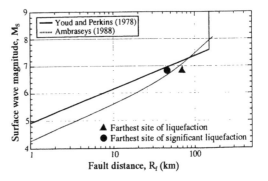

Figure 4 Comparison between results obtained from the 1995 earthquake and relationships proposed for Grade-1 zonation work (added to TC4, ISSMFE, 1993)

137

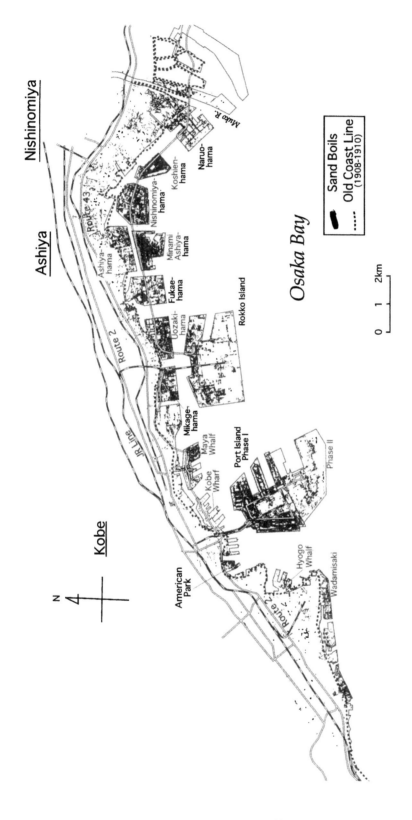

Figure 5 Zone of liquefaction in Kobe, Ashiya and Nishinomiya (Hamada et al.1995)

138

The Ambraseys' bound was originally developed for moment magnitude M_w which is represented as follows:

$$M_w = 4.68 + 9.2 \times 10^{-3} R_f + 0.90 \log R_f \qquad (5)$$

Substituting $M_w = 6.9$ in the equation (5), $R_f = 64$ km is obtained, which almost agree with the actual distance.

In any case, the maximum distance from the epicenter and earthquake fault is related to geological and geomorphological conditions which are not the same in different directions. In the case of the 1995 event, the maximum distance in the north of the epicenter was only several km but it was about 90 km in the northeast direction depending on the extents of low-lying areas.

4 ESTIMATION OF LIQUEFACTION POTENTIAL BASED ON GRADE-2 METHOD

Significant soil liquefaction occurred in low-lying areas in Kobe, Ashiya, Nishinomiya and Amagasaki, being within 40 km of the epicenter, as shown in Fig. 5. Geomorphology-based technique presented in Grade-2 methods as defined in the TC4 Manual was applied for the above mentioned areas. A geomorphologic land classification map compiled by Midorikawa and Fujimoto (1996), shown in Fig. 6, was used to identify liquefaction susceptible units, which was drawn up based on the Land Condition Maps, Kobe and Osaka-nanseibu at scales of 1:25,000 (Geographical Survey Institute, 1976, 1983) referring to maps compiled in 1880's as well as most recent ones.

The map was then transformed into liquefaction potential map with reference to the criteria listed in Table 1 (Table 5.3 of the TC4 manual). According to the criteria, liquefaction potential for the units of the lowland was classified into three levels; liquefaction likely, possible and not likely under the ground motion at the JMA intensity V (Approximately M.M.I. VIII). In the case of the 1995 earthquake, a seismic intensity of VI-VII was assigned to the studied areas as mentioned before, so two levels of liquefaction susceptibility were applied for the each geomorphologic units; one is a susceptibility just as it is in Table 1 for intensity of V and another is one level higher susceptibility for intensity VI-VII. The results of the mapping are summarized in Figs. 7 and 8.

5 COMPARISON BETWEEN LIQUEFACTION POTENTIAL AND ACTUAL LIQIUEFACTION OCCURRENCE

A couple of liquefaction potential maps shown in Figs. 7 and 8 were compared with the distribution of liquefaction occurrences generated by the Hyogoken-nanbu earthquake shown in Fig. 5. Tables 2 and 3 show summaries of the each comparison. In the tables, the left two columns summarize the geomorphologic units in the studied areas and the geomorphology-based liquefaction potential shown in Figs. 7 and 8. The next columns show actual liquefaction occurrence in each unit during the 1995 earthquake; liquefaction occurrences were classified into three levels in accordance with number of liquefaction occurrence within each unit as explained in the foot note below the each table.

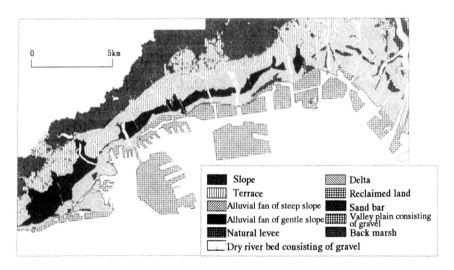

Figure 6 Geomorphological land classification map of Kobe region (Midorikawa and Fujimoto, 1996)

Table 1 Susceptibility of detailed geomorphological units to liquefaction subjected to ground motion of the J.M.A. intensity V or M.M.S. VIII (Wakamatsu, 1992)

Geomorphological conditions		Liquefaction potential
Classification	Specific conditions	
Valley plain	Valley plain consisting of gravel or cobble	Not likely
	Valley plain consisting of sandy soil	Possible
Alluvial fan	Vertical gradient of more than 0.5%	Not likely
	Vertical gradient of less than 0.5%	Possible
Natural levee	Top of natural levee	Possible
	Edge of natural levee	Likely
Back marsh		Possible
Abandoned river channel		Likely
Former pond		Likely
Marsh and swamp		Possible
Dry river bed	Dry river bed consisting of gravel	Not likely
	Dry river bed consisting of sandy soil	Likely
Delta		Possible
Bar	Sand bar	Possible
	Gravel bar	Not likely
Sand dune	Top of dune	Not likely
	Lower slope of dune	Likely
Beach	Beach	Not likely
	Artificial beach	Likely
Interlevee lowland		Likely
Reclaimed land by drainage		Possible
Reclaimed land		Likely
Spring		Likely
Fill	Fill on boundary zone between sand and lowland	Likely
	Fill adjoining cliff	Likely
	Fill on marsh or swamp	Likely
	Fill on reclaimed land by drainage	Likely
	Other type fill	Possible

The comparison was quantified by assigning a point for agreement between the liquefaction potential and the liquefaction occurrences. As shown in the right column in Tables 2 and 3, a point of zero was assigned if the level of the liquefaction potential and liquefaction occurrences agree with each other. If the level of the liquefaction potential is one level higher than that of the liquefaction occurrences, a point of +1 was assigned. If, on the other hand, the level of the liquefaction potential is one level lower than that of the liquefaction occurrences, a point of -1 was assigned. The lowest row of the tables indicate standard deviation of the estimation points.

The standard deviation in Table 2 becomes 0 although a overestimate and a underestimate were made. Whereas that in Table 3 becomes 0.88 because several overestimates were done. The results of the comparison show that the actual performance of the ground agrees better with the result of the assessment shown in Fig. 7. This indicates that liquefaction

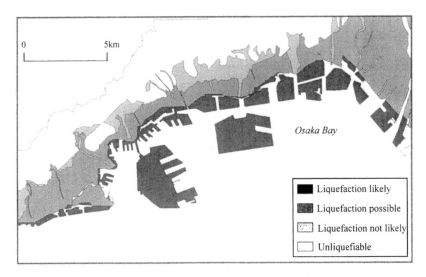

Figure 7 Liquefaction potential map at the JMA seismic intensity V for Kobe region

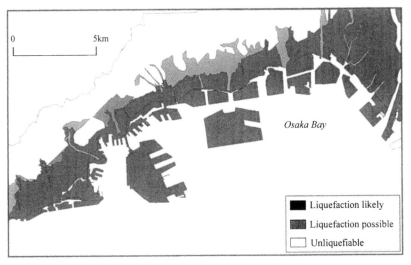

Figure 8 Liquefaction potential map at the JMA seismic intensity VI-VII for Kobe region

potential is more strongly affected by susceptibility of sedimentary deposits to liquefaction in each geomorphological unit than by intensity of ground shaking.

6 CONCLUSION

The magnitude-maximum distance criteria for predicting the maximum extent of a liquefaction susceptible area, presented in the Grade-1 zonation method of the TC4 Manual, were applied for the 1995 Hyogoken-nanbu earthquake. The results of the

studies show that some underestimates were made but they were partially due to errors in converting various kind of magnitude scales such as M_J and M_W into the surface magnitude M_S used in the criteria. Substituting directly in the original equations for magnitude-maximum distance criteria, the discrepancies between the predicted distances to the farthest liquefied sites and actual ones during the 1995 earthquake became to be mostly acceptable range of the assessment as a general guide.

Geomorphological criteria presented in the Grade-2 method was also applied for the 1995 earthquake. The results of the assessment showed that the

Table 2 Comparison between liquefaction potential shown in Figure 7 and actual liquefaction occurrences

Geomorphologic units	Liquefaction potential[1]	Liquefaction occurrences[2]	+: Overestimation -: Underestimation
Alluvial fan of steep slope	×	×	0
Alluvial fan of gentle slope	△	△	0
Natural levee	○	△	+1
Dry river bed consisting of gravel	×	×	0
Back marsh	△	△	0
Delta	△	○	-1
Sand bar	△	△	0
Valley plain consisting of gravel	×	×	0
Reclaimed land	○	○	0
Standard deviation			0

[1]: Liquefaction potential
○: Likely
△: Possible
×: Not likely

[2]: Number of liquefaction occurrences
○: More than a hundred / km²
△ Tens / km²
×: None

Table 3 Comparison between liquefaction potential shown in Figure 8 and actual liquefaction occurrences

Geomorphologic units	Liquefaction potential[1]	Liquefaction occurrences[2]	+: Overestimation -: Underestimation
Alluvial fan of steep slope	△	×	+1
Alluvial fan of gentle slope	○	△	+1
Natural levee	○	△	+1
Dry river bed consisting of gravel	△	×	+1
Back marsh	○	△	+1
Delta	○	○	0
Sand bar	○	△	+1
Valley plain consisting of gravel	△	×	+1
Reclaimed land	○	○	0
Standard deviation			0.88

[1]: Liquefaction potential
○: Likely
△: Possible
×: Not likely

[2]: Number of liquefaction occurrences
○: More than a hundred / km²
△ Tens / km²
×: None

geomorphological criteria listed in Grade-2 method of the TC4 Manual is mostly consistent with the actual performance of the ground during the 1995 Hyogoken-nanbu earthquake although the assumed seismic intensity in the criteria was smaller than the actual ones.

REFERENCES

Architectural Institute of Japan (AIJ) 1996. *Data on the strong motion records for the 1995 Hyogoken-nanbu earthquake* (in Japanese)

Geographical Survey Institute, the Ministry of Construction 1976. *Land Condition Map, Kobe* (in Japanese).

Geographical Survey Institute, the Ministry of Construction 1983. *Land Condition Map, Osaka-nanseibu* (in Japanese).

Kikuchi, M. 1995. Source processes of the Kobe earthquake of January 17, 1995, *Chishitsu News* 486: 12-15 (in Japanese).

Hamada, M., Isoyama, R. and Wakamatsu, K. 1996. *The 1995 Hyogoken-nanbu (Kobe) earthquake, liquefaction, ground displacement, and soil condition in Hanshin area.* Tokyo: Association for Development of Earthquake Prediction.

Hamada, M. and Wakamatsu, K. 1996. Liquefaction, ground displacement and their caused damage to structures. *Special Report on the Hyogoken-nanbu earthquake -Investigation into damage to civil engineering structures-*: 45-91. Japan Society of Civil Engineers.

Midorikawa, S. and Fujimoto, K. 1996. Isoseismal map of the Hyogo-ken nanbu earthquake in and around Kobe city estimated from overturning of tombstones, J. *Struct. Constr. Eng., Architectural Institute of Japan* 490: 111-118 (in Japanese).

National Land Agency (eds.) 1996. *Disasters White Paper*, the Printing Bureau, the Ministry of Finance (in Japanese).

Shibata, T., Oka, F. and Ozawa, Y. 1996. Characteristics of ground deformation due to liquefaction, *Special Issue of Soils and Foudations: 65-79* Japanese Geotechnical Society.

TC4, the International Society for Soil Mechanics and Foundation Engineering 1993. Manual for Zonation on Seismic Geotechnical Hazards, Geotechnical Society of Japan.

Uemura, Y. 1996. *Houses damage and its causes in the area surrounding Kyoto city from the 1995 Hyogoken-nanbu earthquake,* Osaka: Data Center for Faults Research (in Japanese).

Seismic Behaviour of Ground and Geotechnical Structures, Sêco e Pinto (ed.) © 1997 Balkema, Rotterdam, ISBN 90 5410 887 8

A study on the adaptability of two zoning methods for slope instability

S. Yasuda
Tokyo Denki University, Saitama, Japan

ABSTRACT: Two zoning methods which are introduced in the Manual for Zonation on Seismic Geotechnical Hazards were applied to three areas where many slope failures occurred during recent earthquakes. Adaptability of the methods was discussed and some comments were made.

1 INTRODUCTION

In Japan, a lot of slope failures have been occurred at many areas during past earthquakes. Among them, three areas where slope failures occurred during recent major earthquakes, the 1995 Hyogoken-nambu(Kobe) earthquake, the 1993 Hokkaido-nansei-oki earthquake and the 1987 Chibaken-toho-oki earthquake, were selected in this study. Then two zoning methods which are introduced in the Manual for Zonation on Seismic Geotechnical Hazards (ISSMFE, TC4, 1993), were applied in the three areas to study the applicability of the zoning methods.

2 STUDIED AREAS AND ZONING METHODS

The 1995 Hyogoken-nambu earthquake with a magnitude of 7.2 in JMA scale brought many slope failures in Rokko mountain in and around Kobe City. Rokko mountain is located just behind Kobe City with steep slope. The maximum height of the mountain is 931 m. Locations of the failed slopes were investigated by Okimura. According to his research, the studied area was selected as shown in Fig.1.

Many slope failures were occurred also in Okushiri Island, which is located in the north of Japan, during the 1993 Hokkaido-nansei-oki earthquake with a magnitude of 7.8 in JMA scale. Okushiri Island is about 20 km in length and 10 km in width. Almost all area of the island is mountain with fairly steep slopes. The maximum height of the mountain is 585 m. Locations of failed slopes were investigated by the author and his colleagues by comparing two sets of aerophotographs taken before and after the earthquake. The studied area was selected based on the investigation as shown in Fig.2.

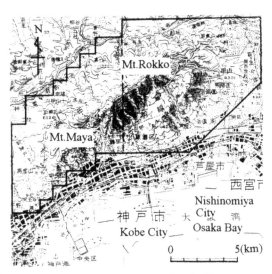

Fig.1 Studied area in and around Kobe City

In 1997, Chibaken-toho-oki earthquake with a magnitude of 6.7 in JMA scale occurred at about 100 km southeast from Tokyo. Many slope failures occurred at the slopes of terraces. Failed sites were investigated precisely by Chiba Prefectural Government. The studied area was selected according to the investigation as shown in Fig.3

Among the Grade-2 methods which are introduced in the Manual for Zonation on Seismic Geotechnical Hazards (ISSMFE, TC4, 1993), two methods, (1) the method proposed by the Kanagawa Prefectural Government (hereafter called Kanagawa Method), and (2) the method proposed by Mora and Vahrson (hereafter called Mora and Vahrson's Method), were applied to the areas. The studied areas were divided

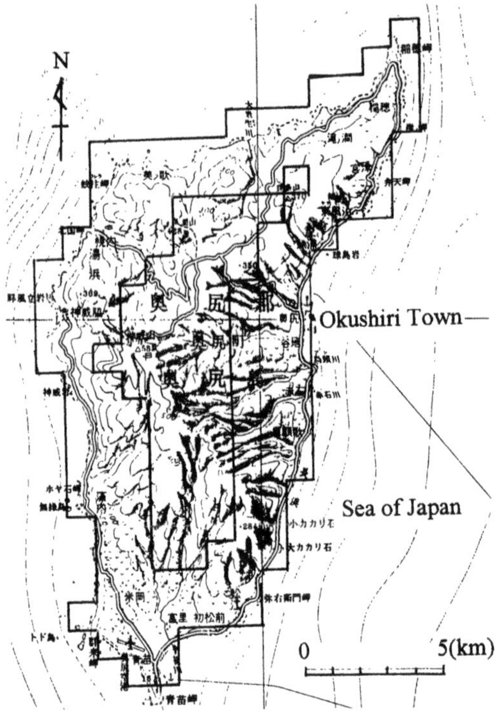

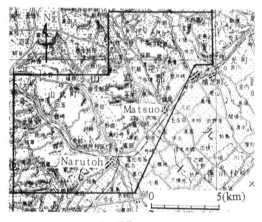

Fig.2 Studied area in Okushiri Island

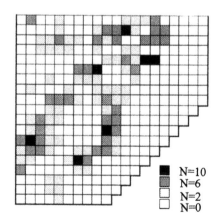

Fig.3 Studied area in Chiba

into small mesh areas of 500 m by 500 m and 1 km by 1 km in the first and second method, respectively. The maximum surface acceleration of each mesh area was estimated by recorded accelerations in Kobe because many records were obtained. On the contrary, few acceleration records were obtained in Okushiri and Chiba. Then the maximum surface acceleration was estimated by Fukushima and Tanaka's formula for Okushiri and by Joner and Boore' formula for Chiba.

3 ESTIMATED ZONING MAPS IN KOBE, OKUSHIRI AND CHIBA

(1) Kobe

Figures 4 and 5 show the number of slides in each mesh area estimated by Kanagawa Method and the number that occurred during the 1995 Hyogoken-nambu earthquake, respectively. These figures show only the map in east part of the Kobe area. By comparing Figs.4 and 5, it can be said that many meshes are estimated to have 6 or 10 slides, though not so many slides were induced during the 1995 Hyogoken-nambu earthquake. This difference is attributed to the effect of the maximum surface acceleration on the estimated number of slides. As

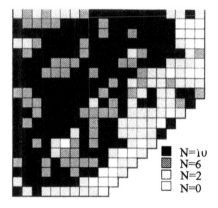

Fig.4 Number of slides in each mesh in Kobe estimated by Kanagawa Method

Fig.5 Number of slides in each mesh which occurred during the 1995 Hyogoken-nambu earthquake

show in Table 4.2 in the manual, the weight W_1 increases rapidly with the maximum surface acceleration. As the acceleration in the area is estimated as more than 400 gals, the weight W_1 reaches 2.754, which is a very high value. Therefore the estimated numbers of slides are overestimated in this case.

Figures 6 and 7 show the estimated landslide hazard index, $H\ell$, estimated by Mora and Vahrson's Method, and the number of slides in each mesh area that occurred during the 1995 Hyogoken-nambu earthquake, respectively. By comparing the two figures, it is seen that almost all meshes are classified in the same grade in Fig.6. Therefore the meshes of $H\ell=3$ must be subdivided.

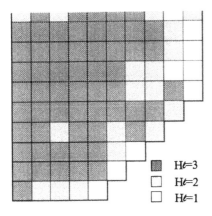

Fig.6 A zoning map estimated in Kobe estimated by Mora and Vahrson's Method

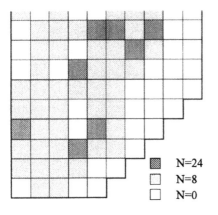

Fig.7 Number of slides in 1 km by 1 km mesh during the 1995 Hyogoken-nambu earthquake

(2) Okushiri

Figures 8 and 9 show the number of slides in each mesh area estimated by Kanagawa Method and the number that occurred during the 1993 Hokkaido-nansei-oki earthquake, respectively. By comparing Figs.8 and 9, it is seen that both figures coincide fairly well with each other in northwest part but do not coincide with each other in southeast part of the studied area. This difference is attributed to the difference of the maximum surface acceleration in two parts. As the epicenter was located in northwest direction from Okushiri Island, the estimated maximum surface acceleration in northwest part is higher than that in southeast part. Then the weight W_1 was 2.306 and 1.004 in northwest part and southeast part, respectively.

Figures 10 and 11 show the estimated landslide hazard index, $H\ell$, estimated by Mora and Vahrson's method, and the number of slides in each mesh area that occurred during the 1993 Hokkaido-nansei-oki earthquake, respectively. By comparing the two figures, it is seen that many meshes are classified in the second grade in Fig.10. Therefore the meshes of $H\ell=2$ are better to be subdivided.

(3) Chiba

Figures 12 and 13 show the number of slides in each mesh area estimated by Kanagawa Method and the number occurred during the 1987 Chibaken- toho-oki earthquake, respectively. By comparing Figs.12 and 13, it can be said that the estimated numbers of slides are overestimated in south part, and underestimated in north part. This difference is attributed also to the difference of the maximum surface acceleration in two parts. As the epicenter was located in southeast direction from the studied area, the estimated maximum surface acceleration in south part is higher than that in north part. Then the weight W_1 was 2.754, 2.306 and 1.004 in south, central and southeast parts, respectively.

Figure 14 and 15 show the estimated landslide hazard index, $H\ell$, estimated by Mora and Vahrson's method, and the number of slides in each mesh area that occurred during the 1987 Chibaken-toho-oki earthquake, respectively. As show in Fig.14, no slope failure was estimated by this method. This result is attributed the parameter for relative relief, Sr. There are low terraces only with the height of less than 75m in this area. Then the parameter Sr became zero in all mesh. Therefore, the range for Sr must be subdivided in this area.

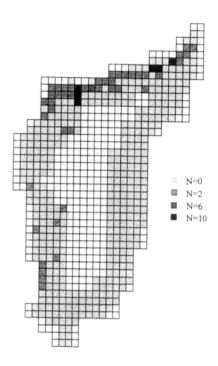

Fig.8 Number of slides in each mesh in Okushiri
estimated by Kanagawa Method

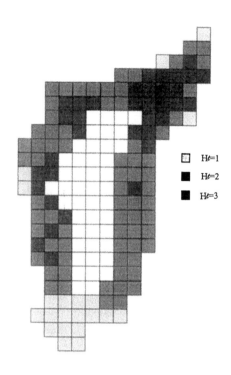

Fig.10 A zoning map estimated in Okushiri
estimated by Mora and Vahrson's Method

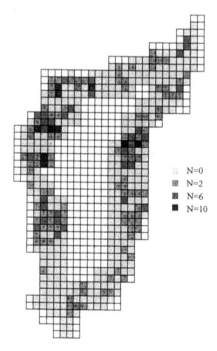

Fig.9 Number of slides in each mesh which
occurred during the 1993 Hokkaido-
nansei-oki earthquake

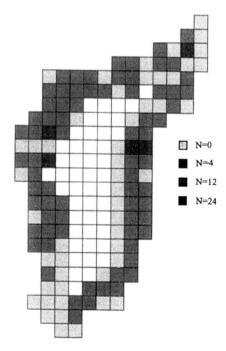

Fig.11 Number of slides in 1 km by 1 km mesh
during the 1993 Hokkaido-nansei-oki
earthquake

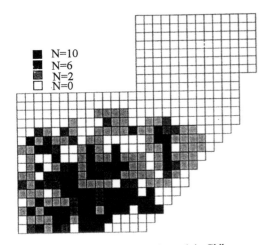

Fig.12 Number of slides in each mesh in Chiba
estimated by Kanagawa Method

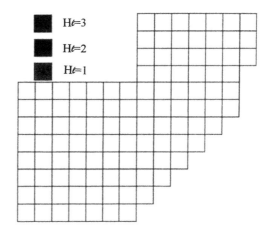

Fig.14 A zoning map estimated in Chiba estimated
by Mora and Vahrson's Method

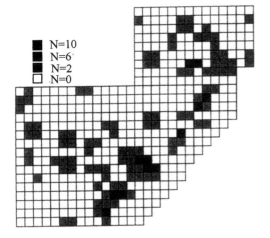

Fig.13 Number of slides in each mesh which
occurred during the 1987 Chibaken-toho-oki
earthquake

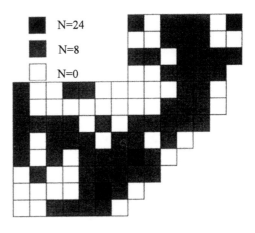

Fig.15 Number of slides in 1 km by 1 km mesh
during the 1987 Chibaken-toho-oki
earthquake

4 SOME COMMENTS ON THE APPLICATION OF THE ZONING METHODS

(1) Kanagawa Method

Figures 16 and 17 compare the number of slides estimated Kanagawa Method with the number of slides which occurred during the 1995 Hyogoken-nambu and 1987 Chibaken-toho-oki earthquakes. As shown in the average lines, solid lines in the figures, estimated number of slides are almost several times compare with the failed number. As mentioned above, the effect of weight W_1 was too intense in Kobe and south part of Chiba.

The maximum surface acceleration for Kobe was estimated by the records on the hard grounds shown in Fig.18. Records were obtained on rocks also during the earthquake as shown in Fig.18. The maximum surface on the rock was only 200 to 300 gals in the studied area. Then the zoning was carried out also under this acceleration. Figure 19 shows the number of slides estimated under the acceleration of 200 to 300 gals. By comparing Figs.4, 5 and 19, it can be said that the appropriate acceleration to apply Kanagawa Method may be a mean value between the acceleration on hard ground and the acceleration on rock.

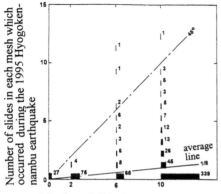

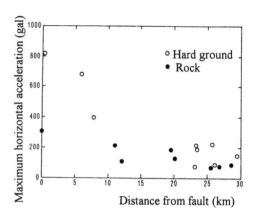

Fig.16 Comparison of the number of slides which occurred during the Hyogoken-nambu earthquake with the estimated number of slides

Fig.18 Relationship between recorded acceleration and distance from fault

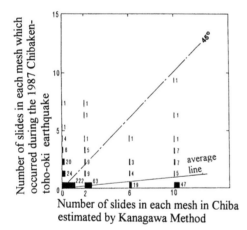

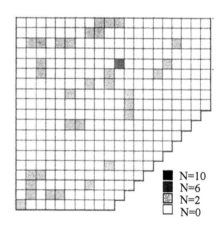

Fig.17 Comparison of the number of slides which occurred during the Chibaken-toho-oki earthquake with the estimated number of slides

Fig.19 Number of slides in each mesh in Kobe estimated by Kanagawa Method under 200 to 300 gals of acceleration and distance from fault

Maximum surface acceleration estimated by Joner and Boore's formula at middle and south parts in Chiba was 300 to 500 gals. These values may be greater than the acceleration on rock because Joner and Boore's formula is derived mainly from the records on hard grounds. If W_1 is assumed 1.004 which means 200 to 300 gals in acceleration, no slope failure is estimated in the studied area. If W_1 is assumed 2.306 the number of slope failures is overestimated. Then, if a mean value, $W_1=1.66$ is assumed, estimated number of slides are fairly coincided with number of slides as shown in Figs.20 and 13. Therefore, it can be said also that the appropriate acceleration may be a mean value between the acceleration on hard ground and the one on rock.

150

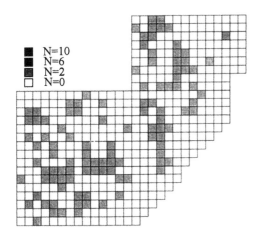

Fig.20 Number of slides in each mesh in Chiba
estimated by Kanagawa Method under
$W_1=1.66$

N=10
N=6
N=2
N=0

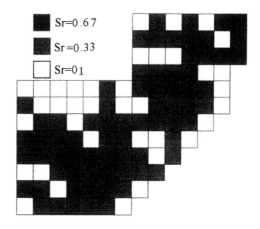

Fig.22 Modified zoning map estimated in Chiba
estimated by Mora and Vahrson's Method
by subdividing Sr

Sr=0.67
Sr=0.33
Sr=0.1

(2) Mora and Vahrson's Method

As mentioned above, ranges for total index, $H\ell$ was too rough in Kobe and Okushiri in Mora and Vahrson's Method. In Chiba, the range for the relative relief value, Sr, was too rough. Then these ranges were subdivided and tried to estimate again. Figure 21 shows the modified zoning map in Kobe by subdividing the ranks of $H\ell=2$ and 3. Figure 22 shows the modified map in Chiba by subdividing the Sr=0 into three ranges. By comparing Fig.21 with Fig.7 and Fig.22 with Fig.15, it can be said that these modified zoning maps are better than the original zoning maps shown in Figs.6 and 14.

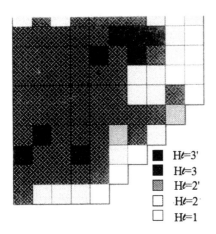

$H\ell=3'$
$H\ell=3$
$H\ell=2'$
$H\ell=2$
$H\ell=1$

Fig.21 Modified zoning map estimated in Kobe
estimated by Mora and Vahrson's Method
by subdividing $H\ell$

5 CONCLUSIONS

Two zoning methods which are introduced in the Manual for Zonation on Seismic Geotechnical Hazards were applied in three areas where slope failures occurred during recent three major earthquakes, to study the applicability of the methods. The followings conclusions were derived:
(1) Number of slides estimated by Kanagawa Method was apt to overestimate. Main reason was that the maximum surface acceleration on the hard ground was used. Appropriate acceleration for Kanagawa Method may be a mean value between the acceleration on hard ground and that on rock.
(3) Zoning maps estimated by Mora and Varhson's Method were slightly too rough. Some parameters were better to be subdivided.

ACKNOWLEDGMENT

Zoning maps were made with the assistance of Messrs. T.Motoi, M.Wada and N.Nagao who are former and present undergraduate students at Tokyo Denki University and Mr. S.Arima who is former student of Kyushu Institute of Technology. The author would like thank them for their assistance.

REFERENCE

TC4, ISSMFE : Manual for Zonation on Seismic Geotechnical Hazardz, 1993.

2 Seismic behaviour of geotechnical structures

Seismic Behaviour of Ground and Geotechnical Structures, Sêco e Pinto (ed.)© 1997 Balkema, Rotterdam, ISBN 90 5410 887 8

Liquefaction mechanisms and countermeasures

Korhan Adalier
Civil Engineering Department, Rensselaer Polytechnic Institute, Troy, N.Y., USA

Mourad Zeghal
Impact Forecasting, L.L.C., Chicago, Ill., USA

Ahmed-W. Elgamal
Civil Engineering Department, Columbia University, New York, N.Y., USA

ABSTRACT: Two free-field downhole array seismic records are now available to document the underlying site liquefaction response mechanisms. Records of acceleration and pore pressure at the Wildlife Refuge (Imperial County, CA, USA) reflected liquefaction during the 1987 Superstition Hills earthquake. In 1995, liquefaction at Port Island (Kobe, Japan) was documented by a four-accelerometer downhole array during the Hyogoken-Nanbu earthquake. Using these records, the associated shear stress-strain response is identified and employed as a guideline for development of liquefaction countermeasure techniques. A number of these techniques are discussed in view of the identified response mechanisms. Furthermore, results of a recent centrifuge testing program are employed to illustrate the beneficial effects of four different countermeasure techniques applied to a liquefiable foundation layer underlying an earth embankment.

1 INTRODUCTION

Recent major seismic events such as the 1964 Niigata, 1989 Loma Prieta, and the 1995 Kobe earthquakes, continue to demonstrate the damaging effects of liquefaction-induced loss of soil strength and associated lateral spreading (Seed 1966, Ishihara et al. 1987, Seed et al. 1990, Soils and Foundations 1996). Experimental laboratory research on soil liquefaction has provided valuable insight concerning excess pore-pressure buildup in saturated loose granular soils (National Research Council 1985). However, for engineering applications, there remains a need to further understand and identify the mechanisms of seismically induced soil deformation due to liquefaction, and associated stiffness and strength degradation.

In-situ seismic records of site liquefaction are scarce. Currently, downhole seismic records are only available for: (1) the Wildlife Refuge site (Imperial County, CA) during the 1987 Superstition Hills earthquake (accelerations and pore-pressures), and (2) the Port Island site (Kobe, Japan) during the 1995 Hyogoken-Nanbu earthquake. Such downhole acceleration and excess-pore-pressure records provide direct information on the mechanisms of site liquefaction; and associated stiffness degradation and lateral spreading. This valuable information is of paramount importance to the development of liquefaction countermeasure techniques.

A brief summary of the Wildlife Refuge and Port Island case histories is presented below. In view of the observed mechanisms, a number of liquefaction countermeasures are also briefly discussed. In addition, the results of a recent centrifuge testing study compares four different techniques for remediation of a liquefiable foundation layer underlying an earth embankment.

2 SHEAR STRESS-STRAIN RESPONSE

A simple identification procedure, proposed earlier in basic form for shake-table studies (Koga and Matsuo 1990), was developed and used to study the downhole earthquake response at Lotung (Zeghal and Elgamal 1993, Zeghal et al. 1995, Elgamal et al. 1996a). Using a shear beam model to describe site seismic lateral response, shear stresses at levels z_i and $(z_{i-1} + z_i)/2$ may be evaluated as follows:

$$\tau_i(t) = \tau_{i-1}(t) + \rho_{i-1}\frac{\ddot{u}_{i-1} + \ddot{u}_i}{2}\Delta z_{i-1}, i = 2,3,... \quad (1)$$

$$\tau_{i-1/2}(t) = \tau_{i-1}(t) + \rho_{i-1}\frac{3\ddot{u}_{i-1} + \ddot{u}_i}{8}\Delta z_{i-1}, i = 2,3,... \quad (2)$$

in which subscripts i and $i - 1/2$ refer to levels z_i (of the ith accelerometer) and $(z_{i-1} + z_i)/2$ (halfway between accelerometers i and $(i - 1)$) respectively, $\tau_i(t) = \tau(z_i; t)$, $\tau_1 = \tau(0;t) = 0$ at the stress-free ground surface, ρ_{i-1} = mass density of the z_{i-1} to z_i soil layer,

$\ddot{u}_i = \ddot{u}(z_i, t)$ is absolute acceleration at level z_i, and Δz_i spacing interval between accelerometers. These stress estimates (Eqs. 1 and 2) are second order accurate. The corresponding second-order accurate shear strains may be expressed as:

$$\gamma_i(t) = \frac{1}{\Delta z_{i-1} + \Delta z_i} \left((u_{i+1} - u_i)\frac{\Delta z_{i-1}}{\Delta z_i} + (u - u_{i-1})\frac{\Delta z_i}{\Delta z_{i-1}} \right)$$

for $i = 2, 3, \ldots$ (3)

$$\gamma_{i-1/2}(t) = \frac{u_i - u_{i-1}}{\Delta z_{i-1}}, i = 2, 3, \ldots \quad (4)$$

in which $u_i = u(z_i; t)$ is absolute displacement (evaluated through double integration of the recorded acceleration history $\ddot{u}(z_i, t)$).

3 WILDLIFE-REFUGE, CALIFORNIA USA

The Wildlife Refuge site is located on the west side of the Alamo river, Imperial County in Southern California. Evidence of liquefaction was observed at or near the site following the 1930, 1950, 1957, 1979, and 1981 Imperial Valley (Youd and Wieczorek 1984) earthquakes. These observations triggered an interest in Wildlife which in an insightful effort, was instrumented in 1982 (Fig. 1, Youd and Wieczorek 1984) by the United States Geological Survey (USGS). On November 24, 1987, the Wildlife site was shaken by the Superstition Hills earthquake ($M_W = 6.6$), causing a sharp increase in recorded pore-water pressure (Holzer et al. 1989). In addition, subsequent field investigations showed evidence of site liquefaction and ground fissures. Figure 2 depicts the NS and EW components of the recorded accelerations at ground surface and 7.5 m depth; and the associated excess pore-water pressure measured at 2.9 m depth (piezometer P5, Fig. 1). As shown in Fig. 2, the surface records displayed peculiar acceleration spikes (Holzer et al. 1989) associated with simultaneous instances of pore-pressure drop.

Figure 3 displays the shear stress-strain history during the Superstition Hills earthquake evaluated using Eqs. 1-4 (Zeghal and Elgamal 1994, Zeghal et al. 1996). A dramatic loss of soil stiffness due to liquefaction is depicted by this stress-strain history. This was further manifested by the effective stress histories at 2.9 m depth (location of piezometer P5, Fig. 1), evaluated from the acceleration and excess pore pressure records of Fig. 2. Shear stress versus effective vertical stress ($\sigma'_v = \sigma_v - p$, where p is excess pore pressure measured by P5, and σ_v is total vertical stress at P5) may be interpreted as an effective stress path.

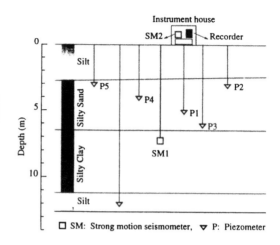

Figure 1   Cross-section and instrumentation at the Wildlife Refuge site (after Bennett et al. 1984).

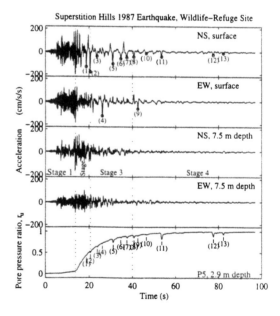

Figure 2   Wildlife Refuge site NS and EW surface and downhole (at 7.5 m depth) accelerations, and associated pore water pressure (at 2.9 m depth) during the Superstition Hills 1987 earthquake.

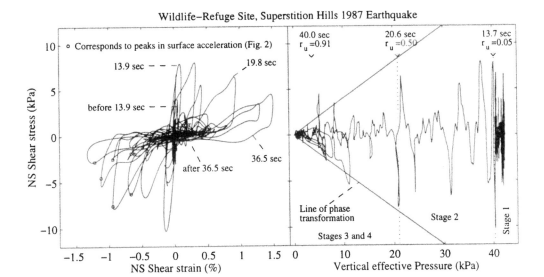

Figure 3   Wildlife Refuge site NS shear stress-strain and effective stress-path histories during the Superstition Hills 1987 earthquake (Zeghal and Elgamal 1994).

During the initial phase (stage 1 [0 s to 13.7 s], Fig. 2), the site showed no evidence of significant stiffness degradation (Fig. 3); and no appreciable rise in pore pressure was recorded by piezometer P5 (Fig. 2). During the strong shaking phase (stage 2 [13.7 s to 20.6 s], Figs. 2 and 3), the site experienced a clear and gradual stiffness degradation associated with a sharp increase in pore water pressure. Soil stiffness and yield strength continued to decrease during stages 3 and 4 ([20.6 s to 96.0 s], Figs. 2 and 3). Cycles of large shear strain were developed during these stages, with a peak strain of 1.5 % at 36 s (Fig. 3). During stages 2-4, site response was marked by shear stress-strain hardening at large strains (Fig. 3). This hardening was more pronounced in the negative direction where acceleration spikes occurred along with instantaneous pore pressure drops (Fig. 2). Such asymmetric hardening was most probably associated with the presence of a nearby free slope towards which permanent ground deformations were observed (Holzer et al. 1989). At low effective confining pressures (high excess pore pressures, stage 4), the effective stress-path (Fig. 3) clearly exhibited a reversal of behavior from contractive to dilative as the line of phase transformation was approached (National Research Council 1985).

Thus, this case history clearly showed (for the first time), an in-situ mechanism of shear stress-strain hardening at large strain excursions during liquefaction. Such a mechanism has been observed in a number of experimental studies (e.g., Koga and

Matsuo 1990, Arulmoli et al. 1992, Taboada and Dobry 1993, Adalier 1996), and is a consequence of soil dilation at large strain excursions, which results in associated instantaneous pore-pressure drops (Vucetic and Dobry 1988).

4   PORT ISLAND, KOBE JAPAN

Port Island is a reclaimed island located on the southwest side of Kobe, Japan. In the phase completed by 1981, 436 ha were reclaimed by bottom-dumping from barges (Nakakita and Watanabe 1981). Soil in the artificial reclaimed layer (O'Rourke 1995, Sitar 1995) consisted of decomposed weathered granite fill (Masa soil mined from the nearby Rokko mountains). A downhole accelerometer array was installed at the North-West corner of Port Island (Fig. 4) in August 1991 (Iwasaki 1995a). The array consisted of triaxial accelerometers located at the surface, 16 m, 32 m, and 83 m depths. All instruments were linked to a common triggering mechanism, and hence the recorded earthquake data were synchronized.

As shown in Fig. 4, the downhole array site consists of: (1) an artificial, reclaimed, loose surface layer down to about 19 m depth, (2) an alluvial clay layer between 19 m and 27 m depth, (3) sand and sand with gravel strata interlayered with clay between 27 and 61 m depth, (4) a diluvial clay layer between 61 m and 82 m depth, and (5) sand with gravel layers interlayered with clay starting at about

82 m depth. The water table was located at a depth of 4 m approximately. A Standard Penetration Test (SPT) profile of the soil strata around the downhole array are also shown in Fig. 4 (Iwasaki 1995a, b). In the upper 20 m layer (Fig. 4), low Standard Penetration Test (SPT) blow counts prevailed (average uncorrected N-values of about 6 blows/ft). Such low values in a granular fill are indicative of high liquefaction susceptibility (Seed et al. 1983).

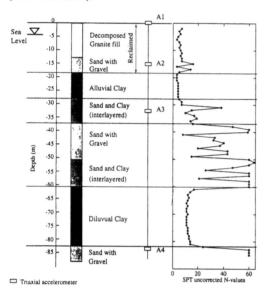

Figure 4   Soil profile and instrumentation at the Port Island site (after Iwasaki 1995a).

Using the recorded downhole accelerations (Fig. 5) and Eqs. 1-4, the shear stress-strain response was evaluated (Elgamal et al. 1996b). Selected representative cycles of this response are shown in Fig. 6. Two remarkably different response patterns were exhibited at the site. Below 32 m depth, the shear stress-strain histories showed an essentially linear soil response, with no appreciable reduction in soil stiffness. On the other hand, at shallow depths, the stress-strain histories indicated: (1) a noticeable reduction in stiffness with a slight shear strain hardening at elevation 24 m (Fig. 6), and (2) an abrupt sharp loss of stiffness and reduction of yield strength near the surface at 8 m depth.

5   MECHANISMS OF SITE LIQUEFACTION

The Wildlife Refuge and Port Island sites exhibited two remarkably different response mechanisms. In first approximation, the responses may be contrasted along the lines presented in Table 1.

Following liquefaction, the Port Island site response was marked by large strains and small stresses, while at the Wildlife site the response was associated with a regain in shear stiffness and strength at large strain excursions. Such regain is due to a tendency for dilation at large strains during liquefaction (cyclic mobility), and is of importance in restricting the extent of accumulated lateral cyclic deformation. This phenomenon is an essential ingredient in liquefaction remediation techniques involving densification, as mentioned below.

Table 1   Response parameters of Wildlife Refuge and Port Island sites.

Parameter	Wildlife Refuge	Port Island
Level of accel. towards loss of stiffness	0.15g	0.7g
Duration of shaking towards loss of stiffness	7 sec	2 sec
Peak input acceleration and duration after liquefaction	0.06g/15 sec	0.2g/7 sec
Liquefied zone	within upper 7 m	within upper 16 m
Cyclic strains	1.5 %	1.5 %
Site condition	nearby slope	level
Downslope strains	> 1.5 %	---
Post-liquefaction shear mechanism	cyclic mobility	---
Duration of liquefied state	>15 min. (Holzer et al. 1989)	>4 min. (Elgamal et al. 1996b)
Vertical settlement	few cms (Holzer et al. 1989)	> 0.5 m nearby (Soils and Found. 1996)

6   LIQUEFACTION COUNTERMEASURES

6.1   Liquefaction countermeasure techniques

Soil improvement against liquefaction hazards can be divided into four main categories: (1) densification, (2) drainage, (3) solidification, and 4) inclusions (Adalier 1996).

Densification: Densification increases liquefaction resistance due to the associated decrease in void ratio or increase in relative density, and mean effective stress (increase in lateral stresses). Densification has been the most popular technique of improving loose saturated liquefiable soils.

Densification can be achieved using several methods such as vibro-compaction, vibro-replacement, deep dynamic compaction, deep blasting, compaction grouting, and displacement piles. A comprehensive description of these methods can be found in a state-of-the-art paper by Mitchell (1981), in Broms (1991), and in Adalier (1996).

Port Island, Kobe (Japan). Hyogoken–Nanbu EQ. Jan. 17, 1995

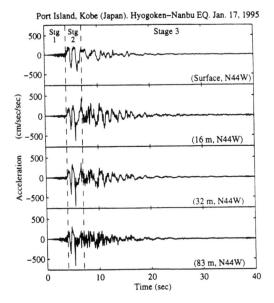

Figure 5   N44W accelerations at ground surface and downhole stations (at 16 m, 32 m and 83 m depths, after Iwasaki 1995a).

Port Island, Kobe (Japan). Hyogoken-Nanbu Earthquake, Jan. 17, 1995·

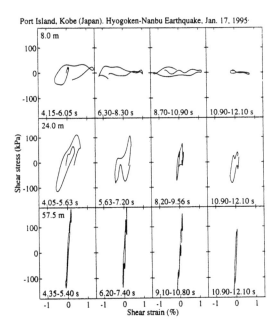

Figure 6   Selected N44W shear stress-strain cycles midway between accelerometers, at 8.0 m, 24.0 m and 57.5 m depths (Elgamal et al. 1996b).

Drainage: Relief wells by gravel drains (stone columns) or prefabricated drains are typical means of increasing drainage in the soil system. It is believed that installation of stone columns (gravel drains) by vibro-replacement mitigates liquefaction potential by increasing the surrounding soil density; providing drainage to control pore pressures; and introducing stiff elements (stone columns) that can potentially carry higher stress levels and provide a deformation restricting effect (Priebe 1991, Baez and Martin 1993, Adalier 1996).

Solidification: These methods involve the introduction of stabilizing chemicals (e.g., cement, bentonite, and sodium silicates) by either injection (e.g., injection grouting) or mixing (e.g., jet grouting and deep-soil-mixing). However, they are relatively expensive, and are generally only considered for limited areas to construct stabilized zones or cells within a larger soil mass (improved grid shape); or for situations where other soil improvement techniques cannot be used (e.g., small, difficult-to-access areas, or under existing structures). Comprehensive information regarding different solidification methods is reported in Mitchell (1981), Winterkorn and Pamukcu (1991), and Adalier (1996).

Inclusions: This method involves formation of inclusions in the soil using concrete, timber, or steel piles, and sheet piles. Such techniques usually rely on the ability of inclusions to reduce shear stress and restrict shear strains in the soil. In addition, depending on the installation method (driving or placing), it can densify the surrounding soil and enhance liquefaction resistance. Generally, these techniques have enjoyed little popularity as liquefaction countermeasure methods.   Thorough discussions may be found in Kramer and Holtz (1991), and Adalier (1996).

6.2 *Applicability of Different Remedial Techniques to Wildlife Refuge and Port Island Cases*

In view of the liquefaction mechanisms contrasted above, the beneficial outcomes of each technique may be listed as shown in Table 2.

7 CENTRIFUGE STUDY

A series of model tests (Adalier 1996) were conducted at the Rensselaer Polytechnic Institute (RPI) Centrifuge Research Center.   The RPI centrifuge is an Acutronic 100g-ton machine. On the centrifuge, uniaxial base shaking may be imparted by available servo-hydraulic centrifuge shake tables. In addition to the 1 ton actuator shake table used in this study, a new large shaker 1 m in length with a 14 ton double actuator system is now

Table 2 Application of remedial techniques to Wildlife Refuge and Port Island sites.

Technique	Case	Outcome
Drainage	*Wildlife Refuge*: applicable. *Port Island*: not applicable since shaking was too strong with rapid excess pore pressure build-up (see Table 1).	Control of pore pressures leading to reduction of vertical settlements, and lateral deformations.
Drainage with densification (e.g., stone columns)	Applicable in both cases.	Densification of surrounding soil (increase in relative density), and introduction of stiff elements to increase shear resistance.
Densification	*Wildlife Refuge*: most densification techniques may not be suitable since the soil contains large portion of fines. *Port Island*: All densification methods can be used for this case (note that deep dynamic compaction has an economical treatment depth of about 12 m). At Port Island, areas treated by vibro-compaction did not suffer much deformations as opposed to the surrounding untreated areas (Soils and Foundations 1996). Compaction grouting may not be economical for the large areas involved in both cases.	Reduce liquefaction potential, settlement, and increase cyclic mobility (to reduce lateral deformations).
Solidification	These methods are relatively expensive and uneconomical for both cases. However, limited area application (e.g., sparse stabilized zones or cells within a large soil mass) may not be sufficient for eliminating liquefaction induced deformations at Port Island (since the shaking was too violent).	Elimination of liquefaction in treated soil volume. Reduction in available pore water, reduction in liquefaction potential, increase in overall stiffness in the areas confined by treated zones.
Inclusions	*Wildlife Refuge*: applicable. *Port Island*: applicable but must be designed for a 0.7g level input motion.	Introduction of stiff elements to increase shear resistance, reduce liquefaction potential, and provide resistance against lateral deformation of the ground.

available (Van Laak 1996). Dynamic testing was conducted at a centrifugal acceleration of 75g.

The seismic response of a cohesive highway embankment overlying a layer of loose saturated clean sand was studied (Adalier 1996), with and without liquefaction countermeasures (Fig. 7). In this series of tests, the models were heavily instrumented with accelerometers, pore-pressure transducers, and LVDTs. Each model was subjected to three consecutive shaking events. Sufficient time was allowed between shaking events to allow for full excess pore pressure dissipation. In each event, shaking was about 8 seconds in duration. Input motion consisted of a sinusoidal signal with a peak amplitude of about 0.1g (first shaking event), 0.2g (second shaking event), and 0.3g (third shaking event).

After dynamic excitation, large deformations and cracks were observed in the benchmark model with no liquefaction countermeasures (Model 1 of Fig. 7). Along the embankment crest, a settlement of 0.75 m was measured. In Model 2 with the densified soil columns, settlement was reduced by about 50%; with significant manifestations of increased foundation strength due to cyclic mobility. Nevertheless, the level of deformations was still excessive. In Model 3, an unexpected outward movement of the remediated cement walls was observed. This movement led to settlements of the embankment and unsatisfactory overall response. The berm of Model 4 was found quite effective, particularly in preserving the embankment integrity. Finally, in Model 5, the liquefied foundation soils

were contained below the embankment with no lateral deformations. Additionally, the embankment further preserved its integrity, ironically due to liquefaction of its underlying foundation, which produced an efficient base isolation mechanism.

8 SUMMARY AND CONCLUSIONS

The conducted studies showed the role of downhole acceleration records in assessing the mechanisms of site amplification, stiffness degradation and liquefaction. At the Wildlife Refuge and Port Island sites, the estimated stress-strain histories showed that: (1) site stiffness and strength decreased steadily with excess pore pressure buildup, (2) at high excess pore pressure levels, site response was characterized by large strains and small stresses, and (3) during liquefaction, significant shear strength may evolve at large shear strains, due to the cyclic mobility effect. A number of guidelines were drawn for liquefaction countermeasure strategies. A brief review of liquefaction countermeasures was presented showing the effect of each technique in mitigating the observed detrimental liquefaction effects. The performance of four such techniques was reviewed based on results of a recent centrifuge testing program. In this program, the underlying remediation mechanisms included stiffening, increased vertical confinement, and containment of a loose liquefiable stratum, underlying a cohesive

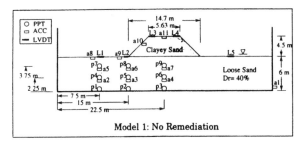

Model 1: No Remediation

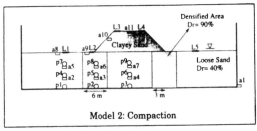

Model 2: Compaction

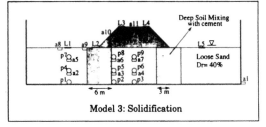

Model 3: Solidification

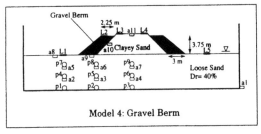

Model 4: Gravel Berm

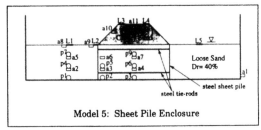

Model 5: Sheet Pile Enclosure

Figure 7 Configurations of centrifuge Models 1 to 5.

earth embankment. All remediation techniques resulted in improved overall dynamic response. Containment and increased confinement were found to be especially effective in preserving the embankment integrity.

ACKNOWLEDGMENTS

This research was supported by the United States Geological Survey (grant No. 1434-94-G-2397), the National Science Foundation (grant No. MSS-9057388), and the National Center for Earthquake Engineering Research (grant No. 90-1503); with matching funds from Earth Mechanics Inc. Port Island downhole acceleration data were provided by the Committee of Earthquake Observation and Research in the Kansai Area (CEORKA), Japan, with the help of Dr. Y. Iwasaki.

REFERENCES

Adalier, K. 1996. *Mitigation of earthquake induced liquefaction hazards*. Ph.D. Thesis, Rensselaer Polytechnic Institute, Troy, NY, 659 p.

Arulmoli, K., Muraleetharan, K. K., Hossain, M. M. and Fruth, L. S. 1992. *Verification of liquefaction analyses by centrifuge studies laboratory testing program soil data report*. Earth Tech. Corp., Irvine, CA.

Baez, J. I., and Martin, G. R. 1993. Advances in the design of vibro systems for the improvement of liquefaction resistance. *Sym. of Ground Improvement*, Vancouver Geotechnical Soceity, Vancouver, B.C.

Bennett, M. J., McLaughlin, P. V., Sarmiento, J. S., and Youd, T. L. 1984. *Geotechnical investigation of liquefaction sites, Imperial Valley, California*. U.S. Geological Survey Open-File Report 84-252.

Broms, B. B. 1991. Deep Compaction of granular soils. *Foundation Engineering Handbook*, 2nd Ed., Von Nostrand Reinhold, NY, NY, pp. 814-832.

Elgamal, A.-W., Zeghal, M., Parra, E., Gunturi, R., Tang, H. T., and Stepp, C. J. 1996a. Identification and modeling of earthquake ground response I: Site amplification. *Soil Dyn. and Earthq. Engrg.*, 15(8), 499-522.

Elgamal, A.-W., Zeghal, M. and Parra, E. 1996b. Liquefaction of an artificial island in Kobe, Japan. *J. of Geotech. Engrg.*, ASCE, 122(1), 39-49.

Holzer, T. L., Youd T. L. and Hanks T. C. 1989. Dynamics of liquefaction during the 1987 Superstition Hills, California, Earthquake. *Science*, 244, 56-59.

Ishihara, K., Anazawa, Y., and Kuwano, J. 1987. Pore water pressures and ground motions monitored during the 1985 Chiba-Ibaragi Earthquake. *Soils and Foundations*, 27(3), 13-30.

Iwasaki, Y. 1995a. Georesearch Institute, Osaka, Japan, Personal communication.

Iwasaki, Y. 1995b. Geological and geotechnical characteristics of Kobe Area and strong ground motion records by 1995 Kobe Earthquake, Tsuchi-to-Kiso. *Japanese Soc. of Soil Mech. and Found. Engrg.*, 43(6), 15-20 (in Japanese).

Koga, Y. and Matsuo, O. 1990. Shaking table tests of embankments resting on liquefiable sandy ground. *Soils and Foundations*, 30(4), 162-174.

Kramer, S. L., and Holtz, R. D. 1991. *Soil improvement and foundation remediation with emphasis on seismic hazards*. A Report of a workshop sponsored by the National Science Foundation and held at the Un. of Washington, Dept. of Civil Eng., Seattle, WA, Aug. 19-21.

Mitchell, J. K. 1981. Soil Improvement State-of-the-Art Report. *Proc., 10th Int. Conf. on Soil Mech. and Found. Engrg.*, Stockholm, Vol. 4, 506-565.

Nakakita, Y., and Watanabe, Y. 1981. Soil stabilization by Preloading in Kobe Port Island. *Proc., 9th Int. Conf. on Soil Mech. and Found. Engrg.*, Tokyo, Japan, 611-622.

National Research Council 1985. *Liquefaction of Soils During Earthquakes*. Committee on Earthquake Engineering, National Academy press, Washington, D.C., 240 p.

O'Rourke, T. D. 1995. *Geotechnical effects: Preliminary report from the Hyogoken-Nanbu Earthquake of January 17, 1995*. National Center for Earthquake Engrg. Research. Bulletin., SUNY, Buffalo, 9(1).

Priebe, H. J. 1991. Vibro Replacement - design criteria and quality control. *Deep Foundation Improvements: Design, Construction, and Testing*, ASTM STP 1089, M. I. Esrig and R. C. Bachus (eds.), Philadelphia, 62-72.

Seed, H. B., Idriss, I. M., and Arango, I. 1983. Evaluation of liquefaction potential using field performance data. *J. of Geotech. Engrg.*, ASCE, 109(3), 458-482.

Seed, H. B. 1966. Landslides during earthquakes due to soil liquefaction. *J. of Soil Mech. and Found. Div.*, ASCE, 94(5), 1053-1122.

Seed, R. B., Dickenson, S. E., Riemer, M. F., Bray, J. D., Sitar, N., Mitchell, J. K., Idriss, I. M., Kayen, R. E., Kropp, A., Hander, L.F. Jr., and Power, M. S. 1990. *Preliminary report on the principal geotechnical aspects of the October 17, 1989, Loma Prieta Earthquake*. Report No. EERC-90-05, Earthquake Engrg. Res. Center., Un. of California, Berkeley,CA.

Sitar, N., Ed. 1995. *Geotechnical reconnaissance of the effects of the January 17, 1995, Hyogoken-Nanbu Earthquake Japan*. Report No. UCB/EERC-95/01, Earthquake Engrg. Res. Cent., Berkeley, CA.

Soils and Foundations 1996. Special issue on geotechnical aspects of the January 17, 1995 Hyogoken Nambu Earthquake. *Soils and Foundations*, Japan Geotechnical Society, January 1996.

Taboada, V. M. and Dobry, R. 1993. Experimental results of Model 1 at RPI. *Proc., Int. Conf. on the Verification of Numerical Procedures for the Analysis of Soil Liquefaction Problems*, Arulanandan, K. and Scott, R. F., eds., Vol. 1, Davis, CA, 3-17, Balkema.

Van Laak, P. 1996. *Development of dynamic capability for geotechnical centrifuge model studies*. Ph.D. Thesis, Rensselaer Polytechnic Institute, Dept. of Civil Engineering, Troy, NY, 195 p.

Vucetic, M. and Dobry, R. 1988. Cyclic triaxial strain-controlled testing of liquefiable sands. *Amer. Soc. Testing Mat.*, Spec. Tech. Publ. 977, 475-485.

Winterkorn, H. F., and Pamukcu, S. 1991. Soil stabilization and grouting. Ch. 9, *Foundation Engineering Handbook*, H-Y. Fang (ed.), 2nd Ed., Van Nostrand Reinhold, NY, NY.

Youd, T. L., and Wieczorek, G. F. 1984. *Liquefaction during 1981 and previous earthquakes near Westmorland California*. U.S. Geological Survey Open-File Report 84-680.

Zeghal, M. and Elgamal, A.-W. 1993. *Lotung site: downhole seismic data analysis*. Rep., Dept. of Civil Engrg., Rensselaer Polytechnic Institute, Troy, NY.

Zeghal, M., Elgamal, A.-W., Tang, H. T., and Stepp, J. C. 1995. Lotung downhole array: evaluation of soil nonlinear properties. *J. of Geotech. Engrg.*, ASCE, 121(4), 363-378.

Zeghal, M. and Elgamal, A.-W. 1994. Analysis of site liquefaction using earthquake records. *J. of Geotech. Engrg.*, ASCE, 120(6), 996-1017.

Zeghal, M., Elgamal, A.-W., and Parra, E. 1996. Identification and modeling of earthquake ground response II: Site liquefaction. *Soil Dyn. and Earthq. Engrg.*, 15(8), 499-522.

Seismic Behaviour of Ground and Geotechnical Structures, Sêco e Pinto (ed.)© 1997 Balkema, Rotterdam, ISBN 90 5410 887 8

Seismic analysis of flexible buried structures

J.P. Bardet
University of Southern California, Los Angeles, Calif., USA

C.A. Davis
Los Angeles Department of Water and Power, Calif., USA

ABSTRACT: A simplified pseudo-static analysis method is proposed to identify the main causes of the transverse buckling of buried structures, such as culverts, which are subjected to earthquake ground motions. The proposed method is based on the case history of a 2.4-m diameter corrugated metal pipe in the Lower San Fernando Dam which failed during the 1994 Northridge Earthquake. The analysis, which considered factors such as static overburden pressure, peak ground acceleration, liquefaction-induced ground displacement, pore pressure buildup, and nonlinear soil response, suggests that the pipe failure was mainly caused by the reduction in moduli of embedding soils due to pore pressure build up, without a complete liquefaction.

1. INTRODUCTION

Until recently, underground structures were thought to be safe during earthquakes as long as they did not cross fault planes. The 1995 Hyogoken-Nanbu earthquake suddenly overturned this belief when it damaged the Daikai Subway Station (Iida et al., 1996), and revealed that underground structures were also vulnerable to earthquake ground motion. A related but less publicized example of the failure of buried structures took place during the 1994 Northridge Earthquake in the Van Norman Complex of the Los Angeles Department of Water and Power (LADWP), in the northern San Fernando Valley in Southern California. In this event, 76 meters of 2.4-m diameter buried pipe were crushed, and 23 meters deformed. As pointed out by Youd and Beckman (1996), the transverse failure of large-diameter conduits is still not well understood because such failures have rarely been observed during past earthquakes. In contrast, numerous longitudinal failures for small-diameter underground pipes have been recorded and extensively studied for transient ground motion (e.g., O'Rourke and Hmadi, 1988) and permanent ground deformations (e.g., Tawfik and O'Rourke, 1985).

This paper summarizes the field investigation results on the pipe collapse in the Van Norman Complex. Based on this case history, the paper proposes a pseudo-static analysis that identifies the main causes for the transverse failure of large-diameter flexible conduits.

2. DESCRIPTION OF DRAIN LINE AND SURROUNDING SOILS

Figures. 1, 2, and 3 show a plan, profile, and cross-section of the pipe (drain line) which is located in the upstream berm of the Lower San Fernando Dam (LSFD) of the Van Norman Complex in Southern California.

2.1 Drain line characteristics

As shown in Fig. 2, the 116-m long drain line is made of 17 segments, which have been numbered 1 to 17. Each segment is 7.3 m long, except for segments 1 and 2, and each is made of 2.4 m diameter, unencased, corrugated metal pipe. At the south end, the line connects to the 2.6 m diameter reinforced concrete tunnel that originally served as the outlet line for the reservoir (LADWP, 1975). The 2.4 m nominal inside diameter, 8 gauge (0.43 cm thick), galvanized, corrugated metal pipe was fabricated according to the specifications of ASTM A 444. The base metal had a minimum yield strength of 230 MPa and minimum tensile strength of 310 MPa (ASTM A 446). The corrugations had a pitch of 6.8 cm and depth of 1.3 cm as defined by ASTM A 760. The pipe segments were joined by butting them together and wrapping a metal strap around the perimeter. The joints were secured tightly but were not sealed with rubber gaskets. As shown in Fig. 2, the slope of the drain line is 0.312%. With the intake set at elevation 314.8 m, this line maintains a relatively

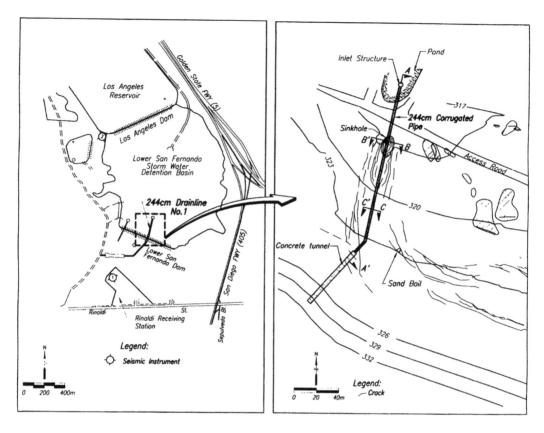

Figure 1. View of Lower San Fernando Dam along Drain Line, and crack patterns and sand boils observed after the 1994 Northridge Earthquake.

constant flow of water throughout the year, and serves as the outlet to empty the storm basin.

The drain line was constructed in 1973, after the LSFD was removed from service. Originally, the LSFD was placed in service in 1915 to impound water for the Lower Van Norman Reservoir which served as a terminal storage and distribution facility for Los Angeles Aqueduct water. The 1971 San Fernando Earthquake caused extensive damage to the LSFD. The hydraulic fill embankment underwent a massive liquefaction induced slide on its upstream face. The construction history and 1971 damage are described by Seed et al. (1973). Following the 1971 earthquake, the reservoir was immediately drained and removed from service. The LSFD was reconstructed to serve as a storm water detention basin. Its upstream slope were rebuilt and its outlet lines were modified to release water from the detention basin.

2.2 Soil characteristics

As shown in Fig. 2, pipe segments 1 to 6 were founded on top of alluvial soil while the remaining segments were founded on weak sedimentary rock (mainly sandstone and siltstone) which is hereafter referred to as bedrock. The northern end of the trench was cut through bedrock along segments 9 to 17, the center through alluvium along segments 6 to 9, and the southern end through hydraulic fill slide debris along segments 1 to 6.

Figure 3 shows a transverse cross-section of the pipe and embedding material. The drain line was placed in the natural ground which was cut to fit the pipe curvature. In 1973, the 5.5 m wide trench was backfilled with compacted sand bedding and overlain by compacted trench fill material. In 1975, during reconstruction of the LSFD upstream slope (LADWP, 1975), an embankment fill was added on the top of the backfilled trench. Its thickness varies from 1.5 m on the north end, to 7.3 m on the south end.

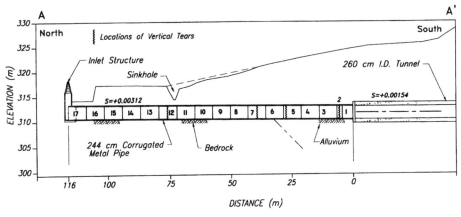

Figure 2. Profile of Drain Line along section AA' of Fig. 1.

Groundwater conditions vary along the pipe. During drain line repairs, groundwater was observed near the trench bottom in bedrock, further south it was slightly higher in the more permeable alluvium, and at the south end it went above the top of the pipe in the hydraulic fill. These observations were made on a free trench surface. Limited observations during excavation for emergency repairs indicated that the water level may be as high as the top of the pipe at the north end.

3. OBSERVATIONS

3.1 Ground motion and field observations at LSFD

Figure 1 shows the location of the strong motion instruments at the Rinaldi Receiving Station (Station 1) and Los Angeles Dam abutment.(Station 2). Station 1, located 500 m south of the drain line, re-

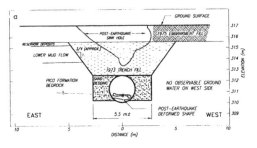

Figure 3. Typical section of drain line at joint of segments 11 and 12.

corded peak horizontal and vertical accelerations of 0.84 g and 0.85 g. Station 2, located 900 m north of the drain line, recorded peak horizontal and vertical accelerations of 0.43 g and 0.32 g, respectively. As pointed out by Bardet and Davis (1996a), the ground motion varied rapidly in the vicinity of the drain line. Figure 5 shows the variation of ground motion accelerations projected onto a vertical plane transverse to the drain line. Horizontal peak accelerations in this plane did not exceed 0.49 g at Station 1 or 0.35 g at Station 2, which provides a range for the level of accelerations impinging on the drain line.

Figure 1 shows the pipe alignment in the LSFD upstream face, and the location of the closest sand boils and ground cracks which were observed following the Northridge Earthquake. As described in Bardet and Davis (1996b), the upstream berm of the LSFD spread laterally in the northern direction, due to the liquefaction of the saturated hydraulic fill which remained after the 1971 San Fernando Earthquake. The south end of the pipe was buried under 12 m of fill encased within the hydraulic fill debris. As shown in Fig. 1, a small sand boil emerged on the ground surface at elevation 323 m. This sand boil was the highest at the LSFD, and was only 24 m east of the pipe alignment, which indicates that the saturated hydraulic fill may have also liquefied along the east side of the pipe. The pipe was therefore located on the western boundary of a large liquefied area, and partially embedded in liquefiable material.

As shown in Figs. 1, 2, and 3, the largest subsidence, which is shaped as a sinkhole, formed over the drain line. This sinkhole was largely responsible for further investigation. According to the post-earthquake elevation contours of Fig. 1, the soil subsidence was the largest on the east side of the pipe alignment. The southern boundary of subsidence is approximately located over the joint between seg-

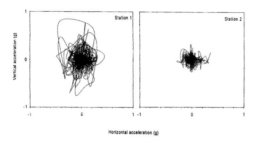

Figure 4. Ground acceleration transverse to the drain line as recorded at the Rinaldi Receiving Station (Station 1) and Los Angeles Dam abutment (Station 2).

ments 6 and 7. Ground cracks follow the pipe alignment and were largest within the subsided area. The sinkhole occurred above a trench which had been excavated in bedrock and was therefore unrelated to the liquefaction of the hydraulic fill.

3.2 Collapse and deformation of drain line

Figure 5 shows a photograph, facing south, of the damaged pipe after the 1994 Northridge Earthquake: 76 meters collapsed laterally, and 23 meters deformed substantially. The excavations that unveiled the pipe were carefully supervised, and the digging operations were performed with extreme caution to avoid any additional damage to the pipe. In other words, the pipe deformation in Fig. 5 resulted solely from earthquake damage, not from digging operations. The deformed sections, which are drawn in Fig. 6, were carefully measured over the entire length of the pipe. They were sketched within an 8% accuracy. This level of accuracy was estimated by comparing the measured perimeters of the deformed and initial sections, and by assuming that the circumference remained equal to πD where D is the diameter of the undeformed pipe.

As shown in Fig. 6, the pipe sections deformed quite differently, although they were similar characteristics. The most common type of collapse was lateral. The largest lateral deformations were observed on segments 7 to 12, and the largest vertical deformation on segments 1, 2, 11, and 12. Segments 13 to 17 deformed laterally with deformation amplitude decreasing from over 60 cm in segment 13, to none in segments 16 and 17. Surprisingly, segment 6 and a part of segment 7 did not deform while adjacent segments completely failed laterally. Only the northern end of segment 7 completely failed. Its southern end was not deformed.

Figure 5. View facing South of the collapsed drain line.

Segments 11 and 12, close to the sinkhole, have complex deformation patterns. segment 11 collapsed in two different ways within its 7.3 m length. Its northern part was compressed laterally and folded over to the east. Its southern part folded in the opposite direction. Segment 12 had a 0.9-m long vertical tear, 2.3 m north from segment 11. The southern part of segment 12 was crushed laterally, pinched tight at its top, and bent over to the west. Its northern part deformed much less around segment 13.

As shown in Fig. 6, there were vertical tears in segments 2, 5, 7, and 12. These vertical tears were less than 1 m long from the top of the pipe and occurred at transitions between different deformation patterns. The tears on segments 2 and 12 took place along manufactured welded seams; no specific observations were made for segments 5 and 7.

Many of the deformation patterns were not continuous across the joint connection. At some joints (e.g., 5:6 and 10:11), the segments deformed completely differently, while at some other joints (e.g., 6:7 and 8:9), they deformed in nearly the same way. The largest difference occurred between segments 11 and 12, which folded over in opposite directions. Throughout the entire pipe length, the base remained semi-round, and left at least 30 cm of space for water to flow through. Many segments were crushed so tight that there was no space left between the walls.

4. ANALYSES

4.1 Causes of collapse

Among all the factors which may have contributed to the failure of the drain line, three factors were immediately excluded by visual observation and field measurement: (1) the degradation of pipe material properties due to corrosion, (2) the effects of axial deformation on the pipe buckling strength, and (3) the longitudinal propagation of transverse buckling. The segments which had a galvanized coating were not even rusted at the locations of local buckling and tears. Corrosion did not change the material properties of the pipe segments, and therefore was not a significant factor of failure. No axial deformation was observed in the pipe segments, or could be detected by measuring the length of each pipe segment and the corrugation spacing. Such an axial deformation, which could have resulted from the upstream berm rubbing against the corrugated pipe as it moved due to liquefaction, would have decreased the flexural stiffness and the buckling resistance. There may have also been a longitudinal propagation of transverse buckling on each segment, but not along the complete length of the drain line which was made of loosely connected segments. In the present analysis, the longitudinal propagation was considered a secondary effect which took place after the onset of transverse buckling and only occurred within limited portions of the pipe (i.e., segments 3 to 5, and 7 to 12).

The remaining factors which may have influenced the performance of the soil-pipe system included (1) the settlement of the upstream berm during shaking, (2) the pore pressure buildup in the embedding materials, (3) the liquefaction of the hydraulic fill, (4) the large peak ground acceleration and velocity, (5) the strain softening of embedding material, and (6) permanent lateral deformation of embedding material.

The most significant factors were difficult to single out based only on field investigation. Their effects may have been combined on some particular segments, and isolated with different intensity on other segments along the 100-m long drain line.

In view of the complexity of the problem, a simplified approach had to be adopted for determining the factors which most influenced the pipe buckling.

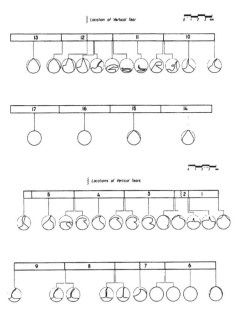

Figure 6. Observed deformed shapes of cross-sections of drain line.

4.2 Static analysis

A review of the design methods for flexible buried pipe is useful to examine the stability of the pipe under gravity loading prior to the earthquake. Moore (1989) thoroughly reviewed the application of buckling theories to buried pipes. He established that the maximum hoop force N in the pipe is the critical load parameter driving failure. In the present analysis, the static loads and the resulting hoop force N were estimated and compared to the critical hoop force N_{cr} required for buckling using the continuum model recommended by Moore (1989).

The approach consists of (1) calculating the hoop force N that results from external loads on the pipe, and (2) examining if the pipe can resist these loads by comparing N with the critical hoop force N_{cr}. The factor of safety FS against buckling is defined as:

$$FS = N_{cr}/N \qquad (1)$$

This approach is obviously simplified because it assumes that the determination of buckling and external loads are not coupled problems.

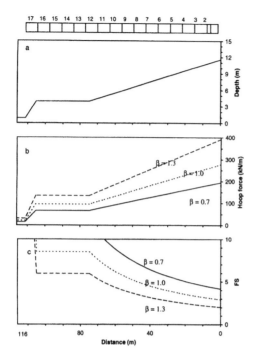

Figure 7. Results of static analysis: (a) burial depth along pipe showing location of pipe segments; (b) static hoop force along pipe for various arching factors; and (c) factors of safety for buckling.

The maximum static hoop force N (kN/m) can be simply evaluated from:

$$N = \frac{F_v}{2} \qquad (2)$$

where F_v is the vertical resultant force acting on the top of the pipe. The exact determination of the maximum hoop force in a flexible buried conduit is a difficult problem due to various factors including geometry of the trench, compaction effects, variation of material properties of the soils around the pipe, interaction between the soil and pipe, effect of sloping ground, and arching effects. Approximate methods such as the limit equilibrium method of Marston (Spangler and Handy, 1982; and Moser, 1990) or the elastic closed form solution of Burns and Richard (1964) give a first order approximation of the hoop force in the pipe. These methods can be used to calculate the load applied to the pipe in terms of a dimensionless arching factor β:

$$\beta = F_v / \gamma HD \qquad (3)$$

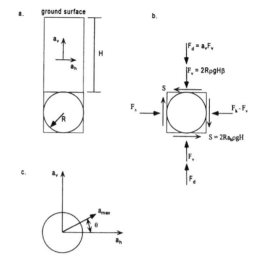

Figure 8. Pseudo-static analysis: (a) horizontal and vertical accelerations; (b) static and pseudo-static forces; and (c) maximum peak ground acceleration..

where γ (kN/m^3) is the soil unit weight, H (m) is the depth of fill over the pipe, and D (m) is the pipe diameter. Moser (1990) recommends using $\beta = 1$; however, depending on the method of analysis and assumptions made, β may be slightly higher or lower than 1 for flexible conduits. Figure 7b shows the variation of N along the pipe length for different values of β.

The critical hoop force N_{cr} (kN/m) is determined from a continuum elastic model for the soil around the pipe (Moore, 1989):

$$N_{cr} = 0.66\left(E_p I\right)^{1/3} \left(\frac{E_s}{1 - v_s^2}\right)^{2/3} \qquad (4)$$

where E_p (kPa) is the Young's modulus of the pipe, I (m^4/m) the moment of inertia of the pipe, E_s (kPa) the soil Young's modulus, and v_s the soil Poisson ratio. In this analysis E_p, I, E_s, and v_s were taken as 2×10^{11} kPa, 9.38×10^{-8} m^4/m, 9000 kPa, and 0.3, respectively.

The static factor of safety FS was calculated using Eq. 1 for different values of β. As shown in Fig. 7, N increases with depth of burial while FS decreases. FS remains greater than 2 for all values of β. FS is largest on the north end of the pipe, while it is the lowest on the south end.

In conclusion, under static conditions, the drain line was definitely stable along its complete length for all types of trench conditions. This result was anticipated because the drain line had never shown signs of distress prior to the 1994 Northridge earthquake.

4.3 Pseudo-static analysis

The design of buried pipes (e.g., Moser, 1990) rarely takes into account seismic factors, except for differential displacement across a fault, and transient and permanent axial deformations (Committee on Gas and Liquid Fuel Lifelines, 1984; Tawfik and O'Rourke, 1985; O'Rourke and Hmadi, 1988). To our knowledge, there are no existing design methods for analyzing the transverse buckling of buried flexible pipes under dynamic loads. Therefore, a pseudo-static analysis had to be developed to account for the combined effect of vertical and horizontal accelerations as shown in Fig. 8a. The proposed pseudo-static analysis consists of (1) finding the additional inertial forces, for which the hoop force N becomes equal to the critical hoop force N_{cr}, and (2) determining the reduction in N_{cr} for which buckling will result. Therefore, the analysis accounts not only for the increase in applied loads caused by the peak horizontal acceleration a_h and peak vertical acceleration a_v, but also for the possible reduction in N_{cr} during shaking.

4.3.1 Pseudo-static forces

Figure 8b shows the loads applied to the square element containing the pipe. When shear waves travel vertically upward, the shear force applied to the pipe is $S = 2\tau R$, as a result from the shear stresses $\tau = \rho\, a_h H$ at depth H where ρ is the soil unit mass. The peak vertical acceleration a_v is assumed to generate the additional vertical load F_d, which is related to F_v through:

$$F_d = \frac{a_v}{g} F_v \qquad (5)$$

The resultant static shear force F_v can also be written:

$$F_v = 2\rho g R H \beta \qquad (6)$$

where β is the dimensionless arching factor defined in Eq. 3. The resultant vertical force is $F_v + F_d$, the horizontal force is F_h, and the shear force is S. The equilibrium of these forces can graphically be represented in a similar way to the Mohr representation of stress. Therefore, one can determine the inclination of the square element for which there is no shear

force S and the normal force is maximum (i.e., $F = F_{max}$). This case corresponds to the maximum principal force:

$$F_{max} = \tfrac{1}{2}\left(F_v + F_d + F_h\right) + \sqrt{\tfrac{1}{4}\left(F_v + F_d - F_h\right)^2 + S^2} \qquad (7)$$

When the pipe fails under dynamic loads, F_{max} is equal to $2N^d_{cr}$, which implies that:

$$XFS = \frac{F_{max}}{F_v} = \tfrac{1}{2}\left(1 + \frac{F_h}{F_v} + \frac{a_v}{g}\right) + \sqrt{\tfrac{1}{4}\left(1 - \frac{F_h}{F_v} + \frac{a_v}{g}\right)^2 + \left(\frac{a_h}{\beta g}\right)^2} \qquad (8)$$

where

$$X = N^d_{cr}/N_{cr} \le 1 \qquad (9)$$

N^d_{cr} is the critical hoop force to cause buckling under earthquake loading, and N_{cr} is the critical hoop force for static conditions (i.e., Eq. 4). The factor of safety FS is for static conditions (i.e. Eq. 1). Equation 8 relates the vertical acceleration, horizontal acceleration, and ratio of initial static force F_h/F_v at the onset of buckling. Rearranging Eq. 8 in terms of the maximum acceleration a_{max} and assuming $F_h = F_v$ gives:

$$\frac{a_{max}}{g} = \frac{2(X\,FS - 1)}{\sin\theta + \sqrt{\sin^2\theta + \left(4\cos^2\theta\right)/\beta^2}} \qquad (10)$$

where $a_v = a_{max} \sin\theta$ and $a_h = a_{max}\cos\theta$ using the sign convention shown in Fig. 8. Spangler (1941) has shown $F_h = F_v$ is a valid assumption for flexible pipes.

In practice, the inclination θ of the ground acceleration is difficult to determine. However, there is a critical inclination corresponding to a minimum value of a_{max}:

$$\theta = \sin^{-1}\left(\frac{\pm\beta}{\sqrt{4 - \beta^2}}\right) \qquad (11)$$

The corresponding minimum of a_{max} is :

$$\frac{a_{max}}{g} = \frac{\beta}{2}\sqrt{4 - \beta^2}\left(XFS - 1\right) \qquad (12)$$

Equation 12 defines the ground acceleration a_{max} which is required to buckle a pipe having a static

169

safety factor FS. The value of a_{max} increases linearly with the static safety factor FS. The slope of this linear relation depends on β and X.

X is a dimensionless factor relating the buckling capacity in static and dynamic analyses. X can be evaluated by using Eqs. 4 and 9.

$$X = \left(\frac{E}{E_s}\right)^{2/3} \tag{13}$$

where E is the soil Young's modulus in pseudo-static analysis, and E_s is the static value used in Eq. 4. Equation 13 assumes Poisson's ratio of embedding soils is the same under static and dynamic conditions, but that its Young's modulus may differ. When $X = 1$, E is equal to E_s, and $X < 1$ when $E < E_s$. Figure 9 shows the maximum acceleration a_{max} required to obtain buckling for $\beta = 1$ and five values of X ranging from 1 to 0.2. For a given value of β and X, the pipe is stable when the point (FS, a_{max}) is below the straight line defined in Eq. 12, and unstable above it. Figure 9 also shows the range of accelerations measured in the pipe vicinity and the static values of FS presented in Fig. 7. Based on Fig. 9, it is possible to conclude that only segment 1 could have failed for $X = 1$ during the Northridge Earthquake, assuming it was subjected to the motions recorded at Station 1. For the other segments to fail, a reduction in E must have taken place during dynamic loading (i.e., X<1). This reduction may have been caused by the buildup of pore pressure in the embedding material, without the need for a complete liquefaction (X = 0).

The present pseudo-static analysis assumes that the buckling of flexible buried pipes is caused by two factors: the peak ground acceleration and the stiffness reduction of embedding soil. According to this theory (e.g., Fig. 9), the pipes which have a static factor of safety larger than 2 are unlikely to buckle without a stiffness reduction. This decrease in stiffness is relative to the static (secant) soil modulus, which is smaller than the soil modulus obtained from shear wave velocity (e.g., Bardet, 1997). Therefore, this simplified theory implies that the decrease of shear modulus with shear strain amplitude during transient shaking is a less significant factor to buckling than the decrease of confining pressure which accompanies pore pressure buildup in saturated soils. This simplified theory clearly predicts that a complete liquefaction, which is associated with a negligible soil stiffness, causes buckling. However, it also points out that buckling can take place without a complete liquefaction. The present analysis is simplified. There is a definite need to analyze this case history with nonlinear finite element methods and constitutive models for soil.

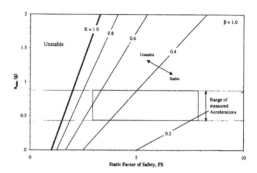

Figure 9. Acceleration required to cause buckling for given static factor of safety and various modulus reduction parameters X.

5. CONCLUSION

Based on the case history of the transverse failure of a 2.4 m diameter flexible pipe of the Lower San Fernando Dam during the 1994 Northridge earthquake, a simplified pseudo-static method of analysis has been proposed to identify the main causes for the transverse failure of buried flexible pipe during earthquakes. This case study demonstrates that flexible buried pipes are not as immune to the effects of earthquake ground motion as previously thought, but can partially or completely collapse during earthquakes. The simplified pseudo-static analysis suggests that the main reason for the transverse failure of buried pipes is the reduction in soil stiffness caused by pore pressure build up, which does not require a complete liquefaction. This simplified method of analysis is useful in geotechnical and pipeline engineering to assess the stability of flexible buried pipes during earthquake ground motions.

ACKNOWLEDGMENTS

The authors thank R. B. Saul and C. C. Plumb for assisting in geological interpretations, J. C. Chen for drafting figures, and M. Harris for proofreading the manuscript. The financial support of the National Science Foundation (grants CMS 9416237 and CMS 9414222) and the contributions of the Los Angeles Department of Water and Power are acknowledged.

REFERENCES

Bardet, J. P., 1997, 'Experimental Soil Mechanics," *Prentice-Hall*, Upper Saddle River, N. J., pp. 277-295.

Bardet, J. P. and C. A. Davis, 1996a, "Engineering Observations on Ground Motion at the Van Norman Complex after the Northridge Earthquake," *Bull. Seism. Soc. Am.*, Vol. 86, No. 1B, pp. S333-S349.

Bardet, J. P., and C. A. Davis, 1996b, "Performance of San Fernando Dams during the 1994 Northridge Earthquake," *Journal of the Geotechnical Engineering Division*, ASCE, Vol. 122, No. 7, pp. 554-564.

Bardet, J. P. and C. A. Davis, 1995. "Lower San Fernando Corrugated Metal Pipe Failure," M. O'Rourke ed., *Proc. Fourth U.S. Conf. on Lifeline Earthquake Engineering*, ASCE, San Francisco, Aug., pp. 644-651.

Burns, J. Q., and R. M. Richard, 1964, "Attenuation of Stress for Buried Cylinders," *Proceedings of the Symposium on Soil Structure Interaction*, University of Arizona, Tucson, Arizona, pp. 378-392.

Committee on Gas and liquid Fuel Lifelines, 1984, "Guidelines for the Seismic Design of Oil and Gas Pipeline Systems," *ASCE*.

Datta, S. K., A. H. Shah, and K. C. Wong, 1984, "Dynamic Stresses and Displacements in Buried Pipe," *Journal of Engineering Mechanics, ASCE*, pp. 1451-1465.

Davis, C. A., and J. P. Bardet, 1996, "Performance of Four Corrugated Metal Pipes during the 1994 Northridge Earthquake" *Proceedings of the 6th Japan-US Workshop on Earthquake Resistant Design of Lifeline Facilities and Countermeasures against Soil Liquefaction*, Tokyo, June, in press.

Iida, H., T. Hiroto, N. Yoshida, and M. Iwafuji, 1996, "Damage to Daikai Subway Station," Soils and Foundations, Special Issue on Geotechnical Aspects of the January 17 1995 Hyogoken-Nambu Earthquake, *Japanese Geotechnical Society*, pp. 283-300.

LADWP, 1975, "Final Construction Report on the Lower San Fernando Dam Alterations" Los Angeles Department of Water and Power, Water Engineering Design Division, Report No. AX-215-42.

Moore, I. D., 1989, "Elastic Buckling of Buried Flexible Tubes - A review of Theory and Experiment," *Journal of Geotechnical Engineering*, ASCE, Vol. 115, No. 3, pp. 340-358.

Moore, I. D., and R. W. Brachman, 1994, "Three-Dimensional Analysis of Flexible Circular Culverts," *Journal of Geotechnical Engineering*, ASCE, Vol. 120, No. 10, pp. 1829-1844.

Moser, A. P., 1990, "Buried Pipe Design," *McGraw Hill*, New York.

O'Rourke, M. J., and K. E. Hmadi, 1988, "Analysis of Continuous Buried pipelines for Seismic Wave Effects," *Earthquake Engineering and Structural Dynamics*, Vol. 16, 917-929.

Seed, H. B, K. L. Lee, I. M. Idriss, and F. I. Makdisi, 1973, "Analysis of the Slides in the San Fernando Dams During the Earthquake of February 9, 1971," *Report No. UCB/EERC 73-2* University of California, Berkeley, California.

Spangler, M. G., 1941, "The Structural Design of Flexible Pipe Culverts," Bul. 153, *Iowa Engineering Experiment Station*, Ames, Iowa.

Spangler, M. G., and R. L. Handy, 1982, "Soil Engineering," Fourth ed., *Harper and Row*, New York.

Tawfik, M. S., and T. D. O'Rourke, 1985, "Load-Carrying Capacity of Welded Slip Joints," *Journal of Pressure Vessel Technology*, Vol. 107, pp. 36-43.

Youd, T. L., and C. J. Beckman, 1996, "Highway Culvert Performance During Earthquakes," *National Center for Earthquake Engineering Research Technical Report*, in press.

171

Seismic Behaviour of Ground and Geotechnical Structures, Sêco e Pinto (ed.)© 1997 Balkema, Rotterdam, ISBN 90 5410 887 8

Lateral-spreading loads on a piled bridge foundation

J.B. Berrill, S.A. Christensen, R.J. Keenan & W.Okada
Department of Civil Engineering, University of Canterbury, Christchurch, New Zealand

J.R. Pettinga
Department of Geological Sciences, University of Canterbury, Christchurch, New Zealand

ABSTRACT: The effect of lateral spreading on the piled foundations of a multispan highway bridge is studied. It is found that by far the greatest lateral load imposed on the bridge substructure comes from the 1.5 m thick unliquefied crust that is carried towards the river on liquefied sand. The buried piers and raked piles resist its displacement, inducing passive failure in this crustal layer. Trenching at two of the piers on the floodplain of the left bank reveals failure surfaces in the crustal soil consistent with passive failure. Soil strengths were measured by *in situ* direct shear testing and by the CPT. In this case, the passive load on the buried portion of the slab piers is estimated at 850 to 1000 kN per pier, compared with roughly 50 kN in drag forces between the liquefied sand and the set of eight, 400 mm square raked piles per pier. The collapse load of the foundation system is estimated to be about 950 to 1150 kN. Thus the load imposed by the unliquefied crust was very close to the ultimate capacity of foundations designed to the standards of the 1960s. The main conclusion is that in lateral spreading, the chief threat to such piled foundations comes from loads imposed by the unliquefied crust, not from the drag forces of the liquefied soil. In this case, clear evidence was found of passive failure as the crust drove against the buried pier, piled through to firm ground.

INTRODUCTION

Lateral spreading of river banks during earthquakes has caused great damage to bridges (Hamada and O'Rourke, 1992; Youd and Hoose, 1978) yet the loading mechanism is not completely understood. Japanese researchers (Tokida *et al.*, 1993; Vargas and Towhata, 1995) have studied drag forces exerted on piles by liquefied soil, and have found that the forces are often small. Another source of loading induced by lateral spreading may arise from a crust of unliquefied soil overlying the liquefied layer as it is driven against buried foundations. This second sort of loading is the principal subject of this investigation.

In most cases of lateral spreading a layer of soil at the ground surface does not liquefy, either because it lies above the watertable or because it is too fine-grained for liquefaction, and this stratum is carried along on the underlying, spreading sand. Where the displacement of this unliquefied crust is resisted by buried foundations, large forces may be generated. In the limit, these forces equal the passive resistance of the unliquefied soil.

This article describes a case study of the effect of lateral spreading on a piled bridge foundation during the 1987 $M_L6.3$ Edgecumbe, New Zealand earthquake. We present evidence of passive failure in the 1.5 m thick unliquefied crust as it was driven against the piers of a multi-span highway bridge. Our main conclusion is that *passive pressure is the correct design load* for the unliquefied part of the soil profile in the presence of lateral spreading. Also by taking the difference between the collapse load of the foundation and estimates of the passive load, we are able to infer that the drag force exerted by the liquefied soil on the buried piles is indeed small, as found by Tokida *et al.*, (1993), Vargas and Towhata (1995) and others.

The case study concerns the Landing Road Bridge, a 13-span highway bridge, across the Whakatane River at the town of Whakatane on the east coast of the North Island of New Zealand (Figure 1). The bridge (Figure 2) crosses the river near the apex of a recent meander. Liquefaction occurred through a strip of land about 300 m wide on the internal, left, bank of the bend. In the free field, about 1.5 m of lateral movement was observed towards the river channel. At the time of the earthquake mounds of soil formed behind the four bridge piers on the left bank, suggestive of passive failure in the crust of unliquefied soil above the watertable. Subsequent

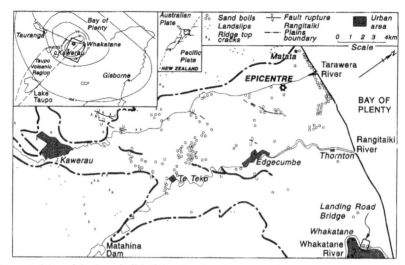

Figure 1 Map of the Rangitaiki Plains showing the ground damage caused by the 1987 Edgecumbe earthquake. Large inset shows Bay of Plenty region and isoseismals from the earthquake.

trenching found failure surfaces confirming the occurrence of a passive failure.

We shall begin by briefly describing the Edgecumbe Earthquake and the response of the bridge itself, then continue with a description of our investigation of the site as a whole and of detailed observations at the piers. These include trenching and *in situ* shear strength measurements. The magnitude of the passive forces is then estimated and compared with loads required for cracking of the precast concrete piles, which was observed in the abutment piles but not in the piles of the internal piers. We conclude that the passive loads are large (of the order of 1 MN per pier), and constitute an important design consideration.

THE 1987 EDGECUMBE EARTHQUAKE

The rupture of 2 March 1987 occurred on a normal fault beneath the Rangitaiki Plains near the north eastern end of the Taupo Volcanic Zone, associated with the boundary between the Pacific and Indian plates, in the North Island of New Zealand. Evidence of liquefaction was seen throughout the Plains (Figure 1) generally near present or recent former river and stream channels (Franks *et al.*, 1989). At least six bridges, ranging in size from a single-span farm bridge to the 240 m long Landing Road Bridge, suffered damage, mostly minor, from convergence of river banks due to lateral spreading.

Figure 1 shows the epicentre near the mouth of the Tarawera River, 17 km from the Landing Road site. Seismological studies show that rupture propagated up and to the south from a hypocentre at a depth of 8 km, to break the ground surface on a pre-existing south-westerly striking fault, running from about 1 km east of the town of Edgecumbe towards the town of Kawerau. The closest point on the surface rupture is 8 km from the site. There was no strong motion accelerograph in Whakatane; the nearest was at the Matahina Dam, due south of the epicentre and also about 8 km from the nearest surface breakage. The instrument at the base of the 80 m high rockfill dam recorded a peak acceleration of 0.33 g. Due to rupture directivity effects, the motion at the bridge site was probably weaker than this. Lowry assigned an intensity of MM to Whakatane (Christensen, 1995).

THE LANDING ROAD BRIDGE AND SITE

The Landing Road Bridge carries State Highway 2 over the Whakatane River. The bridge was constructed in 1962 and is of a standard design used widely in New Zealand. It comprises 13 simply supported spans 18.3 m long, carrying a two-lane concrete deck and two footpaths. The superstructure consists of 5 precast post-tensioned concrete I-beams, bearing on 16 mm rubber pads. The spans are interlinked with bolts through diaphragms over the

Figure 2 The Landing Road Bridge, looking north east to the mouth of the Whakatane River.

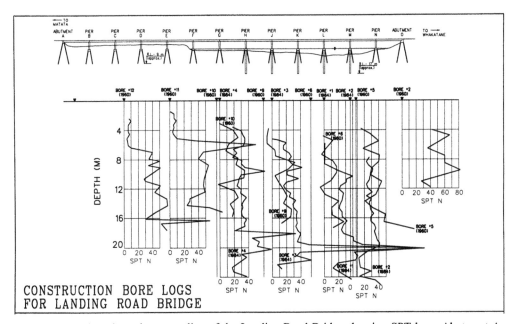

Figure 3 Section along the centre-line of the Landing Road Bridge showing SPT logs. Abutment A is at the north end of the bridge.

piers and through the abutment backwalls. The beams are tied down using holding bolts at all piers and abutments, forming a quite stiff monolithic structure. The substructure comprises concrete slab piers running the full width of the superstructure and supported by eight 406 mm square raked prestressed concrete piles. The abutments are also supported by 406 mm square raked piles, 5 piles on the river side and 3 on the approach side. The abutment backwall is tight-packed and bolted to the beam diaphragm. There are no approach slabs. Five river piers were additionally underpinned with two 1.1 m diameter concrete cylinders each, around 1985 after one pier had been undermined by flooding. The underpinned piers (piers H to M) can be seen in Figure 3.

Figure 4 Aerial view looking towards the north-west of the Landing Road Bridge site. The Landing Road Bridge can be seen in the left foreground. Ground cracking can be seen on the berm beside the bridge and in the fields beside State Highway 2.

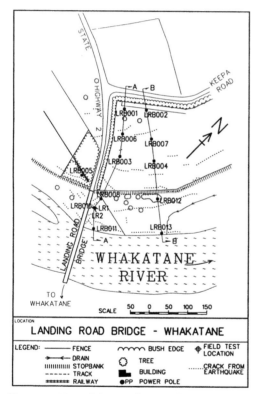

Figure 5 Map of the Landing Road Bridge site showing lateral spreading cracks and the location of *in situ* tests and cross section ties.

Comparison of early and present maps shows that during the past 140 years the river has migrated about 300 m towards the south, leaving behind loose sands and silts to a depth of about 6 m, forming a classical point-bar structure. It is these soils that liquefied. The bridge piles were driven into denser sands and gravels below that depth. These older and denser soils are also found extending to the ground surface on the right bank at the south abutment. Gross features of the soil profile are seen in the SPT logs reproduced in Figure 3. The watertable on the left bank is at about 1.5 to 2 m below ground level, depending on the tidal river level. We note that Tinsley and Dupré (1992) have observed from Californian earthquakes that laterally-accreted geologic structures, such as point bars, are more susceptible to liquefaction than other common geologic structures.

OBSERVATIONS FOLLOWING THE 1987 EARTHQUAKE

Extensive liquefaction, marked by lateral spreading and ejection of sand, occurred around the bridge and for a few hundred metres downstream on the northern (true left) bank of the river. Cracks associated with the lateral spreading show clearly in Figure 4. The cracks tended to be parallel to the river channel except immediately beside the bridge where they swung around to meet the bridge at approximately 45°. Near the bridge, there were about five major crack sequences (Figure 5) which were in excess of 200 mm wide in a strip extending back about 300 m wide from the true left bank of the river. There was no observed evidence of liquefaction on the true right bank of the river.

Particle size distributions of 13 samples of ejecta taken from upstream and downstream beside the bridge had D_{50}'s ranging from 0.2 to 0.5 mm. Some of the ejecta contained small amounts of pumice particles, while others did not.

The stopbank, seen in Figure 4 running through the bridge abutment, had longitudinal cracks which were in excess of 100 m in length in places. Intensive cracking and subsidence occurred in the north-west road embankment, from the abutment to beyond the intersection of SH2 with Keepa Road. An eyewitness states that this approach road was passable immediately after the main earthquake, but one hour later it was no longer passable by car. This suggests that failure took place under static conditions, presumably as a front of pore pressure migrated upwards into the crustal stratum above the watertable and so weakened it that it could no longer support the 4 m high roadway embankment.

At the piers on the left bank (piers B through E in Figure 3), soil had mounded up (Figure 6) on the landward sides of the piers and gaps of up to 600 mm formed on the river side, suggesting passive failures in the soil crust above the watertable. The bridge superstructure did not undergo any significant distress. Compression along the axis of the bridge was indicated in one instance by the buckling of a pair of concrete footpath slabs.

Excavation to about one metre at the northern abutment showed that the front raked piles were cracked on the river side. These cracks extended through 75% of the width of the piles (Figure 7). There was no indication of cracking on the other side of the piles. Piles beneath the internal piers were not inspected at the time, since the pile caps were both buried and below the watertable. Soil at the northern abutment settled 300 to 500 mm, exposing the piles. At the time of the earthquake this abutment was thought to have rotated (pers. comm. L McCallin). This observation was supported in 1994 with the measurement of ½° rotation of the bottom of the abutment face towards the river. The south-eastern abutment was not inspected after the earthquake. In 1992 the authors found that the tops of the first two piers from the northern abutment were leaning towards the river by about 1°, while the remaining piers appeared vertical.

Horizontal cracks near the base of piers H and J, (which are the first two piers additionally underpinned from the left bank of the river), were not noticed by the highway authority until some years after the earthquake but were considered to

Figure 6 Pier C, looking upstream. Note the mound of soil behind pier apparent in background.

Figure 7 Front raked pile of the western abutment of Landing Road Bridge showing settlement of the ground

177

Figure 8 Horizontal cracks through piers H and J of Landing Road Bridge. These cracks have been repaired with epoxy resin. The outriggers constructed to tie the piers to the piles added in 1985 are evident in this photograph.

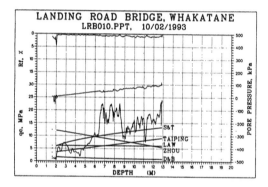

Figure 9 Log of probe LRB010.PPT, with various liquefaction criteria for 2 March 1987 Edgecumbe earthquake.

have occurred as a result of it. This cracking is consistent with the lateral-spreading loads being transmitted both downwards through the raked pile system and upwards through the slab piers to the superstructure. The superstructure would thus carry a compressive load through to the fixed south abutment, constructed in the older and denser sands and gravels which had not been affected by migration of the river channel. Some of this compressive load, originating from the left-bank piers, would be shed to the strengthened piers in mid-stream due to their greater stiffness.

INVESTIGATIONS

Investigations have been carried out at the site over successive summers, first with rotary boring and piezocone probes (Christensen, 1995), trenching at piers C and D (Keenan, 1996) and finally, *in situ* shear strength testing at Pier C (Okada, 1997).

General Investigation

In 1993, five cone penetration tests (CPT), eight piezocone (CPTU) probes and two rotary borings were carried out at the locations shown in Figure 5 at the site. The log of piezocone probe LRB010 is shown in Figure 9, overlain with liquefaction criteria from several authors. Clearly these prediction procedures are highly inconsistent and of little help, at least for the near-field condition of this site, in determining which layer liquefied. Another approach is to compare the particle size distribution (PSD) of ejected soil with that of samples recovered from the borings. However, here this method was not reliable since a precise record was not kept of exactly where samples were taken in 1987. In the event, the most useful evidence came from the comparison of grain sizes in a close up photograph of a sand boil taken in 1987 near pier C with those of samples retrieved from a rotary boring at the same location. This indicated that the loose clean sand strata shown liquefied in the earthquake. shaded in Figure 10, with q_c up to about 8 MPa.

Christensen (1995) used photogrammetry to measure settlements due to the liquefaction and lateral spreading. Ground settlement throughout the liquefied region was found to be of the order of 400 mm, which is consistent with the settlement measured directly at the northern abutment (300-500 mm) shortly after the earthquake in 1987. Liquefaction-induced settlements for probe sites LRB010.PPT and LRB012.PPT were estimated using the procedure of Ishihara (1993) at 170 and 310 mm respectively. These are of the same general order as the average value of 400 mm measured photogrammatically.

Estimation of horizontal displacements was also attempted using photogrammetry but could not be accomplished with any accuracy due to the lack of good markers on the generally featureless floodplain. Using a rough estimate of crack widths, seen in a number of terrestrial photographs from 1987, a total horizontal displacement at the river bank of 1.5 to 2.0 m was estimated. This range compares well with predictions by Hamada's method, outlined in Barlett

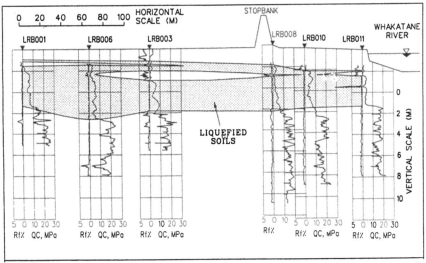

Figure 10 Cross Section AA at Landing Road Bridge showing the estimated liquefied strata. Cross Section BB is generally similar, with the loose sand layer slightly thicker.

and Youd (1992, 1995), which gives an estimate of between 1.2 and 2.0 m for horizontal displacement of the soils along cross section B-B. (We do not have sufficient information to apply the method of Bartlett and Youd itself.)

At this stage, rough calculations were made of the passive load on the buried portion of the piers and pile caps, using soil strengths inferred from cone resistances measured 20 to 30 m from the bridge. These gave loads of the order of 0.5 to 1.0 MN per pier, close to the preliminary estimates of the collapse load of the pile-pier system. This prompted further investigation, carried out by Keenan (1996), who made a more precise estimation of the structural strength of the foundation system and who excavated trenches at piers C and D to look for evidence of passive failure of the soil and to inspect the tops of the piles for damage.

Structural Analysis of the Foundation System

Using the construction plans drawn in 1960, the strength of the pile/pier system was calculated. The superstructure is very much stiffer and stronger than the substructure, making it a redundant component in the structural system for longitudinal loads. The potential collapse mechanism of the substructure is shown in Figure 11. The total horizontal force required to initiate collapse of the substructure was calculated to be 1000–1100 kN per pier. Further details of the structural analysis may be found in Keenan (1996).

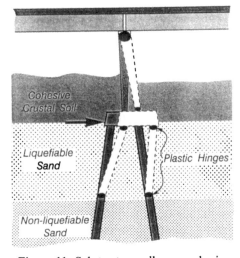

Figure 11 Substructure collapse mechanism.

Since this force was of the same order as the rough estimates of the passive soil load, trenching was undertaken to inspect the tops of the piles for cracking. Since the piers were still very near vertical, it was clear that a full collapse mechanism had not developed, but that it was possible that cracking had started at potential plastic hinges. The most critical point is at the top inside face of the northern piles (north is to the left in Figure 11); the next, the outside face at the top of the southern row of piles.

179

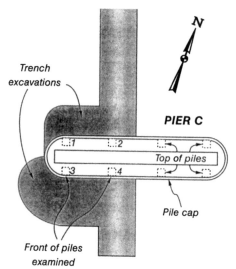

Trench
excavations

PIER C

Top of piles

Pile cap

Front of piles
examined

Figure 12 Plan of excavation at pier C.

Trenching

Since piers C and E had the most prominent mounding of soil behind them, they were selected for investigation. Trenching at both piers told the same story. Here, we will focus on the trench at pier C. Trenches approximately 1.5 m wide were excavated on each side of the pier, with the east face of the trench positioned along the centreline of the bridge, as shown in Figure 12. This enabled close examination of the passive failure surface on the north side of the pier and of lateral spreading cracks in the soil on the south side. Face logs for these trenches are shown in Figure 13.

Soil mounding was evident for about 2.5 m back from the north of the pier face. The elevation difference of the ground surface from one side of the pier to the other was about 450 mm. Upper laminated silt horizons exhibited curvature in the heaved soil zone, clearly showing that the crustal soil had been forced upward. A shear surface could be seen clearly in the sandy silt near the ground surface, and was traced downwards towards the front face of the pile cap, as shown in Figure 13. This surface may have been related to excavation for construction of the pile cap. Other shear surfaces were found running upwards from the intersection of the pier and pilecap, at a slope of 20 to 30°. This slope together with the sharp offset of a lens of sandy silt (material B) suggest that these inner discontinuities were more likely to be the 1987 failure surfaces. The clear mounding of the ground surface and the presence of

the shear surfaces leaves little doubt that passive failure occurred.

After logging of the soil profile, the excavation was deepened and widened in an effort to examine the inner faces of the two upstream piles at their interface with the pilecap (piles 1 and 2 in Figure 12). Since the excavation had now passed below the watertable and into loose, clean sands this proved to be very difficult to accomplish, despite heavy pumping from the excavation. However, the upper 150 mm or so of the front face of piles 1 and 2 could be inspected and they appeared to be undamaged. No cracks could be felt along the front, north face of these piles, nor on the east and west faces through about 50 percent of their width. The inner faces of the piles could not be reached for inspection. Thus it is certain that there was no concrete crushing on the landward face which would have been in compression; however nothing can be said about the inner face of these piles, which would be the first site of tensile cracking in the foundation system.

Trenching south of pier C revealed two vertical lateral spreading cracks in the crustal soil, as shown in the trench log (Figure 13). The fissure closer to the pier was followed by extending the excavation to the west. At the end of the pile cap, it swung to the north and joined up with one of the large lateral-spreading cracks which make an angle of 45° with the length of the bridge and which are still evident as surface depressions.

Soil Properties

For many years, the nearby cardboard mill had been disposing of surplus wood and bark chips on the banks of the Whakatane River, and these had become mixed in with the upper 1 to 2 metres of soil. Although the wood chips comprised only a small percentage of the soil constituents, they made conventional sampling and testing of the generally silty crustal soil difficult. Blocks of the silty material were taken back to the laboratory, but because of the woodchips it was impossible to cut undisturbed samples for triaxial or direct shear strength testing. Therefore, a large *in situ* direct shear testing device was constructed, with a sample area of 0.1 m², and tests carried out in place at the site. Six tests were made at various levels within the failure mass at pier C, down to the level of the top of the pile cap, near the silt-sand interface which coincided more or less with the high-tide watertable. (Okada, 1997)

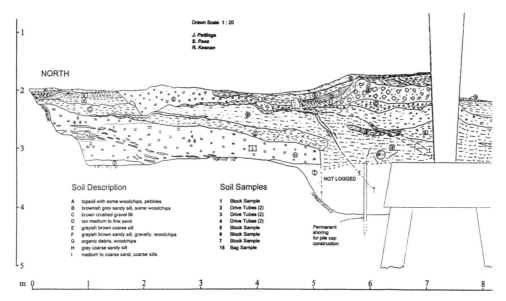

Figure 13 Trench log, pier C, looking downstream

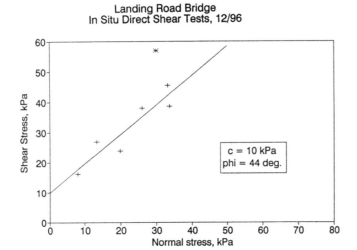

Figure 14 Results of large-scale *in situ* direct shear tests at pier C.

The results are shown in Figure 14. Omitting test 1, which was made in material E (Figure 13) which was not close to the shear surface, the results are fairly consistent, and yield values of c = 10 kPa and ϕ = 44°.

The unit weight of the silty soil was measured at γ = 12 kN/m³. Visual inspection of the soil from a large block sample suggested that the unnaturally low value was due more to the presence of the woodchips rather than the small amount of pumice present.

Estimates of Passive Load

Using the values of c = 10 kPa, ϕ = 44° and γ = 12 kN/m³ in a two-dimensional Rankine analysis yields a passive force of 850 kN for the 8 m wide pier. With the same soil values, a Coulomb analysis gives a total passive force per pier of about 1000 kN, depending on the assumption of wall friction angle.

181

DISCUSSION

Since the bridge is still standing, very close to plumb, and since no major cracking was found at the base of the pier nor at the top of the riverward piles, it is clear that a collapse mechanism did not form. However, with the collapse load estimated to be between 950 and 1130 kN and the imposed load by the unliquefied soil crust somewhere around 850 to 1000 kN, it is likely that collapse was only narrowly avoided. That the foundation did not collapse indicates that the additional lateral load imposed by drag of the liquefied sand against the piles was small. Using the results of Vargas and Towhata (1995), Keenan (1996) estimated a drag force of about 50 kN per pier. The behaviour of the Landing Road Bridge confirms that the drag forces are small compared with those imposed by the unliquefied crust.

Unliquefied crust will be present wherever there is granular material above the watertable, or with cohesive soil regardless of the position of the watertable. In these circumstances, where lateral spreading is likely, passive loads from this unliquefied crust constitute an important design loading.

CONCLUSIONS

The Landing Road Bridge provides a useful case study for a number of aspects of liquefaction and lateral spreading. In particular:

1. Liquefaction occurred in the very recent laterally-accreted loose, clean sands of a point bar which has developed in the Whakatane River over the past 140 years.

2. Average settlement of the surface of the liquefied ground was abut 400 mm. This is of the same order as the predictions of 170 and 310 mm made at CPT probe sites using the method of Ishihara (1993). If allowance were made for the contribution of lateral spreading, not included in the Ishihara procedure, the agreement would be closer.

3. Foundations on piles, passing through liquefied soil to firm ground may attract large loads from lateral spreading. Drag forces exerted by the liquefied soil are apparently quite modest, whereas those imposed by an overlying unliquefied layer can be large, in the limit corresponding to passive failure in the unliquefied crust.

4. At Landing Road, lateral spreading towards the river channel was impeded by the northern four or five piers of the bridge as the buried raked-pile foundations resisted the lateral spreading of the upper 6 m of the soil towards the river channel. Trenching at two piers found clear evidence of passive failure of the unliquefied crust overlying the liquefied sand layer.

5. Using strength parameters obtained from *insitu* direct shear tests, the passive load was estimated at about 1 MN per pier, compared with about 50 kN per pier in drag forces between liquefied sand and the piles. Although these values depend on the particular stratigraphy and structural form, in this case they illustrate the importance of loads imposed by the unliquefied crust. Clearly, these forces constitute a major design consideration.

6. The collapse load of the substructure was calculated to be about 1000 to 1100 kN. Had the unliquefied crust been thicker, the passive failure force would have been greater, and it is likely that the bridge would have failed. Since many bridges were built to similar designs at that time, it is likely that some of them are vulnerable to lateral-spreading damage.

7. The presence of the wood chips in the river bank material suggests a possible remedy for existing (and future) vulnerable bridges; that is, to put in a crushing zone of weak material behind piers likely to be affected by lateral spreading.

ACKNOWLEDGEMENTS

The authors wish to acknowledge the assistance received from staff of the Regional and District Councils and from Messrs Howard Chapman and David Jennings of Works Corporation, as well as the help of Messrs Pasa and Van Dyk and Drs K.J. McManus and R.O. Davis of the University of Canterbury in the field work and Mr Des Bull in estimating pile capacities. Financial assistance from the New Zealand Earthquake Commission and from the University of Canterbury research fund is gratefully acknowledged.

REFERENCES

Bartlett, S.F. and T.L. Youd, (1992) "Empirical Prediction of Lateral Spread Displacement", *Proc. 4th Japan-US Workshop on Earthquake Resistance Design of Lifeline Facilities*, Honolulu, pp. 351-366.

Bartlett, S.F. and Youd, T.L. (1995), "Empirical Prediction of Liquefaction-Induced Lateral Spread", *J. of Geotechnical Engineering, ASCE*, Vol. 121, No. 4, pp. 316-329.

Christensen, S.A. (1995), "Liquefaction of Cohesionless Soils in the March 2, 1987 Edgecumbe Earthquake, Bay of Plenty, New Zealand, and Other Earthquakes", Master of Engineering Thesis, University of Canterbury, Christchurch, New Zealand, 373p.

Franks, C.A., R.D. Beetham and G.A. Salt, (1989), "Ground Damage and Seismic Response Resulting from the 1987 Edgecumbe Earthquake, NZ", *NZ J. of Geol. & Geophysics*, Vol. 32, No. 1, pp. 135-144.

Hamada, M. and T.D. O'Rourke, (1992), "Case Studies of Liquefaction and Lifeline Performance During Past Earthquakes", Technical Report NCEER-92-0001, Vol. 1, National Center for Earthquake Engineering Research, New York.

Ishihara, K., (1993) "Liquefaction and Flow Failure During Earthquakes", (33rd Rankine lecture) *Géotechnique*, Vol. 43, pp. 351-415.

Iwasaki, T., (1986) "Soil Liquefaction Studies in Japan: State of the Art", *Soil Dynamics and Earthquake Engineering*, Vol. 5, pp. 2-68.

Keenan, R.J., (1996) "Foundation Loads due to Lateral Spreading at the Landing Road Bridge, Whakatane", *M.E. Report*, University of Canterbury, 137p.

O'Rourke, T.D. and M. Hamada, (1992), "Case Studies of Liquefaction and Lifeline Performance During Past Earthquakes", Technical Report NCEER-92-0002, Vol. 2, National Center for Earthquake Engineering Research, New York.

Tinsley, J.C. and W. Dupré, 1992, "Liquefaction Hazard Mapping, Depositional Faces, and Lateral Spreading Ground Failure in the Monterey Bay Area, Central California, During the 17 October 1989 Loma Prieta Earthquake", *Proc. 4th Japan-US Workshop on Earthquake Resistant Design of Lifeline Facilities and Countermeasures for Soil Liquefaction,* Honolulu, Vol. 1, pp. 71-86.

Tokida, K., H. Iwasaki, H. Matsumoto and T. Hamada, (1993) "Liquefaction Potential and Drag Force Acting on Piles in Flowing Soils", *Soil Dynamics and Earthquake Engineering*, Vol. 1, Elsevier, pp. 244-259.

Vargas, W. and I. Towhata, (1995), "Measurement of Drag Exerted by Liquefied Sand on Buried Pipe", *Proc. First International Conf. on Earthquake Geotechnical Engineering*, Tokyo, pp. 975-980.

Youd, T.L. and S.N. Hoose, (1978), "Historic Ground Failure in Northern California Associated with Earthquakes", *USGS Prof. Paper No. 993*, 177p.

Seismic Behaviour of Ground and Geotechnical Structures, Sêco e Pinto (ed.)© 1997 Balkema, Rotterdam, ISBN 90 5410 887 8

Liquefaction induced displacements

Peter M. Byrne & Michael Beaty
University of British Columbia, Vancouver, B.C., Canada

ABSTRACT

The key aspect controlling liquefaction induced displacements is the fundamental stress-strain response of the liquefied soil element. Cyclic stresses imposed by the earthquake cause the soil element to experience a large drop in stiffness and strength. A soil structure comprises a collection of such elements, and its response can be computed from the laws of mechanics once the element stress-strain behaviour is known. Characteristic liquefaction response based on cyclic element tests is reviewed. Empirical and mechanics based methods for estimating liquefaction induced displacements are also discussed. The empirical methods are considered to represent a database of field experience, and have been used to verify the mechanics based methods. Two case histories (Mochikoshi Dam No. 1 and La Marquesa Dam) are evaluated using the mechanics approach. Good agreement between the analyses and the field observations is obtained, both in terms of magnitude and pattern of displacement.

1 INTRODUCTION

Earthquake damage to soil-structures can be caused by inertia forces and by soil displacements, and these two effects are usually examined separately. If liquefaction is not triggered, soil displacements will generally be small unless there are steep slopes or weak soils, and any damage that occurs is due to inertia forces. If liquefaction is triggered the displacements may be very large due to the drop in strength and stiffness associated with liquefaction. These displacements have caused major damage to earth structures during past earthquakes, such as at Kobe, Japan in 1995.

Liquefaction induced displacements are particularly important in areas of moderate to high seismicity where structures such as dams, bridges, storage tanks and lifeline facilities are located on liquefaction-prone soils. The rational design of new structures as well the retrofit of existing structures in these areas requires reliable estimates of these displacements.

A flow slide will occur if the residual strength of the liquefied soil is not sufficient for stability. Even if a flow slide does not occur, displacements caused by lateral spreading may still be very large. It is these displacements, and our ability to predict them, that is the topic of this presentation.

Displacements, whether associated with liquefaction or not, depend on the fundamental stress-strain relations of the soil. Before examining methods for predicting such displacements and reviewing case history examples, it is important to briefly examine the characteristic liquefaction response.

2 CHARACTERISTIC RESPONSE

2.1 *Monotonic loading*

Loose sand under undrained monotonic loading can respond in a strain softening manner that is commonly referred to as liquefaction, as shown in Fig. 1(a). Here, the shear stress-strain relation shows a peak with a sharp drop to a significantly lower residual strength value. The stress path is shown in Fig. 1(b), where the pore pressure rise drives the stress path to lower effective normal stresses. The peak shear stress occurs at a

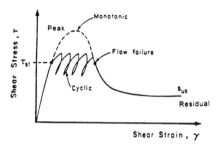

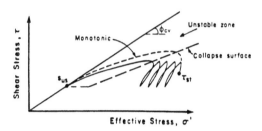

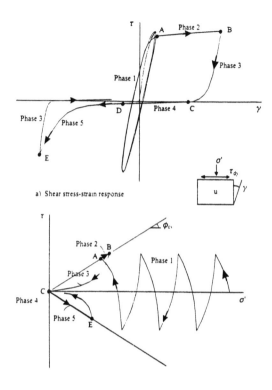

Fig. 1. Response of loose saturated sand under undrained static and cyclic loading.

a) Shear stress-strain response

Fig. 2. Characteristic undrained cyclic shear stress-strain and stress path response for sand.

developed friction angle, referred to as the collapse line, which is below the constant volume friction angle, ϕ_{cv}. The shear stress then drops until it achieves its residual value at the ϕ_{cv} line.

If the material is densified, the unstable zone shown in Fig. 1(b) moves up and to the right, and strain softening under undrained loading no longer occurs at lower stress levels.

2.2 Cyclic Loading

Under cyclic loading the undrained shear stress-strain response of sand remains stiff for a number of cycles and then abruptly changes to a dramatically softer response. This softer response is caused by a rise in porewater pressure during each cycle of load which drives the stress path towards the ϕ_{cv} line or the collapse line. This represents the pre-triggering response or phase 1 as shown on Fig. 2. Once the ϕ_{cv} line is reached, large deformations may occur and the stress path moves up the ϕ_{cv} line (actually slightly above ϕ_{cv}), depicted as phase 2. Upon unloading, phase 3, high pore pressures occur that lead to a zero effective stress state (point C, Fig. 2). Very large strains may then occur at virtually zero shear stress. This state is true liquefaction as the material deforms as a fluid. With further strain, phase 5, the soil skeleton dilates, the pore pressure drops and the stress point moves up the ϕ_{cv} line

again. Upon reversal, the procedure repeats itself with the strain loops getting larger with each cycle.

This behaviour represents level ground conditions where there is no static driving force on the horizontal plane, and is similar to that observed (Fig. 3) at the Wildlife Site in California during a liquefaction event in 1987 (Byrne and McIntyre, 1994). Figure 3 was developed from acceleration recordings above and below the liquefied layer, and essentially represents an in-situ shear test. The average strain in the liquefied layer was obtained by double integrating the acceleration records. The corresponding shear stress is directly proportional to the acceleration of the overlying soil. This field data also demonstrates that strains tend to be small for level ground conditions.

If the applied loading is not symmetric, then the post-liquefaction response is also unsymmetric with strains accumulating in the direction of the load as shown in Fig. 4. It is this behaviour that is responsible for the large liquefaction-induced displacements that can be so damaging to earth structures and is termed lateral spreading.

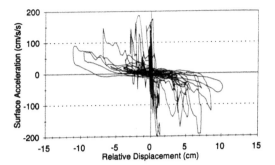

Fig. 3. Wildlife Site. Surface acceleration vs. relative displacement, N-S component, from 1987 Superstition Hills earthquake.

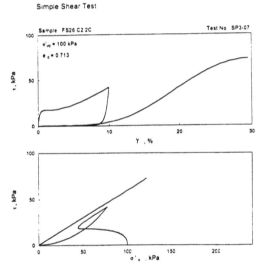

Fig. 4. Undrained response of loose sand to cyclic simple shear loading, Syncrude sand (Vaid, 1996).

3 MECHANICS OF ESTIMATING LIQUE-FACTION RESPONSE

If the number of load cycles are not sufficient to trigger liquefaction then displacements will be small and likely tolerable. The resistance to triggering of liquefaction, termed the Cyclic Resistance Ratio (CRR), depends mainly on the density of the soil, but also upon the confining stress, static bias, and soil type. Although the CRR can be estimated by testing undisturbed samples, it is usually obtained indirectly from penetration resistance and field experience as follows:

$$CRR = CRR_1 \times K_\sigma \times K_\alpha \times K_M \qquad (1)$$

where CRR_1 is commonly obtained from the Seed chart (Seed, et al., 1984), and K_σ, K_α, and K_M are correction factors for confining stress, static bias and Richter Magnitude (Seed and Harder, 1990; Byrne et al., 1994; Pillai and Byrne, 1994).

The emphasis here is on the post-triggering behaviour. The post-liquefaction stress-strain response is characterized by a dramatic drop in shear stiffness. Both laboratory data as well as field observation of earthquake-induced liquefaction indicate a drop in stiffness by a factor of 100 to 500 upon initiation of liquefaction. As straining after liquefaction occurs, the soil stiffens due to skeleton dilation, but eventually the dilation is suppressed by the increasing effective stresses, and the soil reaches a residual strength, s_r.

The residual strength depends on density, confining stress, and soil type. It also depends on the direction of loading relative to the direction of deposition (i.e., inherent anisotropy effects), and can be obtained by direct testing or estimated from field experience (Fig. 5) as suggested by Seed and Harder (1990). It should be noted that their estimates of residual strength may be low because it was assumed that the movements that occurred during the earthquake were sufficient to mobilize the residual strength. It is more likely that deformations were simply enough to mobilize sufficient resistance for stability, and this might explain the large scatter in results.

The potential for a flow slide can be estimated by assigning residual strengths to the liquefied zones and assessing stability using a limit

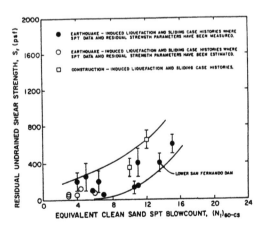

Fig. 5. Relationship between $(N_1)_{60}$ and undrained residual strength (Seed and Harder, 1990).

187

equilibrium analysis. However, this approach may be unconservative if there are also zones that do not liquefy. The strains required to mobilize the residual strength in the liquefied zones could be very large, and the strength in the unliquefied zones at such strains could drop below their peak values. Compatibility of strength values between zones must be considered.

Even if a flow slide is not predicted, the displacements due to lateral spreading could still be very large. The magnitude of these displacements depends mainly on the post-liquefaction stress-strain curves. Ideally, the whole dynamic process of liquefaction should be modeled using an effective stress approach which includes the build-up of porewater pressure, the triggering and post-triggering response. Such analyses have been described by Finn et al. (1986), Byrne and McIntyre (1994), and Byrne and Beaty (1997). However, for practical purpose, it is not unreasonable to assume that all zones trigger simultaneously, and base the predicted deformations on the post-liquefaction stress-strain curves. In this total stress approach, strains occurring prior to liquefaction are assumed to be small and can be neglected (Byrne, 1991).

Idealized pre- and post-liquefaction curves are shown in Fig. 6. Key parameters of the post-liquefaction curve are the residual strength ratio, s_r/σ'_{vo}, and the limiting or failure strain, γ_F, to reach this residual strength. These two parameters depend mainly on the density of the soil and the number of load cycles after liquefaction has been triggered. This effect can also be incorporated through the factor of safety against triggering

(Jitno, 1995). The curves can be obtained from direct testing of undisturbed samples as for Duncan Dam (Byrne et al., 1994), or indirectly from the data base of existing tests and/or field experience. Suggested approximate values are listed in Table 1. These are based on data reported by Seed and Harder (1990) as well as numerous tests on sands carried out under the direction of Dr. Vaid at the University of British Columbia.

Table 1. Post-liquefaction parameters.

$(N_1)_{60}$	s_u/p	γ_F (%)	
		$F_{TRIG} \approx 1.0$	$F_{TRIG} = 0.5$
0 - 4	0.05 - 0.10	25 - 50	>100
4 - 10	0.10 - 0.20	10 - 25	30 - >100
10 - 15	0.15 - 0.4	8 - 15	20 - 35
15 - 20	0.3 - 0.5	5 - 10	15 - 25
>20	> 0.5	< 5	< 15

4 EMPIRICAL EQUATIONS

Two empirical equations are commonly used in practice to estimate liquefaction-induced displacements. Both are based on field observations, together with some knowledge of the soil conditions and the ground surface profile.

4.1 Hamada

Hamada et al. (1987) presented a very simple equation for predicting the permanent lateral ground surface displacements based on field measurements in Japan and California as follows,

$$D = 0.75 \ H^{1/2} \ S^{1/3} \tag{2}$$

where D is the displacement in meters, H is the thickness of the liquefied layer in m, and S is the ground slope or gradient in percent. The data used to develop the equation suggests that actual displacements might range between about half to twice the predicted value.

It should be noted that neither the type of soil (sand or silt), its density, nor the level of shaking, are reflected in this equation. By implication, soils of $(N_1)_{60} = 4$ or 20 give the same displacement assuming that both are triggered to liquefy.

4.2 Youd

Bartlett and Youd (1992) presented an empirical approach which extended the simple Hamada

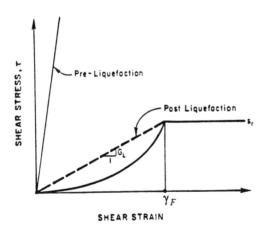

Fig. 6. Idealized post-liquefaction stress-strain models.

equation to consider the soil type and layer thickness, as well as measures of the amplitude and duration of the strong ground shaking. They developed two equations, one for "free face" and a second for "ground slope" conditions. The equations are based on regression analyses to best fit the observed data. They found that while the type of soil, silt or sand, was very important, its density or $(N_1)_{60}$ value was not important provided it was less than 15. We consider their equations as representing an invaluable "smoothed" database of field experience. Any mechanics approach should first be checked and verified with this database.

5 MECHANICS BASED METHODS

Prediction of earthquake-induced movements of earth structures is a difficult problem. Complex effective stress dynamic analyses procedures can be used, but are state-of-art rather than state-of-practice tools. These analyses are appropriate for special cases, particularly where drainage effects are important. Simpler analysis procedures, or total stress methods, are described below.

5.1 *Newmark method*

The simplest and most commonly used method of estimating earthquake induced displacements is that proposed by Newmark (1965). He modeled a potential sliding block of soil as a single-degree-of-freedom rigid plastic system of mass M resting on an inclined plane of slope α as is shown in Fig. 7.

Energy principles require that the work done by external forces (W_{ext}) minus the work done by the stress field (W_{int}) equals the change in kinetic energy of the system. This principle can be expressed as:

$$W_{ext} - W_{int} = 1/2\ M(V_f^2 - V^2) = -1/2\ MV^2 \qquad (3)$$

where V_f is the final resting velocity and is equal to zero, and V is the specified initial velocity.

Applying this concept to Newmark's model for a single pulse, the displacement, D, is given by:

$$D = V^2/(2gN) \qquad (4)$$

where g is the acceleration of gravity , and N is the yield acceleration of the sliding block.

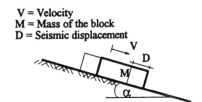

V = Velocity
M = Mass of the block
D = Seismic displacement

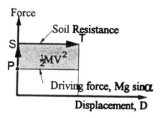

Fig. 7. Newmark rigid plastic model.

5.2 *Extended Newmark method*

As discussed earlier, liquefied soil will not behave in a rigid plastic manner as assumed by Newmark. Based on energy considerations, Byrne (1991) extended Newmark's method to include the very flexible rather than rigid response of liquefied soil.

Idealized pre-cyclic and post-cyclic stress strain curves are shown in Fig. 8. Point P in Fig. 8 is the stress state of a soil element prior to the earthquake. Upon liquefaction, the stress state drops from its static value P to Q. This stress change occurs at very low strain as discussed previously. The soil resistance then increases with strain to a residual value s_r. Since the driving force from the ground slope generally remains constant, the system will accelerate as it deforms and will have a velocity when the strain reaches point R, where the resistance is equal to the driving stress. The strain will continue to increase until an energy balance is reached at point S. If the system also has a velocity from the earthquake shaking, the soil will continue to deform until it reaches point T.

Comparing the rigid plastic Newmark approach with this extension to a general stress-strain relation, it may be seen that the standard Newmark method neglects the displacement from P to S which is caused by the flexible nature of liquefied soil. This could be a considerable displacement since strains of 20 to 50% are commonly required to mobilize the residual strength, s_r.

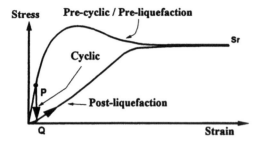

(a) Pre- and post-liquefaction characteristics of loose sand.

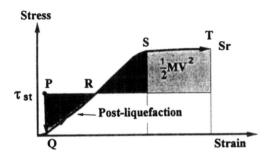

(b) Work-energy principle, extended Newmark.

Fig. 8. Extended Newmark model.

5.3 *Pseudo-dynamic approach*

The displacements of a single degree of freedom system can be computed directly by solving Eq. (3), as described by Byrne (1991). For a multi-degree-of-freedom system, a pseudo-dynamic method incorporating the post liquefaction stress-strain curves can be used with a static finite element code. The pseudo-dynamic approach incorporates inertia forces through use of a seismic coefficient. The magnitude of the seismic coefficient is such that the computed displacements satisfy the total energy balance of the system. The appropriate coefficient is found by an iterative procedure that is generally quick to converge.

The pseudo-dynamic approach has been incorporated into the finite element computer code SOILSTRESS (Byrne and Janzen, 1981) and found to give exact agreement with Newmark's method when conditions correspond to a single-degree-of-freedom rigid plastic system. It also gives good

agreement with liquefaction-induced displacement case histories (Jitno, 1995).

5.4 *Dynamic approach*

The problem can also be solved using a dynamic computer code which directly includes the inertial effects. A computer program such as FLAC (Cundall, 1995) is well suited to this approach because it uses the total element stresses, not just incremental stresses, in the equations of motion for each nodal mass. Therefore, changing the element stresses to a liquefied condition automatically causes a force imbalance under gravity loads, leading to deformations and strains. The predicted displacements account for the energy balance considerations discussed above. A proper plasticity formulation should also be used, so that predicted stresses do not violate the strength criteria as is common in incremental elastic analyses. The use of large displacement theory is also an advantage, since changes in geometry are often required to establish equilibrium.

In a total stress analysis, element properties can be modified to a liquefied state by changing both the stresses and the stiffness. The shear stress, τ_{xy}, is generally set to zero while the horizontal normal stress, σ_x, is set equal to the vertical stress, σ_y (i.e., the Mohr's circle of stress becomes a point). Post liquefaction stress-strain properties and residual strengths are then specified.

The dynamic analysis can be run in two ways. First, the deformations can be assumed to result primarily from the effect of gravity loads. The post-liquefaction stress-strain curve for this case must reflect the expected number of load cycles after liquefaction (i.e., a lower safety factor against liquefaction results in a softer modulus). Alternatively, the effect of the inertia forces from the earthquake can be included by specifying an input base motion and then triggering liquefaction at an appropriate time. In this case it is appropriate to use a bilinear post-liquefaction curve which uses a soft modulus while loading and a stiff modulus during unloading. In this way, the effect of multiple cycles is included directly in the analysis.

6 CASE HISTORIES

The mechanics based approach has been applied to a number of case histories. Some of these are briefly reviewed below:

6.1 *Lateral spreading - North America*

A comparison of predicted and observed displacements for sloping ground, one dimensional conditions, is shown in Fig. 9. The analyses were carried out by Jitno (1995) using a pseudo dynamic approach based on case histories from Bartlett (1992). Very good agreement was achieved.

6.2 *Mochikoshi dam*

This case history is described in detail by Jitno and Byrne (1995). During the Izu-Ohshima-Kinkai earthquakes of January 1978, two tailings dams associated with the Mochikoshi Gold Mining Company failed due to liquefaction of the tailings. Ishihara (1984) estimated the peak ground acceleration at the site was about 0.15 g to 0.25 g.

The cross-section of Dam No. 1 is presented in Fig. 10 and shows the geometry before and after the failure. The dam was built using an upstream construction method and has a maximum height of 28 m, a length of 73 m, and a crest width of 5 m. The water table, during the earthquake, was approximately 3 m below the slope surface.

According to the dam guardian, as reported by Ishihara (1984), the frontal wall of the dam swelled

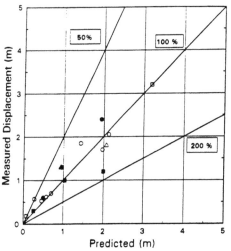

Fig. 9. Predicted vs. measured liquefaction-induced displacements.

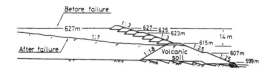

Fig. 10. Geometry of the Mochikoshi Dam No. 1 before and after failure (Ishihara, 1984).

within about 10 seconds of the shaking, causing excessive vertical settlement at the crest. A breach then occurred and a huge mass of tailings were released, carrying away the upper part of the dam.

6.2.1 *Pseudo-dynamic analysis*

The results of the liquefaction analysis using Seed's simplified procedures indicate the full depth of tailings could liquefy with an average factor of safety against liquefaction (F_L) of about 0.80. The final geometry of the tailings after failure, as shown in Fig. 10, reached a slope of about 1:7. This yielded a back-calculated residual strength of about 0.14 σ'_{vo}. The strain at residual strength was estimated to be 22%. This gives a shear modulus number, k_g, of the liquefied materials of 0.6, a reduction of about 1/250.

The predicted deformations are presented in Fig. 11, and the magnitude of the crest displacements are given in Table 2. The analysis shows the liquefied tailings moving up and over the starter dam, causing a bulge at the face of the dam. This mechanism is in reasonable agreement with the observation of the dam guardian.

The computed displacements at the crest were on the order of 6 m (Table 2). Predicted deformations of this magnitude are an indication of a flow slide.

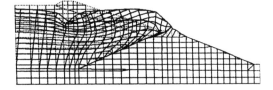

Fig. 11. Deformed shape of Mochikoshi Dam No. 1 (pseudo-dynamic analysis). Mag. factor=1.0.

Table 2. Displacements from pseudo-dynamic analysis.

Upstream Crest (m)		Downstream Crest (m)	
Horizontal Displ.	Vertical Displ.	Horizontal Displ.	Vertical Displ.
6.8	-7.7	6.8	-5.4

Note: Negative sign indicates downward vertical movement or leftward horizontal movement.

6.2.2 *Dynamic analysis*

For comparison, the FLAC computer code was also used to estimate the response to liquefaction. The deformed geometry due to gravity loads (i.e., no base motion) is shown in Fig. 12. These predicted deformations do not represent a final equilibrium position since the analysis was stopped due to excessive distortion in the elements. A complete failure mechanism in plastic flow still exists for the geometry shown. However, the dynamic analysis does confirm the occurrence of large deformations, even without the earthquake loading. The predicted deformations are also consistent with the observations during failure.

6.3 *La Marquesa dam*

La Marquesa Dam is a 10 m high earth dam with a crest length of 220 m and is located about 60 km from Santiago, Chile. The dam geometry, approximate water table, and soil types are shown in Fig. 13 (De Alba et al., 1987). Additional detail has been given by De Alba et al. (1987).

The dam experienced significant deformation during the Chilean earthquake of March 3, 1985. Major movement occurred in the upstream and downstream slopes causing a 1.8 m loss in freeboard. Extensive longitudinal cracks were also observed, particularly in the upstream slope.

De Alba et al. (1987) postulated that, due to the high intensity of the shaking (peak ground acceleration of 0.6 g), a 2 m thick layer of silty sand under the upstream and downstream slopes immediately liquefied at the start of shaking. This layer had average $(N_1)_{60}$ values, converted to equivalent clean sand values, of 6 beneath the upstream shell and 11 beneath the downstream shell. Consequently, the embankment lost its support. The clayey sand shells broke into blocks, moving downward and outward on the liquefied silty sand layer. The deformed dam geometry is also shown in Fig. 13.

6.3.1 *Pseudo-dynamic analysis*

The residual strengths of the silty sand layer were taken directly from the upper bound values of De Alba et al. (1987), 17 kPa and 29 kPa for the upstream and downstream respectively. The limiting strains were based on Seed et al. (1984), and taken to be 100%. The shell was assumed to experience a 50% reduction in stiffness due to the severe earthquake loading, but the strength was not changed because of the clay content.

The kinetic energy due to the earthquake was approximately included in this analysis by assuming a uniform velocity in the dam at the time of liquefaction of 0.6 m/s.

The original mesh (dashed line) and the predicted deformations (solid line) are shown in Fig. 14. The displacements at several locations in the dam are also presented in Table 3. The computed loss of freeboard is about 1.5 m which is in good agreement with the observed value of 1.8 m. The settlement at the upstream and downstream shell are about 3.2 m and 2.5 m, respectively, which are very close to the observed values (about 3.5 m and 2.5 m). In general, the

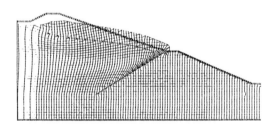

Fig. 12. Deformed shape of Mochikoshi Dam No. 1 (dynamic analysis). Mag. factor = 1.0.

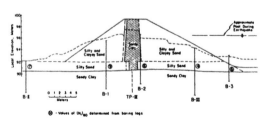

Fig. 13. La Marquesa Dam geometry before and after the earthquake, soil types and water table (after De Alba et al., 1987).

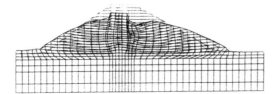

Fig. 14. Deformed shape of La Marquesa Dam (pseudo-dynamic analysis). Mag. factor = 1.0.

Table 3. Vertical displacements from pseudo-dynamic analysis.

Location	Computed Displacement (m)	Observed Displacement (m)
Dam crest	1.5	1.8
U/S shell	3.2	3.5
D/S shell	2.5	2.5

magnitude and the pattern of displacements agree well with the observed values.

6.3.2 *Dynamic analyses*

The analysis was repeated using similar material properties and the computer code FLAC. The analyses were performed both with and without the earthquake inertia forces. A representative earthquake record was input to the base to create the earthquake inertia forces. The post-liquefaction stress-strain curves for this case were modified to reflect a bilinear response.

The deformed shapes for these analyses are shown in Fig. 15, and representative displacement values are given in Table 4. The predicted displacements and deformation pattern from the gravity plus earthquake analysis are in reasonable agreement with the observed deformations and the pseudo-dynamic analysis. The low estimates from the gravity only analysis indicate the potential importance of the earthquake inertia forces.

Table 4. Vertical displacements from dynamic analyses

Location	Gravity Only Displacement (m)	Gravity + EQ Displacement (m)
Dam crest	-0.6	-2.1
U/S shell	-1.3	-3.1
D/S shell	-1.0	-3.0

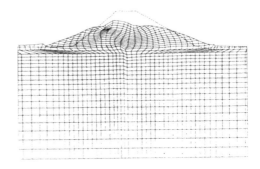

Fig. 15. Deformed shape of La Marquesa Dam (dynamic analysis). Mag. factor = 1.0.

7 SUMMARY

The magnitude of liquefaction induced displacements is a major concern in areas where structures either comprise, or are underlain by, liquefaction-prone soils. Over the past 30 years, field observations have been collected that now form a database allowing predictive models to be developed and verified. Models are of two types:

1) Empirical models that basically capture the observed field performance; and

2) Mechanics or physics based models that consider the stress-strain relations of liquefied soil and the dynamic forces acting on the soil-structure.

The empirical approach can be used directly to estimate movements, particularly for free field conditions, and in many cases may be adequate for deciding if remediation is required. Mechanics based models are needed where heavy structures are present: dams, bridges, piers, caisson walls, large buildings. If these models are calibrated with field experience, they can also be used with some confidence to predict movements. If remediation is required to reduce these movements to tolerable levels, it can be rationally designed using a mechanics based approach.

8 ACKNOWLEDGMENTS

We wish to acknowledge Dr. Hendra Jitno's contribution in that case histories were abstracted from his Ph.D. Thesis, Mr. Ian Manning for the FLAC files for the Mochikoshi dam, and Ms. Kelly Lamb for typing and arranging the paper.

9 REFERENCES

Bartlett, S.F. and Youd, T.L. (1992). Empirical Prediction of Lateral Spread Displacement: Proceedings, 4th US-Japan Workshop on Earthquake Resistant Design Lifeline Facilities and Countermeasures for Soil Liquefaction, NCEER, Honolulu, Hawaii, May.

Byrne, P.M. (1991). A Model for Predicting Liquefaction Induced Displacements Due to Seismic Loading. The Second International Conference on Recent Advances in Geotech. Earthquake Eng. and Soil Dynamics, St. Louis, Missouri, Paper No. 7.14, March.

Byrne, P.M. and McIntyre, J. (1994). Effective Stress Liquefaction Analysis at the Wildlife Site. Invited Lecture, 3rd International Conference on Recent Advances in Geotechnical Earthquake Engineering and Soil Dynamics, St. Louis, Missouri (USA), April 2-7, Vol. 2, pp. 1864-1898.

Byrne, P.M. and Janzen, W. (1981). SOILSTRESS: A Computer Program for Nonlinear Analysis of Stresses and Deformations in Soil. Soil Mechanics Series No. 52, Department of Civil Engineering, University of British Columbia, Vancouver, B.C., Canada, December 1981, updated January 1989.

Byrne, P.M., Imrie, A.S. and Morgenstern, N.R. (1994). Results and Implications of Seismic Performance Studies for Duncan Dam. Canadian Geotechnical Journal, Vol. 31, No. 6, pp. 979-988.

Cundall, P. (1995). FLAC Manual Version 3.3, ITASCA Consulting Group Inc., Thrasher Square East, 708 South Third Street, Suite 310, Minneapolis, Minnesota 55415, U.S.A.

De Alba, P., Seed, H.B., Retamal, E. and Seed, R.B. (1987). Residual Strength of Sand From Dam Failures in the Chilean Earthquake of March 3, 1985. Report No. UCB/EERC-87/11, Department of Civil Engineering, University of California, Berkeley.

Finn, W.D. Liam, Yogendrakumar, M., Yoshida, N. and Yoshida, H. (1986). TARA-3: A Program to Compute the Response of 2-D Embankments and Soil-Structure Interaction Systems to Seismic Loadings. Department of Civil Engineering, University of British Columbia, Vancouver, B.C.

Hamada, M., Towhata, I., Yasuda, S. and Isoyama, R. (1987). Study of Permanent Ground Displacement Induced by Seismic Liquefaction. Computers and Geotechnics 4, pp. 197-220.

Ishihara, K. (1984). Post-Earthquake Failure of a Tailings Dam Due to Liquefaction of the Pond Deposit. Proceedings, International Conference on Case Histories in Geotechnical Engineering, St. Louis, Vol. 3, pp. 1129-11143.

Jitno, H. (1995). Liquefaction Induced Deformations of Earth Structures, Ph.D. Thesis, Department of Civil Engineering, University of British Columbia, Vancouver, B.C., Canada, February.

Jitno, H. and Byrne, P.M. (1995). Predicted and Observed Response of Mochikoshi Tailings Dam. First International Conference on Earthquake Geotechnical Engineering, IS-Tokyo 95, Kenji Ishihara (Ed.), Vol. 2, pp. 1085-1090.

Newmark, N.M. (1965). Effects of Earthquakes on Dams and Embankments. Geotechnique, Vol. 15, No. 2, pp. 139-160.

Pillai, V.S. and Byrne, P.M. (1994). Effect of Overburden Pressure on Liquefaction Resistance of Sand. Canadian Geotechnical Journal, Vol. 31, No. 1, February.

Seed, H.B., Tokimatsu, K., Harder, L. and Chun, R.. (1984). The Influence of SPT Procedures in Soil Liquefaction Resistance Evaluations. Report No. UCB/EERC-84/15, College of Engineering, University of California, Berkeley.

Seed, H.B. and Harder, L.F. (1990). SPT-Based Analysis of Cyclic Pore Pressure Generation and Undrained Residual Strength. Proceedings, H. Bolton Seed Memorial Symposium, Vol. 2.

Seismic Behaviour of Ground and Geotechnical Structures, Sêco e Pinto (ed.)© 1997 Balkema, Rotterdam, ISBN 90 5410 887 8

Nonlinear seismic response of pile foundations during strong earthquake shaking

W.D. Liam Finn & T.Thavaraj
University of British Columbia, Vancouver, B.C., Canada

G.Wu
Agra Earth & Environmental Ltd, Burnaby, B.C., Canada

ABSTRACT

A computationally efficient method is presented for the 3-D analysis of pile foundations under strong earthquake shaking. The key elements of the method are: efficient nonlinear 3-D analysis using a reduced 3-D equation to describe the half-space enclosing the piles, direct analysis of the pile group (no need for interaction factors) including kinematic and inertial interaction, and the ability to establish time-dependent stiffness and damping factors during earthquake shaking. The approach provides a comprehensive understanding of how pile foundations behave during strong seismic shaking, when non-linear soil behaviour is significant. The proposed method has been validated against existing elastic solutions for linear response, against centrifuge test data for single piles and pile groups for strong shaking.

1 INTRODUCTION

Seismic soil-structure interaction analysis involving pile foundations is one of the more complex problems in geotechnical earthquake engineering. A very common example is the 3-D analysis of a pile foundation for a bridge abutment. The analysis involves modelling soil-pile-soil interaction, the effects of the pile cap, nonlinear soil response, and in many cases seismically induced porewater pressures. There are many approaches to solving the dynamic response of pile foundations. Novak (1991) gave an extensive critical review of the more widely accepted methods of analysis. He showed that pile group response cannot be deduced from single pile response without taking dynamic pile-soil-pile interaction into account because the dynamic characteristics of pile groups are strongly frequency-dependent. The dynamic characteristics of piles in a group may be quite different from those of a single pile.

The methods for direct seismic analysis of pile groups are based on linear elastic behaviour. Complete elastic analyses have been conducted using 3-D boundary element formulations (El-Marsafawi et al., 1992a, 1992b), but they require substantial computing time. They are exact for elastic isotropic conditions. However, they cannot take into account the nonlinear behaviour of soil under strong shaking or the effect of seismically induced porewater pressures on dynamic response. The reduction in soil stiffness and the increase in damping associated with strong shaking are sometimes modelled crudely in these analyses by making arbitrary reductions in the shear moduli and arbitrarily increasing the viscous damping. For this reason, the results of these studies have not proved very useful for the response of pile foundations to earthquake loading.

The offshore industry pioneered the seismic design of pile foundations by estimating the nonlinear behaviour of single piles using nonlinear Winkler springs and estimating the group stiffness by using static interaction factors. The nonlinear Winkler springs were obtained from load-deflection curves on prototype piles in various soil conditions. The original basic studies were conducted by Matlock (1963), Matlock et al. (1978) and Reese et al. (1974a, 1974b). Their studies led to the p-y procedure for offshore pile design recommended by the American Petroleum Institute (1991). These curves were derived under static conditions or low-frequency cycling. Therefore, they do not incorporate the frequency effect on damping and stiffness. However, in many low frequency applications, the dynamic stiffness is similar to the static stiffness. El-Marsafawi et al. (1992a, 1992b) developed approximate procedures for estimating elastic dynamic interaction factors based on boundary element analysis. They extended the studies of Kaynia and Kausel (1982), Davies et al. (1985), and Gazetas (1991a, 1991b), using the general 3-D formulation developed by Kaynia and Kausel (1982).

For seismic analysis of pile foundations under strong shaking, engineering practice relies mainly on the p-y approach. There are well recognized problems with this approach; determining the appropriate p-y curves for the site, the approximate nature of the representation of field conditions, and the difficulty of simulating appropriate dynamic interaction between the piles. Traditionally, static interaction factors have been used, but with their wider availability dynamic factors may be adopted in future. Of course both sets of factors are based on elastic analysis. During nonlinear response, the effective stiffness of the pile group is affected strongly by the load on the group. The inertial effects of the superstructure cause additional strains in the ground and hence modify further the effective moduli and damping. This effect was recognized by Matlock et al. (1978) in the development of the computer program SPASM, based on the p-y curve concept. Therefore, it is important to include the forces imposed by the superstructure when evaluating dynamic pile group stiffness.

A procedure for nonlinear dynamic analysis of pile groups, under development at UBC, will be described here. It is a quasi-3D method which permits dynamic nonlinear analysis of pile groups in layered soils. By relaxing some of the boundary conditions associated with a full 3D analysis, the computing costs can be substantially reduced and the analysis is feasible on a Pentium PC. The procedure is validated here by data from centrifuge tests on a single pile and a 2×2 pile group.

The method is presently applicable to situations where high porewater pressures are not expected. An extension to dynamic effective stress analysis is under development.

2 QUASI-3D LATERAL ANALYSIS OF PILES

Under vertically propagating shear waves (Fig. 1) the soil undergoes primarily shearing deformations in xOy plane except in the area near the pile where

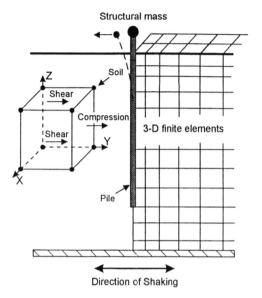

Fig. 1. Quasi-3D model of pile-soil response.

extensive compressional deformations develop in the direction of shaking. The compressional deformations also generate shearing deformations in yOz plane. Therefore, the assumptions are made that dynamic response is governed by the shear waves in the xOy and yOz planes, and the compressional waves in the direction of shaking, Y. Deformations in the vertical direction and normal to the direction of shaking are neglected. Comparisons with full 3-D elastic solutions confirm that these deformations are relatively unimportant for horizontal shaking. Applying dynamic equilibrium in Y-direction, the dynamic governing equation under free vibration of the soil continuum is given by

$$\rho_s \frac{\partial^2 v}{\partial t^2} = G \frac{\partial^2 v}{\partial x^2} + \theta G \frac{\partial^2 v}{\partial y^2} + G \frac{\partial^2 v}{\partial z^2} \qquad (1)$$

where G is the shear modulus, ρ_s is the mass density of the soil, and θ is a coefficient related to Poisson's ratio of the soil. A similar equation can be written for vertical motion.

Piles are modelled using ordinary Eulerian beam theory. Bending of the piles occurs only in the yOz plane. Dynamic pile-soil-pile interaction is maintained by enforcing displacement compatibility between the pile and soils.

A quasi-3D finite element code PILE3D (Wu and Finn, 1994) was developed to incorporate the dynamic pile-soil-pile interaction described previously. An 8-node brick element is used to represent soil, and a 2-node beam element is used to simulate the piles. Using the standard procedure of the finite element method, the global dynamic equilibrium equations are written in matrix form as

$$[M^*]\left\{\frac{\partial^2 v}{\partial t^2}\right\} + [C^*]\left\{\frac{\partial v}{\partial t}\right\} + [D^*]\{v\} = \{P(t)\} \quad (2)$$

in which P(t) is the dynamic external load vector, and $\{\partial^2 v/\partial t^2\}$, $\{\partial v/\partial t\}$ and $\{v\}$ are the relative nodal acceleration, velocity and displacement, respectively. $[M^*]$, $[C^*]$ and $[K^*]$ are the mass, damping and stiffness matrices of the soil-pile system in the direction of vibration for either horizontal or vertical vibration. $[K^*]$ is a complex stiffness matrix which includes the hysteretic damping of the soil.

Direct step-by-step integration using the Wilson-θ method is employed in PILE3D to solve the equations of motion in eqn. (2). In nonlinear analysis, the hysteretic behaviour of soil is modelled by using a variation of the equivalent linear method used in the SHAKE program (Schnabel et al., 1972). Compatibility between shear strains and moduli and damping, is enforced at selected times during shaking, rather than at the end of shaking as in SHAKE. Additional features such as tension cut-off and shearing failure are incorporated in the program to simulate the possible gapping between soil and pile near the soil surface and yielding in the near field.

The loss of energy due to radiation damping is modelled using the method proposed by Gazetas et al. (1993). A velocity proportional damping force F_d per unit length along the pile is given by

$$F_d = c_x \frac{dv}{dt} \quad (3)$$

where the radiation dashpot coefficient, c_x, is given by

$$c_x = 6\rho_s V_s d\left(\frac{\omega \cdot d}{V_s}\right)^{-0.25} \quad (4)$$

in which V_s is the shear wave velocity of the soil, d is diameter of the pile, and ω is the excitation frequency of the external load. Radiation damping associated with vertical motion of the piles is handled in a similar way using a radiation dashpot coefficient, c_z, with

$$c_z = \rho_s V_s d\left(\frac{\omega \cdot d}{V_s}\right)^{-0.25} \quad (5)$$

3 ELASTIC PILE HEAD IMPEDANCES

3.1 Pile head impedances

The impedances K_{ij} are defined as the complex amplitudes of harmonic forces (or moments) that have to be applied at the pile head in order to generate a harmonic motion with a unit amplitude in the specified direction (Novak, 1991). The translational, the cross-coupling, and the rotational impedances of the pile head are represented by K_{vv}, $K_{v\theta}$, $K_{\theta\theta}$, respectively. There is also a vertical stiffness K_{zz}.

Since the pile head impedances K_{vv}, $K_{v\theta}$, $K_{\theta\theta}$ are complex valued, they are usually expressed by their real and imaginary parts as,

$$K_{ii} = k_{ij} + i C_{ij} \quad ..or.. \quad K_{ij} = k_{ij} + i \omega c_{ij} \quad (6)$$

in which k_{ij} and C_{ij} are the real and imaginary parts of the complex impedances, respectively; $c_{ij} = C_{ij}/\omega$ = coefficient of equivalent viscous damping; and ω is the circular frequency of the applied load. K_{ij} and C_{ij} are usually referred to as the stiffness and damping at the pile head. All the parameters in eqn. (6) are dependent on the frequency ω, and in the case of strong shaking on the current distribution of effective moduli and damping ratios.

3.2 Rocking impedances of a pile group

The rocking impedances of a pile group is a measure of the complex resistance to rotation of the pile cap due only to the resistance of each pile in the group to vertical displacements. The rocking impedance K_{RR} of a pile group is defined as the summation of the moments of the axial pile forces around the centre of rotation of the pile cap for a harmonic rotation with unit amplitude at the pile cap. This definition is quantitatively expressed as

$$K_{RR} = \Sigma r_i \times F_i \quad (7)$$

where r_i are distances between the centre of rotation and the pile head centres, and F_i are the amplitudes of the axial forces at the pile heads.

In the analysis, the pile cap is assumed to be rigid. For a unit rotation of the pile cap, the vertical displacements w_i^p at all pile heads are determined according to their distances from the centre of rotation r_i.

The lateral impedance, K_{vv}, and the rocking impedance, K_{RR}, of a 4-pile group with a pile spacing, $s/d = 5$, where s is the centre to centre spacing, and d is the pile diameter, will now be evaluated using the proposed model, assuming elastic response. The results will be compared with the impedances from complete 3-D analyses to check the computational adequacy of the quasi-3D model. The piles have an L/d ratio >15; the ratio of pile modulus to soil modulus, $E_p/E_s = 1000$; the Poisson's ratio $\mu = 0.4$, and the critical damping ratio, $\lambda = 5\%$. The pile cap is rigid and rigidly connected to the pile heads.

In order to show the pile group effect, dynamic impedances of the pile group are normalized to a stiffness of the pile group expressed as the static stiffness of a single pile times the number of piles in the group. This normalized lateral impedance of the pile group, α_{vv}, is the dynamic interaction factor for lateral loading for the given pile head conditions and is defined as

$$\alpha_{vv} = \frac{K_{vv}}{N \cdot k_{vv}^0} \qquad (8)$$

where k_{vv}^0 is the static lateral stiffness of an identical single pile placed in the same soil medium, and N is the number of piles in the pile group. The computed lateral interaction factors for stiffness, α_{vvs}, and damping, α_{vvd}, are compared with those by Kaynia and Kausel (1982) in Fig. 2(a) and Fig. 2(b), respectively. There is good agreement between both sets of factors.

Because the piles are rigidly connected to the pile cap at the pile heads, the total rotational impedance of the pile cap $K_{\theta\theta}^{cap}$ consists of both the rocking impedance K_{RR} of the pile group and the summation of the rotational impedances, $K_{\theta\theta}$, at the head of each pile

$$K_{\theta\theta}^{cap} = K_{RR} + \Sigma K_{\theta\theta} \qquad (9)$$

The total rotational impedance of the pile cap $K_{\theta\theta}^{cap}$ is normalized as $K_{\theta\theta}^{cap} / (N \cdot \Sigma r_i^2 k_{zz}^0)$, in which k_{zz}^0 is the static vertical stiffness of an identical single pile placed in the same soil medium. The computed rotational interaction factors for stiffness and damping are compared with those by Kaynia and Kausel (1982) in Fig. 3(a) and Fig. 3(b),

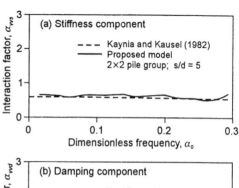

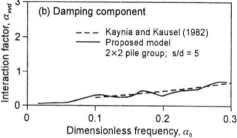

Fig. 2. Comparison of dynamic interaction factors for (a) α_{vvs}, and (b) α_{vvd}, with solution by Kaynia and Kausel (1982).

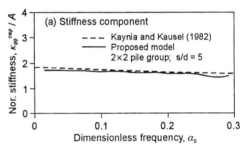

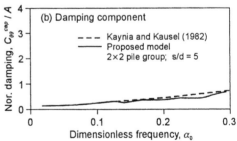

Fig. 3. Rotational interaction factors for (a) stiffness, and (b) damping, compared with those of Kaynia and Kausel (1982).

respectively. The results obtained using the quasi-3D model are quite satisfactory.

4 NONLINEAR SEISMIC RESPONSE ANALYSIS OF A SINGLE PILE

PILE3D was used to analyze the seismic response of a single pile in a centrifuge test conducted at the California Institute of Technology (Caltech) by Gohl (1991). Figure 4 shows the soil-pile-structure system used in the test. The system was subjected to a nominal centrifuge acceleration of 60 g. A horizontal acceleration record with a peak acceleration of 0.158 g is input at the base of the system.

The sand deposit is modelled by 11 layers. Layer thickness is reduced as the soil surface is approached to allow more detailed modelling of the stress and strain field where lateral soil-pile interaction is strongest. The pile is modelled using 15 beam elements including 5 elements above the soil surface. The superstructure mass is treated as a rigid body.

The finite element analysis was carried out in the time domain. Nonlinear analysis was performed to account for the changes in shear moduli and damping ratios due to dynamic shear strains.

The computed time-history of moments in the pile at a depth of 3 m (near point of maximum moment) is plotted against the recorded time-history in Fig. 5. There is satisfactory agreement between the computed and measured moments in the range of larger moments. The computed and measured moment distributions along the pile at the

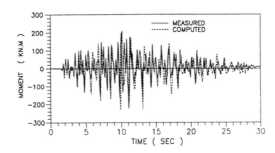

Fig. 5. The computed versus measured moment response at depth D = 3 m.

instant of peak pile head deflection are shown in Fig. 6. The computed moments agree quite well with the measured moments. The moments increase to a maximum value at a depth of 3.5 diameters, and then decrease to zero at a depth around 12.5 diameters. The moments along the pile have same signs at any instant of time, suggesting that the inertial interaction caused by the pile head mass dominates response, and the pile is vibrating in its first mode. The peak moment predicted by the quasi-3D finite element analysis is 344 kNm compared with a measured peak value of 325 kNm.

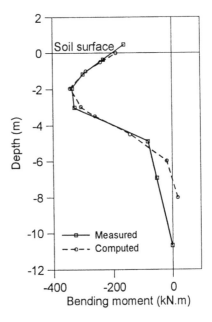

Fig. 6. Computed versus measured moment distribution in the pile at peak pile deflection.

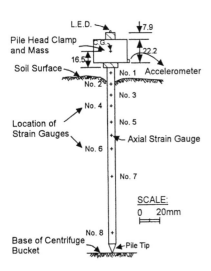

Fig. 4. The layout of the centrifuge test for a single pile.

5 SEISMIC RESPONSE ANALYSIS OF A PILE GROUP

The seismic response of a 4-pile group in a centrifuge test (Gohl, 1991) was analyzed using the program PILE3D. The piles are set in a 2×2 arrangement at a centre to centre spacing of 2 diameters. The finite element mesh used to analyze the group is shown in Fig. 7. A refined mesh is used around the piles near the surface. This region contributes most to the lateral resistance of the piles, and the shear strains are greatest here. The properties of the piles are identical to those of the single pile described earlier. The group piles were rigidly clamped to a stiff pile cap and four cylindrical masses were bolted to the cap to simulate the inertia of a superstructure.

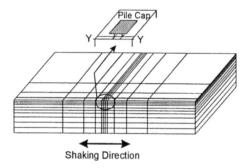

Fig. 7. Finite element model of pile group.

At selected times during the horizontal mode analysis, the rocking stiffness and damping is computed using PILE3D in the vertical mode. This impedance calculation is made using the current values of strain dependent moduli and damping. The current rocking impedance is then transferred to the pile cap as rotational stiffness and damping. The accuracy of the representation of rocking impedance depends on the frequency with which it is updated.

The distributions of computed and measured bending moments along the pile at the instant of peak pile cap displacement are shown in Fig. 8. The computed moments agree reasonably well with the measured moments especially in the region of maximum moment. The computed time-history of dynamic moments for the pile group shows the same kind of agreement with the recorded moments as noted for the single pile test.

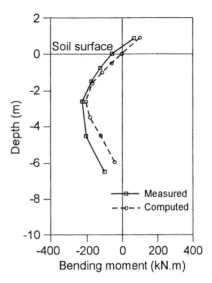

Fig. 8. Comparison between measured and computed bending moments at peak pile cap displacement.

6 PILE IMPEDANCE IN PRACTICE

Commercially available software for the analysis of structures does not incorporate time-dependent stiffnesses. Therefore, for the seismic structural analysis of bridges and buildings on pile foundations, discrete springs and dampers associated with the different degrees of freedom of the pile foundations are assigned to the structural analysis model. The reduction in soil stiffness and the increase in damping associated with strong shaking are sometimes modelled crudely in these analyses by making arbitrary reductions in shear moduli and arbitrary increases in viscous damping. Another approach is to use nonlinear p-y springs and to calculate the stiffness at an arbitrary displacement of the pile head.

The 3D response analysis program, PILE3D, provides a method for the direct computation of pile impedances taking into account nonlinear behaviour of the soil, pile-soil-pile interaction, hysteretic and radiation damping, and structural inertia effects. This capability allows a more rational selection of effective spring and damping constants for use in commercial software.

The distribution of shear moduli at a depth of 2.1 m in the soil around the pile in Fig. 4 at a time T = 12.58 secs is shown in Fig. 9. The distribution of shear moduli and damping are both time- and

Fig. 9. Distribution of shear moduli around pile at a depth of 2.1 m at time 12.58 s into earthquake.

space-dependent. This dependence results in a corresponding time-dependence of stiffness and damping of pile foundations. Dynamic impedances as a function of time were computed using the time- and space-dependent nonlinear shear moduli. Harmonic loads with an amplitude of unity were applied at the pile head, and the resulting equations were solved to obtain the complex valued pile impedances. The impedances were evaluated at the ground surface.

The time-dependent dynamic stiffnesses and associated displacements (real part of the impedance) of the pile are shown in Fig. 10. The dynamic stiffnesses experienced their lowest values between about 10 and 14 seconds, when the largest displacements occurred at the pile head. It can be seen that the lateral stiffness component K_{vv} decreased more than the rotational stiffness $K_{\theta\theta}$ or the coupled lateral- rotational stiffness $K_{v\theta}$. The equivalent damping coefficients (not shown) increased with increasing displacement also because the hysteretic damping of the soil increased with the level of straining.

The time histories of lateral stiffness, K_{vv}, and the coupled lateral-rotational stiffness, $K_{v\theta}$, for the 2×2 pile group subjected to the same input motions as the single pile, are shown in Fig. 11.

The stiffness of a pile group, during seismic shaking, incorporates both kinematic and inertial effects. The inertial effects are due to the mass of the superstructure. Because of the approximate procedures used in practice to evaluate the stiffness of a pile foundation, the inclusion of inertial effects is often ignored. However, the effects of inertial interaction may be very important sometimes.

To demonstrate the effect of inertial interaction, the single pile in Fig. 4 was analyzed, with and without the structural mass at the pile head. The

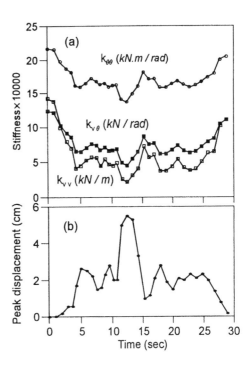

Fig. 10. (a) Time-dependent dynamic stiffnesses, and (b) and associated pile-head displacements for the single pile.

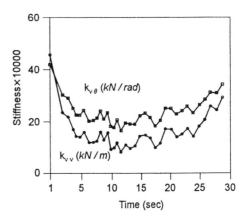

Fig. 11. Dynamic stiffnesses for the pile group.

variations in lateral stiffness, with and without inertial interaction, are shown in Fig. 12. In this case, the inertial interaction has a major effect on the stiffness of the pile. It is clear that inertial interaction should be considered when evaluating pile stiffnesses under strong earthquake shaking.

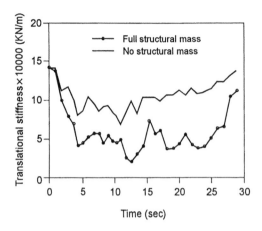

Fig. 12. The effect of the inertia of structural mass on foundation stiffness under strong shaking.

The time histories of lateral and rocking stiffnesses shown in Figs. 10, 11 and 12, show clearly the difficulties in selecting a single spring value to represent the lateral or rotational stiffness of a pile foundation. To make a valid selection, one would need to know which segment of the ground motion was most critical in controlling the seismic response of the structure. A spring based on the minimum lateral stiffness would represent the mobilized stiffness during the period of very strong shaking and would be more critical for longer period structures. Knowing the time variation in stiffness and damping makes it possible to estimate better the appropriate effective discrete stiffness and damping for use in a structural program.

7 FOUNDATION STIFFNESSES OF THE PAINTER STREET BRIDGE BENT USING PILE3D

The Painter Street Overpass located near Rio Dell in Northern California, is a two-span, prestressed concrete box-girder bridge that was constructed in 1973 to carry traffic over the four-lane US Highway 101. The bridge is 15.85 m wide and 80.79 m long. The deck is a multi-cell box girder, 1.73 m thick and is supported on monolithic abutments at each end and a two-column bent that divides the bridge into two spans of unequal length. One of the spans is 44.51 m long and the other is 36.28 m long. The east and west abutments are supported by 14 and 16 piles, respectively. Longitudinal movement of the west abutment is allowed by means of a thermal expansion joint.

Each column of the bent is 7.32 m high and supported by 20 concrete friction piles in a 4×5 group.

Finn et al. (1995) described studies of the seismic behaviour of the abutments and centre bent of this bridge including system identification analyses and quasi-3D analyses of the pile foundation.

The model of the foundation soils was based on data from field tests conducted by the Lawrence Livermore Laboratory in the U.S., and was supplied by Heuze (1994). The shear modulus of the soil varied parabolically with depth with an average value of 100 MPa in the surficial soil layer.

The bridge was subjected to the accelerations of the main shock of the Cape Mendocino-Petrolia earthquake of 1992 with magnitude M = 6.9.

The elastic stiffnesses of the pile foundation are not affected by the inertia of the superstructure because the moduli are not dependent on displacements. The stiffnesses of the real soil under strong shaking are affected by the inertia of the superstructure because the stiffnesses are functions of displacements. Therefore, the proportion of the inertial mass of the deck structure adopted by Makris et al. (1994) in their analysis of the bridge was included in the nonlinear dynamic response of the bridge bent.

The four pile cap stiffnesses of the 20-pile foundation of the bridge bent were evaluated for both elastic and nonlinear response, in order to demonstrate the importance of nonlinear effects. The elastic stiffness of a single pile was also determined to allow an estimate of the effect of pile-to-pile interaction on elastic stiffness under dynamic conditions.

The elastic stiffnesses of a single pile are given in column 1 of Table 1. The elastic stiffnesses of the 20 pile foundation are given in column 2. These stiffnesses are only about 30% of the stiffnesses corresponding to 20 times the stiffness of a single pile. This large reduction in stiffness is due to pile-soil-pile interaction under dynamic conditions.

The time-history of lateral stiffness under the strong shaking taking nonlinear soil behaviour into account is shown in Fig. 13. It is typical of the variations in all stiffnesses. What discrete single valued spring stiffness should be selected for use in a commercial structural analysis program that adequately represents this time-dependent stiffness? This is obviously a matter of judgement. The authors' choice, based on the time histories of the stiffnesses are given in column 4 of Table 1. The

Table 1. Dynamic Stiffnesses of a Single Pile and the 20-Pile Foundation Group

Type of Stiffness	Single Pile (elastic response)	20-Pile Group (elastic response)	20-Pile** Group (strong shaking)
Lateral, k_{uu}	77.7 MN/m	520 MN/m	150 MN/m
Cross Coupling, $k_{u\theta}$	36.7 MN/rad	330 MN/rad	170 MN/rad
Rocking, k_{RR}	--	8400 MN.m/rad	2600 MN.m/rad
Vertical, k_{zz}	376 MN/m	3700 MN/m	2300 MN/m

** Effective stiffness selected for use in commercial structural program.

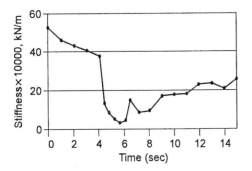

Fig. 13. Time variation of lateral stiffness during strong shaking.

selected stiffnesses are biased towards the values associated with the time of strongest shaking. A comparison of columns 3 and 4 shows that the stiffnesses based on initial in-situ values of moduli were reduced by factors of 3.5 for lateral stiffness and by 3.2 for rocking stiffness by strong shaking. These reductions are of the same order as those deduced from the system identification analysis of the abutment of the bridge during the Cape Mendocino earthquake by Finn et al. (1995), which are shown in Fig. 14.

8 CONCLUSIONS

A validated method for elastic and nonlinear dynamic analysis of pile foundations based on a simplified 3D model of the half space called PILE3D has been presented which can calculate the

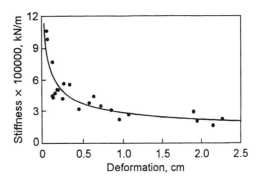

Fig. 14. Variation of lateral stiffness of abutment of Painter Street Bridge with deformation during strong shaking (after Finn et al., 1995).

time-histories of stiffness (and damping) of a pile foundation during an earthquake. The analysis can also include the effects of superstructure inertia on both stiffness (and damping). The time variation of these parameters for a given design earthquake allows a more realistic selection of the representative discrete stiffnesses and damping ratios required by structural analysis programs than the rather arbitrary procedures often used in practice.

9 ACKNOWLEDGEMENTS

Research on seismic soil-structure interaction is supported by grants to the lead author by the Natural Science and Engineering Council of Canada.

10 REFERENCES

American Petroleum Institute (1991). "Recommended Practice for Planning, Designing and Constructing Fixed Offshore Platforms". 19th Edition, Washington, DC.

Davies, T.G., R. Sen, and P.K. Banerjee. (1985). "Dynamic Behaviour of Pile Groups in Inhomogeneous Soil," J. Geot. Eng., ASCE, Vol. 111, No. 12, pp. 1365-1379.

El-Masrafawi, H., A.M. Kaynia and M. Novak. (1992a). "Interaction Factors and the Superposition Method for Pile Group Dynamic Analysis," Research Report, GEOT-1-1992, University of Western Ontario, London, Ontario.

El-Masrafawi, H., A.M. Kaynia and M. Novak. (1992b). "The Superposition Approach to Pile Group Dynamics," Geot. Special Publication No. 34, ASCE, New York, N.Y., pp. 114-135.

Finn, W.D. Liam and Wu, G. (1994). "Recent Developments in Dynamic Analysis of Piles". Proc., 9th Japan Conf. on Earthq. Eng., Tokyo, Japan, Vol. 3, pp. 325-330.

Finn, W.D. Liam, Wu, G. and Thavaraj, T. (1995). "Seismic Response for Pile Foundations for Bridges". Proc., 7th Canadian Conf. on Earthq. Eng., pp. 779-786, May 1995.

Gazetas, G. and M. Makris. (1991a). "Dynamic Pile-Soil-Pile Interaction. Part I: Analysis of Axial Vibration," Earthq. Eng. Struct. Dyn., Vol. 20, No. 2, pp. 115-132.

Gazetas, G., M. Makris and E. Kausel. (1991b). "Dynamic Interaction Factors for Floating Pile Groups," J. Geot. Eng., ASCE, Vol. 117, No. 10, pp. 1531-1548.

Gazetas, G., Fan, K. and Kaynia, A. (1993). "Dynamic Response of Pile Groups with Different Configurations", Soil Dynamic and Earthq. Eng., Vol. 12, pp. 239-257.

Gohl, W.B. (1991). "Response of Pile Foundations to Simulated Earthquake Loading: Experimental and Analytical Results". Ph.D. Thesis, Dept. of Civil Eng., University of British Columbia, Vancouver, B.C., Canada.

Hauze, F.E. (1994). Private Communication.

Kaynia, A.M. (1982). "Dynamic Stiffness and Seismic Response of Pile Groups," Research Report, R82-03, Dept. of Civil Eng., Cambridge, Mass.

Makris, N., Badoni, D., Delis, E., Gazetas, G. (1994). "Prediction of Observed Bridge Response with Soil-Pile-Structure Interaction", J. of Structural Eng., ASCE, Vol. 120, No. 10, October.

Matlock, H. (1963). "Applications of Numerical Methods to Some Structural Problems in Offshore Operations," J. of Petroleum Tech., September.

Matlock, H., S.H.C. Foo, and L.M. Bryant (1978). "Simulation of Lateral Behaviour Under Earthquake Motion," Proc., Geot. Division Specialty Conf. on Earthq. Eng. and Soil Dynamics, American Soc. of Civil Engrs., Pasadena, CA, June, pp. 601-619.

Novak, M. (1991). "Piles Under Dynamic Loads". State of the Art Paper, 2nd Int. Conf. Recent Advances in Geot. Earthq. Eng. and Soil Dynamics, University of Missouri-Rolla, Rolla, Missouri, Vol. III, pp. 250-273.

Reese, L.C., W.R. Cox, and F.D. Koop (1974a). "Analysis of Laterally Loaded Piles in Sand," 6th Annual Offshore Tech. Conf., Houston, Texas, May, Paper No. 2080.

Reese, L.C., W.R. Cox and F.D. Koop (1974b). "Field Testing and Analysis of Laterally Loaded Piles in Stiff Clay," 7th Annual Offshore Tech. Conf., Houston, Texas, May, Paper No. 2312.

Schnabel, P.B., Lysmer, J. and Seed, H.B. (1972). "SHAKE: A Computer Program for Earthquake Response Analysis of Horizontally Layered Sites". Report EERC 71-12, University of California at Berkeley.

Ventura, C.E., Finn, W.D. Liam and Felber, A.J. (1995). "Dynamic Testing of Painter Street Overpass", Proc., 7th Canadian Conf. on Earthq. Eng., June.

Wu, G. (1994). "Dynamic Soil-Structure Interaction: Pile Foundations and Retaining Structures". Ph.D. Thesis, Dept. of Civil Eng., University of British Columbia, Vancouver, B.C., Octobert, 198 pgs.

Wu, G. and Finn, W.D. Liam. (1994). "PILE3D - Prototype Program for Nonlinear Dynamic Analysis of Pile Groups" (still under development). Dept. of Civil Eng., University of British Columbia, Vancouver, B.C.

Wu, G. And Finn, W.D. Liam. "A New Method for Dynamic Analysis of Pile Groups". Proc., 7th Int. Conf. on Soil Dynamics and Earthq. Eng., Chania, Crete, Greece, May 24-26, Vol. 7, pp. 467-474 (1995).

Seismic Behaviour of Ground and Geotechnical Structures, Sêco e Pinto (ed.)© 1997 Balkema, Rotterdam, ISBN 90 5410 887 8

Damage characteristics of independent houses and life line structures caused by Kobe earthquake and its ground motions

Yoshinori Iwasaki
Geo-Research Institute, Osaka, Japan

Abstract

This paper describes the structural damages of Japanese wooden houses and water supply pipe lines in the near fault region caused by Kobe Earthquake of 1995. The damages of wooden house are shown to depend upon mainly three factors of geotechnical condition, distance from the fault, and house ages since construction. The damages of water supply pipe lines are also shown as to be understood as a function of geotechnical conditions and the distance from the fault. The most severe damage was found in soft soil ground for independent structures. However, the damage of continuous line structures is found much severe in the hard ground of terrace land than the soft ground.

1.Damage of Independent Structure-*Wooden House*

Among various structures, Japanese wooden type structures are selected here as a typical index to show the effect of the ground motions to independent structure. To study damage analysis for earthquake disaster mitigation study for Shiga-Prefecture, number of damages of building structures were collected from local city governments. The term "*collapse* is used if the level of the damage is identified as "*useless to live in* and/or "*so severe and the cost of repair is the same as to build new one.*

The collapse ratio[=Nc/Nt] (Nc;number of the collapsed house, Nt; total number of houses in a town block) of wooden house is shown in **Fig.1**. **Fig.2** shows geological condition in the area. The most severe damage is found in the low land alluvial deposit area in front of Rokko mountain, where the area is termed "*damaged belt zone.* In the preliminary analysis, the damage ratio is found strong correlation with the constructed year. The damage ratio increases continuously with the elapsed years

form the construction. If the houses become older, the damage rates increase. The wooden houses are grouped into four groups as

very old house; house age more than 30years after construction (before 1965)

old house; house age 20-30 years (1966 to 1975)

used house; house age 10-20 years (1976 to 1985)

new house; house age 0-10 years (1986 to 1995).

Fig.3 shows that the collapsed ratio increases with house age under the same geotechnical ground condition of soft soil ground.

The damages are analyzed and the collapsed ratio are computed for different ground types of rock, hard ground, soft ground, and liquefied ground and are shown in **Figs.4**. The liquefied area resulted in smaller ground motion and the collapsed ratio is less than soft ground and is almost the same as hard ground.

It is clearly shown that damage ratio depends upon these important three factors as

distance from the fault

ground type

house age since construction

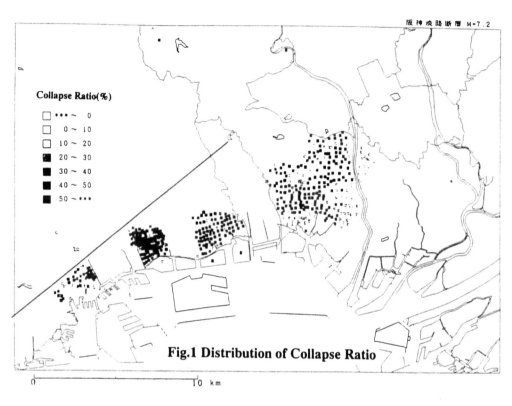

Collapse Ratio(%)

    ••• ~ 0
    0 ~ 10
    10 ~ 20
    20 ~ 30
    30 ~ 40
    40 ~ 50
    50 ~ •••

Fig.1 Distribution of Collapse Ratio

0   10 km

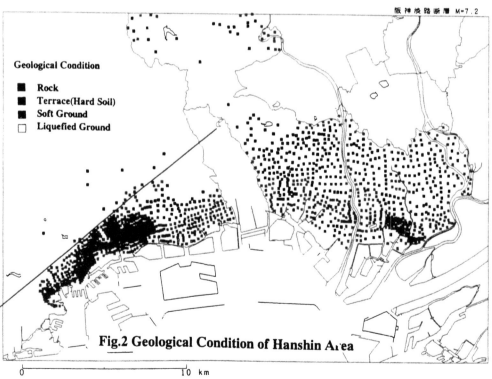

Geological Condition

    Rock
    Terrace(Hard Soil)
    Soft Ground
    Liquefied Ground

Fig.2 Geological Condition of Hanshin Area

0   10 km

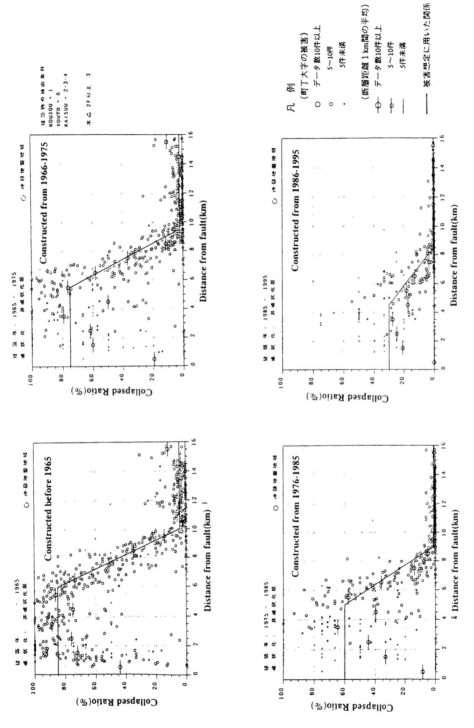

Fig.3 Collapse Ratio of Wooden Houses on Soft Ground

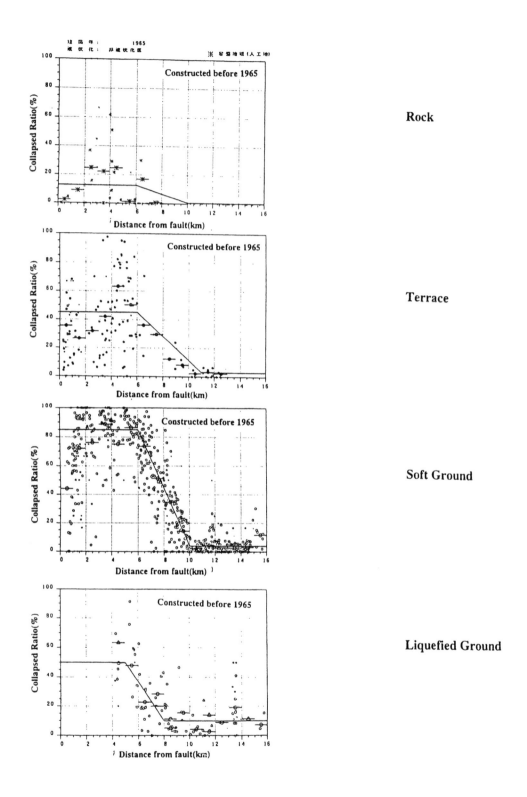

Rock

Terrace

Soft Ground

Liquefied Ground

Fig.4 Collapse Ratio of Wooden Houses on Different Ground Constructed before 1965

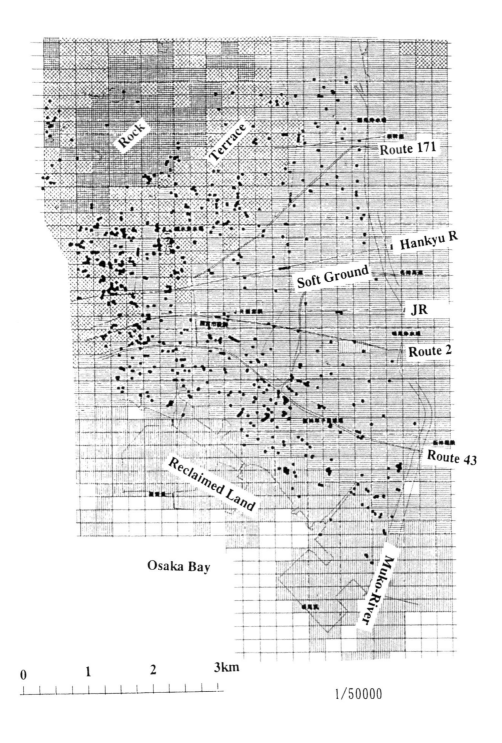

Fig.5 Damage Distribution of Water Supply Pipe in Nishinomiya

209

If the house was built on soft ground and is located within 6-7km from the earthquake fault, the expected damage ratio varies from very high of 80-90% for very old house to medium high of 30% for new house. If the distance from the fault becomes more than 6km, the damage ratio decreases with distance and becomes minimum level at 10km.

2.Damage of Continuous Structure- *Water Supply Pipe Lines*

It is interesting to see damages of pipelines in the ground. **Fig.5** shows the distribution of damaged points as black dots of water system in Nishinomiya based upon the report by Japanese Association of Water Supply. This figure also shows geological conditions in the city. Granite rock is found at north west area which accompanies Pleistocene deposit at the south eastern border, which has been lifted by reverse fault movements in the past and is identified terrace deposit at present. In the central and southern area, low flat land of alluvial deposit, which consists of alluvial soft ground in the central and reclaimed area in the southern area. Liquefaction sites are identified in the southern area especially south of national road of route 43.

There are two damaged zones of pipe lines in Nishinomiya One is the damage concentration in the terrace zone. Another concentration is found in the liquefied zone. In the terrace zone, topology in this area is rather complicated with small valleys and hills. The development of this area required cut and fill earthworks in small scale. The boundaries of cut and filled area were potentially identified main factor to cause damages of pipe lines.

The area is subdivided into 250m grid mesh and the damage ratio(number of damaged points per pipe length of one kilometer) for each mesh is calculated.

The intensity distribution of the damage ratio among each geotechnical ground condition based upon the mesh data is shown in **Table.1.**

The averaged damage ratio is 4 (points per one km length)for rock, 16 for terrace, 6 for low alluvial, and 18 for liquefied areas.

In rock area, the damage ratio is 4, which is caused in the boundary between cut and filled zone and/or boundary of fault block.

Table-1 Damage Ratio in Different Ground Condition

Damage Ratio (number/km)	Rock (%)	Terrace (%)	Alluvial (%)	Liquefied (%)
greater than 6.0	4	6	0	9
4-6	0	6	0	6
2-4	23	7	19	
1-2	4	22	27	23
0-1	82	44	66	43

If the terrace in Kobe and Nishinomiya area were more flat and were developed with less cut and filled earth work, the damage ratio might show almost the same ratio as shown in rock area.In average, the ratio in the liquefied area is three times larger than those in the alluvial area.

Table-2 Averaged Damage Ratio of Pipe Lines for Different Ground

Damage Ratio (number/km)	Rock	Terrace	Alluvial	Liquefied
	4	16	6	18

In the alluvial plain, the damage ratio could be divided by several zones based upon distance from the earthquake fault. The ratio is obtained as shown in **Table-3**.The damage ratio decreases with increase of the distance from the fault.

Table-3 Damage Ratio of Water Supply lines in Alluvial Plain

Distance from fault(km)	6	7	8	9	10
Damage Ratio(number/km)	6	6	4	3	2

3.Near Fault Ground Motion

The maximum ground motion in the near faults on rock site was recorded 305gals and 55kines at Kobe University which locates within one kilometer from the fault. The recorded motions on hard ground was 818gals and 91kines at JMA, Kobe observatory and 635gals and 138kines on soft ground at Takatori JR station.

The attenuation of the maximum horizontal ground motions from the fault were shown in **Fig.6** in log-log scales. In the figure, an experimental curve proposed by Fukushima(1994), was also plotted. They are found rather in good correspondence each other in near fault distance. It might be said there is a tendency for the acceleration to reach a some maximum limit near the fault and become rather constant. However, **Fig.7** shows the same relationship in normal scales. The acceleration increases rather abruptly and shows several times larger values than those of outside of the near fault region. This very strong ground motion is the one of the special characteristics of the near fault ground motion.

The Attenuation of Near Fault Earthquake

There are several factors which control the amplitude of ground motions from the fault. If the attenuation is assumed as inversely proportional against distance from a point source, the attenuation of the ground motion depends upon the source depth. **Fig.8** shows the computed attenuation curve near the fault.

Fig.9 shows that the attenuation of observed Peak Horizontal Acceleration (PHA) against distance from fault for different ground conditions. In these figures, there are two dotted lines, which is a computed attenuation curve for a single point source at the depth of 5km and 10km taking the averaged observed values at 20km as a reference point.

Fig.10 also shows the observed Peak Horizontal Velocity(PHV). The observed values are plotted within the estimated values of attenuation curves for the source depth of about 5-10km from the surface. If the depth of the source becomes shallower, the more rapid decay is expected to have been observed. It is understood that the near site from the earthquake fault has greater potential to receive very strong earthquake if the source becomes very shallow.

Fukui Earthquake(M=7.3) of 1948 was another good example of damaged distribution caused by very near earthquake fault.

Fig.11 shows the distribution of damaged ratio with circle in the Fukui Plain, where the thickness of the deposit to the Tertiary base rock is about a few hundreds meters. Soft alluvial surface layer covers top surface of the basin.

Fig.12 shows the attenuation of collapsed ratio and the distance from the Fukui fault. The damage ratio reached almost 100% at near sites within about 6-7km from the fault. The damage ratio decreases with distance from the fault. This shows the same characteristics of damages caused by near fault earthquake. The damage concentrates within the narrow area from the fault because of the very large intensity of the ground motion.

4.Conclusions;

Based upon case study of damage of wooden house and water supply pipe line, several lessons learnt are

1. Within the near earthquake fault region, the ground motions become very strong and the seismic design should be considered based upon the characteristics of effects of ground motions caused by the specified active fault.

2. The ground motion was very severe especially in the near fault region within 6-7km from fault and the motions were two to three times higher level than those expected by the existing seismic codes.

3. Characteristics of structural damages differs according to geotechnical conditions as well as structural types. Damages of independent structures depend mainly upon the ground motion intensity itself, on the other hand, those of continuous line structure depend not only ground motion but also the differential local displacements.

4. Tendency of the damages of independent structure differs from continuous line structure among the geotechnical conditions as follows,

   independent structure
   soft alluvial>hard terrace>liquefied ≒rock
   continuous line structure

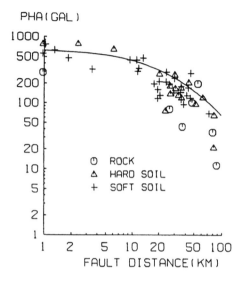

Fig.6 Attenuation of
the Observed Ground Motion
(Log-Log)

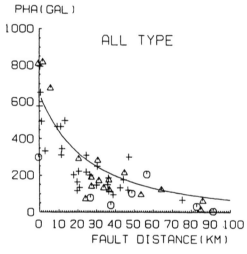

Fig.7 Attenuation of
the Observed Ground Motion
(Normal)

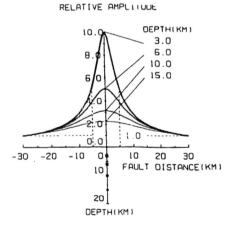

Fig.8 Attenuation Curve
for Point Source

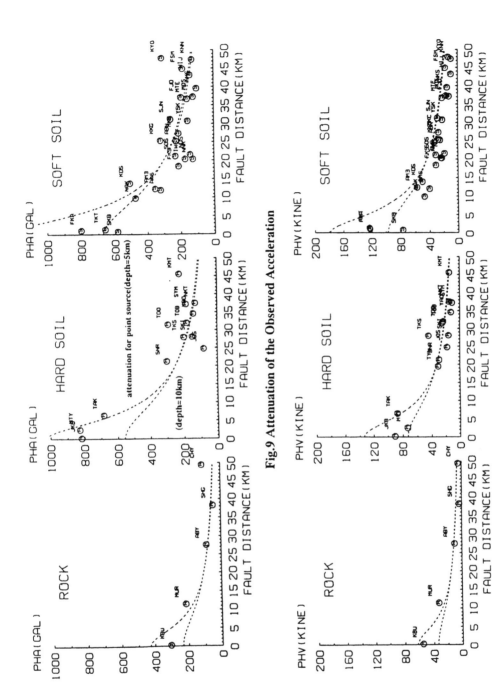

Fig.9 Attenuation of the Observed Acceleration

Fig.10 Attenuation of the Observed Velocity

213

liquefied>hard terrace>soft

degree of damage   greater   >>   smaller

in diluvial>≒rock

This is because the development of the terrace zone in small scaled cut and filled earthwork which resulted in buring the pipe lines crossing the boundaries many times between different geotechnical stiffness.

Reference

Fukushima(1994)Experimental Prediction of Input Earthquake Ground Motion based on the Theoretical Base of Source and Propagation Theory, Ohsaki Research Inst.

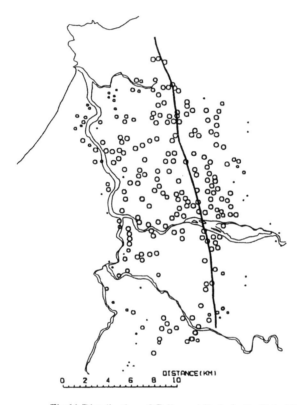

Fig.11 Distribution of Collapsed Ratio in the Fukui Basin of Soft Ground by Fukui Earthquake of 1948

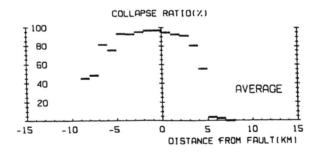

Fig.12 Collapsed Ratio in Soft Ground and Distance from the Fault

Seismic Behaviour of Ground and Geotechnical Structures, Sêco e Pinto (ed.) © 1997 Balkema, Rotterdam, ISBN 90 5410 887 8

The shearing of foundations on an active fault

J.L.Justo & C.Salwa
Department of Continuum Mechanics, University of Seville, Spain

ABSTRACT: The paper describes the failure of a estate of 20 houses built on the verge of a slope near Granada in the ravine of river Gorja. Fissures appeared before the end of construction, and distress little time after the end. Somes houses were demolished. A throurough investigation showed that old landslides had occurred near the site. The slopes of two terraces were actually active faults, as shown by the vertical and horizontal movements measured later along them. The stability of the slope was calculated by the Plaxis finite element program. The strength was below the drained value due to the strain softening produced by the tectonic shear stresses. Actually creep movements, probably combined with tectonic displacements affected the houses.

1 INTRODUCTION

Many houses are placed at the site of San Andreas fault, and it is difficult and expensive to enforce pleople to abandon them. Notwithstanding in the new codes there is an increasing concern with construction near active faults.

At U.S.A. the Applied Technology Council (ATC, 1978) prescribes that "no new building or existing building which is, because of change in use, assigned to Category D shall be sited where there is the potential for an active fault to cause rupture of the ground surface at the building". We must advise that category D is assigned to provide the highest level of design performance criteria.

Part 5, "Foundations, retaining structures and geotechnical aspects", of Eurocode 8 "Design provisions for earthquake resistance of structures", contains in its paragraph 4.1, "Siting" several paragraphs related with the construction in the proximity of seismically active faults:

"The site of construction and the nature of the supporting ground shall be such as to minimize hazards of rupture, slope instability... in the event of an earthquake". "The possibility of ocurrence of these adverse phenomena shall be investigated as specified in the following clauses". ""Buildings of importance categories (I, II, III) shall generally not be erected in the inmediate vicinity of tectonic faults..." "Special geological investigations shall be carried out for urban planning purposes... near potentially active faults in areas of high sismicity, in order to determine the ensuing hazard in terms of ground rupture..."

The importance of these statements has been enhanced by the failure of a estate of 20 houses, called "los Naranjos", with sport fields and gardens built on the verge of a slope near Granada, in the ravine of river Gorja, in the place called "Barranco Hondo". The zone is seismically active and at the other side of the ravine several landslides were visible. Notwithstanding, the site was included in the revision of the urban plan of the zone as able for construction. No geological study was undertaken before construction. A light geotechnical investigation was carried out when one half of the houses were built. Although there was a warning respect to the landslides no special provisions were taken. Fissures appeared before the end of construction, and distress little time after the end. Some houses were demolished. A thorough investigation showed that old landslides had ocurred near the site. The slopes of two terraces were actually active faults, as shown by the vertical and horizontal movements measured later along them.

2 GEOLOGY AND TECTONICS

They have been described by Justo and Escrig (1992).

The zone lies in the south western border of the Tertiary Granada basin, in the foot-hills of Sierra

Nevada. In the region of study the Tertiary deposits are covered with the alluvial cone of Zubia (fig. 1). This alluvial cone consists of detritic sediments, varying from silt to gravel, partly cemented, forming calcareous crusts, with clayey levels. The layers are nearly horizontal with a dip of 10°, and with a structure of palaeochannels. The stones derive from the Triassic limestones and dolomites that crop out in the neighbourhood. According to Chacon (v. Justo and Escrig, 1992), below the cone, at depths ranging from 6 to 20 m, lie conglomerates with silt and clay of Pliocene-Quaternary age. Figure 2 shows a cross-section of the ravine.

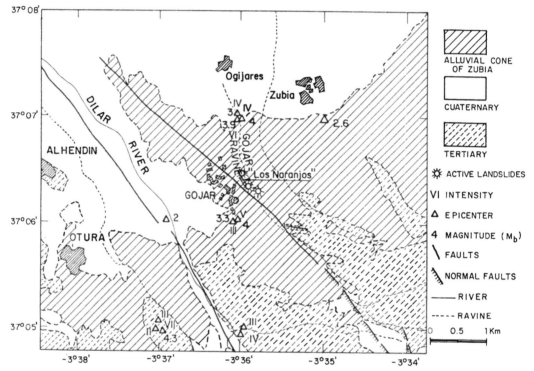

Figure 1. Geology and Tectonics

The whole area is affected by neotectonic processes starting in the upper Miocene. Usually the space extent of neotectonic deformation is mirrored by the distribution of instrumentally recorded seismicity. The areas characterized by neotectonic movement are expressed as seismically active zones, commonly coincident with the tectonic structures. According to this, the Granada basin is a good example of neotectonic activity, with high seismic activity of small magnitude, and faults which affect the superficial sediments. The predominant system of faults has the direction NW-SE in the whole Granada basin, which is coincident with the direction of many scarps. Within this system, the faults responsible for the seismic activity are the vertical and normal faults with a predominant direction N30°W to N60°W (fig. 1). In the surroundings of the Gojar ravine, there is an identified fault which might be responsible for the landslides. As a matter of fact, the direction of the Barranco Hondo is within the range of the predominant faults in the area, and the waters might have driven its way along a fracture. Table 1 collects the earthquakes registered in the surroundings from 1909 till 1989 (when damage in Los Naranjos started). The epicenters have been drawn in figure 1.

3   AERIAL PHOTOGRAMMETRY AND LANDSLIDES

Aerial photographs of the zone from 1956 till 1990 have been carefully examined (v. Justo and Escrig, 1992). We observe a series of landslides in a length of 900 m in the right bank of river Gorja, and, in both banks, a net of sub-vertical fractures, which have become the gullies for the access of water to the ravine. The large landslide that is shown in figure 3 is located in the right bank inmediately upstream of the housing estate and has a width of 200 m; the slip at this site may go back to the 19th century, but towards 1946 was activated anew and ruined the construction that existed there (v. figure 3).

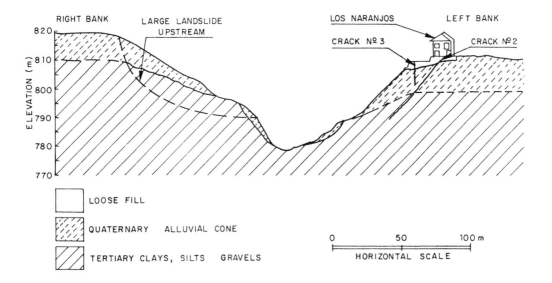

Figure 2. Cross-section of the ravine.

Table 1. Earthquakes registered in the surroundings of "Los Naranjos" from 1909 till 1989.

Date	Longit.	Latit.	Prof. (km)	M_b	I
29/9/1909	-3°36'	37°05'	-	-	IV
10/10/1953	-3°36'	37°06'	-	4	V
29/1/1964	-3°36'	37°07'	5	4	IV
9/9/1964	-3°37'	37°05'	-	-	-
9/9/1964	-3°37'	37°05'	-	-	III
9/9/1964	-3°37'	37°05'	5	4.3	VII
10/9/1964	-3°37'	37°05'	-	-	II
10/9/1964	-3°37'	37°05'	-	-	II
3/3/1977	-3°36'	37°06'	33	3.3	III
29/4/1979	-3°36'	37°07'	5	3	IV
31/7/1979	-3°36'	37°07'	5	3.9	VI
1/8/1985	-3°35'	37°07'	5	2.6	-
7/9/1987	-3°36'	37°05'	5	3.3	III
22/4/1989	-3°37'	37°06'	9	2	-

In the left bank, in the site of Los Naranjos, the scar of an old landslide could be been in the position of the present crack No. 1; in 1987 a fill was dumped on it. A small scar appears to the north of the estate, between houses No. 1 and 2 (figure 4). Stories are told about slides and cracks at the site in the past.

4   OCCURRENCE OF THE EVENTS

Figure 4 shows the plan of the housing estate. The former fissures appeared in May 1989, when the structures of houses 1 to 11 were ended. The construction finished in the middle of 1990; by this time cracks many centimeters wide had opened in walls and gardens. House No. 1 was underpinned with micropiles in September 1990, and houses No. 7, 8, 9 and 10 in April 1991; in all cases the underpinning failed.

Figure 4 shows the state of the cracks the 1st May 1991. Houses 1, 2, 7, 8, 9 and 10, that were crossed by the cracks, were in very bad condition (figure 5). Most of the walls, the tennis court and the swimming pool presented widespread damage. Three families of cracks could be distinguished. Crack No. 1 coincides with the scarp of the old landslide that could be seen in the aerial photographs, and was refilled in 1987. Crack No. 2 is the longest crack; to the east it coincides with crack No. 1 (gardens of houses No. 16 and 15), but further on it comes apart from it: crosses the gardens of houses No. 14, 13, 12 and 11,

cuts the houses from No. 10 to No. 7, approaches the façades of houses No. 6 and 5, and follows to the street and other houses not belonging to this estate. Finally crack No. 3 is paralell to crack No. 2; it starts in the garden of house No. 6, continues through the gardens of houses No. 5 to 2, where it affects somewhat the building, cuts through house No. 1, joins to the outer crack in the street, and finally closes again on the slope, forming the scarp of a landslide. House No. 1 was demolished.

Figure 3. "Los Naranjos" to the right. Large landslide in front.

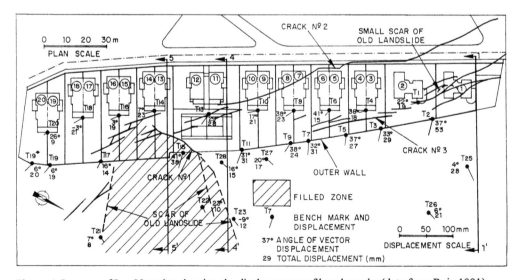

Figure 4. Lay-out of Los Naranjos showing the displacements of bench marks (data from Ruiz 1991).

5 TOPOGRAPHIC MEASUREMENTS

Bench marks were placed in the terrace (figure 4), and three-dimensional displacements measured from the opposite side of the ravine by optical methods. The initial reading was taken the 1st May 1991, when the cracks were well estalished. In figure 4 the horizontal displacement of each bench mark from the 1st May 1991 till the 28th July 1991 has been drawn. We have put at the side of each bench mark the angle that forms the vector displacement with the horizontal and the total displacement in mm.

Figure 5. House No. 10 crossed by the cracks.

The following coments can be made:
- The displacements measured above crack No. 2 dip an average angle of 8° below the horizontal and are nearly paralell to the crack.
- The displacements measured on the slope dip an average angle of only 10.6° below the horizontal and are, in most cases, perpendicular to the cracks.
- The displacement of the outer wall of the estate is directed towards the west and dips an angle of 27.7° below the horizontal, not far from the average angle of the slope.
- The displacements measured near the façade of the houses or in the tennis court have a large component in the direction of the cracks.

All this indicates that there are important displacements that do not seem to be gravitatory, but would correspond better to strike-slip displacements in the direction of the cracks. The importance of horizontal displacements is, in any case, too large for the head of a landslide, and could correspond to the stress relief produced by the carving of the bed of the ravine. The rate of total displacement during the observation period was 0.25 mm/day.

6 RELATIONSHIP BETWEEN THE TOPOGRAPHY BEFORE THE CONSTRUCTION AND THE CRACKS

Figure 6 shows the topography (simplified) before construction. The cracks have been superimposed on it. Before construction the ground was divided into terraces separated by declivities. We may see that the declivity separating the upper and second main terraces nearly coincides with crack No. 2. On the other hand, the declivity at the foot of the second main terrace is near to crack No. 3. All this, together with the stories about past slides and cracks indicated in §3 suggest that the declivities are actually scarps of old landslides. We may see in figures 4 and 6 that the predominant direction of cracks 2 and 3 is within the range of the faults in the zone (v. §2). This, together with the direction of the measured displacements, suggest a tectonic origin for the scarps. The identification between declivity and scarp for crack No. 2, and a detailed examination of the aerial fotographs, suggest a length of at least 450 m for this crack, half of which is beyond both ends of the housing estate, exceeding quite what would be a landslide produced only the construction operations of the estate.

7 GEOTECHNICAL PROPERTIES

The following soil types may be distinguished in the ground:
1. Very soft fill and top soil
2. Clayey and silty sand, from medium to dense (SC, SM)
3. Clayey and silty gravel and sand, from dense to very dense (GC, SM, GP)
4. Red clay with calcareous stones, from firm to hard (CL)

The alluvial cone is formed by intermixed layers of soil types 2 and 3 (prevailing) and some layer of soil type 4, with thicknesses ranging from less than 1 m to a few meters. Below appears the Pliocene-Quaternary, formed by type soil 4.

Table 2 collects the average properties of the soils.

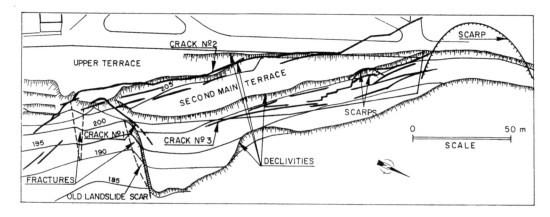

Figure 6. Comparison between the topography before construction and the cracks.

Table 2. Average properties of the soil types

Soil type	N	w_L	I_p	w %	ρ_d kg/m³	q_u kPa	c' kPa	φ
1	2	-	-	12.4	1730	-	-	-
2	35	29	6	17.2	-	-	-	-
3	44	25.3	7.8	-	-	-	-	-
4	95	34.4	14	17.5	1760	189	23	20.8°

N (blows/30 cm)

As soil type 2 is lightly cemented with lime, we often find there honeycombs.

The water level appears at depths ranging from 11.85 up to 22.00 m.

The fill placed on the scar of the old landslide (figure 4) is very loose. There are fills below the construction and gardens of houses No. 1 to 4, at least in some cases very loose. In the rest of the state there are old consolidated fills. Apart form the filling of the old scar (made a little before construction), the construction operations cut soil from the upper part of the estate and dumped it, loose, in the lower part, as the fill of the outer wall and to level the gardens.

8 STABILITY CALCULATIONS

The stability of the slope was calculated by the Plaxis finite element method. The method is elasto-plastic and Mohr-Coulomb properties are assumed for the soil. The parameters used in the calculations are collected in table 3, and drained conditions were supposed. No tension strenght is assumed in the fill.

Sections 1, 4 and 5 (figure 4) have been

calculated. Rupture was reached by a decrease of the shear strength parameters, and the factor of safety, calculated in this way, is shown in table 4. We see that the strength is below the drained value, due to tectonic or gravitational shear stresses that have produced previous deformations that have surpassed the peak strength.

Table 3. Parameters used in finite element calculations

Layer	c' kPa	φ'	γ kN/m³	ψ	G kPa	υ
Fill	1	25°-40°	19.5	0	1000	0.35
Alluvial cone	1-10	34°	22	5°	12000	0.45
Lower clay	23	21°	21.3	0	8000	0.30

Table 4. Calculated factor of safety

Section	F	φ' fill	c' (kPa) alluvial cone
1[1]	1.4	25°	1
4[2]	1.29	40°	10
5[3]	1.104	36°	1

The tension cut off points in the fill appear not far from the place where cracks exist.

In section 5 a stability calculation was carried out with the topografy existing at the site before construction and before the scar of the old lanslide was filled (v. figure 4). The factor of safety was 1.16. The construction operations produced a reduction of

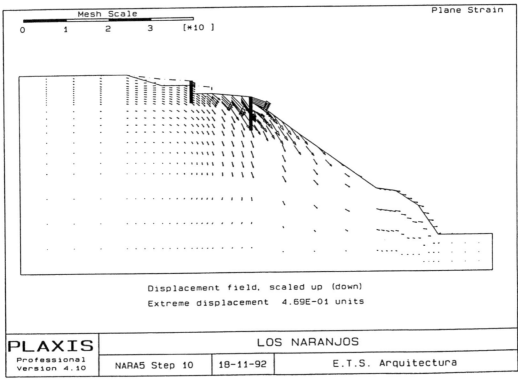

Mesh Scale
0   1   2   3   [*10]

Plane Strain

Displacement field, scaled up (down)
Extreme displacement 4.69E-01 units

PLAXIS	LOS NARANJOS		
Professional Version 4.10	NARA5 Step 10	18-11-92	E.T.S. Arquitectura

Figure 7. Total displacements produced by the construction operations (section 5).

the factor of safety to 1.104, a displacement of the outer wall of 47 cm (figure 7) and a tilt of this wall similar to the observed one. Figure 8 shows the plastic points and the tension cut-off points (points where it was necessary to redistribute the tension stresses in the no tension strength fill). We see that many of them are in the platform near to the tennis courts, where actually cracks appeared.

9 SHEARING OF THE FOUNDATIONS

As indicated above, house No. 1 was demolished. Figure 9 shows, to the left, crack No. 3 crossing the foundation. The crack has teared the tie-beams from the dwarf pillars, due to the settlement and, above all, the tensile displacement of the foundation. Horizontal deformations prevail.

Figure 10 shows another view of figure 9. A defective micropile from the underpining carried out in September 1990 may be seen at the centre of the figure.

10 CONCLUSIONS

The failure of a housing estate in the verge of the slope of a ravine, in a seismically active zone, shows the danger of constructing in places where tectonic and gravitational loads converge. It has been shown the need for investigating possible active faults or the scarps of old lanslides, that may be masked by agricultural works. Some places should be considered not able for building in the urban planning.

Leakage from an underground irrigation channel in the upper street, parallel to the housing estate, may have also contributed to failure.

Crack No. 1 is produced by settlement of the uncompacted fill dumped there before construction.

The ground was treated by pressure grouting. The works ended the 1st April 1992. From the 11th May there was no increase in the ground displacements, measured by means of inclinometers up to July 1992. Notwithstanding some opening of capillary fissures was produced.

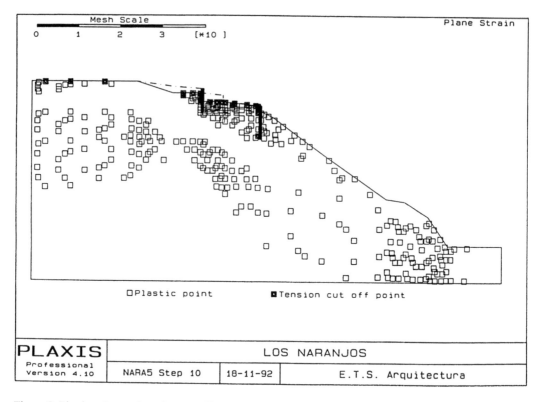

Figure 8. Plastic points and tension cut-off points in cross-section 5.

Figure 9. Shearing of foundations produced by crack No. 3

Figure 10. Another view of figure 8 showing defective micropile

REFERENCES

ATC. 1978. *Tentative provisions for the development of seismic regulations for buildings*. Washington: National Science Foundation.

Justo, J.L. & Escrig, F. 1992. *Dictamen sobre la influencia del terreno en los daños sufridos por la Urbanización los Naranjos*. Granada: Juzgado de Instrucción n° 6. Unpublished report.

Ruiz, J.L. 1991. *Informe sobre la dinámica de laderas que afecta a la margen izquierda del Barranco Hondo, en las inmediaciones de la Urbanización "Los Naranjos"*. *Gojar, Granada*. Granada. Unpublished report.

Seismic Behaviour of Ground and Geotechnical Structures, Sêco e Pinto (ed.) © 1997 Balkema, Rotterdam, ISBN 90 5410 887 8

Natural frequency response of structures considering soil-structure interaction

Sanjeev Kumar
Geotechnology, Inc., St. Louis, Mo., USA

Shamsher Prakash
University of Missouri-Rolla, Mo., USA

ABSTRACT: One of the important parameters in the analysis and design of structures is their natural frequency response. Natural frequencies of a structure fixed at the base are different from the frequencies of a similar structure supported on soil i.e. flexible based structure. Results of a comprehensive parametric study, conducted using a simplified model to perform soil-structure interaction analysis, are presented. It is shown that there is a significant effect of considering soil-structure interaction on the *fundamental* natural frequency of the system. However, there is a negligible effect of considering soil-structure interaction on the natural frequencies in *higher* modes. Results are also presented to show that there is a negligible effect of the number of stories or height of the structures, founded on soil, on the natural frequencies of structures in the *highest and second highest* modes for similar foundation characteristics (stiffness and damping). The *fundamental* natural frequency of the structures is observed to increase nonlinearly with the increase in soil shear modulus.

INTRODUCTION

Structures are generally supported on soils unless rock is very near to the ground surface. The behavior of structures under static or dynamic loads, founded on soils, is different from that of similar structures founded on rock. This difference in the behavior is because of the phenomenon commonly referred to as *Soil-Structure Interaction (SSI)*. Problems involving dynamic loads differ from the corresponding static problems in two important aspects: the time varying nature of the excitation; and the role played by inertia. The response of a structure supported on soil and excited at the base by a dynamic force, estimated on the basis of static soil-structure interaction models or rigid foundation assumption, cannot authentically match actual performance. Therefore, the *Dynamic Soil-Structure Interaction (DSSI)* effect on the seismic response of a structure should not be overlooked (Luan et al., 1995).

The exact analysis considering soil-structure interaction requires that the structure be considered to be a part of a larger system which includes the foundation and the supporting medium and that the effect of the spatial variability of the ground motion and the properties of the soils involved be taken into consideration. However, because of the complexity of the problem, performing such a detailed analysis for all types of structures is not always possible in practice. Therefore, soil-structure interaction analysis under seismic excitation can be conveniently performed in three steps (Gazetas et al., 1992, Veletsos et al., 1988):

1. Obtain the motion of the foundation in the absence of the superstructure. The resulting motion is called "foundation input motion".

2. Determine the dynamic impedances of the foundation (spring and dashpot coefficients) associated with translation, rocking, and cross-rocking oscillations.

3. Compute the seismic response of the superstructure supported on springs and dashpots of Step 2, and subjected at its base to the foundation input motion of Step 1.

For Steps 1 and 2 various formulations have been developed and published in the literature (for Step 1: Idriss and Sun, 1992; Faccioli, 1991; Rosset, 1977; Schnable et al., 1972 and for Step 2: Dobry and Gazetas, 1985, 1988; Gazetas, 1984, 1991; Gazetas et al., 1992; Novak, 1974; Novak and El-Sharnouby, 1983; Roesset and Angelides, 1980).

The response of a structure is governed by the equation of motion (1).

$$[M]\{\ddot{x}\} + [C]\{\dot{x}\} + [K]\{x\} = -[M_F]\{1\}\ddot{x}_g \qquad (1)$$

where [M], [C], and [K] are mass, damping, and stiffness matrices, respectively. $[M_F]$ is the matrix to calculate force at each degree of freedom.

SIMPLIFIED MODEL TO PERFORM DSSI

In practice the mass, stiffness, and damping matrices of a combined foundation-structure system are assembled from the corresponding matrices of the two subsystems (foundation and superstructure) as shown in Figure 1 (Chopra, 1995). The portion of these matrices associated with common degrees-of-freedom at the interface between the two subsystems include contribution from both subsystems.

Kumar (1996) and Kumar and Prakash (1997) have shown that the formulation of matrices as shown in Figure 1 may not be appropriate particularly when rotation of the base/foundation is considered. Kumar (1996) developed a simplified model to perform soil-structure analysis of shear type structures with flexible foundations. For an n-story, shear type structure, response can be calculated using Eqn.(2).

Development of the model is discussed in detail elsewhere (Kumar, 1996 and Kumar and Prakash, 1997). Equation (2) is the simplified equation of motion for an n-story shear type structure, which also includes secondary forces (forces due to P-Δ effect) and time dependent external forces. The form of this equation is similar to (1). For problems where the effect of secondary forces (P-Δ effect) is not considered, the terms associated with mg can be set to zero. Similarly, if the analysis is required to be performed for only earthquake base motion, the external force vector on the right hand side of Eqn.(2) should be set to zero.

SOIL AND STRUCTURE PARAMETERS USED

Three structures, 2-, 4-, and 6-story, supported on pile foundations, are considered in the present study to perform parametric study. The structures considered consist of: story height = 3.5 m; size of each column in a story = 300 x 300 mm (2 columns in each story); damping coefficient (ξ) for structure = 0.02; mass of each story = 19.9 Mg/m^3; and mass moment-of-inertia of each story about its own axis = 38.2 kN-m-sec^2.

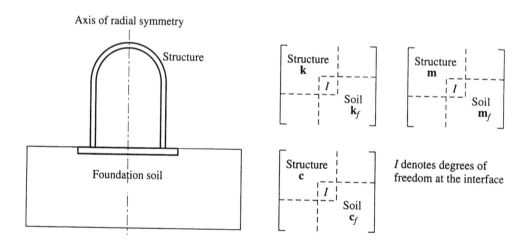

Fig. 1 Assembly of mass, stiffness, and damping materices using corresponding materices of subsystems (after Chopra, 1995)

$$
\begin{bmatrix}
m_b & 0 & 0 & 0 & \text{----} & 0 \\
0 & \rho_b + \sum_{n-1}^{n}\rho_n & 0 & 0 & \text{----} & 0 \\
0 & 0 & m_1 & 0 & \text{----} & 0 \\
0 & 0 & 0 & m_2 & \text{----} & 0 \\
| & | & | & | & & \\
0 & 0 & 0 & 0 & & m_n
\end{bmatrix}
\begin{Bmatrix}
\ddot{x}_b \\ \ddot{\theta}_b \\ \ddot{x}_1 \\ \ddot{x}_2 \\ | \\ \ddot{x}_n
\end{Bmatrix}
$$

$$
+
\begin{bmatrix}
c_x+c_1 & c_1\ell_1+c_{x\phi} & -c_1 & 0 & \text{---} & 0 \\
c_1\ell_1+c_{\phi x} & c_\phi+\sum_{n-1}^{n}c_n\ell_n^2 & c_2\ell_2-c_1\ell_2 & c_3\ell_3-c_2\ell_2 & \text{---} & -c_n\ell_n \\
-c_1 & c_2\ell_2-c_1\ell_1 & c_1+c_2 & -c_2 & \text{---} & 0 \\
0 & c_3\ell_3-c_2\ell_2 & -c_2 & c_2+c_3 & \text{---} & 0 \\
| & | & | & | & & \\
0 & -c_n\ell_n & 0 & 0 & & c_n
\end{bmatrix}
\begin{Bmatrix}
\dot{x}_b \\ \dot{\theta}_b \\ \dot{x}_1 \\ \dot{x}_2 \\ | \\ \dot{x}_n
\end{Bmatrix}
$$

$$
+
\begin{bmatrix}
k_x+k_1 & k_1\ell_1+k_{\phi x} & -k_1 & 0 & \text{---} & 0 \\
k_1\ell_1+k_{\phi x}+\sum_{n-1}^{n}m_n g & k_\phi+\sum_{n-1}^{n}k_n\ell_n^2 & k_2\ell_2-k_1\ell_1-m_1 g & k_3\ell_3-k_2\ell_2-m_2 g & \text{---} & -k_n\ell_n-m_n g \\
-k_1 & k_2\ell_2-k_1\ell_1 & k_1+k_2 & -k_2 & \text{---} & 0 \\
0 & k_3\ell_3-k_2\ell_2 & -k_2 & k_2+k_3 & \text{---} & 0 \\
| & | & | & | & & \\
0 & -k_n\ell_n & 0 & 0 & & k_n
\end{bmatrix}
\begin{Bmatrix}
x_b \\ \theta_b \\ x_1 \\ x_2 \\ | \\ x_n
\end{Bmatrix}
$$

$$
= -
\begin{bmatrix}
m_b & 0 & 0 & 0 & \text{----} & 0 \\
0 & 0 & 0 & 0 & \text{----} & 0 \\
0 & 0 & m_1 & 0 & \text{-----} & 0 \\
0 & 0 & 0 & m_2 & \text{----} & 0 \\
| & | & | & | & & \\
0 & 0 & 0 & 0 & & m_n
\end{bmatrix}
\begin{Bmatrix}
1 \\ 1 \\ 1 \\ 1 \\ | \\ 1
\end{Bmatrix}
\ddot{x}_g +
\begin{Bmatrix}
P_{bx}(t) \\ P_{b\theta}(t) \\ P_1(t) \\ P_2(t) \\ | \\ P_n(t)
\end{Bmatrix}
\qquad (2)
$$

The mass and mass moment-of-inertia of each story are calculated from assumed floor dimensions. The pile foundation supporting each column consists of four, HP 12X72 steel H-piles at a center-to-center spacing of 0.9 meters. Each pile cap is assumed to have dimensions of 1.6m X 1.6m X 1.0m. Actual and effective depths of embeddment of pile caps are 0.8 m and 0.6 m, respectively. The response of the structures is calculated for a wide range of soil shear modulii, from 15,000 kPa to 170,000 kPa. Two types of base input motions, sinusoidal and a recorded earthquake time history (El Centro earthquake of 1940, N-S component), are used to study the steady state response and the response under transient loading, respectively. Both the recorded time history and sinusoidal base motions are scaled to a maximum acceleration of 0.1g at rock. The frequency of sinusoidal base motion assumed is 0.92 Hz. Free-field ground motion at the ground surface was computed using computer program SHAKE which models one-dimensional wave propagation through layered media. Stiffness and damping of the structure are calculated using the procedures given by Chopra (1995). Stiffness and damping of foundations is calculated as recommended by Gazetas (1991).

EFFECT OF CONSIDERING *DSSI* ON *FUNDAMENTAL* NATURAL FREQUENCIES

The *fundamental* natural frequency response of the 2-story structure with respect to the low strain soil shear modulus (G_{max}) is shown in Figure 2. The *fundamental* natural frequencies of this structure, fixed at the base, is 2.13 Hz, and for flexible based structure, it varies from 1.50 Hz at G_{max} = 15,000 kPa to 1.98 Hz at G_{max} = 170,000 kPa. Similar results for the 4- and 6-story structures are presented as Figures 3 and 4, respectively. For the 4-story structure (Figure 3) the fixed base fundamental frequency computed is 1.20 Hz and that for flexible based structure the variation is from 0.75 Hz at G_{max} = 15,000 kPa to 1.06 Hz at G_{max} = 1,70,000 kPa. Figure 4 shows that the fundamental natural frequency of the 6-story structure, fixed at the base, is 0.83 Hz, whereas the fundamental natural frequency for the similar structure, flexible at the base, varies from 0.47 Hz at G_{max} = 15,000 kPa to 0.71 Hz at G_{max} = 170,000 kPa. Figures 2 through 4 show that, when soil-structure interaction

is considered, the *fundamental* natural frequency of the system for the flexible based structure is lower than the corresponding value for the fixed base structure. Also, the *fundamental* natural frequency of the flexible based structures increased nonlinearly with the increase in soil shear modulus and approaches the fundamental frequency of a fixed base structure. Thus if a very high value of G_{max} is used, the fundamental natural frequency of the system will be close to the fundamental frequency of the fixed base structure. However, computations were not made because a very high value of G_{max} will make the matrices ill-conditioned and, therefore, may give erroneous results.

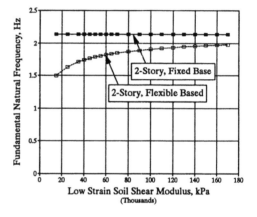

Fig. 2 *fundamental* natural frequency with modulus (2-story structure)

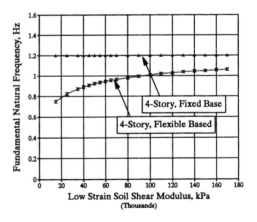

Fig. 3 *fundamental* natural frequency with modulus (4-story structure)

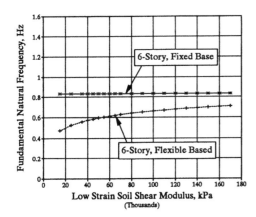

Fig. 4 *fundamental* natural frequency with modulus (6-story structure)

EFFECT OF CONSIDERING *DSSI* ON *ALL* NATURAL FREQUENCIES

In most of the practical problems, *fundamental* natural frequency of the system is of interest. However, frequencies in higher modes may also be of interest depending on the frequency of the base motion. For better understanding of the response of structures, natural frequency response of structures in *higher modes* is also presented. The study is performed assuming the linear and nonlinear behavior of soils.

As stated earlier, the study is performed for a range of low strain soil shear modulii. However, for clarity, the results are presented for a typical G_{max} of 55,000 kPa. Similar observations, as discussed herein for G_{max} of 55,000 kPa, were made at other low strain soil shear modulii (Kumar, 1996).

Linear Elastic Soil. Table I shows the natural frequencies of the 2-, 4-, and 6-story structures (fixed base and flexible based), for *all* modes, at a typical soil modulus value of (G_{max}) 55,000 kPa.

The following observations are made from the natural frequencies listed in Table I.

1. There is a significant effect of dynamic soil-structure interaction on the *fundamental* natural frequencies of the system. Similar observation is made from Figs. 2 through 4.

2. The total variation in the *fundamental* natural frequencies of all the three structures (2-, 4-, and 6-story) when flexible at the base is approximately 200% (0.6 Hz for 6-story structure to 1.8 Hz for 2-story structure) with respect to the lowest fundamental natural frequency (0.6 Hz). This shows that the number of stories in a structure or the total structure height played a significant role in the natural frequency response of the selected structures.

TABLE I. Natural Frequencies of Structures at G_{max} = 55,000 kPa.

Frequency	2-story (Hz)		4-story (Hz)		6-story (Hz)	
	Fix	Flexible	Fix	Flexible	Fix	Flexible
Fundamental natural frequency	2.13	1.80	1.20	0.93	0.83	0.60
2nd natural frequency	5.59	5.56	3.45	3.41	2.45	2.41
3rd natural frequency		12.47	5.29	5.24	3.92	3.88
4th natural frequency		42.60	6.49	6.48	5.17	5.16
5th natural frequency				11.48	6.12	6.10
6th natural frequency				42.33	6.71	6.71
7th natural frequency						10.91
8th natural frequency						42.19

3. There is a negligible effect of considering soil-structure interaction on natural frequencies in *higher* modes.

4. The natural frequencies of all three structures (2-, 4-, 6-story) in the *highest* natural mode, when flexible at the base, are nearly the same, i.e. the total variation is from 42.19 Hz to 42.60 Hz, approximately 1%. Therefore, the number of stories in a structure or the total structure height have negligible effect on the natural frequencies of the structures in the *highest* mode.

5. The variation in the natural frequencies of the flexible based structures for the second highest mode is not significant. The total variation is from 10.91 Hz to 12.47 Hz, approximately 14%. Compared to 200% variation in the *fundamental* natural frequencies. It can, therefore, be inferred that the number of stories in a structure or the total structure height has insignificant effect on the natural frequency in the *second highest* mode.

Nonlinear Soil. In this study, two types of base motions, sinusoidal and recorded earthquake ground motion are considered. Natural frequency response of selected structures for both types of base input motion are presented below. It is understood that the natural frequencies of a system, consisting of soil and structures, depend only on the properties of the soil and structure and not on the base input motion. However, this is true only when soil is assumed to behave as linear elastic. When nonlinear behavior of soil is considered, the stiffness of soil changes with time, depending on the level of strains in the soil. Also, at any particular time, the strains in soil, and thus the foundation stiffness, will not be same for any two different base input motions. The strains in the soil when the top story displacement is maximum, are presented below for sinusoidal base motion and El Centro base motion.

Type of structure	Strain in soil (%)	
	Sinusoidal	El Centro
2-story	0.061	0.253
4-story	0.286	0.129
6-story	0.092	0.144

The strain values listed above show that strain in the soil for different base input motions is significantly different even though the structure characteristics, low strain soil modulus, and peak acceleration at rock are the same.

Table II shows the natural frequencies of the 2-, 4-, and 6-story structures, assuming linear and nonlinear behavior of soil, computed at a typical value of the maximum soil shear modulus (55,000 kPa) for sinusoidal base motion. Similar results for El Centro base motion are presented in Table III. The natural frequencies presented in both the table are at the time when the top story displacement is maximum. However, when displacement in other stories and the foundation are maximum, the difference in natural frequencies from those reported in Tables II and III, if any, is insignificant for all practical purposes (Kumar, 1996).

The following observations are made from the natural frequencies listed in Table II and III.

1. There is a significant effect of considering nonlinear behavior of soil on the *fundamental* natural frequencies and natural frequencies in the *highest* and *second highest* modes. However, there is a negligible effect on natural frequencies in *other* modes.

2. The variation in the natural frequencies from the soil linear case is higher when strain in the soil is higher and vice versa.

3. The total variation in the *highest* natural frequencies of all the three structures for sinusoidal base motion (Table II) is from 20.44 Hz to 29.52 Hz (approximately 44 percent with respect to the lowest value). Similarly, for El Centro base motion (Table III), the variation is from 21.11 to 34.81 Hz (variation of approximately 65 percent with respect to the lowest value).

4. The effect of the number of stories or total height of the structure on the natural frequencies in the *highest* and *second highest* modes is significantly less compared to the effect on *fundamental* natural frequencies.

TABLE II. Natural Frequencies for Sinusoidal Base Motion at G_{max} = 55,000 kPa

Frequency	2-story (Hz)		4-story (Hz)		6-story (Hz)	
	Linear	Nonlinear	Linear	Nonlinear	Linear	Nonlinear
Fundamental natural frequency	1.80	1.60	0.93	0.67	0.60	0.49
2nd natural frequency	5.56	5.55	3.41	3.39	2.41	2.40
3rd natural frequency	12.47	10.40	5.24	5.16	3.88	3.84
4th natural frequency	42.60	29.52	6.48	6.48	5.16	5.15
5th natural frequency			11.5	9.02	6.10	6.08
6th natural frequency			42.3	20.44	6.71	6.70
7th natural frequency					10.91	9.43
8th natural frequency					42.19	25.76

TABLE III. Natural Frequencies for El Centro Base Motion at G_{max} = 55,000 kPa

Frequency	2-story (Hz)		4-story (Hz)		6-story (Hz)	
	Linear	Nonlinear	Linear	Nonlinear	Linear	Nonlinear
Fundamental natural frequency	1.80	1.36	0.93	0.87	0.60	0.46
2nd natural frequency	5.56	5.53	3.41	3.41	2.41	2.40
3rd natural frequency	12.47	9.13	5.24	5.23	3.88	3.84
4th natural frequency	42.60	21.11	6.48	6.49	5.16	5.15
5th natural frequency			11.5	10.59	6.10	6.08
6th natural frequency			42.3	34.81	6.71	6.70
7th natural frequency					10.91	9.24
8th natural frequency					42.19	23.44

DISCUSSION ON TABLES I, II, AND III

Effect of DSSI. From the natural frequencies listed in Table I, it can be inferred that when the soil is assumed to behave as linear elastic, consideration of dynamic soil-structure interaction only affected the fundamental natural frequencies. The effect on natural frequencies in the *highest* and *second highest* modes cannot be discussed because of the different number of degrees-of-freedom in the fixed base and flexible based cases.

Consideration of nonlinear behavior of soil (Tables II and III) affected the *fundamental* natural frequency, the natural frequency in the *highest* mode, and the natural frequency in the *second highest* mode. The variation in the natural frequencies is observed to depend on the level of strains in the soil. The higher the strain level the higher is the effect of soil structure interaction. Additional results on the effect of soil nonlinearity on the frequency response of structures are presented elsewhere (Kumar, 1996).

Effect of Number of Stories or Structure Height

From the observations made on Table I through III, it is concluded that if the soil behaves as linear elastic (very low strains in the soil), there is a negligible effect of number of stories or height of the structure on the natural frequencies in the *highest* and *second highest* modes. However, when soil is expected to behave nonlinearly (high strains in the soil), number of stories or total height of the structure may have some effect on natural frequencies in the *highest* and *second highest* modes, in addition to *fundamental* natural frequency because of different levels of strains for different number of stories. Therefore, it is the authors' opinion that when frequencies in the higher natural modes are of interest, analysis should be performed considering nonlinear behavior of soil.

It is observed that when the soil is assumed linear, the foundation stiffness coefficients in all the three structures remained the same irrespective of strains in the soil and therefore, natural frequencies in the *highest* and *second highest* modes remained same for all the three structures. When nonlinear behavior of soil is considered, the foundation stiffness coefficients changed in all the three structures, and so are the natural frequencies in the *highest* and *second highest* modes. Variation in the natural frequencies in other modes is negligible.

It is interesting to note that the result of inclusion of a flexible foundation, considered in developing the simplified model, is to increase in the number of degrees of freedom by 2. Any change in foundation characteristics affected *fundamental* natural frequencies and natural frequencies in the *highest* and *second highest* modes only. Therefore, the natural frequencies in the *highest* and *second highest* modes appear to be associated with the foundation characteristics only.

CONCLUSIONS

The *fundamental* natural frequency of the structures increases nonlinearly with the increase in soil shear modulus. It is shown that there is a significant effect of considering soil-structure interaction on the *fundamental* natural frequencies of the system but there is a negligible effect of soil-structure interaction on the natural frequencies in *higher* modes for similar foundation characteristics. The results presented here show that the natural frequencies of flexible based structures in the *highest* and *second highest* modes are associated with foundation characteristics only. For low-rise buildings, the effect of the number of stories in a structure or the total structure height on the natural frequencies in the *highest* and *second highest* modes is insignificant if the foundation characteristics are similar.

ACKNOWLEDGEMENTS

The authors wish to thank Mr. Amlan Sengupta, Doctorate Student, University of Missouri-Rolla for his useful suggestions and discussions during the study. Ms. Marge Sebelius, Geotechnology, Inc., edited the manuscript with great care. All help is acknowledged.

APPENDIX I. REFERENCES

Chopra, A. K. (1995). *Dynamics of structures-theory and applications to earthquake engineering.* Prentice Hall, Englewood Cliffs, New Jersey.

Dobry, R., and Gazetas, G. (1985). "Dynamic stiffness and damping of foundations by simple methods." *Vibration Problems in Geotechnical Engineering,* G. Gazetas and E. T. Selig, eds., ASCE Annual Convention, Detroit, Michigan , 75-107.

Dobry, R., and Gazetas, G. (1988). "Simple method for dynamic stiffness and damping of floating pile groups." *Geotechnique,* 38(4), 557-574.

Faccioli, E. (1991). "Seismic amplification in the presence of geologic and topographic irregularities." *Proc. 2nd Int. Conf. on Recent Adv. in Geotech. Earthquake Engrg. and Soil Dyn.,* St. Louis, Missouri, 1779-1797.

Gazetas, G. (1984). "Seismic response of end-bearing single piles." *Soil Dyn. and Earthquake Engrg.,* 3(2), 82-93.

Gazetas, G. (1991). "Foundation vibrations. "*Foundation Engineering Handbook,* H.Y. Fang, ed., Van Nostrand Reinhold, 553-593.

Gazetas, G., Fan, K., Tazoh, T., Shimizu, K., Kava, and Markis, N. (1992). "Seismic pile-group-structure interaction." *Piles Under Dynamic Loads, Geotech. Engrg. Div.,* S. Prakash, ed., Geotechnical Special Publication No. 34, ASCE, 56-94.

Idriss, I.M. and Sun, J.I. (1992). "SHAKE91: A computer program for conducting equivalent linear seismic response analysis of horizontally layered soil deposits." University of California-Davis, California.

Kumar, S. (1996). "Dynamic Response of Low-Rise Buildings Subjected to Ground Motion Considering Nonlinear Soil Properties and Frequency-Dependent Foundation Parameters", Ph.D. Thesis, University of Missouri-Rolla, Rolla, Missouri.

Kumar, S. and S. Prakash. (1997) "A Simplified Model to Perform Dynamic Soil-Structure Interaction Analysis", International Journal of Numerical and Analytical Methods in Geomechanics (submitted).

Luan, M., Lin, G., and Chen, W. F. (1995). "Lumped-parameter model and nonlinear DSSI analysis." *Paper No. 5.02, Proc. 3rd Int. Conf. on Recent Adv. in Geotech. Earthquake Engrg. and Soil Dyn.*, St. Louis, Missouri, 355-360.

Novak, M. (1974). "Dynamic stiffness and damping of piles." *Can. Geotech. J.*, 11(4), 574-598.

Novak, M., and El-Sharnouby, B. (1983). "Stiffness and damping constants for single piles." *J. Geotech. Engrg.* ASCE, 109(7), 961-974.

Rosset, J.M. (1977). "Soil amplification of earthquakes."*Numerical methods in Geotechnical Engineering*, edited by Desai, C.S. and Christian, J.T., McGraw-Hill, 639-642.

Rosset, J.M. and Angelides, D. (1980). "Dynamic stiffness of piles" *Numerical methods in offshore piling*, London, England, 75-81.

Schnable, P.B., Lysmer, J., and Seed, H.B. (1972). "SHAKE: a computer program for earthquake response analysis of horizontally layered sites." Report EERC 72-12, University of California, Berkeley.

Prakash, S., and Sharma, H. D. (1990). *Pile foundations for engineering practice*. John Wiley and Sons, NY.

Veletsos, A.S., Prasad, A.M., and Tang, Y. (1988). "Design approaches for soil-structure interaction." Technical report NCEER-88-0031

APPENDIX II. NOTATIONS

The following symbols are used in this paper:

c_x	= Damping of the foundation in translation
c_ϕ	= Damping of the foundation in rotation
$c_{x\phi}$ or $c_{\phi x}$	= Damping of the foundation in cross-rotation
c_i	= Damping of the ith story columns
g	= Acceleration due to gravity
k_i	= Stiffness of the ith story columns
k_x	= Stiffness of the foundation in translation
k_ϕ	= Stiffness of the foundation in rotation
$k_{x\phi}$ or $k_{\phi x}$	= Stiffness of the foundation in cross-rotation
ℓ_i	= i^{th} Story height from center-to-center of the masses
m_i	= Mass of ith story
m_b	= Mass of the foundation
$P_i(t)$	= Time dependent external force
x_i	= Total displacement of ith mass with respect to the position of the structure after free-field displacement
x_g	= Free-field displacement of the system
x_b	= Horizontal displacement at the base (foundation displacement) due to soil-structure interaction only. Note that this does not include free-field displacement
θ_b	= Rotation of the base
ξ_i	= Modal damping ratio for i^{th} mode
ρ_i	= Mass moment of inertia of the ith story mass about its center of gravity
ρ_b	= Mass moment of inertia of the base about its center of gravity

Seismic Behaviour of Ground and Geotechnical Structures, Sêco e Pinto (ed.) © 1997 Balkema, Rotterdam, ISBN 90 5410 887 8

Investigation of damaged foundations in the Great Hanshin earthquake disaster

Tamotsu Matsui
Osaka University, Japan

Masahiko Kitazawa, Atsushi Nanjo & Furitsu Yasuda
Hanshin Expressway Public Corporation, Japan

Abstract

In this paper, inspection methods for pile foundation damage which was caused by the 1995 Hyogoken-Nambu Earthquake are described, together with their typical results, focusing on foundations of the Hanshin Expressway structures. Also, residual lateral displacement of the foundations after the Earthquake are described, which were measured by satellite-based global positioning system(GPS).

As the results, it is concluded that all the piles located at inland area were confirmed to be sound, although they had some flexural cracks, while some piles located near waterfront of reclaimed lands suffered cracks widely distributed along pile axis and were significantly affected by liquefaction and lateral flow of ground.

1 Introduction

The 1995 Hyogoken-Nambu Earthquake, which hit the Hanshin area in Japan on January 17th, 1995, gave serious damage to structures of the Hanshin Expressway, especially on the No.3 Kobe route and the No.5 Bay route (see **Fig. 1**).

Details of degree and location of the damage can be referred in the literature (Matsui and Oda, 1996). The majority of the foundations for both routes were pile foundation, mostly cast-in-place bored piles of more than one meter in diameter.

Of these damages, the No.3 Kobe route suffered such heavy damage as falling of elevated highways and collapsing or inclining of bridge piers. Owing to such damages, it is very possible for foundation structures under the ground to suffer invisible damage. Therefore, such investigations as direct visual inspection, indirect visual inspection and nondestructive inspection were carried out on typical pile foundations classified by ground type and geography to confirm their soundness level (Ishizaki et al., 1996).

Along the No.5 Bay route that connects many recently reclaimed islands, liquefaction occurred due to the Earthquake, causing greater residual displacements in piers. For this reason it was very possible that the foundations of those piers might be damaged by the

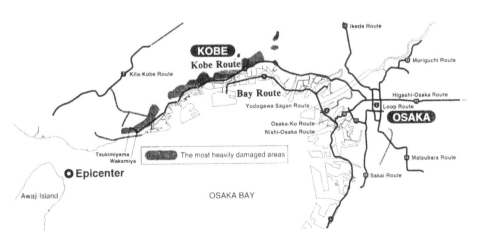

Fig. 1   Location map of the No.3 Kobe route and the No.5 Bay route

235

earthquake force or lateral flow and/or liquefaction of surrounding ground. Therefore, almost similar inspection methods as in the No.3 Kobe route were also applied to pile foundations on the No.5 Bay route to confirm their soundness.

This paper describes not only the outline of the investigation of damaged pile foundations, but also some analysis based on the results of the investigation.

2 Investigation methods for pile foundation damage

2.1 General
Investigation of pile foundation damage should be carried out from the viewpoint of investigating both damage to pile body itself and ground deformation around foundation. The inspection methods listed in **Table 1** are applied to inspect damage to pile foundation, while the investigation methods in **Table 2** are applied to know the overall information of ground deformation around foundation.

2.2 Direct visual inspection
The direct visual inspection method requires that pile surface is exposed to check cracks and other damage of pile by eye inspection. This method is the most reliable way to confirm pile damage because of the direct observation, though only externally. Inspection of orientation and intensity of cracks in piles may be indispensable in determining the degree of damage. On the other hand, the advantage of this method sometimes reduces due to the need for excavation. Inspection of deeper portion of piles needs deeper excavation and sometimes dewatering.

As described above, direct visual inspection is very reliable but has some limitations especially at areas of high ground water level.

The present inspection of damaged piles were carried out with only about one meter excavation of surrounding ground focusing on checking damage of tops of investigated piles located at the four corners of pile groups, as shown in **Fig. 2**.

2.3 Indirect visual inspection
To avoid the disadvantage of direct visual inspection method, the indirect visual inspection method using bore-hole camera was firstly applied to the investigation of damaged piles in the Hanshin Expressway No.3 Kobe route. The outline of bore-hole camera inspection system is illustrated in **Fig. 3**. This method can allow inexpensive and rapid inspection of damage piles even at a deeper portion.

Bore-hole camera system has been used to investigate the condition of joints, cracks and faults in a rock mass by observing a bore-hole wall. In the present inspection of damaged piles, a 66-mm-diameter hole is bored down at the center of investigated pile and a small TV camera was hung down into the hole to inspect cracks on the bore-hole wall.

Table 1   Inspection methods for damage to pile foundations

Kind of inspection	Inspection methods
Direct visual inspection (Eye inspection)	Direct visual inspection of cracks on pile surface.
Indirect visual inspection (Bore-hole camera inspection)	Indirect visual inspection of cracks inside pile, using a small camera inserted into a bore-hole in a pile.
Nondestructive inspection (Pile integrity test)	Nondestructive inspection of cracks and defects in pile by analyzing reflected impact wave.

Table 2   Investigation methods for ground deformation around foundations

Kind of investigation	Investigation methods
Visual survey (Eye observation)	Visual survey of cracks and settlement around piers by eyes.
GPS survey (Global Positioning System)	Satellite survey of residual displacement of piers located near waterfront or at inland area.

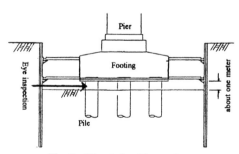

Fig. 2   Direct visual inspection

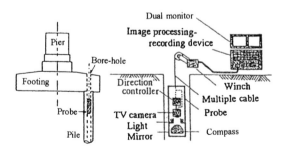

Fig. 3   Indirect visual inspection (Bore-hole camera inspection)

The inspection plan for the No.3 Kobe route was originally considered to use direct visual inspection method as the main method but this was later switched to the indirect visual inspection method in order to save the inspection time and cost. The indirect visual inspection method was also applied as the main inspection method for the No.5 Bay route.

2.4 Nondestructive inspection

The basic idea of a nondestructive inspection is to analyze a reflected elastic wave, which can be achieved rapidly with a significantly low cost. It might be a very useful and effective way of roughly assessing damage to a pile. The principle of the impact wave method, which is widely known as pile integrity test, is to generate an impact wave at pile top with a hammer and to measure the reflected wave with an accelerometer, as shown in **Fig. 4**. The measured wave contains information on changes of pile cross section and cracks in pile body. Analyzing disturbances in the measured wave, the presence of cracks and other defects can be judged, leading to an estimation of the degree of damage. Although this method has advantage on saving time and cost, it requires an evaluation method including correction of an effect by footing. As shown in **Fig. 5**, by excluding outstanding frequency band from the Fourier spectrum which is analyzed from the original wave based on the fast Fourier transformation, the modified wave without an effect by footing is obtained.

This method was mainly used for inspecting damaged piles on the No.5 Bay route. **Fig. 6 a)** shows a typical result of impact wave inspection of damage pile. In this figure, some of crack damage can be seen as irregularity. In **Fig. 6**, indirect visual inspection result is also shown,

comparing with the result of nondestructive inspection. Nondestructive inspection result tends to underestimate the number of multiple adjacent cracks. However, both inspections show similar results in terms of the locations of damage.

3 Investigation result of damage piles on the No.3 Kobe route

3.1 Typical inspection results

The No.3 Kobe route passes over the natural inland ground, on which almost no liquefaction was observed along this route after the Earthquake. At some sites of ground around piers, small-scale cracks and settlement were observed on the ground surface. **Photo 1** shows one of the typical examples.

In the restoration project of elevated highway structures on the No.3 Kobe route, the possibility of reusing the pile foundation was a major subject for discussion. In the following way, inspection results of damaged piles supporting a piltz-type bridge in which super-and sub-structures fell down by the Earthquake are described. The elevated, 18-span continuous concrete bridge collapsed due to flexural-shear failure of the columns. **Photo 2** shows a worst example of damaged pile inspected by the direct visual inspection. As the foundation was totally exposed around pile top in this case, all the piles could be inspected visually. Cracks on the pile surface were found in all the pile tops, the biggest of which was about 2 mm in width. While, no separating of concrete was caused on any pile surface, and hammer tests on pile surface revealed that only normal concrete acoustics could be heard. A typical example of bore-hole camera inspection result is shown in **Photo 3**. It is confirmed

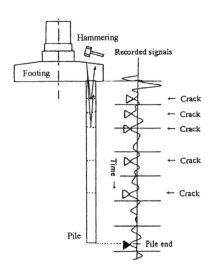

Fig. 4 Nondestructive inspection (Pile integrity test)

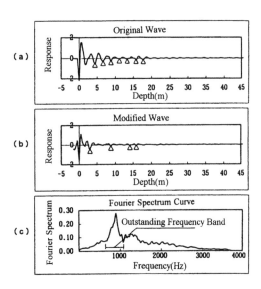

Fig. 5 Modified wave without an effect by footing

237

Photo 1　An example of ground deformation just after
　　　　the Earthquake (the No.3 Kobe route)

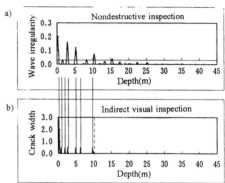

Fig. 6　Comparison between results of indirect visual
　　　　inspection and nondestructive inspection of a
　　　　damaged pile

Photo 2　A worst example of damaged pile inspected
　　　　by the direct visual inspection

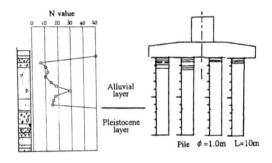

Fig. 7　Crack distribution on piles of an investigated
　　　　foundation of collapsed piltz-type bridge

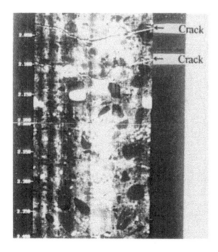

Photo 3　An typical example of bore-hole camera
　　　　inspection result on a damaged pile

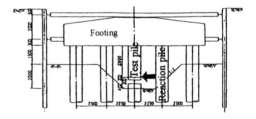

Fig. 8　Outline of bending load test

that some cracks are clearly recognized by this inspection
method.　**Figure 7** shows the crack distribution on piles
of an investigated foundation of collapsed piltz-type bridge.
It is seen from this figure that cracks are concentrated at
around pile top.

The pile damage inspection project of the Kobe route
was carried out by two phases.　The first phase was
carried out for over 50 foundations among about 800 pile
foundations which were selected considering inspection
area and super-structure type.　According to the factor

analysis of the results, the damage may seem greater on foundations where the settlement and cracks were found, N values in the surface ground were fairly small, the overburden on the pile cap was comparatively small, and/or the damage of pier is extremely great or non. It may seem also greater on foundations located near rivers or buried river sites. In the secondary phase, 123 pile foundations in total were checked. In almost all the investigated piles on the No.3 Kobe route, cracks are concentrated at around pile top where the maximum moment occurs, in an almost similar way as shown above. For a few piles, however, some cracks were found around position at which the density of reinforcement bars changes or interface zone between soft and hard soil layers. No cracks larger than the above-mentioned one were found and some of the foundations inspected confirmed to be completely free of cracks. Therefore, no greater damage than the one at the piltz-type bridge site was found in the No.3 Kobe route.

3.2 Loading test on damaged piles

In order to confirm the possibility of reusing damaged piles, loading tests were carried out for the piles of the piltz-type bridge where a highest degree of damage was found on the No.3 Kobe route. Three kinds of loading tests (bending load test, vertical load test and horizontal load test) of pile were carried out (Yasuda et al.,1997). Only bending load test is described and discussed below.

A bending test of pile was carried out on a test pile specimen cut from the centered pile at 2.8 m below the bottom end of the footing, as shown in **Fig. 8**. Multicycle loading was applied in one direction. The pile specimen was loaded up to the maximum of 25 tf, which is equivalent to 60% of the calculated yielding bending moment at the pile top section. **Figure 9** shows the horizontal load-displacement curves of the test pile at the point where the load was applied. They are almost linear when the applied load increases, and little residual displacement is observed after the applied load is removed. This fact suggests that the damaged pile has not been subjected to the ultimate load during the Earthquake. The rigidity obtained by the M-ϕ relationship shown in **Fig. 10** is smaller than the calculated rigidity at the yielding point, indicating that the reinforcements at the pile top has reached the yielding range.

As the next step, horizontal load-displacement curve of the whole pile foundation system was analyzed by considering the non-linear behavior of damaged pile, in order to evaluate its soundness in the ultimate stage. In the analysis, the rigidity of the damaged pile at the pile top over 1.5 m was reduced to the level obtained in **Fig. 9**. **Figure 11** shows the analyzed horizontal load-displacement curves in both cases with sound and damaged pile tops. It is seen from this figure that the degree of deformation of the damaged pile foundation is slightly greater than that of the sound one, while there is almost no decrease in the ultimate strength. Therefore, it is concluded that even if the rigidity of each pile decreases due to a displacement over the yielding point, the whole pile foundation as a pile group can keep little change in strength and displacement in the ultimate stage, with slight decrease in the initial rigidity.

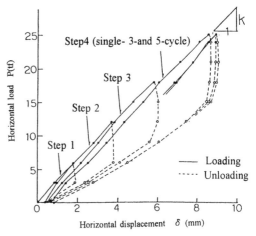

Fig. 9 Horizontal load-displacement curves of the test pile

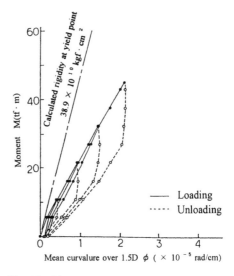

Fig. 10 Moment-mean curvature at the pile top section

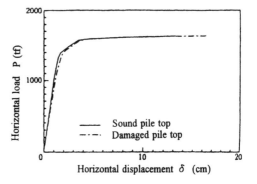

Fig. 11 Horizontal load displacement curves

239

4 Investigation result of damaged piles on the No.5 Bay route

4.1 Typical inspection results of damaged piles

The No.5 Bay route connects many recently reclaimed islands, on which liquefaction occurred along this route after the Earthquake, which caused greater displacements in piers, especially ones located along waterfront of reclaimed lands, as shown in **Photo 4**. Many piers near waterfront had moved laterally toward the sea and had settled around a half meter. The nearer the site is located to the seismic epicenter, the greater the settlement. According to the GPS survey result, it can be seen that the movement of the foundations near waterfront was fairly great in comparison with those in other areas.

Primary inspection on the No.5 Bay route was carried out for some foundations which exhibited by far the greatest amount of displacement. **Photo 5** shows an example of the direct visual inspection results. In spite of a large residual lateral displacement of about one meter, no major damage such as spalling of concrete and buckling of reinforcements at the pile top was observed. From this fact, it is deduced that cracks are distributed widely along the piles. After the primary inspection, further investigation on the No.5 Bay route, in which groundwater level was high and the foundations were large, was carried out, using nondestructive and indirect visual inspection methods, because the other methods may be time and cost consuming.

Figure 12 shows the distribution of average crack density of pile foundations along the No.5 Bay route in the recently reclaimed area, which is obtained by pile integrity tests, and also shows the relative residual displacement obtained by the GPS survey. According to this figure, almost all the piles in the reclaimed lands have cracks over the reclaimed layers. The more the amount of relative residual displacement is, the more the cracks density is.

Figures 13 a) and **b)** show the typical inspection results of cracks on piles located both near waterfront and at inland area. The waterfront piles suffered cracks that were distributed along the piles widely in reclaimed layer, in alluvial deposits, and even in Pleistocene deposits. The inland piles had cracks that were distributed in reclaimed layer and alluvial layer as well as at pile top, however, the crack density was less than that of waterfront piles.

Photo 4 An example of ground deformation just after the Earthquake (the No.5 Bay route)

Photo 5 Exposed surface of the pile top that exhibited greater amount of shift

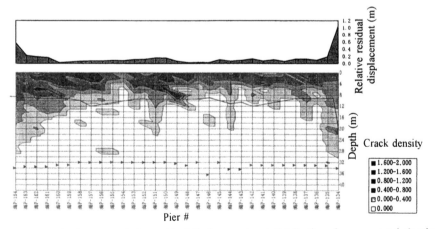

Fig. 12 The relation between relative residual displacement and distribution of average crack density

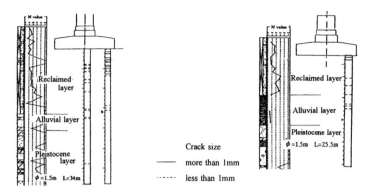

a) Waterfront piles that showed greater amount of displacement   b) Inland piles that showed relatively no displacement

Fig. 13   Crack distribution of damaged piles (the No.5 Bay route)

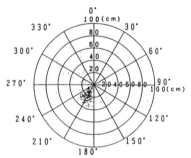

Fig. 14   Direction and amount of movement of foundations shown in vector diagram

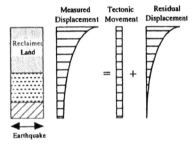

Fig. 15   Schematic diagram of residual lateral displacement of foundation

4.2 Lateral displacement of the foundations

GPS survey which was carried out right after the Earthquake showed that elevated highway foundations on reclaimed lands moved approximately 20 cm in average towards south-west as shown in **Fig. 14**.

The direction and amount of the movement of foundations coincide with those of the movement of fault in the Earthquake (i.e. right-lateral fault) that was obtained by the GPS survey of national bench marks after the Earthquake. Then, the movement must represent a tectonic one of earth and such deep seated movement would not be caused any damage to foundations. Therefore, the residual lateral displacement should be defined as the actually measured movement subtracted the above mentioned tectonic one, as schematically shown in **Fig. 15**. The elevated highway foundations located near waterfront of reclaimed lands moved significantly towards the sea due to the failure of revetment. This indicates that the lateral flow of ground near waterfront is responsible for the movement of the foundations, as the direction and amount of the movement are quite different from those in other locations. **Figure 16** shows a relationship between the distance from shoreline and the lateral displacement of

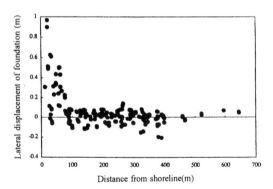

Fig. 16 Relation between distance from shoreline and lateral displacement of foundation

foundations, in which the lateral displacement is shown by the one after subtracting the average tectonic movement (about 20 cm). The figure indicates that the area affected by the lateral flow of ground is approximately within 100 m from shoreline.

241

5   Conclusions

In this paper, the inspection methods for pile foundation damage from the Great Hanshin Earthquake Disaster were described, together with their typical results, focusing on foundations of Hanshin Expressway.   The main conclusions are summarized as follows :

1) Some inspection methods were firstly applied to the investigation of damaged piles in the No.3 Kobe and the No.5 Bay routes in success.   Especially, the indirect visual inspection method and the nondestructive inspection method were established as very useful and effective methods for assessing damage to a pile.

2) All the piles located at inland area were confirmed to be sound, although they had some flexural cracks, while some piles located near waterfront of reclaimed lands suffered cracks widely distributed along pile axis.

3) Loading tests on the most heavily damaged piles of the No.3 Kobe route suggest that these damaged piles have maintain sufficient capacity for lateral resistance.

4) Some piles of the No.5 Bay route near waterfront of reclaimed lands were significantly affected by liquefaction and/or lateral flow of ground.   The area affected by the lateral flow of ground was within about 100 m from shoreline.

References

1) Matsui T, and Oda K : "Foundation damage of structures", Special Issue of Soils and Foundations, pp.189-200, 1996

2) Ishizaki H, Nanjo A, Yasuda F, Adachi Y, Kawakami Y, and Egawa N : "Inspection and restoration of damaged foundation due to the Great Hanshin Earthquake", Proceedings, 3rd Workshop on Seismic Retrofit of Bridges, Japan, 1996

3) Yasuda F, Nanjo A, and Kosa K : "Investigation and checking load bearing capacity of damage piles", KIG Forum'97 Kobe, Japan, pp.249-257,1997

Seismic Behaviour of Ground and Geotechnical Structures, Sêco e Pinto (ed.) © 1997 Balkema, Rotterdam, ISBN 90 5410 887 8

Earthquake geotechnical engineering aspects of the protection dikes of the Costa Oriental of Lake Maracaibo, Venezuela

J. Murria

Venezuelan Foundation for Seismological Research (FUNVISIS), Caracas, Venezuela

ABSTRACT: Oil production from shallow, unconsolidated reservoirs stated in 1928 in the flat, swampy Costa Oriental (eastern coast) of Lake Maracaibo in western Venezuela having caused as much as 5,90 m of subsidence, necessitating the staged construction of three earthen coastal dikes to protect the population and the facilities of the oil industry from Lake waters in the oil fields of Lagunillas, Tia Juana, and Bachaquero.

Extensive studies have shown the potential of liquefaction of the foundation soils in this region of low to moderate seismicity. Mitigative measures, consisting of downstream berms, with on without compaction piles, as well as extension lakeward of the upstream riprap protection, are being implemented and are expected to be completed late in 1997.

This paper presents in detail the earthquake geotechnical engineering studies carried out that led to the design of the mitigative measures being implemented, the problems encountered and the solutions adepted.

1. INTRODUCTION

Oil production from relatively shallow (300-1000 m) reservoirs on the eastern coast (Costa Oriental, formerly Bolívar Coast) of Lake Maracaibo began in 1928 in Lagunillas.

Due to the reservoir characteristics (shallow, unconsolidated, highly compressible, highly porous), ground subsidence has been taking place in an area of some 1600 Km^2, resulting in three separate subsidence basins, corresponding to the oil fields of Tia Juana, Lagunillas and Bachaquero, known collectively as the Costa Oriental oil fields (Figure 1). Subsidence was first detected in Lagunillas in 1929 and has reached as much as 5.90 m (Figure 2) in some points (Van der Knapp and Van der Vlis, 1967; Núñez and Escojido, 1976; Abi Saab and Murria, 1985; Mendoza and Martínez, 1988; Mendoza and Murria, 1989).

The combination of geomorphology (low lying, swampy coastal land) and subsidence, prompted the need for the continual construction of three coastal dikes (Tia Juana, Lagunillas, and Bachaquero) for protection against flooding from Lake Maracaibo waters, as well as three elaborate drainage systems to allow runoff water to be collected and pumped into the lake (Irazábal et al., 1986). Inner (diversion)

dikes had also to be constructed to avoid runoff waters coming into the subsidence area. Figure 3 shows the progressive raising of a typical coastal dike section.

At first, small dikes were constructed with little or no control of the fill materials. Later, it became apparent that, due to the continued rates of subsidence, some higher levels of engineering control were necessary. Furthermore, due to advances in the soil dynamics theory and practice, it was also suspected that the foundation soils were potentially susceptible to liquefaction, based on expected levels of ground shaking for the region. Given the importance of the area in terms of oil production and the value of the existing infrastructure, a series of studies were initiated to evaluate the risks associated with a possible dike failure as a result of liquefaction of the foundation soils (Woodward Clyde Consultants, 1987; Murria and Abi Saab, 1988; Murria, 1991).

The studies performed in the three subsidence areas provide better understanding of the foundation soils under seismic loading and are essential for an appropiate the evaluation of the mitigative measures necessary to increase, to an acceptable level, the integrity of the coastal protection system in case of an earthquake.

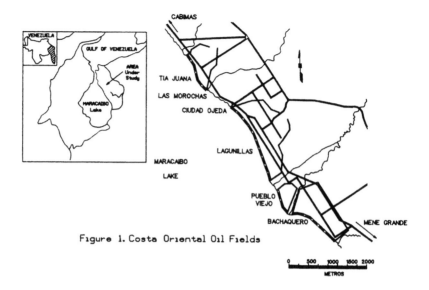

Figure 1. Costa Oriental Oil Fields

2. STUDIES PERFORMED

The design of the dikes had been based on static concepts with little or no consideration given to the dynamic aspects of the problem (NEDECO, 1968; Nuñez and Escogido, 1976; Abi Saab and Murria, 1985). In various opportunities, the susceptibility of liquefaction of the granular foundation soils was evaluated, with little or no conclusive results being obtained (Woodward Clyde and Associates, 1969; Murria and Abi Saab, 1988; Murria 1991). This was due partly to the sparsity of seismic data for the area, which did not allow for maximum design acceleration to be specified. In addition, not much information was available as to the distribution of these soils in the areas of subsidence. As the dikes continued to extend both vertically and laterally (dictated by the development of the subsidence basins) it became more and more important to evaluate the integrity of the system when subjected to earthquake forces. The Venezuelan oil industry, through MARAVEN and LAGOVEN, two of its three operating companies, and INTEVEP its research and development company, began the process of contracting specific studies aimed at resolving uncertainties in the seismicity of the

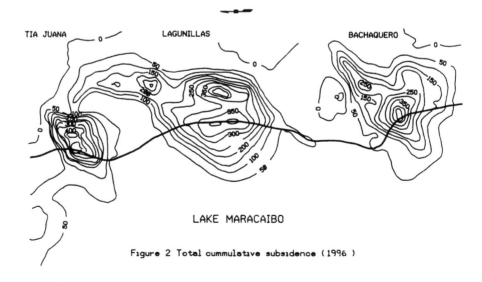

LAKE MARACAIBO

Figure 2 Total cummulative subsidence (1996)

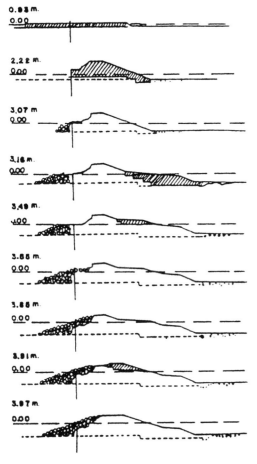

CUMMULATIVE
SUBSIDENCE

0.03 m.
0.00

2.22 m.
0.00

3.07 m
0.00

3.16 m.
0.00

3.49 m.
4.00

3.66 m.
0.00

3.66 m.
0.00

3.91 m.
0.00

3.97 m.
0.00

Figure 3 Progressive raising of the dyke

region and the geotechnical reliability of the dike-foundation system (NEDECO, 1986; Woodward Clyde Consultants, 1987).

The chronological sequence of the work performed at the seismic and geotechnical level can be summarized as follows:

- Review of the existing geotechnical information in order to generate typical soil profiles along the length of the dikes.

- Definition of the geological and tectonic setting of the region and evaluation of the seismic sources capable of producing levels of ground shaking that could affect the stability of the dike-foundation system.

- Review of the available information regarding the historical and instrumental seismicity of the area in order to provide preliminary estimates of the possible design accelerations.

- Initiation of a new neotectonic study to determine the characteristics of the major faults in the area and to evaluate their associated activity.

- Execution of deep seismic refraction studies along the east coast of Lake Maracaibo to provide a crustal model for the area with the objective of defining the attenuation law and to improve the knowledge of the seismicity of the region.

- Testing of scale models in a geotechnical centrifuge for evaluating trends in the dynamic behavior of the dike-foundation system and determining possible failure mechanisms in the presence of liquefiable foundation soils.

- Installation of a permanent seismometer network in the region, providing direct data transmission to an acquisition center and real time data processing.

- Installation of a strong-motion accelerometer network along the coastal dikes.

- Completion of a second stage of geotechnical studies, including specialized in situ and laboratory testing, to provide a better characterization of the foundation soils with emphasis on the determination of dynamic soil parameters for pre- and post-liquefaction analyses.

- Installation of a portable microseismic network to evaluate background levels of seismicity.

- Definition of the appropiate input motion.

3. GEOTECHNICAL CHARACTERIZATION

The field data used for the geotechnical characterization included the results of Standard Penetration Tests (SPT), in situ vane tests (FVT), piezocone tests (CPTU) and downhole and crosshole shear wave velocity determinations (DH-XH V_s). Laboratory test data included both classification and strength and deformation parameters (Zalzman et al., 1993).

For a given soil profile, the average wave velocity, V_s^*, can be calculated if the variation of V_s with depth is known. The average shear wave velocity can then be used to evaluate the fundamental period, T_N, of the soil deposit.

$$V_s^* = \frac{\sum (V_s)_i * H_i}{\sum H_i} \qquad (1)$$

$$T_N = 4H/V_s^* \qquad (2)$$

In an area where the thickness of the soil profile is approximately constant (as can be assumed for the

245

Costa Oriental area), the dynamic soil response can be characterized by the average shear wave velocity alone, rather than the fundamental period of the profile. This was the initial criterion used to determine "similar-type response" profiles in the area. Since the soil response is controlled by the soil characteristics, each representative soil profile was also evaluated in terms of the percentage of different soil types in each. As demonstrated by Vucetic and Dobry (1991), modulus degradation can be correlated broadly with soil plasticity and so this was also used as an index parameter for the initial characterization (Sully et al., 1995).

In general terms, the dike foundation soils can be divided into three principal units:
- an upper layer of loose to medium dense silty sand and non-plastic silts, 2.5 m to 4.5 m thick.
- a middle layer of very soft to soft silty clay, occasionally organic, to a depth of about 13 m, and
- a lower layer of firm to stiff clay, interbedded with dense sand (Sully et al., 1993b).

At various locations along the dike, in situ measurements of shear wave velocity by the crosshole technique have been performed. In order to extend this information to other area where V_s measurements are not available, the data have been correlated with index parameters such as N_{SPT}, S_u or q_c (Echezuría and Sully, 1986). Typical relationships derived for Costa Oriental data are:

$$G_{max} = (2000 \text{ to } 2300)S_u \qquad (3)$$

$$V_s = 77(N_1)^{0.33} \qquad (4)$$

Equation (4) is compared with other published relationship in Figure 4 (Sully et al., 1995).

4. DEFINITION OF INPUT MOTION

The input motions for the microzonification studies were determined from seismic hazard studies performed in the area (Woodward-Clyde Consultants, 1987). For the Costa Oriental region, the variation of maximum bedrock acceleration as a function of return period is shown in Figure 5. From this figure, the value of maximum base acceleration

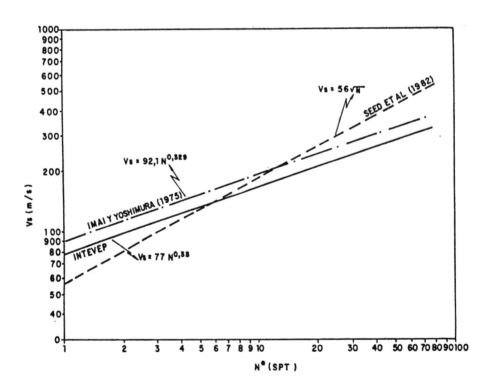

Fig. 4 Comparison of published correlations between V_s and $(N_1)_{SPT}$

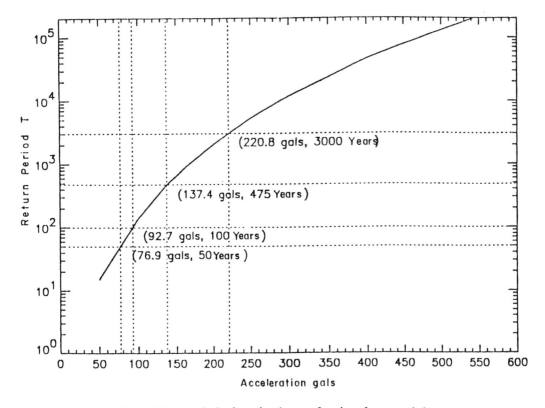

Fig 5 Maximum bedrock acceleration as a function of return period

for any useful life and probability of excedence can be determined (typical values are indicated on the figure). For the COLM dikes, a 3% probability of exceedence in a 100 year expected life has been chosen as the design condition and corresponds to a bedrock acceleration of 0.225 g.

In order to adequately define the input motion not only the maximum acceleration is required, but also information regarding the duration and frequency content of the generated earthquakes that may affect the area. This information was determined during the seismotectonic investigations (Woodward Clyde Consultants, 1987). Based on these results, a duration of around 20 seconds was selected as representative with epicentral distances of the order of 200 Km to 150 Km. Both synthetic and real acceleration records were then evaluated to provide input acceleration records representative of both near- and far-field sources.

The depth of application of the basal acceleration was determined from a sensitivity study based on information from deep borings (down to 500 m) previously performed in both Bachaquero and Tia Juana (NEDECO, 1986). The variation of the small-strain shear modulus with depth was evaluated from the results of in situ and laboratory resonant column test data. Both lower and upper bound limits were defined. A series of response analyses were then performed to study the effect of the variation of "bedrock" depth on the surface accelerations and frequency content. "Bedrock" was defined at depths varying from 150 m to 500 . The modulation of the selected input motions was then evaluated at a depth of 50 m, these time history records being the output at 50 m depth obtained from each of the response analyses with differing input bedrock depths.

In order to reduce the thickness of the soil profile to be analyzed, the modified 50 m depth records are used as input records for this study. In this way, the new "bedrock" depth is assigned at 50 m and only the soil variations of interest in the geotechnical study (between the ground surface and 50 m depth) need to be modelled in the analysis. The 50 m record is an unbiased average based on both soil modulus variations and depth to bedrock. the frequency content variation of the different records is also evaluated and selected to represent the average response as a result of propagation between the

247

depth of application and the 50 m level. The maximum acceleration for the input record (to be applied at 50 m depth below each profile) was determined to be 0.084 g for the 3000 year design earthquake condition. Once the 50 m input acceleration time histories had been evaluated, the dynamic response for the representative near-surface (0 m to 50 m) soil profiles was evaluated (Sully et al., 1995).

5. DYNAMIC SITE RESPONSE ANALYSES

The site response analyses were performed using the modified version, SHAKE91 (Idriss and Sun, 1992), of the well-known program SHAKE (Schnabel et al., 1972). The SHAKE91 program performs an equivalent linear elastic analysis which incorporates modulus reduction according to average strain levels induced by the applied ground motion. The Vucetic and Dobry (1991) degradation curves were used as were the curves for variation in damping ratio. Parametric studies were again used to evaluate the range of response according to variations in the input data and soil properties. Typical variations in the average shear wave velocity in the Tia Juana region, as well as soil type variations for the defined representative soil profiles are shown in Figure 6. As stated earlier, both real and synthetic time histories were used representing near- and far-field sources. Results obtained from these type of sensitivity analyses have demonstrated the importance of soil type and parameter variations on the response for the COLM soils (Fernández, 1993).

5.1 Evaluation of liquefaction potential

According to preliminary results, the majority of the near-surface granular soils in foundation the soils of

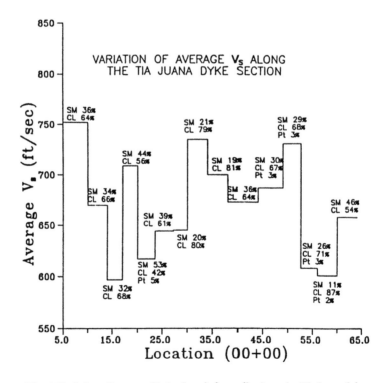

Fig. 6 Variation of average V_s for foundation soils along the Tía Juana dyke

the coastal dikes will undergo initial liquefaction. Whether or not the behavior will be contractive with subsequent strength loss is being presently evaluated by laboratory tests. However, for first phase studies, a residual strength of 1 t/m^2 or a strength ratio, S_{uss}/S_v', equal to 0.12 (whichever was the largest) was assigned to the liquefied soils for evaluation of pseudo-static stability.

5.2 Evaluation of post-seismic deformations

Post-seismic deformation analyses were performed using the computer code DYNARD (Sully et al., 1993a), developed by Woodward Clyde Consultants, and licensed to MARAVEN/INTEVEP. Additional analyses were also contracted to provide validation of the obtained results using the program TARA-3FL (Finn, 1992).

6. DEFINITION AND IMPLEMENTATION OF REMEDIAL MEASURES

Stability and deformation analyses were performed for each characteristics dike and foundation system for both pre- and post-seismic conditions. The assumptions and parameters for these static and pseudo-static analyses have been discussed by Sully et al. (1993). Back analyses of case histories of failures due to liquefaction have been performed to verity the approach and parameters used for both stability and deformation analyses (Murria et al., 1992; Sully et al., 1993a).

When it became apparent that certain sections of the dike would not meet the stability and deformation criteria established for the dike-foundation system, an evaluation of alternative types of mitigation measures was performed (Fernández and Sully 1992). A long list of remedial measures was prepared from which the following were finally chosen:

- Construction of downstream (land side) berms with a variable thickness (between 2 m and 4 m). A maximum length of 40 m was chosen as a practical limit. Berms were constructed in about 23 of the 47 Km of coastal dikes.
- Where the additional berm did not guarantee the required safety factor of 1.2, ground improvement was implemented. After two pilot tests were performed (Finn, 1992; Fernández and Sully, 1992; Villegas and Murria, 1994), the first in Bachaquero and the second in Lagunillas, stone piles 0.60 m in diameter and between 13 m and 16 m in length were constructed, 3 m on centers, in 7 rows parallel to the dike in a total length of 2.3 Km.
- Extension of the upstream (lake side) rip rap in selected sections of the coastal dikes with a total length of about 3 Km. Figure 7 shows schematically the mitigation measures being implemented. Completion date is late 1997.

7. CONCLUSIONS

The methodology and results of the earthquake geotechnical studies in the Costa Oriental protection dikes have been presented with emphasis on the evaluation and remediation of a linear earth structure, critical to the continued oil production in the Costa Oriental of Lake Maracaibo. However, the

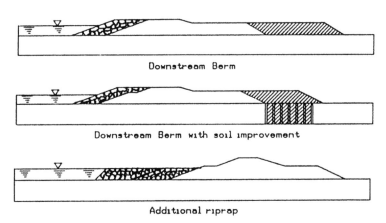

Downstream Berm

Downstream Berm with soil improvement

Additional riprap

Figure 7. Mitigative Measures

results of this study have also been used for implementing contingency plans (Murria y Angarita, 1992) and planning the social and economic development of this area of important industrial activity. Although in this instance the results have only been discussed with respect to the coastal dike system, they can be easily extrapolated to other areas away from the dike (Sully et al., 1993b). A detailed geological and geotechnical evaluation is presently underway to consolidate the available information and make it readily accessible (De Santis et al., 1993; Sully et al., 1995).

The application of the results obtained has demonstrated the usefulness of appropiate studies for understanding and evaluating the controlling geotechnical and seismotectonic parameters, as well as allowing for the implementation of timely and cost-effective mitigation measures. The studies were useful for correctly understanding the possible dynamic response variations due to localized subsoil variations; an important factor in the Costa Oriental area where the geological constraints on deposition have given rise to a vertically and laterally variable sequence of soft sediments.

Studies are still in progress to provide additional data to reduce uncertainties in the present state of knowledge. The permanent strong motion and microseismic networks are also nearing completion and will provide further data for verification of the tectonic model and attenuation law being used. In specific areas of the dike, local seismicity monitoring stations are being installed (accelerometers and piezometers) for calibration of the dynamic response analyses. In addition, the results of the microseismicity study performed by the CETE - FUNVISIS - INTEVEP - MARAVEN (FUNVISIS-CETE Méditerranée, 1994) group will also provide information that may be used to calibrate the small-strain dynamic response of typical soil profiles.

8. ACKNOWLEDGMENTS

Although many people have contributed to the studies described in this paper the author would like to single out Dr. John P. Sully, formerly of INTEVEP, who for many years has been involved in the seismic geotechnical aspects of the Costa Oriental protection dikes and who has published extensively on this subject.

9. REFERENCES

Abi Saab, J. and J. Murria (1985). Origen y desarrollo del sistema de proteccion costanera, Costa Oriental del Lago de Maracaibo. I Jornadas de Tecnología de Producción, INTEVEP, Los Teques, Venezuela, 3/1-3/15.

De Santis, F., A. Singer and J.P. Sully (1993). Caracterización geológica y geotécnica de los sitios del proyecto de microzonificación sísmica experimental CETE-FUNVISIS-INTEVEP. Presented at First Int. Seminar on Seismic Microzona-tion, Cardón, Venezuela.

Echezuría, H. y J.P. Sully (1986). Discussion of the geotechnical properties and liquefaction potential of the COLM dyke foundation soils. Informe Técnico (Confidencial) INTEVEP, S.A., INT-01662,86 Sept., 142 pp.

Fernández, A. (1993). Análisis de sensibilidad de la respuesta dinámica de perfiles geotécnicos. Trabajo Final de Grado, Magister en Ingenería Civil, Universidad Simón Bolívar, 156 p.

Fernández, A. and J.P. Sully (1992). Estudio de deformaciones por efectos dinámicos en la Sección Piloto de Lagunillas. Informe Técnico INTEVEP, S.A., INT-TEIG-0051,92, Oct., 42 pp.

Finn, W.D.L. (1992). Flow deformation analyses-pilot section of Lagunillas dykes. PEACS report to INTEVEP, S.A., August, 20 pp.

FUNVISIS-CETE Méditerranée (1994). Estudio experimental de microzonificación sísmica en los sitios de la COLM y El Vigía. Proyecto INTEVEP 92-174.

Idriss, I.M. and J. Sun (1992). User's for SHAKE91. Center for Geotechnical Modeling, Univ. of California, Davis, 13 p.

Irazábal, A.; J. Abi Saab; J. Murria; J. Groot (1986). Drainage problems in areas subjected to subsidence due to oil production. Proc. of the 2nd Int. Conf. on Hydraulic Design in Water Resources Eng. and Land Drainage, Southampton University, U.K., April, pp. 545-554.

Mendoza, H. and C. Martínez (1988). Ground subsidence modelling in Costa Oriental Oilfieds, Venezuela. 5th Int. Symp. on Deformation Measurements and 5th Canadian Symp. on Mining Survey and Rock Deformation Measurements. Fredericton, N.B., 6-9 June 1988.

Mendoza, H. And J. Murria (1989). Ground subsidence modeling in western Venezuela. Symp. on Land subsidence, Dhanbar, Bihar, India.

Murria, J. (1991). Subsidence due to oil production in western Venezuela: Engineering problems and solution. Fourth Int. Symp. on Land Subsidence, Houston, TX, May 1991.

Murria, J. and J. Abi Saab (1988). Engineering and construction in areas subjected to subsidence due to oil production. 5th International (FIG) Symp. on Deformation Measurements, Fredericton, N.B., Canada, 367-373.

Murria, J. y J.L. Angarita (1992). Plan de Contingencia contra riesgos de inundación en la Costa Oriental del Lago de Maracaibo (Plan COLM), Primer Taller, Planes de Contingencia, CEPET, Maracaibo, 28 February 1992.

Murria, J., E. Gajardo and J.P. Sully (1992). Seismic zonation for COLM dikes. Workshop on Seismic Zoning Methodologies for Geotechnical Hazards, Lisboa, Portugal.

NEDECO (1968). Bolívar Coast Dykes, Venezuela. Internal Report to Compañía Shell de Venezuela. March 1968.

NEDECO (1986). Internal report to MARAVEN.

Nuñez, O. and Escojido, D. (1976). Subsidence in the Bolívar Coast. Publication N° 121 of the Int. Assoc. of Hydrological Sciences, Proc. of the Anaheim Symposium, December 1976, pp. 257-266.

Schnabel, P.B., J. Lysmer and H.B. Seed (1972). SHAKE: A computer program for earthquake analysis of horizontally layered sites. Univ. of California at Berkeley, Earthquake Engineering Research Center, Report No. EERC 72-12, Dec., 88 pp.

Sully, J.P., A. Fernández and S. Zalzman (1993). Definition of soil parameters for use with DYNARD for flow deformation analyses. Informe Técnico INTEVEP, S.A., in preparation.

Sully, J.P., A. Fernández, S. Zalzman and M. González (1993a). Backanalysis of two case histories to validate modulus values for liquefied soils for use with DYNARD. Informe Técnico INTEVEP, S.A., INT-TEIG-0074,93, Agosto, 13 pp.

Sully, J.P., E. Gajardo, J. Murria and J. Abi Saab (1993b). Seismic Zonation for COLM Dykes, Proc. Caribbean Conf. on Natural Hazards: Volcanoes, Earthquakes, Windsforms, and Floods, St. Augustine, Trinidad.

Sully, J.P., O. Morales, E. Gajardo, J. Murria and J. Abi Saab (1995). Microzonation Studies for Lake Maracaibo coastal protection system. Proc. of "Recent Advances in Geotechnical Engineering". St. Louis, MO.

Villegas, B. and J. Murria (1994). Preliminary evaluation of the use of compaction piles for improvement of the foundation soils of the Coastal dikes of Lake Maracaibo, Venezuela. Proc. from the Fifth U.S. Japan Workshop on Earthquake Resistant Design of Lifeline Facilities and Countermeasures Against Soil Liquefaction, NCEER Technical Report NCEER, 94-0026, November 7, 1994.

Van Der Knapp, W. and Van Der Vlis, A.C. (1967). On the cause of subsidence in oil producing areas. Proc. of the 7th World Petroleum Congress, México City, 70.85-95.

Vucetic, S. and Dobry, R. (1991). Effect of soil plasticity on cyclic response. Journal of Geotech. Eng., ASCE, 117:1.

Woodward-Clyde & Associates (1969). Seismicity and seismic geology of northwestern Venezuela. Report to Compañía Shell de Venezuela.

Woodward Clyde Consultants (1987). Geotechnical Studies, Costa Oriental Dikes, Volumes 1 and 2, Prepared for MARAVEN, S.A., Lagunillas, Venezuela.

Zalzman, S., Nuñez, M., Sgambatti, J. y Echezuría, H. (1992). Caracterización geotécnica de la fundación de los diques de la COLM. Informe Técnico INTEVEP, S.A., en edición.

Seismic Behaviour of Ground and Geotechnical Structures, Sêco e Pinto (ed.)© 1997 Balkema, Rotterdam, ISBN 90 5410 887 8

Seismic design procedure for kinematically stressed piles

Aspasia Nikolaou
State University of New York at Buffalo, N.Y., USA

George Gazetas
National Technical University, Athens, Greece

ABSTRACT: Based on results of an extensive parameter study, a relatively simple procedure is proposed for estimating peak values of bending moments in a pile embedded in a two-layered soil and subjected to S-wave seismic excitation.

1 INTRODUCTION

Seismically loaded piles are usually designed to withstand only the inertial forces generated from the oscillation of the superstructure. Recent observations, however, have associated pile damage with deformations induced by the passage of seismic waves through the surrounding soil, particularly in the presence of sharp stiffness discontinuities in the soil profile (Mizuno, 1987). This type of pile distress is called *kinematic* to distinguish it from the distress generated by the *inertial* forces of the superstructure.

The importance of kinematic loading has been recognized in the recently published Eurocode 8, dealing with the seismic design of civil structures. Part 5 of that code states:

"Piles shall be designed for the following two loading conditions:

(a) inertia forces from the superstructure.......

(b) soil deformations arising from the passage of seismic waves which impose curvatures and thereby lateral strain on the piles along their whole length... Such kinematic loading may be particularly large at interfaces of soil layers with sharply differing shear moduli. The design must ensure that no "plastic hinge" develops at such locations..."

In this paper, the results of a comprehensive parameter study are summarized with a goal towards developing an inexpensive design procedure for computing the peak bending moments in a pile under an arbitrary earthquake time-history excitation.

Specifically, first, a number of simple closed-form expressions are fitted to a comprehensive set of numerical results to estimate the harmonic (steady-state) kinematic pile bending moments at the depth, or in the vicinity, of soil discontinuities, for a two-layered profile. The maximum harmonic moments are then correlated to the corresponding peak values during actual earthquakes. A set of simple design formulae and charts are presented for the design of piles against these peak values of maximum moments.

2 METHOD OF ANALYSIS

The response to vertical S-wave excitation of a single pile embedded in a layered soil can be obtained numerically in a single step using either a suitable finite-element (FE) formulation with "wave--transmitting" boundaries (as for instance, the one described by Blaney et al, 1976), or a boundary-element-type code (as the one described by Mamoon & Banerjee, 1990). Instead, a beam-on--Winkler-foundation formulation is used herein, sketched in Figure 1. In this, the role of soil-pile interaction is played through a set of continuously-distributed springs and dashpots, the frequency-dependent parameters of which $[k = k(\omega)$ and $c = c(\omega)]$ have been calibrated against results of finite-element and boundary-element analyses. Such springs and dashpots connect the pile to the free-field soil; the wave-induced motion of the latter (computed with any available method: e.g., Schnabel et al (1973), Roesset, (1978)) serves as the support excitation of the pile-soil system.

A frequency-domain method of solution to this problem, presented in Makris & Gazetas (1992) and Kavvadas & Gazetas (1993), has been extended to the time domain using the Discrete Fourier Transform (DFT) method. This can accommodate precisely the frequency dependence of k and c, contrary to methods that require frequency--independent parameters to obtain the response directly in the time domain. In addition, whereas in the earlier aforementioned publications the problem is solved by forming 4n algebraic equations (n = the total number of layers penetrated by the pile) relating pile to free-field displacements (and their first three spatial derivatives), the method used in this work (Mylonakis, 1995) follows a layer transfer matrix approach. This is computationally more efficient, as it involves a sequence of transfer-matrix multiplications and is known in the literature as the Haskell-Thomson (1950) or the Holzer method (Clough & Penzien, 1975). Furthermore, the often unrealistic *"rigid bedrock"* assumption of that earlier formulation has been eliminated, and the motion can be prescribed at the surface of an outcropping ("elastic") rock.

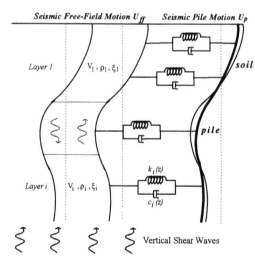

Figure 1 The Beam-on-Dynamic-Winkler-Foundation (BDWF) model used in this paper

One of the relative merits of the developed dynamic Winkler method is its versatility in handling *approximately* moderate levels of nonlinearity in the soil surrounding the pile. Such nonlinearity arises from the larger stresses induced in the immediate neighbourhood of the pile and could be modelled approximately with a linear analysis of a radially inhomogeneous soil (Veletsos & Dotson (1986), Dotson & Veletsos (1987), Michaelides et al (1997)). The springs and dashpots resulting from such a (plane-strain) analysis would reflect the nonlinearities due to pile-soil interaction, rather than the (additional) nonlinearities due to shear waves propagating in the free field soil.

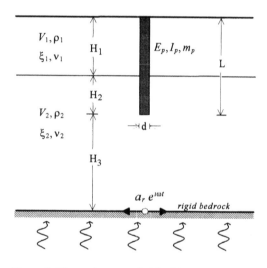

Figure 2. The problem studied in this paper

The results of such a beam-on-dynamic--Winkler-foundation (BDWF) analysis presented in this article refer to a single pile embedded in a two-layer soil deposit and subjected to vertical S-waves (Figure 2). The excitation is described through the base rock acceleration: $a_r \exp(i\omega t)$ for harmonic excitation and $a_r(t)$ for the general case. To illustrate the capability of the utilized BDWF method, Figures 3 and 4 compare its results with those of two rigorous formulations: a boundary-element-type formulation (by Kaynia & Kausel, 1982), and a finite-element formulation (by Blaney et al, 1976). Figure 3 plots the ratio of the pile displacement U_p over the free-field soil displacement U_{ff} (known in the literature [e.g. Gazetas et al 1992, Fan et al 1991] as *kinematic ratio* I_u), while Figure 4 plots the distribution with depth of pile displacements and bending moments; the performance of the simplified method is very satisfactory.

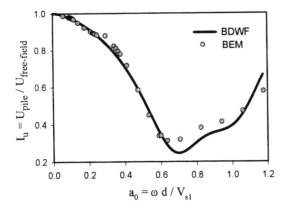

Figure 3. Comparison of the ratio $I_u = U_p/U_{ff}$ calculated with the Boundary Element Method (BEM) and the proposed BDWF Method for a pile with $L/d=12$, $E_p/E_{s1}=1000$, $V_1/V_2=1/5$, and $\rho_1/\rho_2=3/4$

3 SUMMARY OF RESULTS: STEADY-STATE BENDING MOMENTS

A brief summary of the main findings of these studies is given herein, with the help of Figures 5 and 6. They refer to a single pile in a two-layer soil profile, subjected to harmonic steady-state excitation, which consists exclusively of vertically-propagating S-waves. The most important conclusions that have emerged are as follows:

(1) For a given excitation, the kinematic bending moments depend mainly on:

- the stiffness contrast between any two consecutive soil layers in the deposit; for the examined profile this contrast would be measured with the ratio V_1 / V_2 of their respective S-wave velocities
- the boundary conditions at the head of the pile or the pile cap; the results presented herein consider only the two extreme cases, i.e. of fixed-head piles with no cap rotation and of free-head piles with no rotational constraint at their top
- the proximity of the excitation frequency, ω, to the fundamental (first) natural frequency, ω_1, of

the soil deposit and, to a lesser degree, to the second natural frequency, ω_2, of the deposit

- the relative depth, H_1 / ℓ_a, measured from the top of the pile down to the interface of the layers with the sharpest stiffness contrast normalized with respect to the active length ℓ_a of the pile.

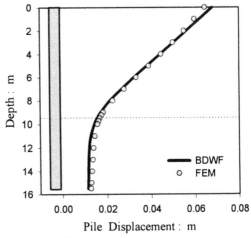

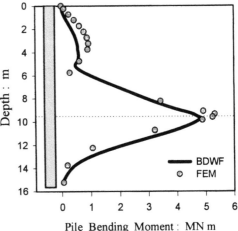

Figure 4. Comparison of kinematic displacement and bending moment derived using the Finite Element (FEM) and the BDWF methods (pile and soil properties same as in Figure 3)

(2) The bending moments are largest either at the pile head, or in the vicinity of the interface of soil layers with the sharpest stiffness (one pile diameter away from the interface). The moments at the interface for free and fixed head piles are almost identical, except when the pile is "short" and

"rigid" (meaning, when $H_1 < \ell_a$). Atop the fixed-head piles the moment is generally of the same order of magnitude as, or smaller than, the moment created at the interface of the two layers. In some cases though, the moment at the top may be much higher than at the interface. These are the cases for which the active pile length ℓ_a, well-known from the inertial interaction studies as

$$\ell_a \approx 1.75 \, d \, (E_p / E_{s1})^{0.25}$$

is larger than the height of the first soil layer. Apparently, these will be the cases where relative "stiff" piles (e.g., $E_p/E_{s1} > 5000$) with relative "small depth" to the interface ($H_1 / L < 1/2$) are forced to develop large bending moments at the pile head in order to satisfy the quite severe no-rotational top boundary condition.

(3) In most cases, the maximum steady-state bending moment occurs at the fundamental natural period of the soil deposit. The pile moment transfer functions display a very rapid decline when moving away from resonance. The ratio of the maximum pile moment at resonance (whether it occurs at the top of the soil deposit or at the layer interface) divided by the respective static moment [i.e., the ratio max $M(T_1) / M(T)$] follows, more or less, the

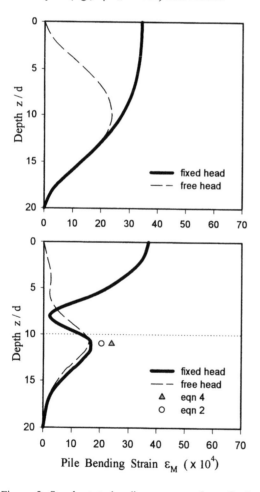

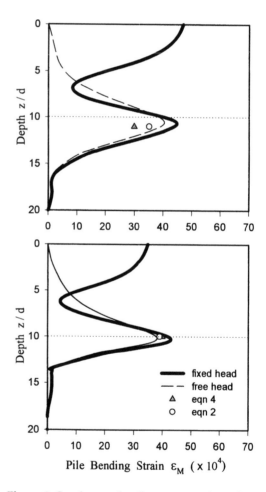

Figure 5. Steady-state bending moments for a fixed and a free head pile with $L/d = 20$ and $E_p / E_{s1} = 5000$ in a homogeneous profile and in a 2-layer profile with $V_1/V_2 = 1/2$

Figure 6. Steady-state bending moments for a fixed and a free head pile with $L/d = 20$ and $E_p / E_{s1} = 5000$ in a 2-layer profile with $V_1/V_2 = 1/4$ and in a 2-layer profile with $V_1/V_2 = 1/10$

free-field amplification of acceleration (i.e., the ratio: a_{ff} / a_r). This shows the great influence of the first mode of vibration on the magnitude of the bending moment and contradicts some earlier statements in the literature that higher modes would produce larger kinematic moments. Indeed, while higher frequencies *do* indeed tend to develop "wavy" shapes of deflection and thus have the potential for inducing relatively large curvatures at the soil interface, the actual curvature is also affected by the overall drift between the top and the bottom of the pile. This drift becomes maximum at the first natural mode and consequently produces the largest moments at the first resonance.

(4) Pile moment in the vicinity of the interface is also influenced by the pile and soil characteristics as well as the differences between the two layers. The relative pile-soil stiffness is of great importance: the stiffer the pile with respect to the soil, the larger the developing moment. The stiffness contrast between the two layers is also very significant: high values of kinematic moment occur mainly when the stiffness changes sharply across the interface ($V_1 / V_2 < 0.25$).

(5) Two closed-form expressions have been developed for approximately computing the maximum steady-state bending moment at the interface between the two layers:

• the first expression is based on a rough estimate of the shear stress τ_{interf} that is likely to develop at the interface as a function of the free-field acceleration of the surface, $a_{surface}$:

$$\tau_{interf} \approx a_{surface} \, \rho_1 \, H_1 \qquad (1)$$

The fitted formula is written:

$$M_{max} = 0.042 \; \tau_{interf} \; d^3 \left(\frac{L}{d}\right)^{0.30} \left(\frac{E_p}{E_{s1}}\right)^{0.65} \left(\frac{V_1}{V_2}\right)^{0.50} \qquad (2)$$

• the second expression has been motivated by the maximum strain ε_M due to a bending moment M acting in a pile cross-section of diameter d:

$$\varepsilon_M = \frac{M}{E_p I_p} \frac{d}{2} = \frac{M}{E_p \left(\frac{\pi d^4}{64}\right)} \frac{d}{2} \approx \frac{10 M}{E_p d^3} \qquad (3)$$

The corresponding formula is written:

$$M_{max} = \frac{2.7}{10^7} \, E_p \, d^3 \left(\frac{a_r}{g}\right) \left(\frac{L}{d}\right)^{1.3} \left(\frac{E_p}{E_{s1}}\right)^{0.7} \left(\frac{V_1}{V_2}\right)^{-0.3} \left(\frac{H_1}{L}\right)^{1.25} \qquad (4)$$

Predictions of the two formulae are shown as dots on Figures 5 and 6. We have found them to provide reasonably successful results in a large number of cases.

4 TIME VERSUS FREQUENCY-DOMAIN BENDING MOMENTS

Under transient seismic excitation, described through actual and artificial accelerograms applied at the base of the deposit, we have found that the above conclusions are still valid --- with one exception: the peak values of the transient bending moments are smaller than the steady-state amplitudes. An example is given in Figure 7, for a relatively rigid fixed-head concrete pile (diameter 1.3 m and length 15.5 m) penetrating a 9.5 m thick top layer of soft clay and socketed 6 m into a deep layer of dense sand. Eight actual accelerograms and one artificial motion, all scaled to a 1 m/s² (0.10 g) peak acceleration, were used as excitation at the rock level. It is evident from this figure that the envelope of peak moments (in the "time domain") has a distribution with depth which is of the same shape as the distribution of steady-state amplitudes (in the "frequency domain"). But the values of the latter are about 3 - 5 times larger than the former, depending on the excitation.

A correlation can be proposed between the largest values of bending moments, in the time and frequency domains:

$$\max M(t) = \eta \; \max M(\omega) \qquad (5)$$

where: max M(t) is the maximum pile bending moment in the time domain and maxM(ω) is the maximum steady-state pile bending moment, which usually corresponds to a frequency equal to the fundamental frequency of the deposit. The frequency-to-time reduction factor, η, takes values between about 0.15 and 0.50, depending on:

• the duration of the accelerogram in terms of the number of strong motion cycles N_{cycles}
• the relative frequency characteristics between the excitation "signal" and the pile moment

transfer function: as a first approximation, this relation is expressed in terms of the ratio T_p / T_s (i.e. "predominant" earthquake period / fundamental period of the soil deposit), and

- the sharpness of the pile-moment transfer function, expressed by the ratio $maxM(T) / M(T)$ which is directly related to the effective damping ratio ξ_{eff} of the pile-soil system.

Quantification of the variables N_{cycles} and T_p requires pertinent seismological input. For design purposes we present Figure 8, in which η is plotted as a function of the number of significant cycles, N_c, for $\xi_{eff} = 10\%$, depending on the intensity of the seismic excitation. The earthquake ground acceleration records (real and artificial) used as excitation from the plotted data points are described in Table I.

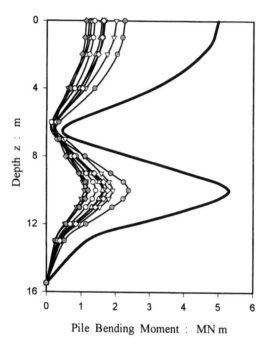

Table I. The ground motions used in the time-domain analyses. Symbol N_c stands for the effective number of cycles in the record, T_p is the range of dominant periods, and PGA is the peak ground acceleration.

Earthquake	Record	PGA (g)	N_c	T_p (sec)
Arificial	AASHTO S1 soil	0.12	>10	0.1-0.5
Northridge 1994	Pacoima downstr.	0.42	2.5	0.15 - 0.5
Pyrgos 1993	Pyrgos trans.	0.45	0.5	0.12 - 0.45
Whittier 1987	LA116 Ch. 1	0.38	6	0.1 - 0.25
	Pacoima Ch. 1	0.15	3.5	0.1 - 0.3
	Tarzana Ch. 3	0.40	5	0.3 - 0.4
Loma Prieta 1989	Anderson downstr.	0.24	4	0.15 - 0.3
Kobe 1995	Kobe JMA	0.82	3	0.3 - 0.9
Mexico 1985	La Villita 0°	0.12	3	0.5 - 0.6

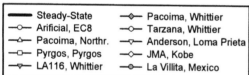

—— Steady-State	—◇— Pacoima, Whittier
—○— Arificial, EC8	—○— Tarzana, Whittier
—△— Pacoima, Northr.	—▽— Anderson, Loma Prieta
—□— Pyrgos, Pyrgos	—◇— JMA, Kobe
—▽— LA116, Whittier	—●— La Villita, Mexico

Figure 7. Distribution with depth of peak steady-state moments and envelope of time-domain moments (using the earthquakes of Table I), for a pile with E_p = 25 GPa, d =1.3 m, L = 15.5m, ρ = 2.5 Mg/m^3. The 2-layer profile has a fundamental period of 0.52 sec, total height of 30m, and shear wave velocities of 80 m/sec (first layer, H_1= 9.5m) and 330 m/sec (second layer, H_2 = 6m and H_3 = 14.5m)

Hence, the dependence:

$$\eta = \eta \ (N_{cycles}, T_p / T_s, \xi_{eff}) \qquad (6)$$

It is worth mentioning that although η increases as the number of cycles increases and as the soil natural period gets closer to the dominant period of the shaking, similar results were not observed for periods away from resonance. In such cases, the factor η remains practically unaffected by the increase in strong motion duration. This fact is taken into account in the design plot of Figure 8, the data of which are combined in two clusters: one for *"near resonance"* response (i.e. the profile natural period is within the range of the signal's dominant periods), and one for *"non resonance"* response (i.e., the soil natural period is substantially smaller or larger than that of the earthquake). The solid data points shown in this figure correspond to the first (*"resonance"*) case, while the open points to the second, ("non resonance") case.

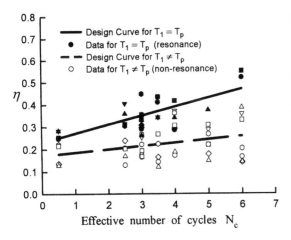

Figure 8. Proposed design curves for the peak moment frequency-to-time reduction factor η as function of the effective number of cycles N_c and resonance conditions

5 CONCLUSION

The algebraic expressions (1) - (4) along with the chart of Figure 8 can be easily used for obtaining quick estimates of the peak values of maximum kinematic pile bending moments in an essentially two-layer soil stratum. Such large moments tend to develop at the interface between two soil layers when their stiffnesses differ markedly.

For more general soil profiles the information provided in the paper can still be (cautiously) used as a guide in preliminary design.

6 ACKNOWLEDGMENTS

Financial support for this work has been provided by the National Center for Earthquake Engineering Research (NCEER) and by the EU Human Capital and Mobility Programme.

7 REFERENCES

AASHTO (1983). "Guide specifications for highway bridges", Washington DC

Barghouthi, A.F. (1984). "Pile Response to Seismic Waves", Ph.D. Dissertation, Univ. of Wisconsin, Madison

Blaney, G.W., Kausel, E. & Roesset J.M. (1976). "Dynamic Stiffness of Piles", Proc. 2nd Int. Conf. Num. Methods Geomech. Virginia Polytech. Inst. & State Un. Blacksburg VA II, pp. 1010-1012

Clough R.W. & Penzien J., Dynamics of Structures, McGraw-Hill, 1975

Dotson, K.W. & Veletsos, A.S. "Vertical and Torsional impedances for radially inhomogeneous viscoelastic soil layers", Report. NCEER-87-0024, National Center for Earthquake Engineering Research, State University of New York, Buffalo, 1987

Eurocode EC8 (1990). "Structures in seismic regions, Part 5: Foundations, retaining structures, and geotechnical aspects", First Draft. Brussels: Commission of the European Communities

Fan, K., Gazetas, G., Kaynia, A., Kausel, E., & Ahmad, S. (1991). "Kinematic seismic response of single piles and pile groups", J. Geotech. Engng Div., ASCE, 117, No. 12, 1860-1879

Florres-Berrones, R. & Whitman, R. V. (1982). "Seismic Response of End-Bearing Piles", J. Geotech. Engng Div., ASCE 108, No. 4, 554-569

Gazetas, G., Fan, K., Tazoh, T., Shimizu, K., Kavvadas, M., & Makris, N. (1992). "Seismic response of soil-pile-foundation--structure systems: Some recent developments", Piles Under Dynamic Loads, Geotech. Special Publ. No. 34, ASCE, S. Prakash, ed., 56-93

Gazetas, G., Mylonakis, G., & Nikolaou, A. (1995). "Simple Methods for the Seismic Response of Piles Applied to Soil-Pile-Bridge Interaction", State of the Art Paper, 3rd Intern. Conf. on Recent Advances in Geotechnical Earthquake Engineering & Soil Dynamics, Vol. 3, pp. 1547-1556, St. Louis, April 2-7, 1995

Kavvadas, M. & Gazetas, G. (1992). "Kinematic Seismic Response and Bending of Free-Head Piles in Layered Soil", Geotechnique, Vol. 43, No. 2, pp. 207-222

Kaynia, A. M. & Kausel, E. (1982). "Dynamic Behavior of Pile Groups" Proc. 2nd Int. Conf. Numer. Meth. in Offshore Piling, Austin, pp. 509-532

Makris, N. & Gazetas, G. (1992). "Dynamic Pile-Soil-Pile Interaction. Part II: Lateral and

Seismic Response", *Earthq. Engng. & Struct. Dynamics*, Vol. 21, No. 2

Mamoon, S.M., & Banerjee, P.K. (1990). "Response of Piles and Pile Groups to Traveling SH-waves", *Earthq. Engng. & Struct. Dynamics*, Vol. 19, No. 4, pp. 597-610

Michaelides, O., Gazetas, G., Bouckovalas, G., & Chrysikou, E. (1997) "Approximate Non-linear Dynamic Axial Response of Piles", Geotechnique, 1997 (in press)

Mizuno, H. (1987) "Pile Damage during Earthquakes in Japan", *Dynamic Response of Pile Foundations*, ed. T. Nogami, ASCE Special Publication, pp. 53-78

Mylonakis, G. (1995). "Contributions to Static and Seismic Analysis of Piles and Pile-Supported Bridge Piers", Ph.D. Dissertation, State University of New York at Buffalo.

Mylonakis, G., Nikolaou, A., & Gazetas, G. (1997). "Soil-Pile-Bridge Seismic Interaction: Kinematic and Inertial Effects. Part I: Soft Soil", *Earthq. Engng. & Struct. Dynamics*, Vol. 26, No. 3, pp. 337-359

Nikolaou, A. (1995). Seismic Distress of Piles, M.S. Thesis, State University of New York at Buffalo

Nikolaou, A., Mylonakis, G., and Gazetas, G. "Kinematic Bending Moments in Seismically Stressed Piles", Report NCEER-95-0022, National Center for Earthquake Engineering Research, State University of New York, Buffalo, 1995

Pender, M. (1993). "Aseismic Pile Foundation Design Analysis", *Bulletin of the New Zealand National Society for Earthquake Engineering*, Vol. 26, No. 1, pp. 49-160

Wolf, J.P. (1985). *Dynamic Soil-Structure Interaction*, Prentice-Hall

Veletsos, A.S. & Dotson K.W. (1986). "Impedances of Soil Layer with Disturbed Boundary Zone", *Journal of Geotechn. Engng.*, Vol. 112, No. 3

Velez, A., Gazetas, G., & Khrishnan, R. (1983). "Lateral Dynamic Response of Constrained - Head Piles", *Journal of Geotech. Eng. Div., ASCE*, Vol. 109, No. 8, pp. 1063-1081

Seismic Behaviour of Ground and Geotechnical Structures, Sêco e Pinto (ed.) © 1997 Balkema, Rotterdam, ISBN 90 5410 887 8

Analytical formulae for the seismic bearing capacity of shallow strip foundations

A. Pecker
Géodynamique et Structure, Bagneux, France

ABSTRACT: Following the observed foundation failures in Mexico City during the 1985 Michoacan earthquake, extensive studies have been carried out on the seismic bearing capacity of shallow foundations. These studies have thrown new light on the main factors affecting the seismic behavior of shallow foundations at failure and new analytical formulae have emerged to compute the seismic bearing capacity under pseudo-static conditions. These formulae account for the load inclination and eccentricity as well as for the inertia forces developed within the soil medium; they are valid for cohesive and dry cohesionless materials.

1 INTRODUCTION

The evaluation of the seismic bearing capacity of foundations has not received much attention from the earthquake engineering community. If the cyclic behavior of foundations has been extensively studied, with the development of impedance analyses in the last decade, their behavior at failure did not initiate much research. The major reason probably lies in the few observations of foundations failures during earthquakes (Auvinet and Mendoza, 1986; Romo and Auvinet, 1992; Pecker, 1994). This situation changed with the 1985 Michoacan Guerero earthquake and significant improvements in the foundation design have been obtained ever since. The object of this paper is to review the recent developments in the evaluation of the seismic bearing capacity of shallow foundations for which rational methods of analyses seem to emerge in the last few years. These methods, based on limit analyses theories for the pseudo-static problem, have lent themselves to the derivation of analytical formulae which properly account for the load inclination and eccentricity of the forces acting on the foundation (inertia forces from the superstructure) and for the seismic forces developed within the soil medium by the passage of the seismic waves.

2. FACTORS AFFECTING THE BEARING CAPACITY OF SURFICIAL FOUNDATIONS

Many factors affect the seismic bearing capacity of foundations; knowledge and proper evaluation of these parameters are essential prerequisites for a correct evaluation of the foundation stability.

2.1 Pre-earthquake Conditions

Observations of foundation behavior during the 1985 Michoacan-Guerero earthquake clearly evidenced that the initial static pressure and load eccentricity have pronounced effects on the seismic behavior of foundations. Foundations with low static safety factors (high contact pressures) or with significant load eccentricities behaved poorly whereas well-designed foundations appear not to have suffered significant damages.

2.2 Seismic loads

A key aspect of any soil-structure interaction problem and hence of bearing capacity evaluations is the calculation of the dynamic forces acting at the soil foundation interface. It must be recognized that the dynamic loads acting on the foundation have six components:

- a vertical force which, in most common situations, can be neglected since its magnitude is small with respect to the static permanent vertical load;
- two shear forces (T) acting in two orthogonal, horizontal directions; these forces arise from the inertia forces developed in the structure by the shaking; they induce an inclination with respect to the vertical of the force acting on the foundation;
- two overturning moments around horizontal axes which are closely related to the inertia forces (T); they arise from the elevated position of the center of gravity of the structure above the foundation level; these moments induce an eccentricity of the load acting on the foundation;

- a torsional moment, if the center of mass of the structure is not aligned with the geometric center of the foundation; this situation occurs for instance, when the permanent loads present an initial eccentricity with respect to the center of the foundation. This moment can be disregarded for bearing capacity evaluations, although the permanent eccentricity must be accounted for in the preearthquake conditions (paragraph 2.1) and in the evaluation of the dynamic shear forces and overturning moments, since it can affect their values due to dynamic coupling between the degrees of freedom of the system.

For strip foundations, these forces reduce to the vertical one, the shear force and the overturning moment; only the maximum value of these forces are needed but, in some instances, when permanent displacements are to be computed, their variations in time are also required.

2.3 Soil Strength

Even for static conditions, the correct evaluation of the soil strength is a challenge for the geotechnical engineers. It depends on many factors such as the soil fabric, stress and strain history, stress path, drainage conditions, anisotropy, ... Additional factors must be considered for seismic conditions, depending on the soil type which will be broadly characterized as cohesionless soils or cohesive soils:

- the rate of loading may significantly affect the value of the strength measured in conventional laboratory test usually carried out at slower rates than those anticipated in the field during an earthquake. The strength of dry cohesionless soils is not affected by the rate of loading but plastic cohesive soils may exhibit undrained strengths 30% to 60% higher than the conventional strength. For Mexico city clay, for instance, an average increase of 40% is quoted by Romo (1995).

- degradation under cyclic loading. The repetition of alternate cycles of loading may cause a cyclic degradation of the material and a subsequent decrease in its strength. However, not all materials are subjected to degradation: dry cohesionless soils are insensitive to it and cohesive soils will only experience degradation when they are strained beyond a given strain threshold which is material dependent. For instance, permanent cyclic deformations start to accumulate for Mexico city

clays only when the cyclic strain is higher than approximately 3% (Romo, 1995);

- pore pressure build-up and drainage conditions. Saturated cohesionless soils usually experience an increase in their pore water pressure due to cyclic loading under undrained conditions, which may lead to a liquefaction condition unless the drainage conditions allow for a rapid dissipation of these pressures. Whether the soil is initially in a dense or in a loose state will lead to completely different situations: a loose sand loses its whole strength and gives rise to large deformations and catastrophic failures; dense sands, owing to their dilatant behavior, can mobilize very large undrained strengths, provided drainage is prevented, and cannot develop dramatic failures.

2.4 Inertia Forces in the Soil Medium

During an earthquake, the passage of seismic waves gives rise to inertia forces within the soil medium, which are equilibrated by dynamic stresses (mainly horizontal shear stresses). These stresses mobilize a fraction of the available soil strength and consequently, the strength available to balance the inertia forces arising from the superstructure is not necessarily the full soil strength. As an extreme case, when the ground acceleration at a given depth is too large, the induced seismic stresses may cause failure: this phenomenon called fluidization (Richards et al, 1990), has been noticed in numerous theoretical studies (Pecker - Salençon, 1991; Budhu and Al Karni, 1993; Richards and Shi, 1994, Paolucci and Pecker, 1997).

However, it must be noticed that the inertia forces within the soil mass and the forces acting on the foundation, arising from the inertia forces of the superstructure, both vary in time and there is no obvious reason why they should be in phase and maximum, with the same direction, at the same instant. Moreover, inertia forces within the soil profile decrease with depth due to the attenuation of accelerations, which is not reflected in the aforementioned studies. To properly account for the influence of the soil inertia forces, they must be treated as an independent parameter (Pecker - Salençon, 1991). To consider the same seismic coefficient k_H for the soil inertia force ($F_X = \rho\, g\, k_H$) and for the structural inertia forces ($T = k_H\, m\, g$) leads to erroneous conclusions with respect to the influence of the soil inertia forces; the major reduction in the foundation bearing capacity outcomes from the load inclination and eccentricity on the foundation and not from the soil inertia forces (Dormieux - Pecker, 1995).

3. GENERAL FRAMEWORK FOR THE EVALUATION OF THE FOUNDATION BEARING CAPACITY

Assuming that all the factors listed under paragraph 2 are properly accounted for, the dynamic bearing capacity of surficial foundation can be examined from two different approaches.

Probably, the most rigorous approach would be to develop a global model (finite element model) including both the soil and the structure. Obviously, if the analysis is meant to be significant, a realistic non-linear constitutive soil model must be used. Owing to this last constraint, to computer limitations and also to the fact that development of a global model would require competence in geotechnical engineering, structural engineering, soil-structure interaction, numerical analysis, such an approach is seldom used in everyday practice.

The alternative approach, which represents the state of practice today, is to uncouple the evaluation of dynamic loads (a structural engineer task), from the verification of the bearing capacity (a geotechnical engineer task). This is a so-called substructure approach which suffers the following shortcomings which, up to now, have not clearly been evaluated:

- the evaluation of the dynamic loads is based on an elastic analysis of the soil-structure system; at most, some degree of non-linearities can be accounted for in an approximate manner, but how the dynamic loads are affected by yielding of the foundation is usually not evaluated; recently, Paolucci (1997) has shown that the base shear transmitted by the superstructure may differ from that predicted from a classical linear elastic soil structure interaction analysis, if soil yielding is accounted for;

- the bearing capacity is checked using a pseudo-static approach, in which only the maximum loads acting on the foundations are considered.

3.1 Pseudo-Static Evaluations

Up to very recently, the seismic bearing capacity of shallow foundations was checked using classical bearing capacity formulae in which the seismic action is regarded as an equivalent static force and load eccentricity and inclination are treated as correction factors (S and i) to the N_γ, N_c and N_q bearing capacity factors. The ultimate bearing capacity writes:

$$q_u = \frac{1}{2} \gamma B S_\gamma i_\gamma N_\gamma + C S_c i_c N_c + q S_q i_q N_q \quad (1)$$

Recently, methods based on limit equilibrium analyses (Chen 1990, Salençon 1983) have emerged taking into account the soil inertia foces (Sarma, Isossifelis 1990, Budhu, Al-Karni 1993, Richards et al 1993). All these methods, although they present significant improvements on the preceding bearing capacity equation, suffer limitations which restrict the significance of their findings:

- the horizontal accelerations of the soil and of the structure are assumed to have the same magnitude and to act in the same direction; this is a severe limitation since there is no reason why both the structure and the soil should respond in an identical manner to a given motion;

- results are based on an assumed failure surface which is similar to an asymmetric Prandtl's mechanism (Sarma, Iossifelis) or a simplified version of it (Richards et al). This mechanism does not allow for any uplift of the foundation which can be significant for high horizontal accelerations;

- all these methods are upper bound solutions to the true bearing capacity problem, but no indication of their goodness is given by comparison with lower bound estimates.

Based on their results, all authors conclude that the incorporation of inertia forces in the soil results in a dramatic reduction in the bearing capacity of a foundation. However, as shown by Dormieux - Pecker (1995), the major reduction outcomes from the load inclination and eccentricity on the foundation; the incorporation of the inertia forces in the soils only contributes for an additional small reduction (for a Coulomb's material). This result may have important implications for seismic building codes, since only few solutions are available incorporating soil inertia forces; could they be neglected, it will make the situation simpler for designers.

A more elegant solution to the problem is provided by the concept of bounding surface in which the loading parameters are treated as independent parameters.

To this end, the yield design theory provides a rigorous treatment of the problem (Salençon, 1983, 1990). This theory belongs to the category of limit analysis methods. Alike any limit analysis method, the derivation of upper bound and lower bound solutions allows to bracket the exact solution and possibly to determine it exactly when both bounds coincide. A proper application of the yield design theory requires the knowledge of :

- the problem geometry; in the following, the foundation is assumed to be a strip footing resting on the surface of an homogeneous halfspace;

- the materials strengths; they refer to the soil strengths which are represented by a Tresca strength criterion (cohesive soil) or a Mohr-Coulomb strength criterion (dry cohesionless soil) and to the soil-foundation interface which is assumed without tensile strength to allow for uplift between the soil and the foundation;

- the loading parameters; five independent loading parameters are considered in the derivation of the bounding surface: the normal force N, the horizontal shear force T, the overturning moment M and the soil inertia forces F_x (= ρ g k_H) and F_y (= ρ g k_v) in the horizontal and vertical direction.

The set of admissible loads is located within a surface, defined in the loading parameters space, called the bounding surface.

$$\phi\ (N,\ T,\ M,\ F_x,\ F_y) \leq 0 \qquad (2)$$

In the case where F_x = F_y = 0, experimental evidence of equation (2) has been given by Butterfield - Gottardi (1994) and Kitazume - Terashi (1994).

This approach has been followed by Pecker Salençon (1991), Salençon Pecker (1995 (a) and (b)), Paolucci Pecker (1997).

3.2 Dynamic approach

As noted previously, the forces acting on the foundation or within the soil mass vary with time. They can exceed the available resistance of the foundation soil-system for short periods without leading to a general failure of the foundation. This is an essential difference between static, permanent loading and dynamic, time-varying loading. In the first instance, an excessive load generates a general failure, whereas the second situation induces permanent, irreversible displacements which superimpose to the cyclic displacements. Failure can therefore be no longer defined as a situation in which the safety factor drops below 1.0. It must rather be defined as excessive permanent displacements which impede the proper functioning of the structure. This definition, first introduced by Newmark (1965) has been successfully applied to the design of dams, gravity retaining walls assimilating the potentially unstable soil mass to a rigid sliding block. It has also been used for the bearing capacity of foundations (Sarma - Iossefilis, 1990; Richards et al, 1993).

This method has been further extended, relaxing the condition of rigid soil blocks and considering a more realistic deformable body, as it is actually assumed in the computed failure mechanisms. The soil foundation system is assumed to behave as an elastic perfectly plastic system, in which the bounding surface defined previously, is adopted as the boundary for the apparition of plastic deformations. Using the kinetic energy theorem, the angular velocity of the foundation and its permanent displacement can be computed (Pecker - Salençon, 1991, Pecker 1996). Under the assumptions spelled above, the method permits a rigorous definition of failure in terms of unacceptable permanent displacements. The reader is referred to these references and also to Paolucci (1997) for additional details on this topic which is not further developed herein.

4. SIMPLIFIED FORMULA FOR COHESIVE SOILS

Solutions to the bearing capacity of shallow strip footings resting on the surface of a cohesive halfspace have been obtained by Pecker - Salençon (1991), Salençon - Pecker (1995 a and b) for a soil obeying a Tresca strength criterion with or without tensile strength. These solutions were derived from the static and the kinematic approaches of the theory

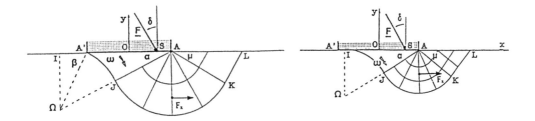

Without uplift

With uplift

Figure 1. Example of Kinematic Mechanisms - Cohesive soils

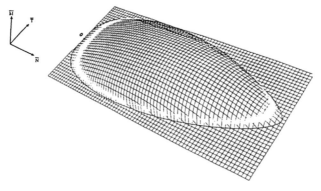

Figure 2. Skeletal View of the Bounding Surface for Cohesive Soils

and it was shown that both, the lower bound and the upper bound solutions, were very close to each other, giving therefore an almost exact solution to the problem. The most prominent kinematic mechanisms used are presented in figure 1 in two situations: without uplift of the foundation and with uplift. The first situation is prevailing for small load eccentricities or inclinations whereas the second one governs when these two parameters become significant. These mechanisms depend upon three geometric parameters for which the optimum values, which minimize the maximum resisting work, are numerically determined.

A simple dimensional analysis shows that the results can be expressed in terms of the adimensional parameters:

$$\overline{N} = \frac{N}{CB}, \quad \overline{T} = \frac{T}{CB}, \quad \overline{M} = \frac{M}{CB^2}, \quad \overline{F} = \frac{F_x B}{C}$$

where C is the soil undrained shear strength, B the foundation width and N, T, M, F_x, the four independent loading parameters, since for a Tresca strength criterion F_y does not play any role.

In the case where $\overline{F}$ is nil and for a soil without tensile strength, the bounding surface is presented in figure 2; only the upper part of the surface corresponding to $\overline{M} \geq 0$ is presented in figure 2.

In the general case, the following equation has been found appropriate to define the bounding surface:

$$\frac{\left[(1 - e\,\overline{F})\,\beta\,\overline{T}\right]^2}{(\alpha\,\overline{N})^a\left[1 - \alpha\,\overline{N} - e\,\overline{F}^g\right]^b} + \frac{(1 - f\,\overline{F})\,(\gamma\,\overline{M})^2}{(\alpha\,\overline{N})^c\left[1 - \alpha\,\overline{N} - e\,\overline{F}^g\right]^d} - 1 = 0 \qquad (3)$$

in which the parameters a ... g and α ... γ have determined numerically:

a = 0.70, b = 1.29, c = 2.14, d = 1.81, e = 0.21, f = 0.44, g = 1.22, $\alpha = \dfrac{1}{\pi + 2}$, $\beta = 0.5$, $\gamma = 0.36$

Equation (3) is valid under the constraints: $0 < \alpha\,\overline{N} \le 1$, $|\overline{T}| \le 1$.

When $\overline{F} = \overline{M} = \overline{T} = 0$, equation (3) reduces to the well known bearing capacity formula:

$$\alpha\,\overline{N} = 1 \quad \text{i.e.} \quad N = (\pi + 2)\,C.B \qquad (4)$$

When $\overline{F} = 0$ and $\overline{M} \ne 0$, $\overline{T} \ne 0$, equation (3) can be viewed as an extension of Meyerhoff's formula for eccentric-inclined loads.

With non zero values of $\overline{F}$ (the soil inertia forces), the following conclusions can be derived (Pecker Salençon 1991, Pecker et al. 1995):

- for normally encountered ground accelerations, characterized by a value $\overline{F} \le 2$, and for foundations for which $\overline{N} \le 2.5$, i.e. for foundations with a safety factor higher than 2.0 under a vertical centered load, the effect of the soil seismic forces can be neglected without loss of accuracy. For foundations with lower safety factors, the soil seismic forces induce a dramatic reduction in the bearing capacity. This is illustrated in figure 3 which presents cross-sections of the bounding surface for various values of $\overline{F}$

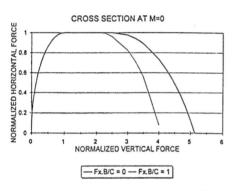

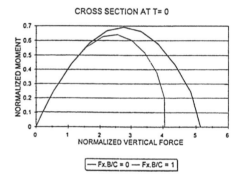

Figure 3. Bounding Surfaces

5.0 SIMPLIFIED FORMULA FOR DRY COHESIONLESS SOILS

Solutions for the cohesionless soil, obeying Coulomb's strength criterion with or without tensile strength, have been obtained within the framework of the yield design theory (Salençon - Josseron, 1994). The kinematic mechanisms of the upper bound approach are similar to those presented in figure 1, provided the arc of circles are replaced by log spirals.

Based on the results of Salençon-Josseron (1994), and on additional results using the same theory (Paolucci Pecker, 1997), a simplified formula has been developed for the equation of the bounding surface preserving the same general expression given by equation (3).

Restricting our presentation to the case of a purely cohesionless soil (C = 0) and introducing the adimensional variables:

$$\overline{N} = \frac{N}{N_{max}} \qquad \overline{T} = \frac{T}{N_{max}}$$

$$\overline{M} = \frac{M}{B\,N_{max}} \qquad \overline{F} = \frac{k_H}{\tan\phi}$$

in which N_{max} is the ultimate load under a vertical centered force

$$N_{max} = \frac{1}{2}\,\gamma\,B^2\,N_\gamma \qquad (5)$$

266

where N_γ is the well known bearing capacity factor which takes values of the order of 25 for $\phi = 30°$, and 52 for $\phi = 35°$, the following equation has been derived for the bounding surface

$$\frac{\left[(1 - e\,\overline{F})\,\beta\,\overline{T}\right]^c}{(\overline{N})^a \left[(1 - g\,\overline{F})^d - \overline{N}\right]^b} + \frac{\left[(1 - f\,\overline{F})\,\gamma\,\overline{M}\right]^c}{(\overline{N})^a \left[(1 - g\,\overline{F})^d - \overline{N}\right]^b} - 1 = 0 \qquad (6)$$

in which:

$a = 0.92$, $b = 1.25$, $c = 1.14$, $d = 0.39$, $e = 0.41$, $f = 0.32$, $g = 0.96$, $\beta = 2.90$, $\gamma = 3.95$

Equation (6) is valid under the following constraint:

$$0 < \overline{N} < (1 - g\,\overline{F})^d$$

In equation (6), opposite to the case of cohesive soils, the gravity force and vertical inertia force in the soil medium do affect the bearing capacity. They can be accounted for in the soil unit weight:

$$\gamma = \rho\,g\,(1 + k_v) \qquad (7)$$

where k_v, the vertical seismic coefficient, can take positive or negative values.

In the case of a zero $\overline{F}$, the bounding surface is presented in figure 4.

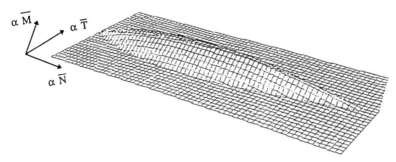

Figure 4. Skeletal View of the Bounding Surface for Cohesionless Soils ($\phi = 30°$)

Generally speaking, the same conclusions as for the cohesive soils were reached (Paolucci Pecker, 1997): in a range of reasonable values of the pseudo-static seismic coefficient ($k_H \leq 0.3$) the bearing capacity reduction due to the effects of the inertia forces in the soil is small and does not exceed 15-20%.

6.0   CONCLUSIONS

Based on the static and kinematic approaches of the yield design theory, the bounding surfaces for the seismic bearing capacity of shallow strip foundations resting at the surface of a cohesive or dry cohesionless soil have been numerically determined.   In this derivation, the loading parameters (vertical force, horizontal force and overturning moment acting on the foundation, as well as soil inertia forces within the soil medium) are treated as independent variables preserving all the necessary flexibility to assign them the appropriate values.

These bounding surfaces have been defined by simple analytical formulae expressed in terms of adimensional variables related to the loading parameters.   These formulae reduce to the well known bearing capacity formulae for vertical centered loads, can be viewed as extensions of Meyerhoff's formulae for eccentric inclined loads and do incorporate the soil inertia forces for the seismic case.   Therefore, they constitute a significant improvement for the evaluation of the pseudo-static seismic bearing capacity of shallow strip foundations.

ACKNOWLEDGMENT

This work was supported by the European Commission under contract CI1*CT92-0069 "Seismic Bearing Capacity of Foundations on Soft Soils" and the Human Capital and Mobility Program under contract ERB-CHRX-CT92-0011.

REFERENCES

Auvinet, G., Mendoza, M.J. (1986).
Comportamiento de diversos tipos de cimentacion en la zone lascustre de la Ciudad de Mexico durante el sismo del 19 de septiembre de 1985. *Proceedings Symposium: "Los Sismos de 1985: Casos deMecanica de Suelos, Mexico.*

Budhu, M., Al-Karni, A. (1993). Seismic bearing capacity of soils. *Geotechnique, 33*, pp.181-187.

Butterfield R., Gottardi G. (1994). A complete three-dimensional failure envelope for shallow footings on sands. *Geotechnique, 44, n° 1*, pp. 181-184.

Chen W.F., Liu X.L. (1990). Limit analysis in soil mechanics. *Developments in geotechnical engineering, vol. 52*, Elsevier Applied Sciences, ed., New York, N.Y.

Dormieux L., Pecker A. (1995). Seismic bearing capacity of foundations on cohesionless soils - *Journal of Geotechnical Engineering, ASCE, 121(3)*, pp. 300-303.

Kitazume M., Terashi M. (1994). Operation of PHRI Geotechnical Centrifuge from 1980 to 1994. *Technical note of the Port and Harbour Research Institute*, Ministry of Transport, Japan.

Newmark N. (1965). Effects of earthquakes on dams and embankments. *Geotechnique*, vol. *XV(2)*, pp. 139-160.

Paolucci R. (1997). Simplified evaluation of earthquake induced permanent displacements of shallow foundations. Accepted for publication in *Journal of Earthquake Engineering.*

Paolucci R., Pecker A. (1997). Seismic bearing capacity of shallow strip foundations on dry soils. Accepted for publication in *Soils and Foundations.*

Pecker A. (1994). Seismic design of shallow foundations. *Proceedings 10th European Conference on Earthquake Engineering, 2,* pp. 1001-1010, Vienna.

Pecker A. (1996). Seismic bearing capacity of shallow foundations. *11th World Conference on Earthquake Engineering*, Acapulco.

Pecker A., Salençon J., (1991). Seismic bearing capacity of shallow strip foundations on clay soils. *Proceedings of the International Workshop on Seismology and Earthquake Engineering.* CENAPRED, Mexico city, pp. 287-304.

Pecker A., Auvinet G., Salençon J., Romo, M.P., Verzura L. (1995). Seismic bearing capacity of foundations on soft soils. *Report to the European Commission. Contract CI1* CT92-0069.*

Richards R., Elms, D.G. and Budhu, M. (1990). Dynamic fluidization of soil. *ASCE Journal of Geotechnical Engineering, 116,* pp. 740-759.

Richards R., Elms, D.G. and Budhu, M. (1993). Seismic bearing capacity and settlements of shallow foundations. *ASCE Journal of Geotechnical Engineering, 119,* pp. 662-674.

Richards Jr., R. and Shi, X. (1994). Seismic lateral pressures in soils with cohesion. *ASCE Journal of Geotechnical Engineering, 120,* n° 7, pp. 1230-1251.

Romo M., Auvinet, G. (1992). Seismic behavior of foundations on cohesive soft soils. *Recent Advances in Earthquake Engineering and Structural Dynamics.* Ouest Editions, Nantes, Chapter III.4, pp. 311-328.

Romo M. (1995). Clay behavior ground response and soil-structure interaction studies in Mexico city. *Proc. 3rd Conf. on Recent Advances in Geotech. Earthq. Engng and Soil Dynamics,* vol. *2,* pp. 1039-1051.

Salençon J. (1983). Calcul à la rupture et analyse limite. *Presses de l'Ecole Nationale des Ponts et Chaussées,* Paris.

Salençon J. (1990). An introduction to the yield design theory and its applications to soil mechanics. *European Journal of Mechanics, A/Solids, vol. 9(5),* pp. 477-500.

Salençon J., Josseron E. (1994). Seismic bearing capacity of footings. *Annual Progress Report, Prenormative research in support of Eurocode 8-Topic 4: Capacité portante des fondations superficielles sous sollicitations sismiques. Contract ERB-CHRX-CT92 -0011.*

Salençon J., Pecker A. (1995a). Ultimate bearing capacity of shallow foundations under inclined and eccentric loads. Part I: purely cohesive soil. *Eur. J. Mech., A/Solids, 14,* n°3, 349-375.

Salençon J., Pecker A. (1995b). Ultimate bearing capacity of shallow foundations under inclined and eccentric loads. Part II: purely cohesive soil without tensile strength. *Eur. J. Mech., A/Solids, 14,* n°3, 377-396.

Sarma S.K. and Iossifelis I.S. (1990). Seismic bearing capacity factors of shallow strip footings. *Geotechnique, 40,* pp. 265-273.

Seismic Behaviour of Ground and Geotechnical Structures, Sêco e Pinto (ed.) © 1997 Balkema, Rotterdam, ISBN 90 5410 887 8

Evaluation of cyclic liquefaction potential based on the CPT

P.K.Robertson & C.E.(Fear)Wride
Geotechnical Group, University of Alberta, Edmonton, Alb., Canada

ABSTRACT: Due to the inherent difficulties and poor repeatability associated with the SPT, several correlations have been proposed to estimate the resistance against cyclic loading for clean sands and silty sands using the CPT. Nevertheless, the SPT has continued to play a major role in the estimation of cyclic resistance due to the need to make corrections to the penetration resistance in silty sands. The SPT provides samples which have been required because the corrections have been based on fines content. However, the amount of field performance CPT data from sites where liquefaction has or has not been observed has increased significantly in the last 10 years. There is now sufficient CPT experience to enable reliable estimates of cyclic resistance to be made directly from CPT data without the need for samples. An integrated CPT method for estimating cyclic resistance is described and an example is presented.

1. INTRODUCTION

Soil liquefaction is a major concern for structures constructed with or on sandy soils. Since 1964, much work has been carried out to explain and understand soil liquefaction. The progress of work on soil liquefaction has been described in detail in a series of state-of-the-art papers, such as Yoshimi et al. (1977), Seed (1979), Finn (1981), Ishihara (1993), and Robertson and Fear (1995).

Several phenomena are described as soil liquefaction. In an effort to clarify the different phenomena, the mechanisms and a set of definitions for soil liquefaction were described by Robertson and Fear (1995). Cyclic liquefaction is defined as the occurrence of zero effective stress due to undrained cyclic loading. At zero effective stress no shear stress exists. When shear stress is applied, large deformations occur due to the large reduction in soil stiffness. This paper describes how the CPT can be used to estimate the resistance to cyclic liquefaction and is a shortened version of a paper prepared for the 1996 NCEER Workshop on Evaluation of Liquefaction Resistance (Robertson and Wride, 1997).

2. CYCLIC RESISTANCE BASED ON THE CPT

The late Professor H.B. Seed and his co-workers developed a comprehensive approach to estimate the potential for cyclic liquefaction due to earthquake loading. The approach requires an estimate of the cyclic stress ratio (CSR) profile caused by a design earthquake. This is usually done based on a probability of occurrence for a given earthquake. A site specific seismicity analysis can be carried out to determine the design CSR profile with depth. A simplified method to estimate CSR was first developed by Seed and Idriss (1971) based on the maximum ground surface acceleration (a_{max}) at the site. The CSR profile from the earthquake can be compared to the estimated cyclic resistance ratio (CRR) profile for the soil deposit. At any depth, if CSR is greater than CRR, cyclic liquefaction is possible. This approach is the most commonly used technique in most parts of the world for estimating soil liquefaction due to earthquake loading.

Currently, the most popular simple method for estimating CRR makes use of the penetration resistance from the Standard Penetration Test (SPT) although, more recently, the Cone Penetration Test (CPT) has become very popular due to its greater reliability and repeatability and the continuous nature of its profile. The methodology based on the SPT has many problems, primarily due to the unreliable nature of the SPT. The main factors affecting the SPT test have been reviewed (e.g. Seed et al., 1985; Skempton, 1986; Robertson et al., 1983). It is highly recommended that the engineer become familiar with the details of the SPT in order to avoid or at least minimize some of the major factors.

Due to the inherent difficulties and poor repeatability associated with the SPT, several correlations have been proposed to estimate CRR for clean sands and silty sands using corrected CPT penetration resistance. Although cone penetration resistance is often just corrected for overburden stress (resulting in the term q_{c1}), truly normalized (i.e. dimensionless) cone penetration resistance corrected for overburden stress (q_{c1N}) is given by:

$$q_{c1N} = \left(\frac{q_c}{P_{a2}}\right)C_Q = \frac{q_{c1}}{P_{a2}} \qquad [1]$$

where:

q_c = measured cone tip penetration resistance

C_Q = correction for overburden stress

= $\left(P_a/\sigma'_{vo}\right)^{0.5}$

= a maximum value of 2 is generally applied to CPT data at shallow depths.

P_a = reference pressure of 100 kPa in same units as σ'_{vo}; P_a =100 kPa if σ'_{vo} is in kPa

P_{a2} = reference pressure of 100 kPa in same units as q_c; P_{a2} = 0.1 MPa if q_c is in MPa.

In recent years, there has been an increase in available field performance data, especially for the

CPT (Ishihara, 1993; Kayen et al., 1992; Stark and Olson, 1995; Suzuki et al., 1995). The recent field performance data have shown that the existing CPT-based correlations to estimate CRR are generally good for clean sands. The recent field performance data show that the correlation between CRR and q_{c1N} by Robertson and Campanella (1985) for clean sands provides a good estimate of CRR, as shown in Figure 1. Based on discussions at the 1996 NCEER Workshop on Evaluation of Liquefaction Resistance, the curve by Robertson and Campanella (1985) has been adjusted slightly. The resulting recommended CPT correlation for clean sand is shown in Figure 1. Occurrence of liquefaction is based on level ground observations of surface manifestations of cyclic liquefaction. For loose sand ($q_{c1N} < 75$) this could involve large deformations resulting from a condition of essentially zero effective stress being reached. For denser sand ($q_{c1N} > 75$) this could involve the development of large pore pressures, but the effective stress may not fully reduce to zero and deformations may not be as large as in loose sands. Hence, the consequences of 'liquefaction' will vary depending on the soil density as well as the size and duration of loading.

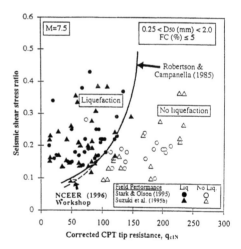

Figure 1. Recommended cyclic resistance ratio (CRR) for clean sands under level ground conditions based on the CPT.

The CPT field observation data used to compile the curve in Figure 1 are apparently based on the following conditions:

Holocene age, clean sand deposits
Level or gently sloping ground
Magnitude M = 7.5 earthquakes
Depth range from 1 to 15 m (3 to 45 ft)
(84% is for depth < 10 m (30 ft))
Representative average CPT q_c values for the layer that was considered to have experienced cyclic liquefaction.

Caution should be exercised when extrapolating the CPT correlation to conditions outside of the above range. An important feature to recognize is that the correlation appears to be based on average values for the liquefied layers. However, the correlation is often applied to all measured CPT values, including low values below the average. Hence, the correlation can be conservative in variable deposits in which a small part of the CPT data could indicate possible liquefaction.

It is important to note that the Seed approach based on either the SPT or the CPT has many uncertainties. The correlations are empirical and there is some uncertainty over the degree of conservatism in the correlations as a result of the methods used to select representative values of penetration resistance within the layers assumed to have liquefied (Fear and McRoberts, 1995). A detailed review of the CPT data, similar to those carried out by Liao and Whitman (1986) and Fear and McRoberts (1995) on SPT data, would be required to investigate the degree of conservatism contained in Figure 1. The correlations are also sensitive to the amount and plasticity of the fines within the sand. One reason for the continued use of the SPT has been the need to obtain a soil sample to determine the fines content of the soil. However, this has been offset by the unreliability of SPT data. With the increasing interest in the CPT due to its greater reliability, it is now possible to estimate fines content and grain size from CPT data and incorporate this directly into the evaluation of liquefaction potential. The following is a summary of the approach suggested by Robertson and Fear (1995), based on the CPT.

For the same CRR, the CPT penetration resistance in silty sands is smaller due to the greater compressibility and decreased permeability of silty sands. Robertson and Fear (1995) recommended an average correction which was dependent on fines content, but not on penetration resistance. The recommended modified corrections, based on both penetration resistance and fines content, can be defined as follows

$$\Delta q_{c1N} = K_{CPT} \, (q_{c1N})_{cs} \qquad [2]$$

where:

Δq_{c1N} = CPT tip correction for silty sands

$(q_{c1N})_{cs}$ = equivalent clean sand normalized CPT tip resistance

$\qquad = q_{c1N} + \Delta q_{c1N} \qquad [3]$

q_{c1N} = measured CPT tip resistance, corrected for overburden and normalized

K_{CPT} = 0   for FC < 5%   [4a]

$\qquad = 0.0267 \, (FC - 5)$

$\qquad\qquad$ for 5% < FC < 35%   [4b]

$\qquad = 0.80$ for FC > 35%   [4c]

FC = fines content, in percent.

Figure 2 presents the CPT correction factor, K_{CPT}, plotted in terms of fines content. The CPT correction for silty sands (Δq_{c1N}) can also be expressed in terms of the measured penetration resistance rather than the clean sand equivalent resistance, as follows:

$$\Delta q_{c1N} = \frac{K_{CPT}}{1 - K_{CPT}} \, q_{c1N} \qquad [5]$$

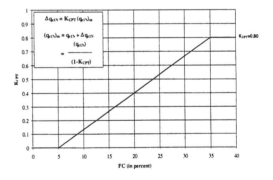

Figure 2.   Recommended fines content correction factor for the CPT.

Although the correction factor shown in Figure 2 is based on fines content, the cyclic resistance of a soil is a function of many factors, such as type of fines, plasticity, and grain characteristics, as well as fines content. Hence, the corrections should be applied with caution. The recommended procedure is to determine the fines content of the soil and apply a correction to the measured CPT q_{c1N} value using Figure 2 and Equation 5 and then to use the corrected penetration resistance with the clean sand curve shown in Figure 1 to estimate CRR. For high fines content soils, the criteria based on water content and liquid limit, as developed by Wang (1979) and interpreted by Marcuson et al. (1990), should also be applied.

2.1 Grain Characteristics from CPT

It is possible to estimate grain characteristics, such as fines content, of sandy soils directly from the CPT and, hence, correct the measured penetration resistance to an equivalent clean sand value. In recent years, charts have been developed to estimate soil type from CPT data (Olsen and Malone, 1988; Robertson and Campanella, 1988; Robertson, 1990). Experience has shown that the CPT friction ratio (ratio of the CPT sleeve friction to the cone tip resistance) increases with increasing fines content and soil plasticity (Suzuki et al., 1997). Hence, grain characteristics, such as fines content can be estimated from CPT data using soil behaviour charts, such as that shown in Figure 3. The addition of pore pressure data can also provide valuable additional guidance in estimating fines content.

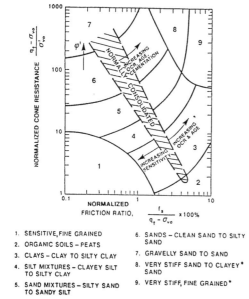

Figure 3.   Normalized CPT soil behaviour type chart, as proposed by Robertson (1990).

Based on field experience, it is possible to estimate grain characteristics, such as fines content and grain size directly from CPT results using the soil behaviour type chart shown in Figure 3. The boundaries between soil behaviour type zones 2 to 7 can be approximated as concentric circles (Jefferies and Davies, 1993). The radius of each circle can then be used as a soil behaviour type index. Based on the CPT chart by Robertson (1990), the soil behaviour type index, I_c, can be defined as follows:

$$I_c = [(3.47 - \log Q)^2 + (\log F + 1.22)^2]^{0.5} \quad [6]$$

where:
Q = normalized penetration resistance,
dimensionless = $(q_c - \sigma_{vo})/\sigma'_{vo}$
F = normalized friction ratio,
in percent = $[f_s /(q_c - \sigma_{vo})]$ x 100%
f_s = CPT sleeve friction stress
σ_{vo} = total overburden stress

The CPT soil behaviour chart (Figure 3) uses a normalized cone penetration resistance (Q) based on a stress exponent n=1.0, whereas the CRR is based on a normalized cone penetration resistance (q_{c1N}) based on a stress exponent n=0.5. Olsen and Malone (1988) correctly suggested a normalization where the stress exponent (n) varies from around 0.5 in sands to 1.0 in clays. However, this normalization for soil type is complex and iterative. Robertson (1990) suggested the simpler linear normalization for soil type based on n=1.0. Currently, this simpler procedure is recommended for soil classification using the CPT.

The boundaries of soil behaviour type are then given in terms of the index, I_c, as shown in Table 1. The soil behaviour type index does not apply to Zones 1, 8 or 9. Along the normally consolidated region in Figure 3, soil behaviour type index increases with increasing friction ratio; this relationship can then be combined with an approximate relationship between fines content and friction ratio to give the following simplified relationship:

$$\text{Fines Content, FC (\%)} = 1.75\, I_c^{3.25} - 3.7 \quad [7]$$

Table 1. Boundaries of soil behaviour
type (after Robertson, 1990)

Soil Behaviour Type Index, I_c	Zone	Soil Behaviour Type (see Figure 3)
$I_c < 1.31$	7	Gravelly sand
$1.31 < I_c < 2.05$	6	Sands: clean sand to silty sand
$2.05 < I_c < 2.60$	5	Sand Mixtures: silty sand to sandy silt
$2.60 < I_c < 2.95$	4	Silt Mixtures: clayey silt to silty clay
$2.95 < I_c < 3.60$	3	Clays
$I_c > 3.60$	2	Organic soils: peats

The range of potential correlations is illustrated in Figure 4, which shows the variation of soil behaviour type index (I_c) with fines content. The recommended relationship given in Equation 7 is also shown in Figure 4. Note that this equation is slightly modified from the original work by Robertson and Fear (1995) in order to increase the prediction of FC for a given value of I_c.

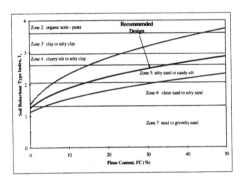

Figure 4. Variation of CPT soil behaviour type index (I_c) with fines content.

The proposed correlation between CPT soil behaviour index (I_c) and fines content is approximate, since the CPT responds to many other factors affecting soil behaviour, such as soil plasticity and mineralogy. Hence, when possible, it is recommended that the correlation shown in Figure 4 (see Equation 7) be evaluated and modified to suit a specific site and project. However, for small projects, the above correlation provides a useful guide. Caution must be taken in

applying the relationship in Equation 7 to sands that fall close to or in Zone 1 in Figure 3 so as not to confuse very loose clean sands with denser sands containing fines.

2.2 Influence of thin layers

Another problem associated with the interpretation of penetration tests occurs when thin sand layers are embedded in softer deposits. Theoretical as well as laboratory studies show that the cone resistance is influenced by the soil ahead of and behind the penetrating cone. The cone will start to sense a change in soil type before it reaches the new soil and will continue to sense the original soil even when it has entered a new soil. As a result, the CPT will not always measure the correct mechanical properties in thinly interbedded soils. In particular, care should be taken when interpreting cone resistance in thin sand layers located within soft clay or silt deposits. Based on studies by Vreugdenhil et al. (1994), Robertson and Fear (1995) suggested a correction factor for cone resistance in thin sand layers embedded in thick fine grained layers. Thin sand layers embedded in soft clay deposits are often incorrectly classified as silty sands based on the CPT soil behaviour type charts. Hence, a slightly improved classification can be achieved if the cone resistance is corrected for layer thickness effects before applying the classification charts.

2.3 Cyclic Resistance from CPT

In an earlier section, a method was suggested for estimating fines content directly from CPT results (Equation 7). Equations 4, 5 and 6 can be combined so that the correction to obtain the equivalent clean sand normalized penetration resistance, $(q_{c1N})_{cs}$, can be estimated directly from the measured CPT data. Using $(q_{c1N})_{cs}$, CRR (M = 7.5) can be estimated using the following simplified equation (which approximates the clean sand curve recommended in Figure 1):

$$CRR = 93 \left(\frac{(q_{c1N})_{cs}}{1000} \right)^3 + 0.08 \qquad [8]$$

where: $(q_{c1N})_{cs}$ is in the range
of $30 < (q_{c1N})_{cs} < 160$.

In summary, Equations 1 to 8 can be combined to provide an integrated method for evaluating the cyclic resistance of saturated sandy soils based on the CPT. The CPT-based method is an alternative to the SPT or shear wave velocity (V_s) based in-situ methods; however, using more than one method is useful in providing independent evaluations of liquefaction potential. The proposed integrated CPT method for evaluating CRR is summarized in Figure 5 in the form of a flowchart. The flowchart clearly shows the step-by-step process involved in using the method and indicates the recommended equations for each step of the process. However, alternative methods could be used in lieu of the proposed equations for any of the steps, provided that evidence is available to justify the application of such alternatives. The process described here includes estimating fines content (FC) from I_c (Equation 7; Figure 4) and then estimating the clean sand equivalent correction (Δq_{c1N}) by first estimating K_{CPT} from FC and then combining it with the measured penetration resistance, q_{c1N}, (Equations 4 and 5; Figure 2). As a result, Δq_{c1N} could be linked directly to I_c by combining Equations 4, 5 and 7, as indicated in the flowchart.

An example of this proposed CPT-based method is shown in Figure 6 for one of the sites from the CANLEX (Canadian Liquefaction Experiment) project (Robertson and Fear, 1995). The measured cone resistance is normalized and corrected for overburden stress to q_{c1N} and the soil behaviour type index (I_c) is calculated. Based on I_c, the fines content, FC (%), is estimated and the correction factor, K_{CPT}, is calculated. The values of K_{CPT} and q_{c1N} are used to calculate Δq_{c1N}. The final continuous profile of CRR at N=15 cycles (M = 7.5) is then calculated based on the equivalent clean sand values of q_{c1N} (i.e. $(q_{c1N})_{cs} = q_{c1N} + \Delta q_{c1N}$) and Equation 8. Included in Figure 6 are results from cyclic simple shear tests on undisturbed samples (obtained using ground freezing and sampling) converted to 15 cycle equivalent values of CRR using the conversions proposed by Seed et al. (1985). A good comparison is seen, although, at this site, the fines content is generally less than 10 % and the corrections to CPT tip resistance for fines content are very small. Note that for $I_c > 2.6$ (i.e. FC > 35%; see Equation 7), the soil is considered to be non-liquefiable; however, it should be checked using other criteria (e.g. Marcuson et al., 1990).

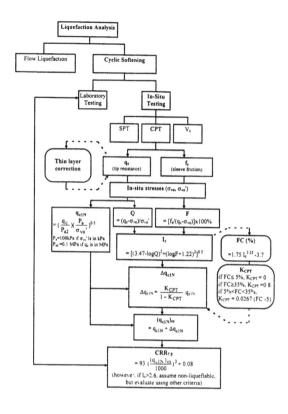

Figure 5. Flowchart illustrating the application of the integrated CPT method of evaluating cyclic resistance ratio (CRR) in sandy soils.

3. RECOMMENDATIONS

For low risk, small scale projects, the potential for cyclic liquefaction can be estimated using penetration tests such as the SPT or CPT. The CPT is generally more reliable than the SPT and is the preferred test, when possible. The CPT provides continuous profiles of penetration resistance which are useful for identifying soil stratigraphy and for providing continuous profiles of estimated cyclic resistance ratio (CRR). Corrections are required for grain characteristics, such as fines content. These corrections may be best expressed as a function of soil behaviour type index, I_c, which is affected by a variety of grain characteristics.

For medium to high risk projects, the CPT can be useful for providing a preliminary estimate of the potential for liquefaction in sandy soils. For higher risk projects, it is also preferred practice to drill sufficient boreholes adjacent to CPT soundings to verify various soil types encountered and to perform index testing on disturbed samples. A site specific correlation between fines content and CPT results is recommended for moderate to high risk projects. A procedure has been described to correct the measured cone resistance for grain characteristics, such as fines content, based on the CPT friction ratio. The corrections are approximate, since the CPT responds to many other factors affecting soil behaviour. Expressing the corrections in terms of soil behaviour index, I_c, may be the best method of incorporating other grain characteristics, in addition to fines content. Hence, when possible, it is recommended that the corrections be evaluated and modified to suit a specific site and project. However, the suggested general corrections provide a useful guide. The CPT is generally limited to sandy soils with limited gravel contents. For soils with a high gravel content penetration may be limited.

Caution should be exercised when extrapolating the suggested CPT correlations to conditions outside of the range from which the field performance data were obtained. An important feature to recognize is that the correlations appear to be based on average values for the liquefied layers. However, the correlations are often applied to all measured CPT values, including low values below the average for a given sand deposit. Hence, the correlations could be conservative in variable deposits where a small part of the penetration data could indicate possible liquefaction.

Finally, as mentioned earlier, it is clearly useful to evaluate CRR using more than one method. The seismic CPT provides a useful technique for independently evaluating liquefaction potential, since it measures both the usual CPT parameters and shear wave velocities within the same borehole. The seismic part of the CPT provides a shear wave velocity profile typically averaged over 1 m intervals and, therefore, contains less detail than the cone tip resistance profile. However, shear wave velocity is less influenced by grain characteristics and few or no corrections are required (Andrus and Stokoe, 1996). Shear wave velocity should be measured with care in order to provide the most accurate results possible since the estimated CRR is sensitive to small changes in shear wave velocity. There should be consistency in the liquefaction evaluation using either method. If the two methods predict different CRR profiles, samples should be obtained to evaluate the grain characteristics of the soil.

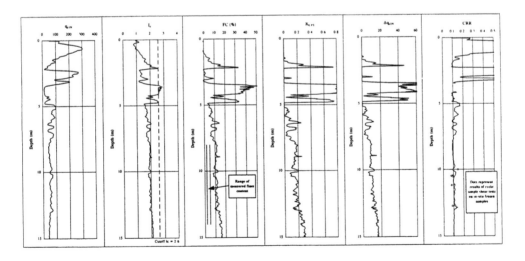

Figure 6.   Example of applying the integrated CPT method to estimate cyclic resistance ratio (CRR) compared with results of cyclic simple shear tests on in-situ frozen samples from CANLEX (Canadian Liquefaction Experiment Phase II (Massey Site).

4. ACKNOWLEDGEMENTS

The authors appreciate contributions by members of the 1996 NCEER Workshop on Evaluation of Liquefaction Resistance (T.L. Youd, Chair). Particular appreciation is extended to the following individuals: I.M. Idriss, J. Koester, S. Liao, W.F. Marcuson III, J.K. Mitchell, R. Olsen, R. Seed, and T.L. Youd.

5. REFERENCES

Andrus, R.D. and Stokoe, K.H., 1996. Guidelines for evaluation of liquefaction resistance using shear wave velocity. *Proc. of the 1996 NCEER Workshop on Evaluation of Liquefaction Resistance (T.L. Youd, Chair)*, Salt Lake City, Utah.

Arango, I., 1996. Magnitude scaling factors for soil liquefaction evaluations. *J. of Geotech. Eng.*, ASCE 122(11): 929-937.

Fear, C.E., and McRoberts, E.C., 1995. Reconsideration of initiation of liquefaction in sandy soils. *J. of Geotech. Eng.*, ASCE 121(3): 249-261.

Finn, W.D.L. (1981). Liquefaction potential: developments since 1976. *Proc. of the 1st Int. Conf. on Recent Advances in Geotech. Earthquake Eng. and Soil Dynamics*, St. Louis, 2, 655-681.

Ishihara, K., 1993. Liquefaction and flow failure during earthquakes, The 33rd Rankine Lecture. *Géotechnique* 43(3): 351-415.

Jefferies, M.G., and Davies, M.P., 1993. Use of CPTu to estimate equivalent SPT N_{60}. *ASTM Geotech. Testing J.*, 16(4): 458-467.

Kayen, R.E., Mitchell, J.K., Lodge, A., Seed, R.B., Nishio, S., and Coutinho, R., 1992. Evaluation of SPT-, CPT-, and shear wave-based methods for liquefaction potential assessment using Loma Prieta data. *Proc. of the 4th Japan-U.S. Workshop on Earthquake Resistant Design of Lifeline Facilities and Countermeasures for Soil Liquefaction*, Technical Report NCEER-94-0019, 1, M. Hamada and T.D. O-Rourke (eds.), 177-204.

Liao, S.S.C. and Whitman R.V., 1986. *A catalog of liquefaction and non-liquefaction occurrences during earthquakes*. Res. Report, Dept. of Civil Eng., M.I.T., Cambridge, MA.

Marcuson, III, W.F., Hynes, M.E., and Franklin, A.G., 1990. Evaluation and use of residual strength in seismic safety analysis of embankments. *Earthquake Spectra*, 6(3): 529-572.

Olsen, R.S., and Malone, P.G., 1988. Soil classification and site characterization using the cone penetrometer test. *Penetration Testing 1988, ISOPT-1*, De Ruiter (ed.), Balkema, Rotterdam, 2: 887-893.

Robertson, P.K., 1990. Soil classification using the CPT, *Can. Geotech. J.*, 27(1): 151-158.

Robertson, P.K. and Campanella, R.G., 1985. Liquefaction potential of sands using the cone penetration test. *J. of Geotech. Div. of ASCE*, March 1985, 22(3): 298-307.

Robertson, P.K. and Campanella, R.G., 1988. *Design manual for use of CPT and CPTu*. Pennsylvania Department of Transportation, (Penn Dot), 200 p.

Robertson, P.K. and Fear, C.E., 1995. Liquefaction of sands and its evaluation. *Proc. of IS Tokyo '95, 1st Int. Conf. on Earthquake Geotech. Eng.*, Keynote Lecture.

Robertson, P.K. and Wride (Fear), C.E. 1997. Cyclic liquefaction and its evaluation based on SPT and CPT. *Proc. of the 1996 NCEER Workshop on Evaluation of Liquefaction Resistance (T.L. Youd, Chair)*, Salt Lake City, Utah.

Robertson, P.K., Campanella, R.G. and Wightman, A., 1983. SPT-CPT correlations. *J. of Geotech. Div. of ASCE,* 109: 1449-1459.

Robertson, P.K., Woeller, D.J., and Finn, W.D.L., 1992. Seismic cone penetration test for evaluating liquefaction potential under cyclic loading. *Can. Geotech. J.*, 29: 686-695.

Seed, H.B., 1979. Soil liquefaction and cyclic mobility evaluation for level ground during earthquakes. *J. of the Geotech. Eng. Div.*, ASCE, 105(GT2): 201-255.

Seed, H.B. and Idriss, I.M., 1971. Simplified procedure for evaluating soil liquefaction potential. *J. of the Soil Mech. and Found. Div.*, ASCE, 97(SM9): 1249-1273.

Seed, H.B., Tokimatsu, K., Harder, L.F., and Chung, R., 1985. Influence of SPT procedures in soil liquefaction resistance evaluations. *J. of Geotech. Eng.*, ASCE 111(12): 1425-1445.

Stark, T.D. and Olson, S.M., 1995. Liquefaction resistance using CPT and field case histories. *J. of Geotech. Eng.*, ASCE 121(12), 856-869.

Skempton, A.W., 1986. Standard penetration test procedures and the effects in sands of overburden pressure, relative density, particle size, aging and overconsolidation. *Géotechnique* 36(3): 425-447.

Suzuki, Y., Tokimatsu, K., Koyamada, K., Taya, Y., and Kubota, Y., 1995. Field correlation of soil liquefaction based on CPT data. *Proc. of the Int. Symposium on Cone Penetration Testing*, CPT'95. Linkoping, Sweden, 2: 583-588.

Vreugdenhil, R., Davis, R., and Berrill, J., 1994. Interpretation of cone penetration results in multilayered soils. *Int. J. for Numerical Methods in Geomechanics*, 18: 585-599.

Wang, W., 1979. *Some findings in soil liquefaction.* Water Conservancy and Hydroelectric Power Scientific Research Institute, Beijing, China.

Yoshimi, Y., Richart, F.E., Prakash, S., Balkan, D.D., and Ilyichev (1977). Soil dynamics and its application to foundation engineering. *Proc. 9th Int. Conf. on Soil Mech. and Found. Eng.*, Tokyo, 2, 605-650.

Seismic Behaviour of Ground and Geotechnical Structures, Sêco e Pinto (ed.) © 1997 Balkema, Rotterdam, ISBN 90 5410 887 8

Earthquake-induced settlements in mat foundations on clay deposits

Miguel P. Romo & Julio A. García
Institute of Engineering, UNAM, Mexico

ABSTRACT: Most seismic designs of mat foundations on clayey deposits are carried out considering only the aspects of bearing capacity, using the concept of load excentricity. Settlements induced by seismic loading are seldom considered mainly because it has been believed that they are of low relevance on the basis of the few case histories reported in the literature. This way of thinking was proved wrong by the significant number of cases that developed in Mexico City during the September 19. 1985 seismic event, where mat foundations on clay deposits underwent large plastic displacements that caused damage beyond any repair to the buildings they supported.

In this paper, the main features of a procedure based on limit equilibrium that allows computation of plastic displacements induced by earthquakes on mat foundations are presented and its capabilities are evaluated throughout comparisons with an actual case.

1 INTRODUCTION

When designing a foundation the engineer must fulfill functionality and stability requirements within a cost-effective framework. The former is achieved by keeping settlements under a maximum allowable limit and the second by providing the foundation with sufficient bearing capacity to carry the building loading.

Most methods to evaluate the seismic bearing capacity of mat foundations are based on simplified procedures that make use of Terzaghi's solution (Terzaghi, 1943) with a reduced foundation width to account for the load excentricity induced by earthquake loading (Meyerhof, 1953) and coefficients that modify the bearing capacity factors to include the load inclination effect (i.e. Sokolovski, 1960; Hansen, 1961; Vesic, 1975). There have been other proposals that include horizontal transient loading (Prakash and Chummar, 1967), inertial effects of the sliding soil wedge (Sarma and Iossifelis, 1990; Richards et al, 1993; Dormieux and Pecker, 1995) and other alternatives to the Prandtl's failure criterion (Pecker and Salençon, 1991; Andersen and Lauritzen, 1988). Centrifuge laboratory tests coupled with numerical studies have shed additional light to improve our knowledge on the dynamic bearing capacity problem (Prevost et al, 1981a; 1981b). Most reported studies on foundation vibration conclude that dynamic loading develops a fairly well defined

failure mechanism involving a broad soil wedge similar to that under static loading. Also, field observations carried out rigth after the September 19, 1985 earthquake supported this conclusion (Mendoza and Auvinet, 1987) although other failure mechanisms that involve soil lateral displacements are not ruled out.

Some of the procedures proposed for dynamic bearing capacity evaluations (Prakash and Chummer, 1967; Sarma and Iossifelis, 1990; Pecker and Salençon, 1991; Richards et al, 1993; Pecker et al, 1994) have been extended to estimate plastic displacements caused by transient loading adopting the Newmark's rigid-block sliding hypothesis (Newmark, 1965). Other approaches that involve uncoupled static and dynamic finite element analyses to evaluate the stress state in the soil-foundation system, and make use of the concept of strain potential (Seed et al, 1974) have been suggested (i.e. Romo, 1991).

The main shortcoming of above referred to procedures is that they consider soil sliding along one-sided failure surface thus the plastic displacements are the results of cumulative movements in only one direction. Earthquake loading induces back and forth shaking to the buildings that transmit oscillating inertia forces to the foundation system causing two complementary sliding surfaces along which permanent displacements build up to produce the differential settlements observed after earthquake shaking

cases. In this paper, the main features of a procedure based on limit equilibrium and Newmark's approach, that considers two complementary sliding surfaces, is presented and evaluated throughout comparisons with reported case histories. The results included in this article and those presented elsewhere (Romo and García, 1993; 1995; 1996) show that the proposed procedure yields reliable results.

2 METHOD OF ANALYSIS

The observation and analysis of a large number of cases of foundations that underwent permanent displacements during the September 19, 1985 earthquake, showed that these movements resulted from the cummulative effects of differential sequential deformations due to foundation rotations, as in a zigzag displacement pattern, as depicted in fig 1. When the building inertia forces ($P(t)$) act in one direction (fig 1a) the soil-foundation contact pressures at that side of the foundation are increased and at the opposit extreme of the foundation are decreased. Owed to the compressibility of the supporting soil, rotations around point O may be generated inducing a building tilting. When the direction of the inertia forces ($P(t)$) change (fig 1b) an entirely similar phenomenon develops at the opposite side of the foundation, causing the soil-building system to rotate around point O'.

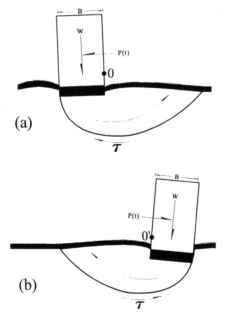

Figure 1. Conceptual model for foundation seismic displacements

Whenever the foundation contact pressures exceed the range of soil elastic behavior, rotations at both ends of the foundation accumulate throughout the duration of the earhquake, yielding as a result permanent differential settlements. Field observations after the September 19, 1985 seismic event pointed out that most failure mechanisms were rotation-like as clearly evidenced by one case where the building toppled due to the rotation shear failure of the soil foundation. On the basis of these observations, it was decided to assume a logarithmic spiral as failure surface in the method of analysis presented herein after.

This modelling, in addition to the variables that are included in the limit equilibrium analysis of the two soil wedges that randomly slide along the two complementary failure surfaces, allows consideration of pre-earthquake foundation differential settlements, the effects of ongoing building tiltings during shaking and the characteristics of the dynamic response of the structure.

2.1 Bearing capacity

Consider a building with height H_e, width B and length L (much larger than B) on a mat foundation having same plan area (BL) and founded at a depth D on an isotropic, homogeneous half-space. Plane strain conditions are assumed. The structure has a mass per unit length m_e and applies a soil pressure q to the half-space which has a unit weight γ_s, an undrained strength C and a friction angle ϕ.

When a sudden displacement is induced to the half-space by a linear acceleration a_t, there appears in the building an inertial force with opposit direction to the half-space movement and having a magnitude equal to $m_e a_e$ where a_e is the building response time-varying acceleration.

It is assumed that the soil has a rigid-plastic behavior and the failure surface conforms a logarithmic spiral with a rotation center at point O (fig 2a) that is located at a height H_o on the vertical that passes through the opposit end from which the spiral starts. The spiral center is defined following a trial and error approach that yields the most critical failure surface (minimum safety factor) for the problem at hand. Studies (Romo and García, 1995; 1996) have shown that the position of the spiral center is affected by the building response acceleration, a_e, in adition to all parameters indicated in fig 2 that define the bearing capacity problem. An example of the effect of a_e, H_e and B on H_o for a frictionless, $\phi = 0$, soil is included in fig 3. The results show that as the building response increases (growing values of the acceleration a_e) the value of H_o becomes smaller. Since the radius of the circular ($\phi = 0$) failure surface varies at a lower rate than H_o does, this means that for larger values of a_e, the failure surface becomes deeper. On the other hand, for a given acceleration a_e, as the height of the building, H_e, increases or the width of the foundation decreases, the failure surface becomes shallower. These aspects have direct influence on the evaluation of the dynamic bearing capacity of mat foundations and should be given proper consideration when using pseudostatic procedures for designing this foundation type.

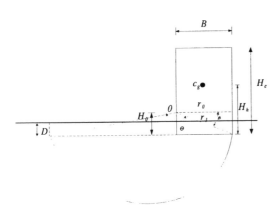

(a) Problem geometry

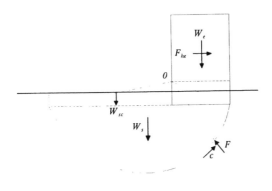

(b) Forces acting on the system

Figure 2. Problem definition

The forces acting on the soil-structure system are indicated in fig 2b. The driving loads are the weight of the structure (including the foundation and live loads), W_e, and the horizontal intertial force, F_e, which is time dependent. It is recomended that H_h be defined from a modal analysis of the structure. The resistant forces are due to soil friction, F, and soil cohesion, C, weight of the sliding soil wedge, W_s, and the weight of the soil layer above the grade foundation depth, W_{sc}.

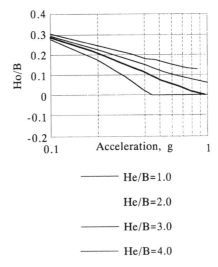

Figure 3. Wedge rotation center for $\phi=0°$

From force moment equilibrium around point O (or O') it is possible to establish the following analytic expression for the bearing capacity (static or dynamic), q, for mat foundations:

$$q = \frac{2}{B}\left(\frac{1}{B+T}\right)\sum M_R \qquad (1)$$

where $\sum M_R$ is the sum of resistant moments due to the cohesion, M_c, the weight of the slinding soil wedge, M_{ws}, and the weight of the soil above the foundation grade level, M_{sc} (Romo and García, 1995; 1996). The parameter T is given by

$$T = \frac{a_e}{g}\left(H_e - 2H_o\right) \qquad (2)$$

if it is assumed that the response acceleration, a_e, is acting at the mid-height of the structure, or

$$T = \frac{2a_e}{g}\left(H_e - 2H_o\right) \qquad (3)$$

281

if the point where it is assumed to be acting the acceleration a_e is defined from a response analysis of the structure using, for example, a dynamic modal procedure.

Once the problem is defined and the building response acceleration evaluated, the mat foundation bearing capacity may be computed by means of eq (1).

2.2 *Permanent displacements*

When the soil-structure system referred to above is acted upon by a horizontal randomly varying time history, then the failure mechanisms of fig 2 will develop alternatively in both directions every time the driving moments exceed the resistant moments. The soil-structure and the soil wedge will rotate as a rigid body around the corresponding rotation center until static equilibrium be restored again. This will repeat itself any time the value of the static safety factor drops below one.

This deformation mechanism, based on a rigid-plastic soil behavior, leads to alternate rotations of the structure whose foundation describes a zigzag-type of movement conducting to both differential and total permanent settlements as indicated in fig 1.

For the conditions explained, the equation of motion for the soil wedge-structure may be developed from establishing moment equilibrium around the rotation point O. Once static equilibrium is overcome, and following the second Newton's law, the soil wedge in its rotation develops an inertial force equal to the soil wedge mass times its acceleration that is considered to be acting at the gravity center of the wedge.

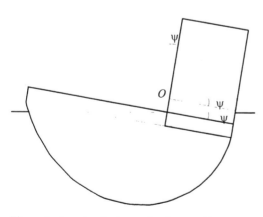

Figure 4. Angulo girado por la cimentación y masa de suelo

It is important to mention that because of the permanent rotations, ψ, the system accumulates during shaking (fig 4), the distances from the forces point applications to the point rotation vary with time, thus the driving and resisting moments also are modified throughout the excitation. In the method presented here the ensuing effects of this variations are accounted for.

Sum of force moments around point O leads to the following differential equation (Romo and García, 1994; 1995; 1996):

$$\frac{\gamma_s}{g} J_o \frac{\partial^2 \psi}{\partial t^2} + \psi \left(W_s \bar{y} - q \frac{BH_e}{2} \right) =$$

$$\left(\frac{qB^2}{2} + \frac{qBa_e}{2g} \left(H_e - 2H_o \right) - \sum M_R \right) \tag{4}$$

where $\dfrac{\gamma_s}{g} J_o$ is the polar moment of inertia of the soil wedge, $\bar{y}$ is the ordinate of the wedge gravity center and all other variables have been already defined. It is worth to recall that a_e is the representative acceleration time dependent building response. This differential equation is solved step-by-step in the time domain following a Newmark-type numerical procedure (Romo and García, 1994).

3 MODEL EVALUATION

The reliability of the results obtained with this model has been evaluated through several comparisons with actual cases in Mexico City that evolved during the September 19, 1985 seismic event. Herein, due to paper length limitations, only one case is presented and discussed.

3.1 *General geotechnical and structural characteristics*

The general geotechnical characteristics of the building, foundation and soil stratigraphy are depicted in fig 5b. The building is a 1445 ton six story, steel reinforced concrete structure on a reinforced concrete mat foundation laid down at 1.10m from the ground surface. The net mat-soil contact pressure was estimated at 5.50 ton/m^2 that with an average soil cohesion of 1.68 ton/m^2 and an average unit weight of 1.2 ton/m^3 yielded a static safety factor of 1.5 (Mendoza, 1987).

The plan view of the building shown in fig 5a depicts an irregular shape of the foundation that precludes a direct application of the method of analysis. To overcome this difficulty, an equivalent foundation with rectangular shape was defined on the basis of having same area and equal minimum inertia moment to those of the actual foundation. Similar approximations have been used in the past for foundation vibration problems with acceptable results (Gazetas and Tassoulas, 1986).

strong motion instrument located near the building site, it was required to use accelerograms recorded nearby the site for more recent earthquakes for the estimation of the movements that the 1985 seismic event could have induced at the building site. To this end, the April 25, 1989 earthquake that had a Richter magnitude of 6.8 and epicenter on the Pacific coast some 380 km from Mexico City was selected on the basis that the frequency content and effective time duration of both events recorded at the hill zone in Mexico City were equivalent.

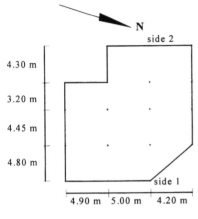

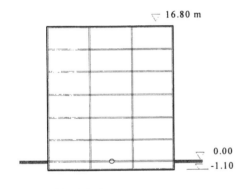

Plain view

Frontal view

a) Geometry characteristics

Figure 5. Foundation-building general characteristics an soil profile

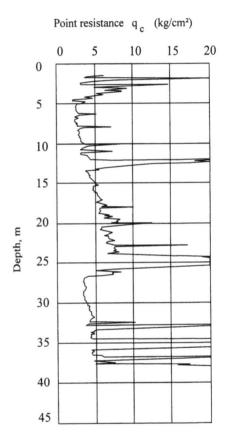

b) Cone penetration resistance

3.2 *Definition of a_e (t)*

To evaluate the effects of the ground movements on the foundation-building during the earthquake of September 19, 1985 it was neccesary to define first such movements. Since at that time there was not a

To determine the ground motions that likely developed during the 1985 event the following procedure was used. First, using a one dimensional wave propagation model (Bárcena and Romo, 1994) the motions recorded nearby the building site during the 1989 earthquake were reproduced using as excitation the accelerogram recorded in this event at a nearby outcrop. This assured that the model was reliable enough to produce satisfactory results for the 1985 earthquake. Second, the site response analysis was carried out for the September 19, 1985 using the same model and the motions recorded at the outcrop during this event. In view of the differences between both earthquakes in terms of intensity, in the response computed for the 1985 event some effects due to the nonlinear soil behavior were manifested. The acceleration response spectrum of the surface ground motions at the building site is shown in fig 6.

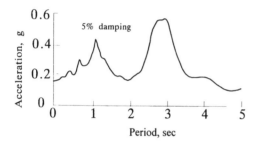

Figure 6. Ground motion at building site

Once the free field motions were determined, the representative building response acceleration time history, a_e (t), was estimated using a modal analysis of the structure. In the calculations of the plastic displacements of the soil foundation, a_e (t) was assumed to be acting at the so called model height. The resulting accelerogram is depicted in fig 7. The maximum response acceleration was 0.34 g and the duration of the intense part (90% of the energy content according to Arias' intensity) was 50 sec.

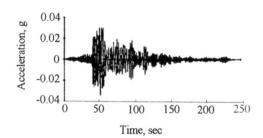

Figure 7. Building response, a_e (t)

3.3 Observed versus computed displacements

The model proposed in section 2.2 has been compared to a number of cases that resulted in Mexico City as a consequence of the September 19, 1985. In this section one of these comparisons is included to show the capabilities of the analysis

procedure and its degree of approximation. In fig 8 the displacements induced by the earthquake are compared to the corresponding displacements computed with the procedure. It may be observed that the theoretical differential displacements reproduce adequately well the acutal movements. Other comparisons presented elsewhere (Romo and García, 1994; 1995; 1996) show similar results.

Sensitivity studies with this procedure have permitted the evaluation of the relative importance of all parameters on the permanent displacements induced by seismic forces on mat foundations (Romo and García, 1995; 1996; Romo and Díaz, 1997). An example of the effect of the maximum building response acceleration and the soil cohesion is also depicted in fig 8. It is seen that the plastic displacements increase with the building inertia forces and with decreasing soil strenghts.

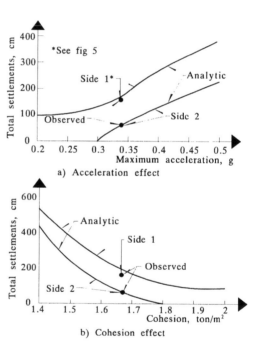

Figure 8. Observed and computed total settlements

4 CONCLUSIONS

The procedure described in this paper allows the evaluation of settlements and tiltings of foundation-building systems induced by earthquake loading. This method has been compared with several case histories developed in Mexico City during the

September 19, 1985 seismic event, yielding results with sufficient accuracy for practical applications. In fact, in a recent study (Romo and Díaz, 1997) this method was used to develop graphs and simple equations to compute straight-forwardly earthquake-induced differential settlements in raft foundations.

The studies carried out with this procedure show that the most relevant parameters are: weight and width of the building (both related to the soil-foundation contact pressure), height of the structure and the horizontal inertia forces, the shear strength of the soil and the grade foundation depth.

From the analyses it was observed that there exists a critical magnitude of each of these parameters from which the total settlements start having an important differential component that leads to the building toppling.

5 ACKNOWLEDGMENT

The authors gratefully recongnize the finantial support received from the Federal District and the valuable comments provided by Professors Gabriel Auvinet and Manuel Mendoza throughout the investigations. To Mr. Arturo Paz and Mr. Roberto Soto for their willingness and skills in editing this paper.

REFERENCES

Andersen, K. H. y Lauritzsen, R. 1988. *Bearing Capacity for Foundations with Cyclic Loads*, Boletín N° 175, Norwegian Geotechnical Institute.

Bárcena, A. y Romo, M. P. 1994. *RADSH: Programa de computadora para analizar depósitos de suelo estratificados horizontalmente sujetos a excitaciones dinámicas aleatorias*, Informe del Instituto de Ingeniería, UNAM, México.

Dormieux, L. y Pecker, A. 1995. Seismic Bearing Capacity of Foundation on Cohesionless Soil, *Journal of Geotechnical Engineering*, ASCE, 121, N° 3, pp. 300-303.

Gazetas, G. And Tassoulas, J. L. 1986. Horizontal stiffness of arbitrarily shaped embedded foundations, ASCE Journal of the Geotechnical Engineering, Vol. 113, No. 5, May

Hansen, J. B. 1961. *A General Formula for Bearing Capacity*, Boletín N° 11, Danish Geotechnical Institute, Copenhague.

Mendoza, M. J. 1987. Foundation Engineering in Mexico City: Behavior of Foundations, *Proceedings of the International Symposium of Geotechnical Engineering of Soft Soils*, México, SMMS, Vol. 2, pp. 351-367.

Mendoza, M. J. y Auvinet, G. 1987. *Comportamiento de cimentaciones de edificios en la Ciudad de México durante el sismo del 19 de Septiembre de 1985*, Informe del Instituto de Ingeniería para el Departamento del Distrito Federal, UNAM.

Meyerhof, G. G. 1953. The Bearing Capacity of Foundations under Eccentric and inclined Loads, *Proceedings of the Third International Conference on Soil Mechanics and Foundation Engineering*, Zürich, Vol. I, pp. 440-445.

Newmark, N. M. 1965. Effects of Earthquakes on Dams and Embankments, *Géotechnique*, 15(2), Londres, pp. 139-160.

Pecker, A. y Salençon, J. 1991. Seismic Bearing Capacity of Shallow Strip Foundations on Clay Soils, *Proceedings of the International Workshop on Seismology and Earthquake Engineering*, CENAPRED, México, pp. 287-304.

Pecker, A., Auvinet, G., Salençon, J. and Romo, M. P. 1995. Seismic bearing capacity of foundations on soft soil, Final Report, Commission of the European Communities, Directorate General for Science Research and Develpment, Contract CI1*CT92-0069

Prakash, S. y Chummar, A. V. 1967. Response of Footings to Lateral Loads, *Proceedings of the International Symposium on Wave Propagation and Dynamic Properties of Earth Materials*, University of New Mexico Press, Albuquerque, New Mexico, pp. 679-691.

Prevost, J. H., Cuny, B. y Scott, R. F. 1981a. Offshore Gravity Structures: Centrifugal Modeling, *Journal of the Geotechnical Engineering Division*, ASCE, 107, N° GT2, pp. 125-141.

Prevost, J. H., Cuny, B., Hughes, T. J. R. y Scott, R. F. 1981b. Offshore Gravity Structures: Analysis, *Journal of the Geotechnical Engineering Division*, ASCE, 107, N° GT2, pp. 143-165.

Richards, R. Jr., Elms, D. G. y Budhu, M. 1993. Seismic Bearing Capacity and Settlements of Foundations, *Journal of Geotechnical Engineering*, ASCE, 119, N° 4, pp. 662-674.

Romo, M. P. 1991. Comportamiento dinámico de la arcilla de la Ciudad de México y sus repercusiones en la ingeniería de cimentaciones, *Sismodinámica* 2, pp. 125-143.

Romo, M. P. y García, J. A. 1994. Procedimiento para el cálculo de desplazamientos permanentes inducidos por sismo en cimentaciones profundas, *2° Simposio de consultores y constructores de cimentaciones profundas*, CENAPRED, México.

Romo, M. P. And García, J. 1995. Análisis de

movimientos permanentes por sismo en cimentaciones sobre arcilla blanda, Report to the Federal District, Institute of Engineering, UNAM, November

Romo, M. P. And García, J. 1996. Procedimiento para el análisis de movimientos permanentes por sismo en cimentaciones superficiales y profundas, Series of the Institute of Engineering, 581, September

Romo, M. P. And Díaz, R. 1997. Movimientos permanentes por sismo en cimentaciones: Un método simplificado de cálculo, Report to the Federal District, Institute of Engineering, UNAM, January

Sarma, S. K. e Iossifelis, I. S. 1990. Seismic Bearing Capacity Factors of Shallow Strip Footings, *Géotechnique*, 40(2), pp. 265-273.

Terzaghi, K. 1943. *Theoretical Soil Mechanics*, John Wiley & Sons, Inc., New York.

Vesic, A. S. 1975. Bearing Capacity of Shallow Foundations, *Foundation Engineering Handbook*, H. F. Winterkorn y H. Y. Fang editores, Van Nostrand Reinhold Co., Inc., New York, pp. 121-147.

Seismic Behaviour of Ground and Geotechnical Structures, Sêco e Pinto (ed.)© 1997 Balkema, Rotterdam, ISBN 90 5410 887 8

Seismic bearing capacity of rigid deep strip foundations

S.K. Sarma & Y.C. Chen
Civil Engineering Department, Imperial College of Science, Technology and Medicine, London, UK

ABSTRACT: Bearing capacity for rigid deep foundations such as caissons under seismic loading are evaluated using the limit equilibrium technique. It is found that the bearing capacity depends on the earth pressures and the soil-shaft interface friction angles on the side of the foundation, particularly on the side of the passive resistance. The analysis includes the inertia of the involved soil mass and the structure induced in an earthquake. The solutions which shows stresses in the soil mass violating the failure criterion of the soil are rejected. The variation of the seismic accelerations within the depth of the foundation is not taken into consideration.

1. INTRODUCTION

The ultimate bearing capacity of a foundation reflects the capacity of the underlying soil to withstand the load of the structure at incipient failure. The effect of an earthquake is to introduce a horizontal inertia load to the structure as well as to the soil. The inertia load on the structure can be treated as an inclined load while that on the soil has to be included in the analysis. The bearing capacity of the foundation is computed by finding the most critical failure surface and the corresponding load which brings the soil to an incipient failure state gives the ultimate bearing capacity.

The bearing capacity of a shallow strip foundation is usually expressed in the well known form:

$$Q = q N_q + c N_c + 0.5 \gamma B N_\gamma \qquad (1)$$

where q = surcharge load

c = cohesion of the soil

γ = unit weight of the soil

B = base width of the foundation.

N_q, N_c and N_γ are the bearing capacity factors which depend on the friction angle of the soil. These factors may also be modified to include the effect of load eccentricity, load inclination, size of the foundation, effect of sloping ground, soil inertia, etc. Reference may be made to Meyerhof (1951, 1953), Brinch Hansen (1970), Vesic (1975), Caquot and Kerisel (1953), Sarma and Iossifelis (1990), Sarma and Chen (1995, 1996) and many others.

Seismic bearing capacity of a rigid deep strip foundation has not been studied thoroughly in the past. The inertia effect of the soil is neglected and the seismic effect on the structure is considered as an inclined load.

The bearing load of a deep foundation (P) is usually separated into two parts (Equation 2) : the base resistance (P_b, Equation 3) and the shaft resistance (P_s, Equation 4). They are considered separately. However, the effect of the skin friction and adhesion on the base resistance is usually neglected.

$$P = P_b + P_s \qquad (2)$$

$$P_b = Q_b \times A_b \qquad (3)$$

$$P_s = \tau_s \times A_s \qquad (4)$$

where Q_b = end bearing pressure at foundation base

A_b = the area of the base

τ_s = average shear strength on the shaft wall

A_s = the surface area of the shaft.

The simplest practice for estimating the end bearing capacity of a deep foundation is to consider it as a shallow foundation at the depth and then include the soil above the foundation base as a surcharge load ($q = \gamma D$, where D is the depth of the foundation). In this practice, the failure surface does

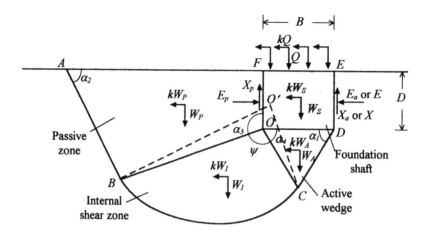

Fig. 1 The failure mechanism of a deep strip foundation in horizontal ground.

not daylight and therefore the resistance of that part of the failure surface is not taken into account. Further more, the effect of the skin friction and adhesion between the foundation side and the soil is also neglected. This practice produces a very conservative estimate.

In order to reduce the conservatism in this estimate, the effect of the skin friction and adhesion and the effect of the missing part of the slip surface is taken into account by considering the thrusts on the vertical sections. This method is however approximate.

Vesic (1965) gave an overview of the bearing capacity of deep foundations in cohesionless soil. However, it is on the static vertical loading only. Meyerhof (1953) provided a solution for the inclined load on deep footings which is an extension of his method (1951) for vertical load. The method is based on a possible failure surface arrived at by using a fixed active block, a passive block and a sheared zone in between. He did not look for any alternative failure mechanism. Furthermore, for foundations subjected to seismic loading, the inertia of the soil mass can not be included. Meyerhof and his coworkers (Meyerhof, et al., 1973; Meyerhof, et al., 1981; Vansangkar, et al., 1983) have performed many model tests on rigid deep vertical piles subjected to inclined loads as well as on battered piles and have derived various design graphs. Brinch Hansen (1970) and Vesic (1975) proposed modifying factors to be applied on the bearing capacity factors to account for the inclined load conditions. But so far, the authors were unable to find any published works on the seismic bearing capacity of deep

foundations which includes the inertia of the soil mass.

2. MODEL

In this paper, the bearing capacity of a deep foundation is derived by using the limit equilibrium technique, which is in principle similar to that used by Sarma and Iossifelis (1990), and Sarma and Chen (1995, 1996) for shallow strip foundations. The end bearing pressure (Q_b) in Equation 5 does not include the term resulting from surcharge as in the case of shallow foundations (Equation 1).

$$Q_b = c\,N_c + 0.5\,\gamma B\,N_\gamma \qquad (5)$$

Fig. 1 shows the assumed failure mechanism which consists of an active wedge under the foundation, a passive zone and an internal shear zone between the two. B and D are the width and depth of the foundation; k is the seismic coefficient; Q is the foundation bearing pressure; W_P, W_I, W_A and W_S are weights of the passive zone, the internal shear zone, the active wedge and the foundation shaft; E ($= 0.5\,\gamma D^2 K$) and X ($= E\tan\delta$) are the normal and shear forces on both sides of the foundation shaft in static case, while K and δ are the corresponding earth pressure coefficient and soil-shaft interface friction angle; E_a ($= 0.5\,\gamma D^2 K_a$) and X_a ($= E_a\tan\delta_a$) are the normal and shear forces on the active side (the right hand side) of the foundation shaft in seismic case, while K_a and δ_a are the corresponding earth pressure coefficient and soil-shaft interface friction angle; X, δ, X_a and δ_a are positive in the upward

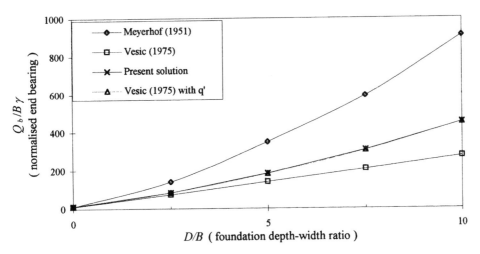

Fig. 2 Comparison of static end bearing (Q_b) solutions for foundations with different D/B ratio (for ϕ=30, δ=30, k=0, K=1) with those of Meyerhof (1951) and those based on Vesic's (1975) factors (q' includes shaft friction; angles in degrees).

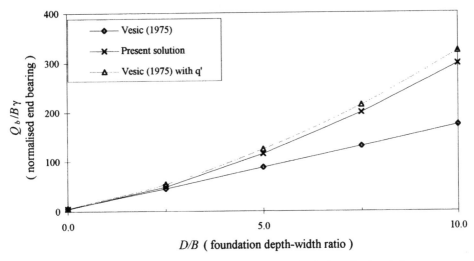

Fig. 3 Comparison of seismic end bearing (Q_b) solutions for foundations with different D/B ratio (for ϕ=30, k=0.2, δ_a=δ_p=30, K_a=K_p=1) with those based on Vesic's (1975) factors (q' includes shaft friction; angles in degrees).

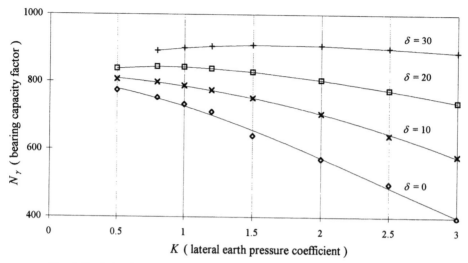

Fig. 4 Static bearing capacity factor N_γ versus K (for ϕ=30, D=10B, k=0)
for deep foundations with different interface friction angle ,δ (angles in
degrees).

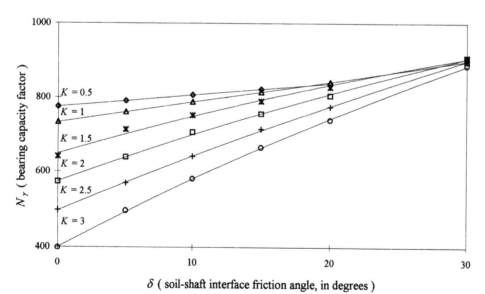

Fig. 5 Static bearing capacity factor N_γ versus δ (for ϕ=30, D=10B, k=0)
for deep foundations subjected to different level of lateral earth pressure,
quantified by K (angles in degrees).

direction ; α_2, α_3, α_1 and α_4 are the angles defining the passive zone and the active wedge; O' is the centre of the log-spiral curve defining the internal shear zone, and $\psi = (1.5 \pi - \alpha_3 - \alpha_4) \geq 0$.

In this analysis, it is necessary that the failure surface ABCDE (see Fig. 1) daylights on both ends so that a kinematically permissible slip surface may develop. Therefore, it becomes essential to interpret the earth pressures on both sides of the foundation.

The failure mechanisms of a deep foundation subjected to static load and to seismic load are different. Under static vertical load, it is necessarily symmetric about both sides of the foundation and the pressures are the same on both sides (defined by K). Under seismic loading, the failure mechanism is not symmetric and the pressures on both sides are different.

For seismic case, the authors call the earth pressure on the left hand side of the foundation as the passive state (defined by K_p) while that at the right hand side as the active state (defined by K_a), even though they may not reach the state of active or passive failure (defined by K_A and K_P as usually considered in earth pressure problems). It should be noted that the direction of the shear force (X_a) on the active side of the foundation may not be the same as that of the active failure state perceived in classical earth pressure theories.

The failure surface may daylight on the active side either by following the side of the footing or by producing a slip surface through the soil. If the slip surface passes through the soil, then an active failure state develops in the soil, while if it passes along the side of the foundation, then an active state may not develop in the soil. Furthermore, even if a failure surface develops through the soil, kinematically, the soil mass in the active zone may not move down relatively to the foundation, unlike the active earth pressure problem. Therefore, it is likely that more complex failure mechanism takes place in the active side. The assumed shear strength on the side of the foundation may imply actual failure or only partially mobilized shear strength.

At the passive side of the foundation, it is very unlikely that passive failure state will develop in the soil during an earthquake.

The pressures on the foundation shaft are indeterminate and can only be confirmed either by experiments or by rigorous analytical solution such as using the finite element method. Even in this case, the solution will depend on the assumed stress strain characteristics and may not represent the field condition.

It is therefore necessary to perform parametric

study, which covers possible variation of the soil-shaft interface friction angles (δ for static case, δ_a and δ_P for seismic case) and lateral earth pressure coefficients (K for static case; K_a and K_p for seismic case) on the foundation shaft. In this study, the sign-convention of the friction angles are positive when the mobilized shear resistance from the soil is in the upward direction, while the range of lateral earth pressure coefficient considered is : $K_A \leq (K \text{ or } K_a \text{ or } K_P) \leq K_P$. The coefficients K_A and K_P is calculated from Mononobe-Okabe equations (1926, 1929) in this study.

3. RESULTS AND DISCUSSION

This paper presents the solutions for $\phi = 30°$ only. Also, the effect of cohesion is not considered.

Fig. 2 and Fig. 3 show the comparison of the present solutions for the end bearing pressure of a deep foundation with Meyerhof's (1951) solutions and those obtained by using Vesic's (1975) bearing capacity and modification factors. Vesic's depth factor does not include the effect of shaft friction; however, it is reasonable to consider the shaft friction as a uniform surcharge Δq ($= 0.5 \gamma D^2 K \tan\delta / L$, where L is the horizontal length of the passive wedge for a shallow foundation) acting on the failed soil mass. Therefore, Equation 1 can be applied with q' ($= q + \Delta q$) instead of q. The comparisons show good agreement between the present solutions with those based on Vesic's factors with modified surcharge q'. However, Meyerhof's solutions are almost twice the value as the present solutions, when the foundation is deeply embedded. This is due to his assumption of failure mechanism where the far end of the failure surface turns towards the shaft as the foundation sits deeper.

Fig. 4 and Fig. 5 show the dependence of the static solutions on the interface friction angle (δ) and lateral earth pressure coefficient (K). In the range of earth pressure coefficient studied ($0.5 \leq K \leq 3$), the solutions reach the peak values at different level of earth pressure, depending on the interface friction angle. In general, the higher the interface friction angle (δ), the higher the end bearing capacity the foundation can provide.

Fig. 4 also indicates that higher earth pressure does not necessarily increase the base resistance of a deep foundation. In most of the cases, it reduces the end bearing capacity. However, it is a function of the interface friction angle. Within the studied range ($0.5 \leq K \leq 3$), the base resistance drops dramatically as

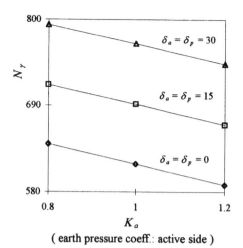

(earth pressure coeff.: active side)

Fig. 6 Seismic bearing capacity factor N_γ versus K_a for deep foundations with different interface friction angle $\delta_{a,p}$ (for ϕ=30, D=10B, k=0.1, K_p=1.2, δ_a=δ_p, angles in degrees).

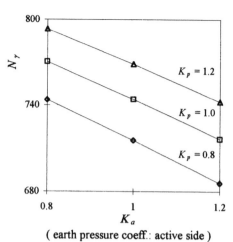

(earth pressure coeff.: active side)

Fig. 7 Seismic bearing capacity factor N_γ versus K_a for deep foundations subjected to different level of K_p (for ϕ=30, D=10B, k=0.1, δ_a=δ_p=30, angles in degrees).

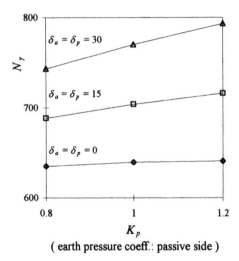

(earth pressure coeff.: passive side)

Fig. 8 Seismic bearing capacity factor N_γ versus K_p for deep foundations with different interface friction angle $\delta_{a,p}$ (for ϕ=30, D=10B, k=0.1, K_a=0.8, δ_a=δ_p, angles in degrees).

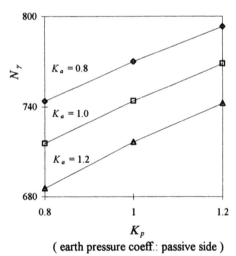

(earth pressure coeff.: passive side)

Fig. 9 Seismic bearing capacity factor N_γ versus K_p for deep foundations subjected different level of K_a (for ϕ=30, D=10B, k=0.1, δ_a=δ_p=30, angles in degrees).

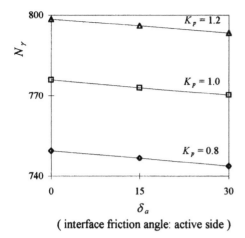

(interface friction angle: active side)

Fig. 10 Seismic bearing capacity
factor N_γ versus δ_a (for ϕ=30,
D=10B, k=0.1, K_a=0.8, δ_p =30)
for deep foundations subjected to
different K_p (angles in degrees).

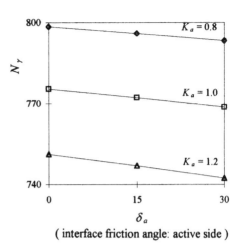

(interface friction angle: active side)

Fig. 11 Seismic bearing capacity factor
N_γ versus δ_a (for ϕ=30, D=10B,
k=0.1, δ_p=30, K_p=1.2) for deep
foundations subjected to different K_a (
angles in degrees).

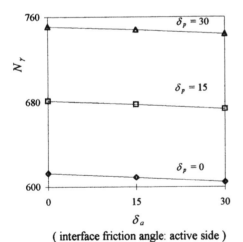

(interface friction angle: active side)

Fig. 12 Seismic bearing capacity
factor N_γ versus δ_a (for ϕ=30,
D=10B, k=0.1, K_a=K_p=1.0) for
deep foundations with different δ_p (
angles in degrees).

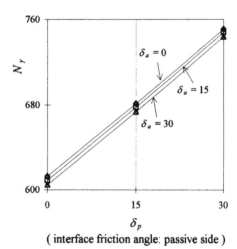

(interface friction angle: passive side)

Fig. 13 Seismic bearing capacity
factor N_γ versus δ_p (for ϕ=30,
D=10B, k=0.1, K_a=K_p=1.0) for
deep foundations with different δ_a (
angles in degrees).

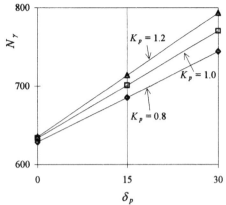

(interface friction angle: passive side)

Fig. 14 Seismic bearing capacity factor N_γ versus δ_p (for ϕ=30, D=10B, k=0.1, K_a=0.8, δ_a =30) for deep foundations subjected to different K_p (angles in degrees).

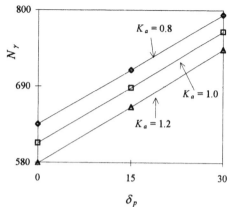

(interface friction angle: passive side)

Fig. 15 Seismic bearing capacity factor N_γ versus δ_p (for ϕ=30, D=10B, k=0.1, δ_a=30, K_p=1.2) for deep foundations subjected to different K_a (angles in degrees).

the earth pressure increases towards that of the passive failure state, especially when the interface friction angle is zero (at which $K_P = 3$).

Fig. 5 shows that a higher interface friction angle provides a higher bearing capacity factor, N_γ, in static case. This effect is more significant at a higher level of earth pressure.

Fig. 6 to Fig. 15 show the dependence of the bearing capacity factor, N_γ, on the lateral earth pressure coefficients, K_a and K_p, and on the interface friction angles, δ_a and δ_p , when the soil-foundation system is subjected to 0.1g of horizontal acceleration.

It is very likely that δ_a and δ_p may be different from each other. However, in Fig. 6 to Fig. 9 they are assumed equal. These figures indicate that K_a and K_p have opposite effects on the bearing capacity factor, within the studied range ($0.8 \leq [K_a$ and $K_p] \leq 1.2$). The increase of K_a reduces the factor, while the increase of K_p increases it. The effect of K_p is larger with higher interface friction angles; however, for smaller $\delta_{a,p}$ it is generally insignificant when compared with that of K_a (Fig. 8 and Fig. 6). Unlike the case of K_p, the effect of K_a as measured by the gradient of the curves is almost independent of $\delta_{a,p}$ (Fig. 6). In general, the effects of K_a and K_p seem to be independent of each other (Fig. 7 and Fig. 9).

Although K_a is smaller than K_p in seismic cases, the solutions are provided for the full range of K_a and K_p in Fig. 7 and Fig. 9.

Fig. 10 to Fig. 12 show that δ_a does not have significant effect on the bearing capacity factor, N_γ, although the increase of δ_a does reduce it. However, it is not the case with δ_p, whose increase provides much higher capacity (see Fig. 13 to Fig. 15).

Fig. 14 also shows that the combined effect of δ_p and K_p on the bearing capacity factor is greater than their individual effect.

4. CONCLUSION

Limit equilibrium technique appears to be robust in determining the bearing capacity factors for rigid deep foundations. The limited results show that the earth pressures and the mobilised interface friction angles have different levels of effect on the bearing capacity. Neglecting the effect of the skin friction on the end bearing capacity produces conservative result.

REFERENCES:

Brinch Hansen, J. (1970) "A revised and extended formula for bearing capacity." *Bulletin No.28, Danish Geotechnical Institute*, 5-11.

Caquot, A. and Kérisel, J. (1953) "Sur le terme de surface dans le calcul des fondations en milieu pulverulent." *Proc. 3rd ICSMFE*, **1**,336-337.

Meyerhof, G.G., Mathur, S., Valsangkar, A. (1981) "The bearing capacity of rigid piles and pile groups under inclined loads in layered sand." *Canadian Geot. Journal*, **18**, 514-519.

Meyerhof, G.G., Ranjan, G. (1973) "The bearing capacity of rigid piles under inclined loads in Sand. Pile groups." *Canadian Geot. Journal*, **10**, 428-438.

Meyerhof, G.G. (1951) "The ultimate bearing capacity of foundations." *Geotechnique*, **2**(4), 301-332.

Meyerhof, G.G. (1953) "The bearing capacity of foundations under eccentric and inclined loads." *Proc. 3th ICSMFE*, **1**, 440-445.

Mononobe, N. (1929) "Earthquake-proof construction of masonry dams." *Proc. World Engineering Conference*, **9**, 275.

Okabe, S. (1926) "General theory of earth pressure." *J. Japanese Soc. of Civil Engineers*, Tokyo, Japan, **12**(1).

Sarma, S.K. and Chen, Y.C. (1995) "Seismic bearing capacity of shallow strip footings near sloping ground." *Proc. 5th SECED Conf. on European Seismic Design Practice*, Chester, UK, 505-512.

Sarma, S.K. and Chen, Y.C. (1996) "Bearing capacity of strip footings near sloping ground during earthquakes." *Proc. 11th WCEE*, Paper No. 2078.

Sarma, S.K. and Iossifelis, I.S. (1990) "Seismic bearing capacity factors of shallow strip footings." *Geotechnique*, **40**(2), 265-273.

Valsangkar, A. and Meyerhof, G.G. (1983) "Model studies of collapse behaviour of piles and pile groups." *Development in Soil Mechanics and Foundation Engineering No.1*, Ed. Banerjee and Butterfield, Applied Science Publishers.

Vesic, A.S. (1965) "Ultimate loads and settlements of deep foundations in sand." *Proc. of Symposium on Bearing Capacity and Settlement of Foundations*, Duke Univ., US, Ed. Vesic, A.S.

Vesic, A.S. (1975) "Bearing capacity of shallow foundations" *Foundation Engineering Handbook*, 1st edn., H.F. Winterkorn and H.Y. Fang (eds.), Chap. 3, Van Nostrand Reinhold Co. Inc., New York, 121-147.

Seismic Behaviour of Ground and Geotechnical Structures, Sêco e Pinto (ed.)© 1997 Balkema, Rotterdam, ISBN 90 5410 887 8

Damage to Higashinada sewage treatment plant by the 1995 Hyogoken-Nanbu earthquake

Yasushi Sasaki
Hiroshima University, Japan

Junichi Koseki
University of Tokyo, Japan

Katsuhisa Shioji
City Bureau, Ministry of Construction, Japan

Makoto Konishi
Japan Sewage Works Agency, Japan

Yoshihiro Kondo
Construction Bureau, Kobe City, Japan

Toshiro Terada
Nippon Koei Co. Ltd, Japan

ABSTRACT: The 1995 Hyogoken-Nanbu earthquake inflicted significant damage to Higashinada Sewerage Treatment Plant, Kobe City. The revetment bordering the plant was thrust approximately 2 *m*, the land behind subsided by a maximum of 2 *m*, and, according to direct inspection during the reconstruction work of the aeration tank and the final sedimentation tank for the main plant, their piles suffered major cracks at a depth of about 2 or 3 *m* from the pile head. Based on the results of back-analyses, the main cause of damage to the piles was thought to be both the inertial force of the superstructure and the reduction of subgrade reaction in the almost completely liquefied soil layers. The lateral spreading of the liquefied soil layers were judged to have caused minor cracks near the bottom of the layers for the aeration tank and the final sedimentation tank, whereas not for the primary sedimentation tank.

1 INTRODUCTION

The 1995 Hyogoken-Nambu earthquake caused structural and/or mechanical damage to 43 sewerage treatment plants, as indicated in Fig. 1. At that time, a total of 103 plants were in operation in Hyogo, Osaka, and Kyoto Prefectures. Among them, the operation was interrupted by the earthquake at 8 plants as listed in Table 1.

In particular, Higashinada Treatment Plant constructed at a reclaimed area in Higashi-nada Ward, Kobe, was most severely damaged, by which the treatment performance was impeded over a long term. Possible effects of liquefaction of the reclaimed subsoil layer on the damage to Higashinada plant were discussed by Tohda et al. (1996) in contrast to the good performance of Port Island Treatment Plant where the reclaimed subsoil layer had been improved before the earthquake. Occurrence of permanent ground displacement and damage of facilities associated with the ground displacement at Higashinada plant were reported by Hamada and Wakamatsu (1996).

This paper describes the results of an investigation conducted after the earthquake to reveal the extent and the mechanism of damage to tank facilities of Higashinada plant. The preliminary results were presented elsewhere (Sasaki et al.,

1996), while they are modified in this paper considering additional information which was obtained by direct inspection of piles conducted during restoration work of the tank facilities.

2 SURVEY OF DAMAGE

Plan of the whole plant is shown in Fig. 2, where the location of the damaged sections is indicated together with that of conduits soaked with water due to breakage of connected pipes.

Damage to pile foundation was surveyed by inspection of pile heads after excavating the surrounding soil layers and by sonic integrity tests measuring the reflection of elastic waves transmitted through the pile, where they were applicable. As a result, a portion of the main plant was evaluated to have suffered damage to the pile foundation. On the contrary, there was practically no damage to the pile foundation of the branch plant and the pumping facility.

The permanent horizontal displacement at the top of the southern revetment of the Uozaki Canal and the permanent settlement of ground surface in the main plant site were evaluated as shown in Fig. 3, where, the maximum thrust of the revetment was about 2 m, and the maximum ground subsidence was

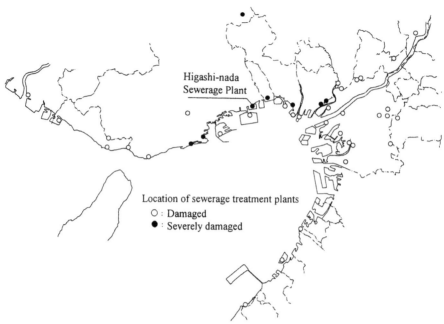

Fig. 1 Location of damaged sewerage treatment plants

Table 1 List of sewage treatment plants where the operation was interrupted

Plants	Max. treatment capacity (m³/day)	Major cause of interruption	Interrupted period
Higashinada, Kobe City	225,000	Joints of sewage conduit as an inlet to primary sedimentation tank were opened by lateral spreading of surrounding liquefied soil.	105 days (with temporary treatment by using a part of adjoining canal)
Chubu, Kobe City	78,000	Some of sludge collectors at primary and final sedimentation tanks fell down.	23 days for the half of secondary treatment system
Seibu, Kobe City	162,000	Sewage pumps were submerged in water flooded by breakage of water conduits.	6 days for No.1 system; no damage to No. 2 system
Ashiya, Ashiya City	73,000	Joints of sewage conduits as inlets to primary sedimentation tank were broken.	14 days
Edagawa, Nishinomiya City	126,000	Pipes connecting between final sedimentation tank and chlorine mixing tank were broken.	7 days for one third of secondary treatment system
Tobu No.1, Amagasaki City	79,000	About half of sludge collectors at primary and final sedimentation tanks were out of order.	16 days for secondary treatment system
Tobu No.2, Amagasaki City	82,000	Conduits connecting primary sedimentation, aeration and final sedimentation tanks were broken.	9 days for secondary treatment system
Mukogawa Upstream, Hyogo Prefecture	55,000	About one fourth of sludge collectors at final sedimentation tank were out of order.	2 days for sand filtration system

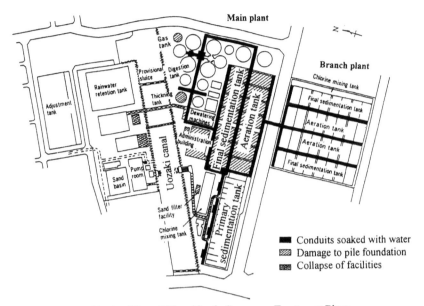

Fig. 2 Plan of Higashinada Sewerage Treatment Plant

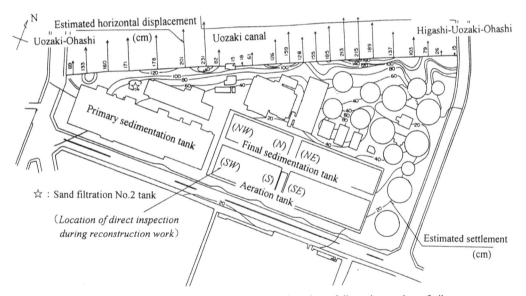

Fig. 3 Estimated permanent displacement and location of direct inspection of piles

larger than 1 m. These are consistent with the ground displacements evaluated by aerial survey (Hamada et al., 1995).

The estimated ground profile in the main plant site is shown in Fig. 4, which consists of a reclaimed Masa soil (decomposed granite soil) layer, Holocene sandy deposits, Holocene clayey deposits, and Pleistocene deposits. The southern revetment for the

Uozaki Canal was L-shaped RC retaining walls underlain by a replaced sand layer.

Because the SPT N-values of the reclaimed soil layers and the Holocene sandy deposits were generally less than 10, they were estimated to have liquefied during the earthquake. Furthermore, based on Figs. 3 and 4, it was estimated that the thrust of the southern revetment caused lateral spreading of

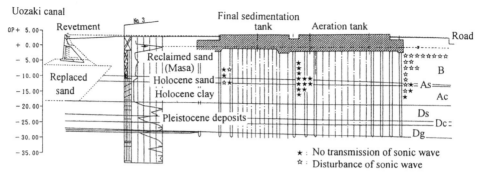

Uozaki canal

OP+ 5.00 — Revetment   No.3   Final sedimentation tank   Aeration tank   Road

Reclaimed sand (Masa)
Holocene sand
Holocene clay
Pleistocene deposits

Replaced sand

★ : No transmission of sonic wave
☆ : Disturbance of sonic wave

Fig. 4 Estimated soil profile and results of sonic integrity tests

Table 2 Results of sonic integrity tests on foundation piles of tank facilities

Facility	Pile (diameter in *mm*)	Number of piles (*ratio to subtotal in %*)			
		Damaged	Intermediate	Undamaged	(Subtotal)
Primary sedimentation tank for main plant	Prestressed concrete (ϕ 400)	2 (*18*)	6 (*55*)	3 (*27*)	11
Aeration and final sedimentation tanks for main plant	Prestressed concrete (ϕ 350)	44 (*44*)	29 (*28*)	28 (*28*)	101
Aeration and final sedimentation tank for branch plant	Prestressed concrete (ϕ 500)	0 (*0*)	4 (*45*)	5 (*55*)	9
	Cast-in-place concrete (ϕ 1000, 1200, 1500)	0 (*0*)	1 (*1*)	22 (*96*)	23
Sand filtration No.2 tank	Prestressed concrete (ϕ 600)	0 (*0*)	1 (*33*)	2 (*67*)	3

Table 3 Results of direct inspection of foundation piles during reconstruction work of aeration and final sedimentation tanks

Location (*refer to Fig. 3*)	Number of piles (*ratio to surveyed piles in %*)					
	Damage to upper part			Damage to lower part		
	Crack on pile surface	Steel wire breakage	Overall breakage	Crack on pile surface	Steel wire breakage	Overall breakage
North-west (*NW*)	17 (*100*)	10 (*59*)	6 (*35*)	7 (*100*)	1 (*14*)	1 (*14*)
North-middle (*N*)	5 (*100*)	3 (*60*)	0 (*0*)	5 (*100*)	0 (*0*)	0 (*0*)
North-east (*NE*)	7 (*100*)	4 (*57*)	0 (*0*)	Not excavated.		
South-west (*SW*)	15 (*100*)	10 (*67*)	0 (*0*)	15 (*100*)	2 (*13*)	0 (*0*)
South-middle (*S*)	4 (*100*)	1 (*25*)	0 (*0*)	4 (*100*)	0 (*0*)	0 (*0*)
South-east (*SE*)	8 (*100*)	2 (*25*)	0 (*0*)	Not excavated.		
(Subtotal)	56 (*100*)	30 (*54*)	6 (*11*)	31 (*100*)	3 (*10*)	1 (*3*)

the liquefied soil layers.

In Table 2, the results of the sonic integrity tests conducted on more than one hundred foundation piles are summarized. About 44 % of piles surveyed at the final sedimentation tank and the aeration tank for the main plant were evaluated to have been damaged, and the depth of the failure corresponded mainly to the boundary between the liquefiable layers and the underlying soft clayey deposits as shown in Fig. 4. It should be noted that the damage ratio of foundation piles of the primary sedimentation tank for the main plant was 18 %, which was less than half of that of the adjoining aeration tank, and piles for the branch plant were evaluated to be almost free from failure.

The final sedimentation tank and the aeration tank had to be reconstructed due to residual tilting of the tanks, while the primary sedimentation tank could be reused after minor repair. The embedded depth of the former tanks was about 4.5 m, while that of the latter tank was about 7.0 m. All of them were supported by pre-stressed concrete piles with a diameter of 35 or 40 cm. These different performance and condition will be discussed in the following analyses.

During excavation work to reconstruct the final sedimentation tank and the aeration tank, some of the original piles were inspected directly. The results are summarized in Table 3 in terms of the occurrence rate of cracks at pile surface, breakage of pre-stressing steel wires and overall breakage of pile body. All the piles suffered surface cracks both at a depth of about 3 m from the pile head and at a depth corresponding to the bottom of the liquefiable layers. In general, the opening width of the upper cracks were larger than that of the lower ones, as reflected by the difference in the occurrence rate of breakage of the steel wires at the respective depth. It should be noted that overall breakage of the pile body was found only at the north-west corner of the final sedimentation tank, which was at the closest location to the Uozaki Canal. This serious damage may be affected by the lateral spreading of the liquefied soil layers caused by the thrust of the southern revetment of the Uozaki Canal.

Although the results of the sonic integrity tests as shown in Fig. 4 were not consistent with the aforementioned observation during reconstruction work with respect to the depth of the major cracks, they may be effective in detecting the occurrence of severe damage because the estimated damage rate (44 %) by sonic integrity tests was not very far from the occurrence rate (54+11=65 % for the upper part)

of breakage of the steel wire or the pile body.

During the reconstruction work of the final sedimentation tank and the aeration tank, it was also observed that almost all the pile heads were not damaged and that additional reinforcing steel bars had been used at the pile heads in order to ensure a firm connection to the bottom beam of the superstructure at the time of the original construction.

3 ANALYSES OF PILE FOUNDATION

In order to evaluate the effects of the response of the superstructure and the lateral spreading of the liquefied soil layers on the foundation piles, dynamic response analyses and back-calculations of the failure mode of the piles were conducted.

3.1 Response of tanks

Two-dimensional equivalent linear response analyses were performed to estimate the horizontal response acceleration of the final sedimentation tank, the aeration tank and the primary sedimentation tank for the main plant. The profiles transverse to the Uozaki Canal were modelled as typically shown in Fig. 5 for the final sedimentation tank and the aeration tank.

The shear moduli of subsoil layers at a strain level of 10^{-6} were evaluated based on the elastic shear wave velocity measured in the borehole as shown in Table 4, Their strain-dependent characteristics were assumed based on the experimental results of similar soils reported by Iwasaki et al. (1980a, 1980b) and Yokota and Tatsuoka (1977). The rigidity of the foundation piles were converted to equivalent one per unit width.

The strong motion record in the N-S direction observed at the Kobe Station of Japan Meteorological Agency was inputted to the base layer, which resulted in the maximum response acceleration of the three tanks equally about 380 cm/sec^2. It was converted to a seismic coefficient (=0.39) to evaluate the maximum inertial force of the superstructures tentatively.

3.2 Modeling of foundation piles

The foundation piles were modeled as a single pile supported by linear Winkler-type springs representing subgrade reaction. As schematically shown in Fig. 6, following four types of failure modes were assumed.

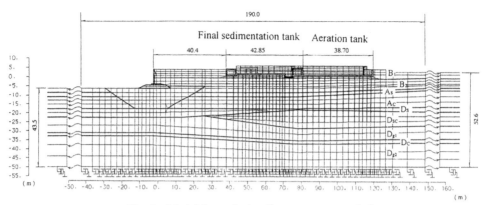

Fig. 5 Model for equivalent linear response analysis

P: Inertial force of superstructure
γ : Saturated unit weight of liquefied soil

α : Coefficient of equivalent earth pressure due to lateral flow
β : Reduction factor of subgrade reaction due to liquefaction

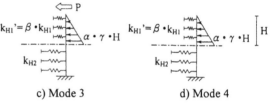

a) Mode 1   b) Mode 2   c) Mode 3   d) Mode 4

Fig. 6 Assumed failure modes

Table 4 Material property assumed for equivalent linear response analysis

Soil layer	Unit weight (kN/m^3)	Go (MN/m^2)
B1	20.0	38.0
B2	21.0	41.7
As	21.5	65.5
Ac	16.5	33.2
Dsc	20.0	92.1
Dg1	22.5	260.7
Dc	19.0	125.1
Dg2	22.5	260.7
Go: shear modulus at strain level of 10^{-6}		
Poisson's ratio for all the layers: 0.49		

a) Mode 1

The maximum inertial force P of the superstructure was solely considered.

b) Mode 2

Together with the maximum inertial force P, reduction of the subgrade reaction from the liquefied soil layers by a factor β was introduced.

c) Mode 3

In addition to the condition assumed in mode B, the effect of the liquefied soil layers which flows laterally through the piles was considered as a hydrostatically distributed earth pressure with an equivalent coefficient α in terms of total overburden stress σ_v; i.e. earth pressure per effective width of the pile (=diameter D) was given by $\alpha \cdot \sigma_v \cdot D$.

d) Mode 4

The maximum inertial force P was eliminated from the condition assumed in mode 3. In this mode, the lateral spreading of the liquefied soil layers was assumed to take place mainly after the earthquake or during the latter part of the earthquake when the amplitude of response acceleration of the superstructure was negligibly small.

The piles of both the aeration tank and the primary sedimentation tank were modeled as shown in Fig. 7, where the coefficient of subgrade reaction k_H was estimated based on SPT N-value and the finite fixity at the pile head was considered by introducing a rotational spring k_R. The pile of the final sedimentation tank was not modeled, because the inertial force P acting on the pile was only slightly smaller than that of the aeration tank with the same pile in diameter and with almost similar subsoil profile.

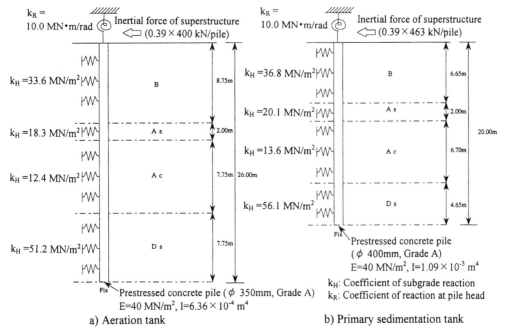

a) Aeration tank
b) Primary sedimentation tank

Fig. 7 Single pile model supported by Winkler-type springs

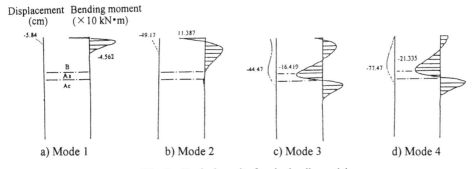

Displacement   Bending moment
(cm)   (×10 kN•m)

a) Mode 1   b) Mode 2   c) Mode 3   d) Mode 4

Fig. 8 Typical results for single pile model

3.3 *Back-analyses of failure modes*

Typical vertical distributions of the calculated deflection and bending moment of the pile for the different failure modes are shown in Fig. 8. The location of the maximum bending moment at middle depth was near the pile head in modes 1 and 2, and near the bottom of the liquefiable layers (B and As) in modes 3 and 4. Based on these results, the major cracks of the piles of the final sedimentation tank and the aeration tank, which were observed at a depth of about 3 m from the pile head by the direct inspection during their reconstruction work, were evaluated to have been caused mainly by the maximum inertial force of the superstructure at a condition similar to

mode 1 or 2 without the effect of lateral spreading of liquefied layers. The cracks observed at a depth corresponding to the bottom of the liquefiable layers, however, may not be explained without assuming a condition similar to mode 4 considering the effect of the lateral spreading.

Results of parametric calculations for modes 1 (β=1.0) and 2 (β<1.0) are shown in Fig. 9 together with the ultimate bending capacity Mu of the pile resulting in the tensile yielding of the steel wire, which was estimated considering the effect of axial load acting on the pile. The calculated maximum bending moment at middle depth exceeded the bending capacity Mu at β=1/100 for both the aeration tank and the primary sedimentation tank.

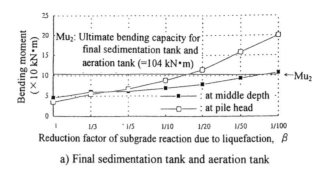

a) Final sedimentation tank and aeration tank

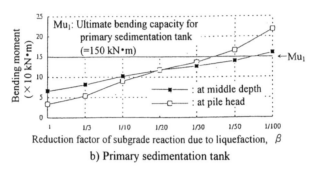

b) Primary sedimentation tank

Fig. 9 Effects of reduction factor β in modes 1 and 2

This may suggest that that the reclaimed soil layer and the Holocene sandy layer liquefied almost completely when the major cracks near the pile head of the aeration tank were formed. On the other hand, the aforementioned good performance of the primary aeration tank may not be explained without assuming a different level of response caused by a different degree of embedment, which could not be evaluated by the equivalent linear response analyses neglecting the effect of liquefaction.

It should be noted that the bending moment at the pile head is much larger than the bending capacity Mu at $\beta=1/100$, as shown in Fig. 9. The pile head may, however, have been without any damage because of the increase in the bending capacity by the aforementioned additional reinforcing steel bars.

Results of parametric calculations for mode 4 are shown in Fig. 10, where the reduction factor β is set zero because there should be no subgrade reaction in the liquefied layers which flows laterally. The calculated maximum bending moment at middle depth exceeded the bending capacity Mu at α slightly larger than 0.04 and 0.07, respectively, for the aeration tank and the primary sedimentation tank. Since the opening width of cracks at lower part of the piles of the final sedimentation tank and the aeration tank was smaller than that at their upper

part, the equivalent earth pressure coefficient α due to lateral spreading of liquefied soil layers, in general, may not have been larger than 0.04. This leads to an estimation that the piles of the primary sedimentation tank may not have been damaged by the lateral spreading, which seems to be consistent with the aforementioned difference in the damage ratio evaluated by the sonic integrity tests as shown in Table 2. At the north-west corner of the final sedimentation tank, however, the pile may have been damaged relatively severely as shown in Table 3 by the lateral spreading because it was at the closest location to the Uozaki Canal.

In order to investigate the effect of the major damage to upper part of the piles which precedes the occurrence of the lateral spreading, another series of parametric calculations for mode D was made by changing the pile head fixity to a hinged condition for simplicity. The results for the aeration tank were shown in Fig. 11, where the calculated maximum bending moment exceeded the bending capacity Mu at α about 0.03. This suggests that the equivalent earth pressure coefficient α due to lateral spreading may be decreased if the possibility of reduction in the fixity of pile near the pile head is considered.

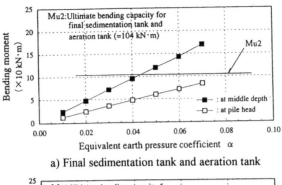

a) Final sedimentation tank and aeration tank

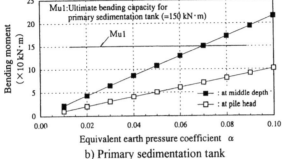

b) Primary sedimentation tank

Fig. 10 Effects of coefficient α in mode 4

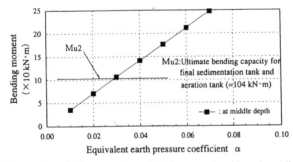

Fig. 11 Effects of coefficient α in mode 4 by assuming hinged condition at pile head

4 SUMMARY

Based on the observation by direct inspection of piles during the reconstruction work of the aeration tank and the final sedimentation tank for the main plant at Higashinada Sewerage Treatment Plant, which faces Uozaki Canal, and on the results of the back-analyses, the major damage to the upper part of the piles was estimated to have been caused by both the inertial force of the superstructure and the reduction of subgrade reaction in the almost completely liquefied soil layers.

The relatively small opening of cracks at the lower part of the piles of the aeration tank and the final sedimentation tank was estimated to have been caused by the lateral spreading of the liquefied soil layers. The equivalent earth pressure coefficient α in terms of total overburden pressure due to the lateral spreading was estimated to be smaller than 0.04 even at the north-west corner of the final sedimentation tank, where the piles were damaged relatively severely by the lateral spreading because it was at the closest location to the Uozaki Canal.

It was also estimated that the good performance of the primary sedimentation tank may have been due to its relatively large embedment depth, which may have resulted in relatively smaller response of the superstructure, and that its piles may not have been damaged by the lateral spreading. ·

The sonic integrity tests employed in the present investigation may be effective in detecting the occurrence of severe damage to piles, whereas their results were not consistent with the observation during the reconstruction work with respect to the depth of the major cracks.

ACKNOWLEDGMENTS

The survey of piles during the reconstruction work of the tank facilities was made jointly by Kobe City, Okumura-gumi Co., Ltd., Mitsui Construction Co., Ltd. and Toa Construction Co., Ltd. Assistance of Messrs. T. Nakai and K. Takenaka at Kobe City, Messrs K. Himeno and H. Shiki at Okumura-gumi Co., Ltd. and Mr. H. Naito at Nippon Koei Co., Ltd. in conducting the investigation is greatly acknowledged.

REFERENCES

Hamada, M., Isoyama, R. and Wakamatsu, K. 1995: The Hyogoken-Nanbu (Kobe) earthquake, liquefaction, ground displacement and soil condition in Hanshin area, *Association for Development of Earthquake Prediction.*

Hamada, M. and Wakamatsu, K. 1996: Liquefaction, ground deformation and thir caused damage to structures, *The 1995 Hyogoken-nambu Earthquake -Investigation into Damage to Civil Engineering Structures-, Committee of Earthquake Engineering, Japan Society of Civil Engineers.*

Iwasaki T. et al. 1980: Experimental study on dynamic deformation characteristics of soils (part 2), *Report of PWRI,* Vol. 153 (in Japanese).

Iwasaki T. et al. 1980: Dynamic deformation and strength property of alluvial clay, *Proc. of 15th annual meeting of JSSMFE* (in Japanese).

Sasaki, Y. et al. 1996: Damage to Higashi-nada sewerage treatment pland by the 1995 Hyogoken-nanbu earthquake, *Symposium on the Great Hanshin-Awaji Earthquake Disaster, Japan Society of Civil Engineers* (in Japanese).

Tohda, J., Yoshimura, H. and Li, L. 1996: Characteristic features of damage to the public sewerage systems in the Hanshin area, *Special Issue of Soils and Foundations on Geotechnical Aspects of the January 17 1995 Hyogoken-Nambu Earthquake.*

Yokota K. and Tatsuoka F. 1977: Shear modulus of undisturbed diluvial clay, *Proc of 32nd annual meeting of JSCE* (in Japanese).

Seismic Behaviour of Ground and Geotechnical Structures, Sêco e Pinto (ed.) © 1997 Balkema, Rotterdam, ISBN 90 5410 887 8

Yodogawa dike damage by the Hyogoken-nanbu earthquake

Y. Sasaki
Department of Civil and Environmental Engineering, Hiroshima University, Japan

K. Shimada
Yodogawa Construction Office, Ministry of Construction, Hirakata, Japan

ABSTRACT: The Hyogoken-nanbu earthquake caused extensive damage to the Yodogawa dike for 5,660 m long at 16 sections. Longitudinal fissures, settlement of crest and collapse of facing of dikes took place. The most severely damaged section about 2 km long were found at the left-bank near its estuary. The main cause of this section was the liquefaction of the foundation ground beneath the dike. The soil condition of the foundation ground and the failure process of dikes were studied.

1 INTRODUCTION

The Hyogoken-nanbu earthquake caused extensive damage to the Yodogawa (Yodo river) dikes which is located about 40 km east from the epicenter as shown in Fig.1.1. Main cause of the damage at severely damaged sections was soil liquefaction beneath the dikes behind which a densely populated urban area is extending.

When soil liquefaction takes place in a layer, the damage aspect of the dike on it varies with the boundary condition such as the depth and the thickness of the liquefied layer (Sasaki et al. 1996).

In the case of the Kushiro earthquake in 1993, the Kushiro river dike on a peat ground was failed due to the liquefaction at the subsided bottom part of the embankment (Sasaki et al. 1993, Finn et al. 1997). Although the thickness of the liquefied part was not so thick, as the liquefaction took place at very shallow layer, the settlement of the crest of a 7 m high dike reached to about 2m.

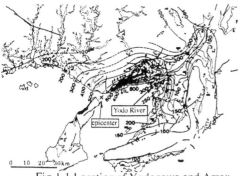

Fig.1.1 Location of Yodogawa and Amax distribution

In the case of the Hokkaido nansei-oki earthquake in 1993, the Shiribeshi-toshibetsu river dike on horizontally deposited loose sand layer failed due to liquefaction (Kaneko et al.). In this case, slip planes grown diagonally within the dike from around the shoulder to bottom were clearly observed (Sasaki et al. 1997).The most severely damaged section of the Yodogawa dike showed resembling aspect to both of these past cases for more than 2 km long.

As the residual height of the damaged dike near the estuary was so low, increase of the subsidence of the dike due to after shocks and a secondary disaster which might be caused by high tide was of concerned. It was essential to establish a restoration program quickly in limited time to meet the strong desire of preventing the anticipated secondary disaster as well as to remedy the damaged sections against future earthquakes.

In order to conduct an efficient investigation to establish the restoration program, the authors first tried to construct an hypothesis on failure process of the dike basing on detected data on damage and existing boring log (Sasaki). The proposed failure mode was later compared with the configuration of blocks divided by cracks within the embankment which was detected by trench excavation during the repair works.

Collected data and proposed framework of the restoration program were examined by a Technical Committee (TCMSERS). A part of the collected data has been reported elsewhere (Matsuo). Present paper describes the damage to the Yodogawa dike and the geotechnical aspects of the foundation ground specially focusing the supposed failure process.

2 EARTHQUAKE MOTION

The distribution of the maximum acceleration during the earthquake is shown in Fig.1.1 (TCMSERS). Although it is located about 40 km apart from the epicenter, more than 200 gals of the maximum acceleration was recorded around the Yodogawa estuary. SMAC type strong motion accelerometers have been installed on and near the dike at the Oyodo observatory at 6.5 km from the estuary. The observatory is located at 45 km from the epicenter in east northeast direction. Fig.2.1 shows the cross section of the dike at the Oyodo observatory. The longitudinal direction of the dike at the observation station is northeast - southwest direction.

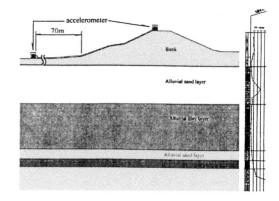

Fig2.1 Oyodo Observatory

Table 2-1 Maximum Acceleration at Oyodo observatory

	Longitudinal	Transverse	Vertical
Near Ground	203 gal	221 gal	239 gal
At Top Crest	199 gal	358 gal	556 gal

Table 2-1 shows the recorded maximum accelerations at the observatory. It is known that the ground around the area was excited by 200-220 gals of maximum acceleration. The maximum acceleration in transverse direction at top of the dike was 1.6 times of that at ground surface near the toe and the acceleration in vertical direction at the top was 2.3 times of the near ground whereas the acceleration in longitudinal direction was not so amplified. Duration time of the strong motion was about 45 seconds as illustrated in Figure 2.2.

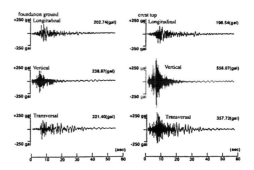

Fig.2.2 Accelaration wave form

Table 3-1 Summary of the damage to the Yodo river

No.	Distance from Estuary	Damage	Length	Remarks
1	L 0+100-2+160 km (Torishima)	Dike Failure, Parapet Wall Collapse	2,130 m	(Emergency Repair 1,800 m)
2	L 2.2+38-2.2+140 km	Dike, Revetment	102 m	
3	R 1.0+100-1.8+80 km (Nishijima)	Dike Failure	750 m	(Emergency Repair 750 m)
4	R 2.4+135-2.6+100 km	Dike Crack	165 m	
5	R 3.4+10- 3.8 km	Dike Settlement, Revetment Crack	393 m	
6	L 3.4+45-3.6+10 km (Takami)	Dike Settlement, Revetment Bulge	165 m	(Emergency Repair 100 m)
7	L 3.8+130-4.0+50 km	Dike Cracks	130 m	
8	L 4.0+120-4.2+15 km	Dike Cracks	105 m	
9	R 4.8 km	Flood Protection Gate at Bridge End		
10	L 4.8+90-5.2+20 km	Dike Cracks, Revetment Collapse	350 m	
11	L6.0+20-6.0+120 km	Dike Cracks	100 m	
12	R 6.6+127-6.8+180 km	Dike Settlement	253 m	
13	R 7.0+60-7.4+85 km	Dike Settlement	406 m	
14	R 7.4+160-7.6+124 km	Dike Settlement, Revetment Cracks	164 m	
15	R 8.4+60-8.4+117 km	Dike Berm Settlement, Revetment Crack	53 m	
16	R 22.8+167-23.2+110 km	Dike, Revetment Cracks	341 m	
17	R 25.2+59-25.2+114 km	Dike Berm Cracks, Revetment Cracks	55 m	
18	L 25.6 km	Water Quality Monitoring Facility		
19	Kanzakigawa R 0.1 km	Revetment	30 m	

3 LOCATION OF DAMAGE TO YODOGAWA DIKES

The location of the damaged sections along the Yodogawa dikes are shown in Table 3-1. There were 19 sites of damaged facilities which include dikes, revetments, and flood protection gate at a bridge end to which disaster restoration fund was allocated. Damage to dike was found at 16 locations and the total length of damaged sections was 5,660 m. Fig.3.1 shows the location of the above mentioned 19 sites. It is known that the damaged sections were mostly located on reclaimed land, delta, former river bed, and former waterway.

Among the 16 sections of dike damage, 14 sections which were 5,270 m long totally were situated in the lowermost 9 km of the river from the estuary where the dikes are located on the land reclaimed in 18 century.

Sections of dike damage are classified into two groups; one is the damage with settlement of top crest and the other is ones with severe cracks although settlements of their crest were not severe. Sections at Torishima (No. 1 in Table 3-1) and Takami (No. 6 in Table 3-1) of the left bank, and at Nishijima (No. 3 in Table 3-1) of right bank are the ones with substantial crest settlement due to dike failure. Those severely damaged and heavily subsided sections of the dikes were found within 4 km from the estuary. The settlements of the top crest of both side dikes are illustrated in Fig.3.2.

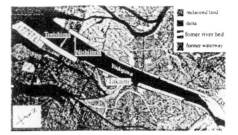

Fig.3.1 Location of damaged section

Photo 3.1 Torishima dike after the earthquake

Photo 3.2 Torishima , sand boiling

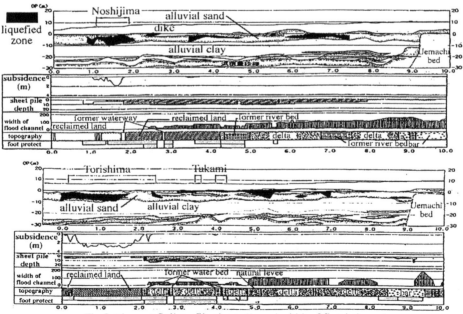

Fig.Settlement,Soil profile,sheet pile,and width of flood channel

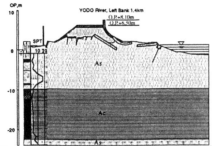

Fig.3.3 Torishima,before & after the earthquake

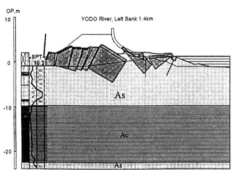

Fig.3.4 Torishima, failure mode

At Torishima section, many longitudinal cracks were seen at crest and land side slope as shown by Photo 3.1 for about 2 km. Concrete parapet wall was slid out towards river and consequently, heavily inclined as shown in Photo 3.2. As the result of this severe failure, asphalt pavement at top crest inclined and the maximum settlement of the top crest was found to be about 3 m.

Figure 3.3 shows the cross sections of the dike at Torishima before and after the event. The appearance of the damaged section implied a failure mode as schematically shown in Fig.3.4.

At Nishijima section, a big fissure in longitudinal direction took place around the center of the top crest, and the land side half of the dike was failed with leaving dislocation of about 1.5 m along this fissure for about 750 m as shown in Photo 3.3 and Fig.3.5. The tentative mode of failure at Nishijima dike which was also proposed from the damage aspect detected just after the earthquake as illustrated in Fig.3.5. At this section, three rows of sand compaction piles had been installed.

At Takami section, a longitudinal crack occurred around the center of the top crest with about 20 cm gap.There was bulging at the slope near shoulder above its berm and at the slope near its toe. This bulging near the toe of the slope pushed up the pavement of the road just adjacent to the dike.

Photo 3.3 Nishijima dike after the earthquake

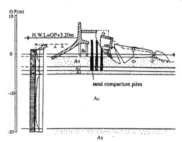

Fig.3.5 Nishijima, failure mode

Photo 3.4 Takami dike after the earthquake

At these three sections, emergency repair works to raise their height were immediately conducted soon after the damage detection survey was completed.

Traces of sand boiling were observed at many locations around the damaged sections, not only at the land side ground, but also at the river bed beneath the water. An example of the sand boiling has been shown in Photo 3.2. Trace of sand boiling

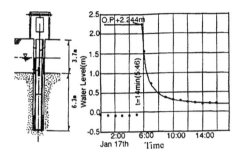

Fig.3.6 Raise of pore pressure

within the collapsed embankment was also observed during an excavation survey of damaged dike at Torishima section.

At the Torishima section, the change of soil condition due to soil improvement work for the " Super Levee " which runs adjacent to the damaged section was being monitored at the construction yard as a pilot study when the earthquake took place. An abrupt rise of pore water pressure was recorded during the earthquake as shown in Fig.3.6.

4 GEOTECHNICAL FEATURES OF THE DIKE AND THE FOUNDATION SOIL

The lower reach of the Yodo river channel was artificially excavated in Meiji era in early 20 century through the area which was previously reclaimed in early 18 century. After the excavation as a floodway was completed, both sides dikes of the Yodogawa at downstream were periodically enlarged in height and width against experienced natural disaster. The dike around Torishima section was first constructed in around 1903, and was enlarged intermittently after the damage caused by 1938 flood, the Nankaido earthquake in 1946, and high tide by the Second Muroto typhoon in 1961. Fig. 4.1 shows the construction history of the dike.

The height of the dike near estuary is currently designed against the high tide during typhoon. Elevation of OP + 8.1 m at top crest is designed against both tidal fluctuation and tidal wave which are anticipated when typhoon passes over the Osaka Bay. Current dike at the time of the earthquake was made of 6.5 m high embankment and 1.6 m high concrete parapet wall at river side shoulder. Top crest and both sides slope were covered by facing. Brown colored fine sand for the lower part and gravely sand for the upper part of the dike have been used as the construction materials for embankment.

In Fig.3.2, the soil profiles of the foundation ground beneath the both side dikes have been shown. At top of it, a layer of fine to medium sand about 10 m thick is deposited underlain by 13-15 m thick cohesive soil layer. The boundary between these layers is ambiguous because of the gradual change of the grain size. The boundary roughly lies between OP - 8 m and OP - 10 m. The SPT blow count N-value for the uppermost sand layer is mostly within 3-10. The N-value for the lower cohesive soil layer is around 3. At the bottom of the alluvial layer beneath these two layers, about 3 m thick sandy layer is deposited underlain by alluvial sandy layer of N>30.

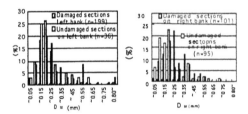

Fig.4.2 Mean grain size of the sand

The soil properties of the uppermost alluvial sand layer are shown in Figs. 4.2 to 4.5.
It is known that the mean grain size (D_{50}) of 0.15-0.2 mm are most frequent at damaged sections in both side dikes, whereas the smaller grain size of less than 0.2 mm is more frequently observed at undamaged sections in right bank.

It is also known that the fines contents (Fc) at damaged sections in both sides dike are smaller than those at undamaged sections where the fines content varies in wider range than at damaged sections.

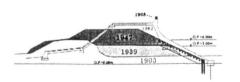

Fig.4.1 Construction history

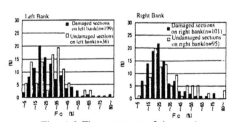

Fig. 4.3 Fines content of the sand

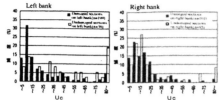

Fig.4.4 Uniformity Coefficient

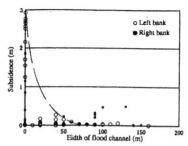

Fig. 4.8 Width of flood channel

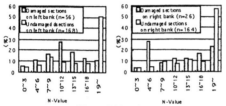

Fig.4.5 N-value

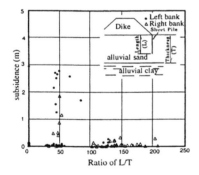

Fig.4.6 Liquefaction susceptibility

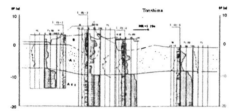

Fig.4.7 Sheet pile depth

Uniformity coefficients (Uc) at damaged sections are noticed to be smaller than those at undamaged sections.

It is noticed that relatively high N-values are more frequently found at undamaged sections than at damaged sections where N-values of 4-6 at right bank and N-values of 10-12 at left bank are most frequently seen.

These findings imply that the uppermost sand layer are generally prone to liquefaction at damaged locations, although the difference of the soil properties between damaged and undamaged sections is not so significant.

The liquefaction susceptibility of the uppermost sand layer against ks=0.22 was examined by a simplified method (Japan Road Association). Fig. 4.6 shows the result of the assessment for Torishima section. It is known that the liquefaction susceptible layer at Torishima section is thicker than at other damaged sections. This coincides with the damage degree.

So the distribution of the susceptible zone to liquefaction in the uppermost sand layer was examined along the dike against ks=0.22. As shown in Fig. 3.2 by solid area, it is known that the sections where the dike subsidence was extensive located at sections where the liquefaction prone layer was comparatively thicker and shallower.

The cutoff wall by steel sheet piles had been placed in near the toe of the Yodogawa dike so as to prevent seepage flow during flood. The width of the flood channel varies at locations along the dike as shown in Fig. 3.2. It should be noted that difference of boundary conditions surrounding the liquefaction zone such as the installation of sheet piles near the toe of dikes and the existence of flood channel as well as the depth and thickness of the layer may affect the subsidence of the dike due to liquefaction of foundation soil.

It is known that dike subsidence was prominent in sections with no flood channel or with a sheet-pile length of less than 4 m.On the contrary, there was almost no subsidence when flood channel width is more than 50 m and sheet piles had been laid out on a loose sand layer with a thickness of more than 10 m as shown in Figs. 4.7 and 4.8 if the soil conditions were the same.

5 DIKE EXCAVATION SURVEY

After a cofferdam was completed as part of permanent restoration program, an excavation survey of the damaged dike was conducted at Torishima section. Five sites were selected based on the extent and type of damage. Fig. 5.1 shows the cross sections of observed damage at these sites.

It should be noted that wide opening of cracks were commonly observed around the dike shoulder right beneath the concrete parapet wall at the five sites. It was also found that many shear planes diagonally tilting toward the water side from the land side slope were found within the dike. The stepwise shear planes are considered to be developed while the dike subsided and tilted toward the water side.

Results of the observation and Swedish sounding reveal that soil blocks of the embankment separated by these slip planes seemed to have kept their original shape even in the underlying liquefied layer, although the bottom of the subsided blocks was not directly observed because of the ground water. Further, intrusion of sand veins were found at several locations right beneath the parapet or closer to the water side.

The development of slip planes inside the embankment on a liquefied layer is due to the stress re-distribution within the embankment brought about by associated loss of the tangential stress at the bottom boundary of the embankment due to the liquefaction. In such a condition, central part of the embankment becomes active Rankine state, accordingly, shear plane takes place around the shoulder at first (Sasaki et al. 1997). The observed wide opening of crack right beneath the parapet is considered to be developed at first when the foundation ground was liquefied.

It should also be noted that the observed aspect of the failed dike looked similar to that of propose one based on a working hypothesis as shown previously in Fig. 3.4.

Based on the results of excavation survey and the detected facts on soil properties mentioned before, the failure process of damaged dike at Torishima is concluded as follows;

(1) Alluvial sandy soil beneath the dike was liquefied by the earthquake shaking.

(2) With the raised pore water pressure in the underlying layer, the stress state in the dike became active Rankine state, and crack was developed around the shoulder.

(3) During high pore water pressure continued in shallower part of the liquefied layer nearby bottom boundary of the dike, central portion of the embankment subsided into the liquefied layer by thrusting the side part of embankment beneath its slope.

(4) Because the ground surface was lower and there was no non-liquefied surface layer, and the parapet provided an additional weight, the water side portion of the dike subsided and the rest of the dike was sheared towards the water side.

(5) Boiled sand intruded into cracks while subsidence proceeded and thrusting was continued.

(6) With the decrease of the raised pore pressure, movement of the soil blocks ceased with leaving

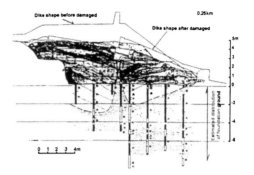

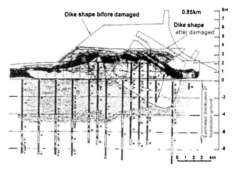

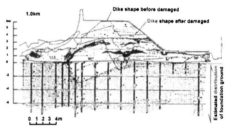

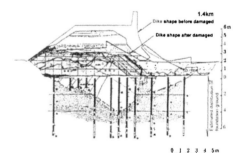

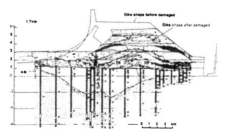

Fig. 5.1 Excavation results

deformation of the dike.
The above mentioned failure mechanism is schematically illustrated in Fig. 5.2.

6 REPAIR WORKS

Since the damage to the dike at Torishima section was so severe that the secondary disaster by high tide was of great concern.Soon after the damage detection was completed, the emergency repair works to raise the dike height tentatively was conducted as shown in Fig. 6.1 and Photo 6.1. The preparatory work for the emergency restoration was initiated on the date of the earthquake and the tentative restoration was completed on January 30, 1995.
From the previously mentioned study on the cause and the mechanism of the dike failure, it was apparent that the soil liquefaction was the main cause. Considering the highly urbanized hinterland of the dike as well as the main cause of the damage, it was decided to conduct a remedial treatment of the foundation soil to prevent the liquefaction induced failure during future earthquakes for the damaged section at Torishima as part of the permanent restoration. Because residences, schools, and factories are densely distributed behind the dike, the treatment method should not cause substantial noise or vibration. Considering the construction period, reliability and past experiences, deep mixing method was adopted. In particular for the region close to residences, it was decided to adopt the earth-removal deep mixing method in order to suppress the unexpected lateral movement of surrounding ground. From the stability of the dike during the anticipated earthquakes, it was decided to improve the soil beneath the dike by grid pattern so as the stabilized area ratio was to be 50 %. The unconfined compression strength of treated soil was designed to be 5 kgf/cm^2.
The extent of the treated area and the

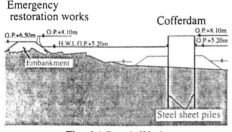

Fig. 6.1 Repair Work

depth were designed to match the enlarged cross section of the dike as shown in Fig. 6.2 and to reduce the future settlement due to consolidation in 13 m thick alluvial cohesive soil layer that lies under the

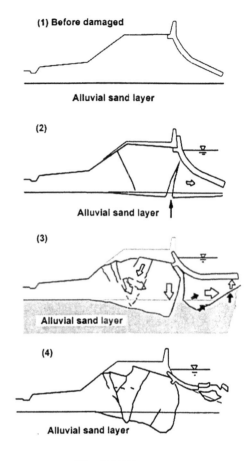

Fig. 5.2 Failure process

Photo 6.1 Emergency raise up of crest height

sandy soil layer. As shown in Fig. 6.2, cross section of the repaired dike has gentler slope in water side and higher embankment without parapet wall so that people can easily enjoy the water front.

It was necessary to construct a temporary cofferdam prior to conducting the remedial treatment of foundation soil so that the hinterland will be secured against possible flood and high tide during the construction. As shown in Fig. 6.1 and Photo 6.2, the cofferdam of double rows of sheet piles was completed on June 16, before the outflow season of the Yodo river.

Photos 6.3 and 6.4 show the views of dike during construction and the completed dike.

Since the concrete facing on water side slope and the parapet wall were not heavily damaged at Nishijima and Takami although the land side halves

Photo 6.3 Remedial treatment of soil

Photo 6.2 Cofferdam construction

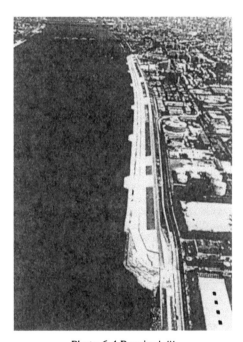

Photo 6.4 Repaired dike

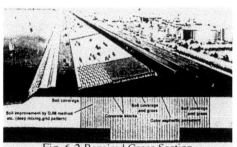

Fig. 6.2 Repaired Cross Section

were collapsed, it was decided to repair those sections by partial re-construction without tentative cofferdam to protect against flood and high tide during construction. This reduced the construction cost. Fig.6.3 and Fig.6.4 show the schematic illustration of the repair at these two sections.

The main part of the restoration program at these three sections was completed by the end of March, 1996. The sections with severe cracks in the dike other than above mentioned three sections were also repaired by March 1996 by re-compaction of the damaged portion of the dike.

315

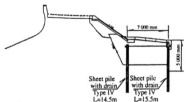

Fig.6.3 Repaired section of Nishijima

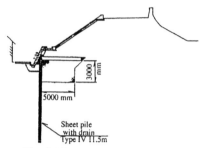

Fig.6.4 Repaired section of Takami

7 CONCLUSION

The Hyogoken-nanbu earthquake caused extensive damage to the Yodogawa dikes at its downstream sections whereas the damage to the dike in upstream portion of the dike was not significant. Most extensive failure took place at left bank of the Torishima dike for about 2 km. It is concluded since traces of sand boiling was apparently observed and the raised pore water level was recorded that the main cause of the damage was the soil liquefaction beneath the dikes,

By correlating the soil conditions and the damage degree, it was found that the liquefaction sand layer was thicker and shallower at more severely damaged sections than the sections where damage was relatively slight, although the soil properties such as grain size distribution was not so different between damaged and non-damaged sections. The tendency of suppression of dike subsidence affected by the existence of wider flood channel and the deeper cutoff sheet piles was also seen.

At the heavily damaged sections during the excavation survey, wide opening of crack was observed within the embankment around the dike shoulder beneath the concrete facing. The excavation survey revealed that the soil blocks separated by fissures which were induced by underlying soil liquefaction tended to keep their shape even in the subsided location in the liquefied layer. Observed configuration of the soil blocks verified basically the proposed hypothesis on failure

mode. The findings learned from this earthquake can be utilized to mitigate the future disaster.

ACKNOWLEDGMENT

The authors are grateful to the collaboration given by the Ministry of Construction, Technical Committee for Mitigation of Seismic Effect on River Structures, ACTEC, and JICE during the damage detection and the establishment of the restoration program. In preparing the figures for this paper, Mr. H.Satoh of Yodogawa construction office, MOC, and Mr. I. Onodera of OYO corporation assisted the authors. Their assistance is gratefully acknowledged.

REFERENCES

1) Sasaki Y., Finn W. D. L., Shibano A. and Nobumoto M. 1996. Settlement of Embankment above a Liquefied Ground which is Covered by Non-liquefiable Surface Layer. *Proc. 6th Japan-US Workshop on Earthquake Resistant Design on Lifeline Facilities and Countermeasures against Liquefaction:*(in Press)
2) Sasaki Y., Oshiki H. and Nishikawa J. 1993. Embankment Failure caused by the Kushiro-oki Earthquake of January 15, 1993. *Special Volume, XIIIth ICSMFE, New Delhi, India:*61-68.
3) Finn W. D. L., Sasaki Y. and Wu G. 1997. Simulation of Response of the Kushiro River Dike to the 1993 Kushiro-oki and 1994 Hokkaido Toho-oki Earthquakes. *Proc. 14th ICSMFE:*(in Press).
4) Kaneko M., Sasaki Y., Nishikawa J., Nagase M. and Mamiya K. 1995. River Dike Failure in Japan by Earthquakes in 1993. *3rd Int. Conf. on Recent Advances in Geotechnical Earthquake Engineering and Soil Dynamics, Vol. I:*495-498,
5) Sasaki Y., Moriwaki T. and Ohbayashi J. 1997. Deformation Process of an Embankment Resting on a Liquefiable Soil Layer. *Proc. Int. Symp. at Nagoya :* (in Press)
6) Sasaki Y. 1995. Diagnostic Report on Damaged Dikes by the Hyogoken-nanbu Earthquake. *Report to Ministry of Construction*
7) TCMSERS (Technical Committee for Mitigation of Seismic Effect on River Structures) 1996. Report on Damage to Yodogawa Dike and its Restoration Work (in Japanese)
8) Matsuo O. 1996. Damage to River Dikes. *Special Issue on Geotechnical Aspects of the January 17, 1995 Hyogoken-Nanbu Earthquake, Soils and Foundation:* 235-240,
9)Japan Road Association 1990. *Highway Bridge Specification.*

Seismic Behaviour of Ground and Geotechnical Structures, Sêco e Pinto (ed.) © 1997 Balkema, Rotterdam, ISBN 90 5410 887 8

Early earthquake warning system for city gas network

Yoshihisa Shimizu
Center for Supply Control and Disaster Management, Tokyo Gas Co., Ltd, Japan

Fumio Yamazaki
Institute of Industrial Science, The University of Tokyo, Japan

ABSTRACT: Tokyo Gas supplies city gas to 8 million customers around Tokyo metropolitan area and has social responsibility to secure safety even after big earthquake. To prevent gas-caused secondary disaster after earthquake, it is necessary to make prompt and reasonable decision. For this purpose, SIGNAL (Seismic Information Gathering Network Alert System) has been developed to support to make decision and has been in operation since June, 1994. One of SIGNAL's function is quick monitoring SI readings and seismic acceleration at 331 locations, acceleration waves at 5 locations and rising ground water levels at 20 locations considered most at risk from ground liquefaction, using a reliable radio network. This real-time earthquake data is linked to data bases of gas pipelines and ground conditions to give an accurate damage assessment, based on pipe damage experiences suffered in previous earthquakes. The assessment of damage to the gas network is useful for the rapid implementation of emergency work and in drawing up efficient and accurate restoration plans.

KEYWORDS

SIGNAL, Earthquake monitoring, Damage estimation, SI sensor, Liquefaction sensor, City gas network, Radio communication network.

1 INTRODUCTION

On January 17, 1995, a magnitude 7.2 earthquake occurred in the Osaka-Kobe and Awaji Island area, causing unprecedented damage especially in the city of Kobe with expressways collapsed in many places, and numerous fires started. The city gas system was also damaged: as shown in the Table 1, medium-pressure leaks occurred in 106 locations, and low-pressure leaks in 26,459 locations (Gas Earthquake Countermeasures Study Group Report, 1996). Gas supply to around 900,000 households was shut off to prevent secondary damage caused by gas leakages. In spite of a massive relief effort involving gas operators from the whole country, restoration work took nearly three months, renewing our awareness of the destructive force of earthquakes.

Shut-off of gas supplies to the worst affected areas

Table 1. Damage to Gas Facilities in Great Hanshin Earthquake

Items	Contents
1. Pipeline Damage	106 Leaks in Medium-Pressure Network 26,459 Leaks in Low-Pressure Network
2. Shut-off Gas Supply	6 to 15 Hours after Earthquake
3. Gas Supply Interruption	To 860,000 Customers
4. Restoration	85 Days Required

was implemented after 6 hours at the earliest, in other cases after as long as fifteen hours. The decision to shut off supplies took so long to reach because collection of information on the extent of damage to gas facilities was extremely difficult after the earthquake. Investigations of damage to underground gas lines were hampered by a lack of staff, damage to roads and enormous traffic jams. Even after damage was found, disruption to the telephone network and traffic jams made it difficult to report the damage. Much attention was focused on the importance and difficulty of collecting damage information after a major earthquake such as the Great Hanshin Earthquake.

Tokyo Gas supplies city gas to eight million customers in an area of 3,100 km² centering on Tokyo, one of the most crowded cities in the world and the focus of a heavy concentration of political, cultural and economic functions (Table 2). We are aware that it is our duty to the community as a gas distributor to ensure public safety in the event of a major earthquake. Tokyo Gas therefore regards preparation for a major earthquake as an important issue, and is focusing on the following three areas:

1) Prevention strategies: Earthquake-resistant design of gas production and supply facilities
2) Emergency strategies: Shut-off supplies to the most heavily affected areas, while maintaining supplies to little damaged areas
3) Restoration strategies: Rapid restoration of service to shut-off areas

In order to prevent secondary gas-related damage at the time of an earthquake, it is necessary to shut off gas supplies to the worst affected areas rapidly. For this reason, Tokyo Gas has set up a emergency control system as shown in Fig. 1 and a system of area blocks, which allow gas supplies to be shut off block by block depending on the extent of damage. (15 medium pressure blocks, 100 low pressure blocks).

Table 2. Outline of Tokyo Gas

Items	Contents	
1. Service Area	3,100km²	
2. Customer Contacts	8 million	
3. Pressure Grades and Pipeline Length	High;	500km
	Medium A;	1,900km
	Medium B;	3,500km
	Low:	39,000km

However, rapid decisions on shutting off supplies depend on accurate information being available on the extent of damage to the gas pipeline network in each block immediately after the earthquake. It is very difficult, though, to gain an overall picture of the damage immediately after a major earthquake, as noted above. In fact there were delays in ascertaining the extent of damage to buried pipes after the Great Hanshin Earthquake, which in turn impeded emergency work and restoration work planning.

Since 1987 Tokyo Gas has been developing a system known as SIGNAL (Seismic Information Gathering & Network Alert System), aimed at providing an overall quantitative assessment of damage to gas pipeline networks immediately after an earthquake. SIGNAL went into operation in June 1994. 356 seismic sensors are installed across the service area, allowing damage to pipes to be assessed by combination of seismic data sent on-line (via radio circuits, the most reliable method during earthquakes) with a database of ground and pipeline information. SIGNAL is capable of completing a damage assessment within 10 minutes of an earthquake, allowing rapid decisions to be made on shutting off gas supplies to the most heavily affected areas as part of the emergency work following a major earthquake. It is also now possible to plan restoration work in a short period of time after an earthquake, based on this damage assessment data, and restoration work can be speeded up by the faster response to outside offers of help, more rapid securing of plant and materials etc.

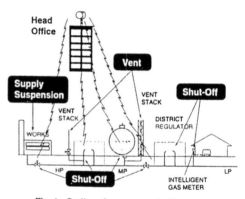

Fig. 1. Outline of emergency shutdown system

Starting from September 1996, information on the SI values and acceleration values at 331 sites has been made available on the Tokyo Gas Internet home page. It is hoped the publication of this data will promote sharing of seismic data and stimulate seismic disaster prevention research.

2 CONFIGURATION OF SIGNAL

SIGNAL is made up of three sub-systems: a seismic motion monitoring system, an epicenter estimation system, and a damage estimation system. It also includes a data bank of basic data on the ground and the pipe system. The overall configuration of the system is shown in Fig. 2.

Overall Configuration of SIGNAL

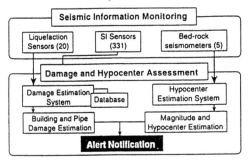

Fig. 2. Overall Configuration of SIGNAL

2.1 Basic Data Base

Basic data on ground conditions, consumers and pipelines have been entered into a data base using our mapping system (GIS) data, based on a grid with cells measuring 250m E-W by 175m N-S (80,000 cells overall). As the size of earthquake vibrations is greatly affected by ground conditions, the data collected by each of the 331 SI sensors (Katayama, et.al., 1988) are made to represent the seismic motion in similar ground around the sensor, with the service area microzoned according to ground conditions and the areas to be represented by each sensor determined in advance. Ground types are divided by topography into upland and alluvial plain types; the latter are further subdivided into three categories according to the natural frequency of the ground, giving a total of four ground types.

Another factor exacerbating damage is ground liquefaction, therefore the depth of the liquefaction layer around each SI sensor is calculated in advance for each grid cell. Data on pipe types, total length of each diameter of pipe, and numbers of consumers are also entered for each cell.

2.2 Seismic Motion Monitoring System

SIGNAL monitors seismic motion in the Tokyo Gas service area using earthquake sensors installed at 356 locations. Fig. 3 shows the locations of earthquake sensors. The system is extremely reliable even in earthquake situations, as data from each sensor are sent by microwave radio transmitters used exclusively for the sensors (Fig. 4). SI sensors installed at 331 locations measure the SI value and maximum acceleration of the seismic motion. SI

values are an index of the level of oscillation of the building during earthquakes. As shown in Fig. 5, SI values show a greater correlation with building damage than do the ground acceleration values. Fig. 6 shows the relationship between actual damage to screwed joint in low-pressure gas pipes during the past major earthquakes, and the SI values for the affected areas. It can be seen that the level of damage increases with the SI value, and measurement of SI values is thus an important step in damage estimation. In order to estimate epicenter and magnitude, ground seismometers are installed at five locations in the surrounding areas. The seismometers are installed at depths of 20–40m in the engineering base rock layer, and the meters continuously measure and transmit 3-directional component acceleration waves. Ground liquefaction also exacerbates damage to buried pipelines, thus in order to assess the occurrence and extent of ground liquefaction, liquefaction sensors (Shimizu, et.al., 1992) are also installed in 20 areas considered to be at high risk from ground liquefaction.

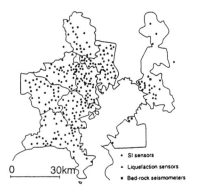

Fig. 3. Distribution of Seismic Sensors

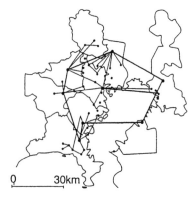

Fig. 4. Radiocommunication Network

319

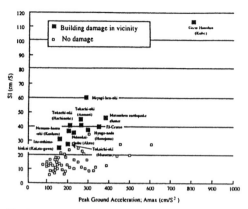

Fig. 5. Correlation between Earthquake Damage, Acceleration and SI values

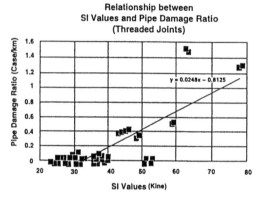

Fig. 6. Relationship between SI values and Pipe Damage Ratio

2.3 Damage Estimation System

Damage in low- and medium-pressure pipelines is estimated on the basis of data sent from the SI sensors as shown in Fig. 7 (Yamazaki, et.al., 1994).

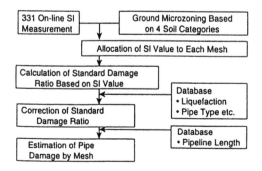

Fig. 7. Damage Estimation System

The first step in damage estimation is the calculation of standard damage ratios and liquefaction depths for each grid cell, based on input of the SI values measured at each of 331 locations. As similar ground types are considered to give identical responses, the SI values for each cell are taken to be the readings from the nearest SI sensor in the same type of ground. Standard damage ratios are calculated from an empirical formula based on records of past earthquake damage, with SI readings as a function. Exacerbation of damage due to liquefaction is taken into account by factoring ground liquefaction into the standard damage ratio, according to the depth of the liquefaction layer. Finally, the number of damage sites for each grid cell is determined by multiplying the total buried pipe length, after correction for pipe type and diameter, by the modified damage ratio. From the estimated number of damage sites for each cell, the number of damage sites in low- and medium-pressure blocks is assessed, forming a basis for decisions on whether emergency shut-off of supply is necessary.

2.4 Epicenter Estimation System

Since data on epicenter location and earthquake magnitude can be obtained within a very short time, such data is ideal for drawing up subsequent emergency response plans. Estimation of the epicenter (Noda, et.al., 1993) is carried out in real time, using acceleration waves transmitted continuously from ground seismometers in five locations. First the initial time for P-waves and S-waves is calculated and compared with the theoretical and observed motion times to minimize error in determining the epicenter location. Magnitude is calculated based on the Japan Meteorological Agency formula.

It is also possible to assess damage from the epicenter location and earthquake magnitude calculated by the epicenter estimation system. In this case, damage is assessed by calculating SI values for each point, using an attenuation damping formula to factor in the distance from the epicenter.

3 OPERATIONAL RECORD OF SIGNAL

Since coming into operation in June 1994, SIGNAL has detected a large number of earthquake motions. A recent example is the Yamanashi-Tobu Earthquake which occurred late at night on March 6, 1996, with its hypocenter near to Lake Kawaguchi. The earthquake intensities recorded by the Meteorological Agency were 5 at Lake Kawaguchi, 3 at Yokohama, and 2 in Tokyo and Chiba. The maximum SI reading recorded in the Tokyo Gas service area was 9 kines, maximum acceleration 222 gals. The PGA distribution is shown in Fig. 8.

It can be seen from Fig. 8 that acceleration in the areas nearest to the hypocenter -- Tama, Machida and Hachioji in Tokyo, Sagamihara in Kanagawa Prefecture -- was close to 200 gals, and in fact many of the microcomputer-controlled meters designed to shut off gas supplies to prevent secondary disasters at this level of acceleration actually cut in to shut off supplies. This resulted in large numbers of telephone inquiries from customers regarding the cut-off in supply, and extra staff had to be brought in to assist the night-duty workers unable to deal with the work load. Following this earthquake, it was discovered that the SIGNAL earthquake motion monitoring PGA distribution results could be used to forecast the likelihood of the microcomputer-controlled meters shutting off supplies, allowing extra staff to be drafted in if necessary immediately after an earthquake. No damage to pipelines occurred during this earthquake. No damage estimate was carried out as the SI readings were too low.

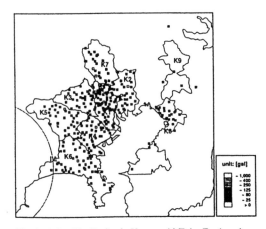

Fig. 8. PGA Distribution in Yamanashi-Tobu Earthquake

4 PUBLICATION OF SIGNAL EARTHQUAKE DATA

In recent years, various organizations and corporations have installed or planned to install earthquake motion monitoring systems. Clearly, integration of all of these systems would result in a highly concentrated network yielding valuable data for seismic disaster prevention research. It was therefore decided, starting from September 1, 1996 (Disaster Prevention Day), to publish SI and acceleration readings from the 331 SIGNAL seismographs via the Internet on the Tokyo Gas home page (http://www.tokyo-gas.co.jp) in the event of an earthquake with a maximum intensity of 3 occurring within the Tokyo Gas service area.

The SI readings are converted to the equivalent on the New Meteorological Agency seismic intensity scale (see Fig. 9; Tong et al., 1996) and displayed on a map color-coded for seismic intensity level. Acceleration data is color-coded according to the gal level. For each of the 331 monitoring points, the maximum SI and PGA readings, together with the location of the seismograph (expressed in latitude and longitude), are also available in text form.

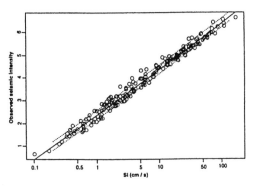

Fig. 9. Relationship between SI Values and New JMA Seismic Intensity

SIGNAL is also linked directly to the SCADA system which controls supply, making protection against intrusion an absolute imperative, and for this reason the system is currently not connected to any Internet server. Publication of seismic data on the Internet therefore involves some time-consuming manual operations, but the aim is to publish the data within 30 minutes.

Publication of seismic data was achieved just after the Ibaragi-ken-oki Earthquake on September 11, 1996. The SIGNAL page on the Internet scored around 10,000 hits on this day, an indication of the level of public interest.

Tokyo Gas will continue to publish seismic data from the 331 SIGNAL monitoring points in the interests of promoting seismic disaster prevention, although it has to be added that the cooperation of other organizations in publishing or setting up a common data base of seismic data would be desirable.

5 CONCLUSION

Rapid and accurate decision-making on shut-off of gas supplies after an earthquake is needed to prevent the occurrence of secondary damage. SIGNAL (Seismic Information Gathering & Network Alert system) was developed to assist in this decision-making. This system comprises 331 SI sensors, 20 ground liquefaction sensors, and 5 ground seismometers which provide data for monitoring of seismic motion after an earthquake. A detailed data base for ground conditions, buried pipes, buildings and structures, liquefaction risk, and damage ratios for buried pipes after earthquakes was prepared to estimate damage to pipes based on the monitored SI values. The system has been running since June 1994.

Starting from September 1996, information on the SI values and acceleration values at 331 sites has been made available on the Tokyo Gas Internet home page. It is hoped the publication of this data will promote sharing of seismic data and stimulate seismic disaster prevention research.

REFERENCES

Katayama, T., Sato, N. and Saito, K. SI sensor for identification of destructive earthquake ground motion. *Proc. of the 9th World Conf. on Earthquake Engineering* VII (1988): 667-672

Noda, S. and Kano, H. Hypocenter estimation system for earthquake countermeasures, *Journal of Natural Disaster Science,* Vol. 15, No. 1, pp 53-90, 1993.

Shimizu, Y., Yasuda, S., Yoshihara, Y. and Yamamoto, Y. 1992. Adaptability Experiments of Liquefaction Sensor. *Proc., 4th Japan-U.S. Workshop on Earthquake Resistant Design of Lifeline Facilities and Countermeasures Against Soil Liquefaction,* 621-638.

Yamazaki, F., Katayama, T. and Yoshikawa, Y. On-line damage assessment of city gas networks based on dense earthquake monitoring. *Proc. of the 5th U.S. National Conference on Earthquake Engineering,* Vol. 4, (1994 a): 829-837.

Gas industry earthquake countermeasures study group report (Supervised by Agency of Natural Resources and Energy) p7-p11, 1996.

Tong, H., Yamazaki, F., Shimizu, Y. and Sasaki, H. Relationship between New JMA Seismic Intensity and Ground Motion Indices. *Proc. of the 51th Annual Conference of the Japan Society of Civil Engineers,* 1-(B); p458-p459, 1996.

Seismic Behaviour of Ground and Geotechnical Structures, Sêco e Pinto (ed.) © 1997 Balkema, Rotterdam, ISBN 90 5410 887 8

Evaluation of stability of dam and reservoir slopes – Mechanics of landslide

Bhawani Singh
Department of Civil Engineering, University of Roorkee, India

R. Anbalagan
Department of Earth Sciences, University of Roorkee, India

ABSTRACT : Detailed stability studies are carried out in landslide hazard prone areas. Initially the mode of failure is identified. Later back analysis is done for determining the strength parameters. Finally forward analysis is made for adjoining slopes in similar geological environment. The computed dynamic settlement of soil and rock slopes during design earthquake is a good criterion for assessing their safety. Hence, in certain cases, a factor of safety of less than one shall also be considered as just safe. Landslide crisis management can be better carried out by computer aided landslide analysis. A package of 18 software programs has been developed and tested extensively in seismic and fragile Himalayan region.

1. INTRODUCTION

A landslide hazard map indicates severity of potential landslides and erosion in a region as well as along various reaches of road/rail line alignments. It is also helpful to understand the probable mode of failures of a rock or soil cut slopes and reservoir slopes. Then the selection and design of preventive measures for stabilization of the landslides becomes much easier. However, dam slopes require particular attention of experts.

With the advent of personal computers, slope analysis is no a longer a boring and time consuming work. The authors have developed a complete package of 18 computer programs for slope stability analysis. It may be used to get the factor of safety of slope or optimum angle of cut slope with or without drainage. Graphic display gives insight into the failure mechanism of slopes.

For efficient use of these programs, it is very important that the basic concepts are understood well. This will in turn help in understanding the mechanics of even complex landslides. It is envisaged that this paper will help to understand the landslide problems and the mechanics so that the recommendations of the experts are not executed blindly.

2. PROBLEM DEFINITION

In the field a slope may fail in a complex manner or the slope failure looks complex. But landslides in general show one dominant and characteristic mode of failure. A careful and detailed examination may indicate one dominant and characteristic mode of failure. Accordingly the landslide may be classified based on the observed mode of failure.

(a) Planar slide along a joint plane on a rock slope,
(b) 3-D wedge slide along two joint planes on a rock slope,
(c) Rotational slide along a cylindrical rupture in soil or rock slope, homogeneous earth dams and mine dumps,
(d) Talus or debris slide along the contact of underlying rock

bed, which is common in Himalaya.
(e) Rock fall due to overtoppling of rock blocks, often seen on steep rock slopes having steep outward dipping joints.

Varnes, (Schuster and Krizek, 1978) classified landslides in more detail based on type of the landslide and the nature of materials involved. At present no realistic mathematical model of such failures in seismic regions is available for the purpose of analysis. This limitation should be kept in mind while applying various theories of landslide in the field.

3. APPROACH TO PROBLEM

There is another difficulty in applying analytical models in the field. It is related to collection of input data. Even in landslide prone areas, it is generally not possible to perform in-situ tests for determining shear strength parameters of discontinuities as well as soil/rock mass. It is also hazardous to guess the future groundwater conditions within the slope during peak rainy season.

Thus a new approach to solution of this problem is required. Past experiences of many landslides show that shear strength parameters along slip surface may be obtained by back analysis of steep slopes in distress. Back analysis also provides a good check on other input parameters e.g. unit weight, groundwater conditions and tension crack. If this work is not done correctly, the factor of safety of even a temporarily stable slope would be found to be less than unity.

The package on computer programs thus also includes programs of back analysis of slopes for each mode of failure.

The long experience has taught us the following approach to solve the problems of landslides (Deoja et al. 1993).

(a) Back analyse slopes in distress or high and steep slopes,

(b) Predict factors of safety of adjoining slopes in the same rock/soil mass (using strength parameters from above analysis),

(c) Select and design the preventive measures for slope if factor of safety is not adequate.

Some idea of strength parameter may be had from published literature. Hoek & Bray (1981) have compiled peak and residual strength parameters along discontinuities. The typical strength parameters of rock mass may be picked up according to state of weathering and joint characteristics. Bieniawski (1981) also suggested strength parameters for rock mass according to their RMR (Rock Mass Rating).

4. LIMITATIONS OF ANALYTICAL MODELS

The following assumptions are generally made in limit state analysis of slopes :

(a) Slope material is homogeneous and isotropic,

(b) It obeys Coulomb's law of friction for both joints and rock/soil mass,

$$\tau = C + (\sigma - u) \tan \phi \quad ...(1)$$

where,

τ = shear strength
C = Effective cohesion
ϕ = Effective angle of internal friction
σ = Normal stress across slip surface
u = Pore water pressure

(c) Strength is uniformly mobilised along entire rupture surface. There is no progressive failure.

(d) Tension crack is vertical.

(e) Earthquake forces are replaced by pseudo-static forces. The horizontal component of earthquake acceleration in case of submerged slope, hill slopes in Himalaya is generally taken as 0.08 in rock and 0.15 in soil. It is taken as 0.10 to account for blasting vibrations. The earthquake acceleration is higher at the top of the slopes,

(f) In case of submerged slope, groundwater level is same as river (reservoir) water level at least up to the slip surface,

(g) In-situ stresses do not affect stability of the slope,

(h) Radius of curvature of slope in plan is infinite compared to height of slope.

5. MYTH OF LUBRICATION EFFECT DURING SATURATION

Terzaghi and Peck (1948) have shown beyond doubt that there is nothing like lubrication between rock/soil particles or joint surfaces during saturation in rainy season. In reality, water pressure (u) developes within pores or joint openings. Then the shear strength of soil/rock mass decreases because of reduction in the effective stress (σ -u) according to Eq. 1. One can easily check that there is no significant difference in coefficient of friction of dry and wet joint surfaces. However, cohesion may be reduced drastically after saturation.

6. SLOPE IN DRY COHESIONLESS MATERIAL

It is a simple case where rock mass is highly fractured or debris is a mixture of sand, gravel and boulders. Fig.1 shows a dry slope and its mechanics of failure.

$$F = \mathrm{Tan}\ \phi / \mathrm{Tan}\ \psi_f \quad (\mathrm{Static}) \qquad \ldots(2)$$

$$F_{dyn} = \mathrm{Tan}\ \phi / \mathrm{Tan}\ (\psi_f + \mu)\mu$$

$$= \mathrm{Tan}^{-1} \alpha_h (\mathrm{Dynamic}) \qquad \ldots(3a)$$

$$\mathrm{or}\ \mu = \mathrm{Tan}^{-1}(\Upsilon\alpha_h / \Upsilon - \Upsilon_w)(\mathrm{Submerged}) \qquad \ldots(3b)$$

Fig.1 Slope in Dry Cohesionless material

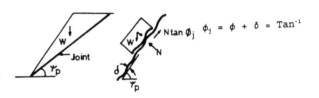

$$F = \mathrm{Tan}\ \phi_j / \mathrm{Tan}\ \psi_p \quad (\mathrm{Static}) \qquad \ldots(4)$$

$$F_{dyn} = \mathrm{Tan}\ \phi_j / \mathrm{Tan}\ (\psi_p + \mu)\ \mu$$

$$= \mathrm{Tan}^{-1} \alpha_h \quad (\mathrm{Dynamic}) \qquad \ldots(5)$$

Fig. 2 Dry Rock Slope with Plane Wedge Sliding

During earthquake, a horizontal force act on the block or element and is equal to the product of weight of block and the horizontal component of earthquake acceleration. It is interesting to note that effect of earthquake is equivalent to increase in the slope angle (ψ_f) by μ, where $\mu = \tan^{-1}\alpha_h$. The slope will fail where slope angle is more than ϕ- μ. It may be noted that reservoir slopes are more vulnerable to failure after earthquake.

Another simple example is of dry rock slopes with plane wedge sliding. In such situation, the dip direction of joint plane is equal to that of the slope face within $\pm 15^{\circ}$. The slope face is steeper than joint plane (Fig.2).

Generally the joint plane is wavy. It is very important to note that waviness of joint plane increases the overall angle of sliding friction by δ, which is the average angle of waviness of asperities with respect to the general dip of the joint plane. It varies from 0 to 20 for smooth to very rough and undulating joints (Barton and Brandis, 1990). They

also find that coefficient of friction ($\tan \phi_j$) along rock joints is about J_r/J_a, where J_r is joint roughness number and J_a is joint alteration number of the joint.

It is also important to note that a slip between wavy joint planes result in further opening of joints ($\tan \delta$). This is called dilatation of joint.

7. ROCK SLOPE WITH PLANE WEDGE SLIDING

A rock slope with plane wedge may develop a tension crack at the time of sliding as shown in Fig.3.

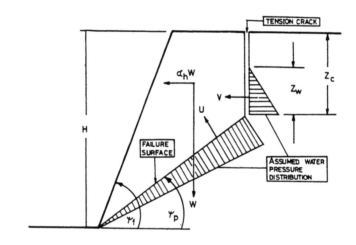

$$F = \frac{C_jA+[W(Cos\ \psi_p - \alpha_h * Sin\psi_p) - U - V*Sin\psi_p]*Tan\phi_j}{W(Sin\psi_p + \alpha_h*Cos\psi_p) + V*Cos\psi_p - P*Cos\ \alpha_p} \qquad ...(6)$$

Where,

$$Z_c = H(1- \sqrt{Cot\psi_f * Tan\psi_p}) < H/2 \qquad ...(7)$$

$$A = (H - Z_c)\ Cosec\ \psi_p) \qquad ...(8)$$

$$W = 1/2\ \ H^2\ [\{1-(Z_c/H)^2\}Cot\psi_p - Cot\psi_f \qquad ...(9)$$

$$U = 1/2\ \gamma_w\ Z_w\ A \qquad ...(10a)$$

$$V = 1/2\ \gamma_w\ Z_w^2 \qquad ...(10b)$$

Fig. 3 Rock Slope with Plane Wedge Sliding
(After Hoek & Bray, 1981)

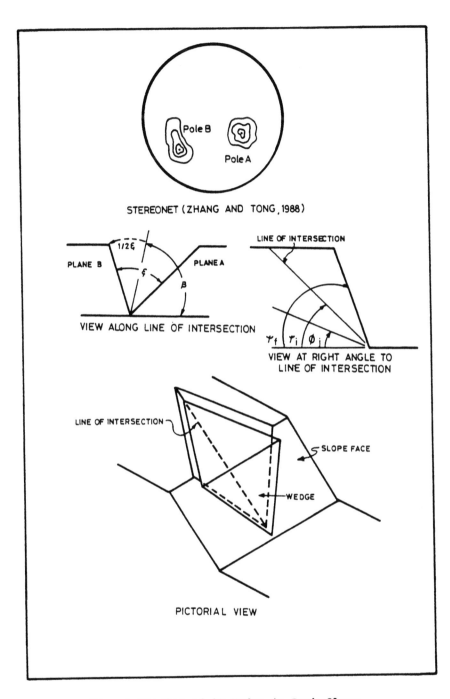

STEREONET (ZHANG AND TONG, 1988)

VIEW ALONG LINE OF INTERSECTION

VIEW AT RIGHT ANGLE TO
LINE OF INTERSECTION

PICTORIAL VIEW

Fig. 4 Sliding of 3D Wedge in Rock Slope

Hoek and Bray (1981) have derived the depth of tension crack (Z_c) as $H[1- \int(Cot\psi_f \cdot Tan\psi_p)]$. However, it is never found to be more than half the height of the slopes even for vertical slopes.

These tension cracks are likely to be filled up with water during rainy season to a depth say Z_w. Then it will exert a hydro-static force V on the plane wedge (Fig. 3). Since the joint plane will act as pipe line or channel after some sliding (when there will be more joint opening due to dilatation), hydrostatic pressure $\gamma_w \cdot Z_w$ will also develop on top most point of the joint plane (Fig.4). However, at its bottom all water pressure may be released. This assumption of groundwater flow has been generally accepted, as rock mass near slope face is more likely to behave as discontinum and not as continum (as in case of soils).

The factor of safety of slope is given Eq.6 in Fig.3. It may be noted that sliding may occur even if $\psi_p < \phi_j$, because of water pressure.

It may be noted that a thin wedge may topple around toe if effective normal reaction across joint (expression within bracket before tan ϕ_j) in Eq. 6 becomes tensile.

Another condition for over-toppling is

$$\psi_f = 90^o - \psi_p + \phi_j - \mu \qquad ...(11)$$

where, ψ_p is the dip of cross joints.

The computer program SASP has been developed to give static and dynamic factors of safety and design of rock reinforcement. For submerged or reservoir slopes, program SARP is available and it accounts for toe cutting.

8. SLIDING OF 3D WEDGE IN ROCK SLOPE

Three dimensional wedge failure occurs when two joint planes A and B strike obliquely across the slope face and their line of intersection dips towards the slope face (Fig. 4).

Hoek and Bray (1981) have given following expression for the factor of safety (F) of 3D wedge sliding along line of intersection.

$$F = [tan\ \phi_j/tan(\psi_i + \mu] *$$
$$[sin\ \beta\ /tan\ (\xi/2)] \qquad ...(12)$$

$$= F\ plane/tan\ (\xi/2) \qquad ...(13)$$
(for symmetric wedge)

where,

ξ = wedge angle (Fig. 4)
ϕ_j = sliding angle of friction
ψ_i = dip of line of intersection
μ = $tan^{-1}h$
β = angle between bisector of joint planes and normal to the line of their inter-section (Fig. 4)
= 90° for symmetric wedge

On comparing Eq. 5 and Eq. 12, Eq. 13 is obtained for symmetric wedge. Thus, for a 90° wedge, the 3D analysis gives factor of safety, which is 1.41 times the factor of safety predicted by plane wedge analysis. Hence there is necessity for a detailed 3D wedge analysis. This is now easily possible with the computer program SASW which is based on simple method proposed by Hoek and Bray (1981).

The input data for SASW includes (i) dip and (ii) dip direction (iii) cohesion, (iv) sliding angle of friction for each joint plane, (v) total number of joint sets, (vi) height of wedge, (vii) reservoir water level, (viii) coefficient of horizontal component of earthquake acceleration and (ix) drainage condition. The output will be in the form of factors of safety of each wedge formed by any two joint planes for static and dynamic conditions. For rigorous analysis of wedge with sloping top terrace and inclined tension cracks, software program RWEDGE is developed on the basis of theory of Hoek & Bray (1981). This program can also be used to design rock reinforcements to stabilise critical wedges.

9. ROTATIONAL SLIDE IN SOIL AND ROCK SLOPES

Rotational slide may take place in a highly jointed rock slope if all joint sets are favourably oriented to induce a slide. In cohesive soil slopes, rotational slide is quite common. It is generally assumed that slip surface is cylindrical. It is further assumed that flow through rock mass is continum type just like in soil mass.

Various forces acting on vertical slices above the cylindrical slip surface are explained in Fig. 5. The pore water pressure is generally accounted for by pore pressure parameter B (Eq.5) in Fig. 5. It may vary from 0 to 0.3 for dry and wet slopes with springs all over. Bishop (1954) suggested a simple method for getting factor of safety of slopes from Eq. 14. It is a transcendental equation and is easily solved by the computer program SARC.

The critical surface is located by calculating factors of safety for many slip surfaces. The selected value represents the slip surface giving a minimum factor of safety.

It may also be pointed out that analysis should be repeated for different depths of tension cracks and reservoir water levels for obtaining a minimum factor of safety. The maximum depth of tension crack and highest RWL may not give lowest factor of safety.

The program SARC may also be used for design of slopes of road embankments. It will also help in development of terraces along hill slope for building complexes. It is also used to design earth or rock fill buttress to stabilise a rotational slide.

10. TALUS/DEBRIS SLIDE

Majority of landslides along hill roads belong to this category.

Talus or debris or highly weathered part of rock mass (regolith) tends to slide along bed rock (Crozier, 1986). The mechanics of the dip slide in hard rock is similar. However, actual slope movement may often look to be different and complex according to the geological conditions.

Coates (1970) analysed this case as shown in Fig. 6. The theory assumes constant thickness of talus equal to Z. During long spell of rains, groundwater table starts building up to depth Z_w below the slope surface. Ranjan et al. (1984) modified Coate's equation to account for earthquake forces (Fig. 6) and Eq. 20a. The following notations have been used.

c, ϕ = strength parameters of talus/debris material,
γ = unit weight of talus,
γ_w = unit weight of water,
ψ_f = slope angle of slope face and bed rock,
α_h, α_v = coefficients of horizontal and vertical components of earthquake accelerations.

329

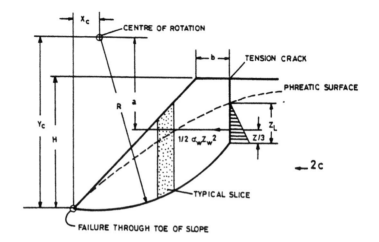

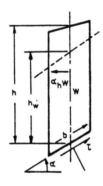

Factor of Safety (Static)

$$F = \sum [X/(1+Y/F)]/\sum Z + Q \qquad \qquad \dots (14)$$

where,

$$X = [c*b + W (1-B) \tan \phi] \sec\alpha \qquad \dots (15)$$
$$B = \gamma_w * h_w / \gamma * h \qquad \dots (16)$$
$$\gamma = \tan\alpha * \tan\phi \qquad \dots (17)$$
$$Z = W * \sin\alpha + \alpha_h * W \cos\alpha \qquad \dots (18a)$$
$$Q = 1/2 \ \gamma_w * Z_w^2 * a/R \qquad \dots (18b)$$

Rock Mass Strength Parameter

$$q_{cmass} = 0.38 * \gamma * Q^{1/3} \qquad \dots (19a)$$
$$\tan\phi = J_r/J_a \qquad \dots (19b)$$

Fig. 5   Bishop's (1954) Simple Analysis of Circular Failure in Slope

330

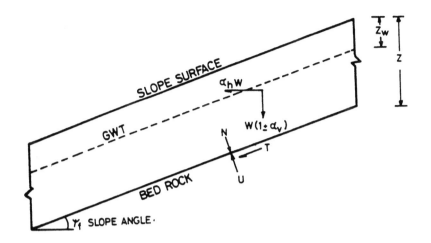

$$F = \frac{c*Sec^2\psi_f/\gamma Z + Tan\phi[1 \pm \alpha_v - \alpha_h*Tan\psi_f - (1-Z_w/Z)\gamma_w/\gamma]}{(1 \pm \alpha_v)* Tan\psi_f + \alpha_h} \quad \ldots (20a)$$

Fig. 6 A Talus/Debris Slope

$$F \simeq F_{dry}/2 \qquad (20b)$$

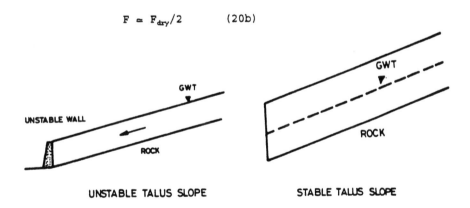

UNSTABLE TALUS SLOPE   STABLE TALUS SLOPE

Fig. 7   Thin Cover of Talus/Debris is Likely to Fail Even on Gental Slope

Surcharge may be taken into account by increasing Z by equivalent soil cover and decreasing Zw by the same amount.

It is important to note that talus/debris slide may take place at slope angle which is much less than the angle of internal friction (Fig. 7). This may happen in case of thin cover of talus with water table close to the slope surface during long spell of rains. According to Eq. 20b, factor of safety drops to half of that in dry weather. So cut slopes and retaining walls will fail again and again. Better drainage is the most suitable solution. A computer program SAST has been developed to get static and dynamic factors of safety. SAST also suggest remedial measures and design of subsurface drains.

A word of caution is needed here that the bed rock at times may be highly undulating and thickness of talus may vary widely. So slope must be divided carefully in several zones for the purpose of analysis and program SANC is used for such analysis.

11. ROCK FALL DUE TO TOPPLING OF ROCK BLOCKS

In case of steep slopes with steep outward dipping joints, slope failure may occur due to toppling of blocks of rock (Fig. 8, Hoek and Bray, 1981 and Zanbank, 1984). The Eq. 11 provides a good criteria for analysis. A thin plane wedge may fail by overturning about the toe due to heavy water pressure in the tension crack. Wedge toppling is also observed frequently. A triangular rock block can topple easily.

Past experience shows that road cutting accelerates the process of toppling, particularly during monsoon, when hydrostatic water pressure develops in the tension cracks and pushes the rock blocks.

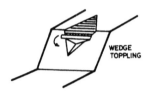

Fig. 8   Failure of Slopes by Overtoppling of Rock Blocks

12. ROCK SLOPE FAILURE DUE TO TOE EROSION

In concave slopes along a river, landslides may occur due to toe cutting. The river water current may cause the toe erosion. In soft soluble rocks like limestone, toe erosion may take place due to groundwater flow and toe softening (Fig. 9).

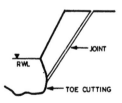

Fig. 9   Plane Wedge Sliding After Severe Toe Cutting

13. STABILITY ANALYSIS OF SOIL AND ROCK SLOPES WITH POLYGONAL WEDGE FAILURE

Slope failures may often show that the slip surface is non-circular and slice boundaries are non-vertical (Fig. 10). Sarma (1979) devised a method of stability analysis of such slopes. Hoek (1987) developed a computer program SARMA,BAS. The authors have also developed another program SANC on the basis of a simple method and theory of Kovari and Fritz (1978). SANC also computes dynamic settlement of a slope during an earthquake. Many complex rock slopes have been analysed using SANC and realistic results obtained.

The software program SANC is ideally suited for stability analysis of non-homogeneous slope materials or earth and rock fill dams in highly seismic regions. The program is also used for stability analysis of debris slide with varying rock and slope profile (Fig. 10,iv).

14. BACK ANALYSIS OF SLOPES

Back analysis of steep slopes in distress discussed earlier will give the much needed strength parameters of slope material.

For that purpose, the mode of failure has to be identified initially. In case of plane failure, F is taken equal to 1.0 in Eq. 6. The cohesion values are calculated for different values of angle of sliding friction (ϕ_j). Computer program BASP is available for this purpose. Judicious choice of C_j and ϕ_j is then made on the basis of past experience.

Similarly Eq. 14 may be used to back calculate strength parameters by substituting F equal to 1.0. Computer program BASC is used for this purpose. However,

one should select realistic values of strength parameters of rock mass. It is very surprising to note that the back analysis of rotational slides gives a linear relationship C and ϕ (Singh & Ramasamy, 1979).

In case of talus or debris slide, Eq. 20a in Fig. 7 is useful for back calculating strength parameters. Computer program BAST is also available to do so. Generally cohesion is small or negligible in talus deposit. So choice of angle of internal friction is not a difficult problem.

Back analysis is a very useful exercise. It saves us from being too conservative as factor of safety cannot be less than 1.0. Nature provides many times indications on slope movement. Tilted trees and tension cracks are often seen at sites of distress.

15. DYNAMIC SETTLEMENT DURING EARTHQUAKE

Experience of computations by various computer programs suggests surprisingly a good and simple correlation for dynamic factors of safety (F_{dyn}).

$$F_{dyn} = F_{static}/(1 + 3.3 \, \alpha_h) \quad \ldots(21)$$

where,

F_{static} = Static factor of safety
α_h = Peak ground acceleration

The constant may vary slightly for different slopes. The critical acceleration (α_{cr}) may be obtained approximately from above correlation for unit dynamic factor of safety.

It is important to understand that slope may not fail even if dynamic factor of safety is less than 1.0. It is because displacement of the wedge may not be significant. Slopes are subjected to much severe vibrations and fracturing during blasting. But significant dynamic displacement does not occur because waves due to blasting contain only one peak, whereas earthquake waves contain many peaks.

Jansen et al. (1990) have developed a simple semi-empirical equation for determining dynamic settlement (meters) :

$$S_{dyn} = 5.8(0.1M)^8 (\alpha_h - \alpha_{cr})/\alpha_{cr} \cdots (22)$$

< 1 meter or
1 percent of height of slope

Where,

M = Magnitude of design earthquake of Richter's scale,

α_{cr} = Critical acceleration at factor of safety of unit,

α_h = Peak ground acceleration.

All computer programs SASP, SARC, SASW and SAST calculate dynamic settlement.

16. DESIGN OF CUT SLOPES

It may be economical to cut the gentle hill slopes along road alignment. These cut slopes should be stable even in adverse weather conditions. Rocks in Himalaya are fragile. Experience alone is the best guide. Table 1 is given here to suggest typical cut slope angles on the basis of past experience in this region.

In case of talus/debris slide, safe cut slope angle does not mean anything as talus/debris is not stable by itself. In such situations, the entire slope should be stabilised by subsurface drainage and/or planting of bushes and trees (Fig. 7).

Table 1 : STABLE CUT SLOPES ON ROAD SIDE WITH HEIGHT (H) LESS THAN 10M

Sl.No.	Type of Soil/Rock	Stable cut slope without any protection work (vertical to horizontal)	Stable cut slope with breast wall or minor protection work
1.	Soil or soil mixed boulder with a. Disturbed vegetation b. Disturbed vegetation overlaid on firm rock	1:1 Vertical for rock portion and 1:1 for soil portion	n:1* Vertical for rock portion and n:1 for soil portion
2.	Same as above but with dense vegetation forests medium rock and shales	1:0.5	5:1
3.	Hard rock, shale or harder rocks with inward dip	1:25 to 0.1 and vertical or overhanged for half tunnels	Not needed
4.	Same as above but with outward dip or badly fractured rock/shale	At dip angle or 1:0.5 or dip of intersection of joint planes	5:1
5.	Conglomerates/very soft shale/sand rock which are easily eroded	Vertical cut to reduce erosion	5:1

* n is 5 for H < 3 m; 4 for H = 3-4 m; 3 for H = 4-6 m

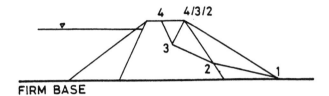

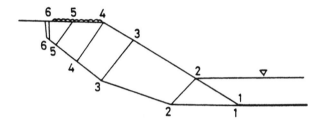

(i) Non-Homogeneous Earth Dam With Slices
     Along Potential Planes Of Weekness or Failure

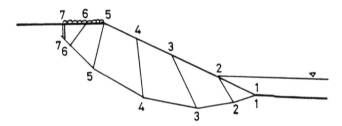

(ii) Rock Slope with Slices Along Preexisting Joints

(iii) Circular Failure with radial To Give Minimum
      Factor of Safety in Homogeneous Slopes

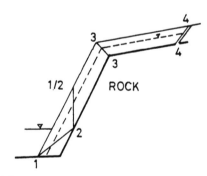

(iv) Debris Slide

Fig. 10   Applications of Proposed Method of Stability Analysis
          of Slope with Non-Circular Slip Surface and Non-
          Vertical Slices.

335

Similarly, in complex geological environment with pre-existing slip surfaces in buried landslide areas, safe cut slope angle does not mean anything. Experts should be consulted for analysing such slopes.

It is important to add here that in steep hills ($\psi_f > 35^O$) it may be more economical to cut the hill slope at 5:1 on road side as mentioned in last column of Table 1 and support the steep cut by a breast wall.

17. CORRECTION FOR RADIUS OF CURVATURE OF CUT SLOPE

Hoek and Bray (1981) have suggested that concave slopes in plan are more stable than convex slopes. Studies of slopes of phyllites in Lesser Himalaya suggest the following correction in slope angle,

$$\psi_{fo} = \psi_f + \theta \, (H/R_o) \qquad \ldots(23)$$

Where,

ψ_{fo} = recommended slope angle,
ψ_f = normal slope angle,
θ = 5° (Phyllites),
H = height of cut slope,
R_o = average radius of cut slope at base of road level (+ve for concave)

Eq. 23 shows that correction may be quite significant i.e. $\pm \, 10^O$ for R = H/2.

It is also interesting to note that compressive stresses develop along concave rim of the slope and help in stabilizing slope. It is due to the fact that arching action takes place like in an arch dam. On the contrary, tensile stresses develop along the convex rim of the slope and destabilize the slope.

It should be mentioned that in case of 3D wedge failure,

correction for curvatures does not appear as there is full freedom for wedge to move down.

18. COMPUTER AIDED DESIGN OF CUT SLOPES

The safe angle of cut slope may be easily obtained by computer programs for a given factor of safety. In rock slopes a static factor of safety of 1.20 is adequate and for soil slopes, factor of safety should be at least 1.50. A dynamic factor of safety of 1.0 is good enough.

In case of plane failures, the computer program ASP should be used to get safe cut slope angle. For 3D wedge sliding, computer program ASW is available and analyses all wedges and all modes of failure before giving safe cut slope angle. Similarly for rotational slide, program ASC is used for any general profile of slope surface.

It should be kept in mind that cut slope angles will be different on both sides where road passes through a mountain, making a trench type cut. This is because dip directions of slopes are 180° opposite to each other.

The programs such as ASP and ASC take into account the radius of curvature of slope. Input data is similar to that for analysis of factor of safety for the respective mode of failure. These programs will be ideally suited for cut slope design of open pit mines.

19. BENCHES

The cut slope in soft rocks should be designed properly (Fig.11). Single or multiple benches are provided for construction and maintenance of slopes in soft rocks. It also serves as protection against fall of rock fragments.

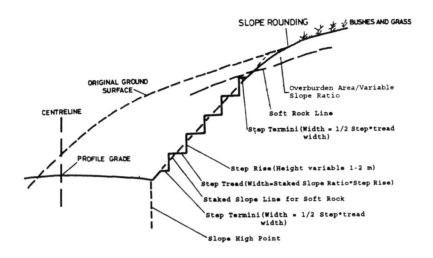

SLOPE ROUNDING — BUSHES AND GRASS

ORIGINAL GROUND SURFACE

CENTRELINE

PROFILE GRADE

Overburden Area/Variable Slope Ratio

Soft Rock Line

Step Termini(Width = 1/2 Step•tread width)

Step Rise(Height variable 1-2 m)

Step Tread(Width=Staked Slope Ratio•Step Rise)

Staked Slope Line for Soft Rock

Step Termini(Width = 1/2 Step•tread width)

Slope High Point

Fig. 11 Idealized Cross Section Showing Stepped Cut Slope Design in Soft Rock

Benches should be provided in such a way that the overall slope angle is equal to the safe cut slope angle.

20. CUT WITH VARIABLE ANGLE

Variable slope angle is recommended for stratified sedimentary rock and if there are layers of different weather properties. For slopes less than 10 m, the Table 1 may be used for cut slope design and for higher slopes, proper stability analysis using suitable computer program should be done.

20.1 Useful Suggestions for Excavation of Cut Slopes (Fig. 11)

(a) Smooth blasting should be adopted in rock slopes having high landslide potential. It may reduce the damage to the rock mass to a minimum.

(b) Soft rocks should be excavated manually with several benches.

(c) The excavation should proceed from top a as it will prevent formation of tension cracks at the top of slope.

(d) Cut should be rounded and blended with terrain to minimize erosion.

21. CONCLUDING REMARKS

A package of 18 computer programs have been developed and used extensively to analyse various types of landslides/failures of slopes in soil and rocks in seismic hilly areas including the Himalayan countries. These programs can be loaded in a PC Notebook and used at the site for an on the spot and quick decision.

The dynamic settlement of slopes with compacted soil and boulders is estimated to be generally within permissible limits in many dam projects. Large number of earthquakes of high intensity in this decade have so far not resulted in unsafe settlement of modern well compacted high earth dams which have done their jobs of retaining high reservoirs very well.

No liquefaction of steeper soil slopes ($\psi_f > 15^\circ$) has been observed in Himalaya because of possible development of negative pore water pressure.

Reservoir slopes of Tehri dam, Koteshwar dam, Lakhwar dam, Kishau dam projects have been analysed extensively using these software programs. The danger due to waves generated by deep seated landslides in the reservoir area is estimated to be negligible to the safety of the dam. Majority of landslides are surficial only and very few are deep seated but far away from dam site.

There is urgent need to use high sophisticated software programs to analyse settlement of soil/rock slopes in complex geological environment.

NOTATIONS USED

A   = Area of joint involved in sliding.
a   = Lever arm of water thrust in tension crack w.r.t. centre of slip circle (Fig. 5)
B   = Pore water pressure coefficient
    = $\gamma_w h_w / \gamma h$
b   = Base length of slice
α = Inclination of base of slice w.r.t. horizontal
C   = Effective cohesion
ϕ = Effective angle of internal friction
C_j = Cohesion along rock joint
F   = Static factor of safety of slope
F_{dyn} = Dynamic factor of safety of slope
H   = Height of slope
h   = Height of slice (Fig. 5)
h_w = Height of groundwater table above slip surface
J_r = Joint roughness number for rock joint

J_a = Joint alteration number for rock joint
M   = Magnitude of design earth-quake on Richter's scale
Q   = Rock mass quality (NGI classification)
q_{cmass} =Uniaxial compressive strength of rockmass in rock slope
R   = Radius of circular slip surface
R_o = Radius of cut slope in plan at road level (Eq. 23)
RWL = Reservoir water level above origin of coordinates of slope profile or
    = Reservoir water level above toe of wedge in rock slope
S_{dyn}= Dynamic settlement of slope in meters during earthquake
U   = Uplift force due to water on joint
u   = Pore water pressure
V   = Water thrust in tension crack
W   = Weight of wedge
Z   = Depth of talus/debris above rock slope (Fig. 6)
Z_c = Depth of tension crack
Z_w = Depth of water in tension crack or depth of ground water from slope surface in debris slide (Fig. 2.6)
α_{cr} = Critical coefficient of earthquake acceleration for unit dynamic factor of safety
α_h = Coefficient of horizontal component of peak ground acceleration
α_v = Coefficient of vertical component of peak ground acceleration
ϕ_j = Angle of sliding resistance along rock joint
γ = Unit weight of soil
γ_w = Unit weight of water
ψ_i = Dip of intersection of two joint planes
ψ_f = Slope angle
ψ_{fo}= Recommended cut slope angle in curved slopes
ψ_p = Dip of joint
μ = $\text{Tan}^{-1} \alpha_h$ for dry slope
    = $\text{Tan}^{-1} [\gamma\alpha_h/(\gamma - \gamma_w)$ for submerged slope
θ = Constant of slope material in Eq. 23.
σ = Normal stress across slip surface
τ = Shear strength
ξ = Wedge angle between two joint planes

REFERENCES

1. Bishop, A.W. (1954) The use of slip circle in the stability analysis of slopes, Geotechnique, V.5,No.1,pp.7-17.
2. Barton, N. and Bandis, S.C. (1990) Review of predictive capabilities of JRC-JCS model in engineering practice. Proc. Int. Symp. on Rock Joints, Norway, pp. 603-610.
3. Bieniawski, Z.T. (1981) Case studies - Prediction of rock masses behaviour by the geo-mechanical classification. Second Australia New Zealand Conference on Geomechanics, Brisbane, pp. 36-41.
4. Coates, D.F. (1970) Rock Mechanical Principle. Department of Energy, Mines & Resources, Monograph 874, Canada, Chap. 6.
5. Crozier, M.J. (1986) Landslides - Causes, Consequences and Environment, Crom Heim, U.S.A.
6. Deoja, B., Dhital, M., Thapa, B. and Wagner, A. (1991) Mountain Risk Engineering Hand book. International Centre of Integrated Mountain Development, Kathmandu, Part I, Chap 10 and 13 : and Part II.
7. Jansen, R.B. (1990) Estimation of Embankment Dam Settlement Caused by Earthquake. Water Power and Dam Construction, December, pp. 35-40.
8. Hoek, E and Bray, J.W. (1981) Rock Slope Engineering, Institute of Mining and Metallury, London, Revised Third Edition, Chap 5 and 7.
9. Hoek, E. (1987) General Two Dimensional Slope Stability Analysis, Analytical and Computational Method in Engineering Rock Mechanics, Ed. E.T. Brown, Allen and Unwin, London,Chap 3, pp. 95-18.
10. Kovari, K. and Fritz, P. (1978) Slope Stability Analysis with Plane, Wedge and Polygonal Sliding Surfaces, Int. Symp. on Rock Mech. related to Dam Foundations, Rio de Janerio, Brazil.
11. Ranjan, G., Singh, B. Saran, S. Viladkar, M.N., and Khazanchi, A.C. (1984) Stability Analysis and Protective Measures for Building Complex on Slope, Int. Symp. on Landslides, Toronto, Canada.
12. Sarma,S.K.(1979) Stability Analysis of Embankments and Slopes, J. of Geotechnical Engineering Division, A.S.C.E., GT.12, pp. 1511-1524.
13. Schuster,R.L., and Krizek,R.J. (1978) Landslides - Analysis and Control, National Academy of Sciences, Washington, D.C., Special Report 176, 234 p.
14. Singh,B. and Ramasamy,G. (1979) Back Analysis of Natural Slopes for Evaluation of Strength Parameters, Int. Conf. on Computer Applications in Civil Engineering, Roorkee, India, V.1, pp. VII-57-62.
15. Terzaghi,K. and Peck,R.B.(1948) Soil Mechanics in Engineering Practice, John Wiley & Sons, Second Ed., Art. 17, 729. p.
16. Zabank,C. (1983) Design Charts for Rock Slopes Susceptible to Toppling, A.S.C.E., J. of Geotechnical Engineering, V. 109, No. 8, pp. 1039-1062.
17. Zhang,S.and Tong.G.(1988)Computerized Pole Concentration Graphs using the Wulf Stereographic Projection, Int. J. Rock Mech. & Geotech. Abs., V.25, No. 1, pp. 45-51.

Seismic Behaviour of Ground and Geotechnical Structures, Sêco e Pinto (ed.) © 1997 Balkema, Rotterdam, ISBN 90 5410 887 8

Seismic response of steep natural slopes, structural fills, and reinforced soil slopes and walls

Nicholas Sitar & Lili Nova-Roessig
Department of Civil and Environmental Engineering, University of California, Berkeley, Calif., USA

Scott A. Ashford
Department of Applied Mechanics and Engineering Science, University of California, San Diego, Calif., USA

Jonathan P. Stewart
Department of Civil and Environmental Engineering, University of California, Los Angeles, Calif., USA

ABSTRACT: Failures of natural slopes, and man-made slopes and embankments commonly occur during major earthquakes and, apart from liquefaction, often constitute the most visible and damaging ground failures. Recently, the 1989 Loma Prieta, 1994 Northridge, 1995 Kobe earthquakes provided a wealth of new data on slope response and slope performance. In particular, the data from Loma Prieta and Northridge provide an excellent opportunity to review site amplification issues. In addition, in all three earthquakes extensive damage occurred as a result of seismically induced deformation of structural fills. The results of recently concluded analytical studies can be used to show the importance of local site conditions on the distribution of ground motions and the resulting ground deformation. Finally, a review of the performance of reinforced soil slopes and walls shows that these types of structures perform very well during earthquakes.

1 INTRODUCTION

The damage caused by failures slopes and fills during major earthquakes can be extensive and at times the failures also lead to a loss of life. Recently, the 1989 Loma Prieta, 1994 Northridge, 1995 Kobe earthquakes provided a wealth of new data on slope response and slope performance. The purpose of this paper is to review three very specific aspects of seismic slope response: topographic site amplification in the vicinity of steep slopes, the behavior of structural fills, and the behavior of reinforced soil slopes and walls.

2 TOPOGRAPHIC AMPLIFICATION IN STEEP NATURAL SLOPES

Topographic amplification is often invoked in order to explain the seismic response of steep slopes. Charles Darwin, in reporting the effects of the February 20, 1835, Chilean earthquake, described narrow ridgetops shattered by ground shaking and suggested that topographic amplification of seismic motions has been a well recognized a phenomenon for some time (Barlow, 1933). Certainly, in the recent past, there have been numerous cases of recorded motions and observed earthquake damage pointing toward topographic amplification as an important effect. Examples include observations

from the 1906 San Francisco Earthquake (Lawson, 1908), 1971 San Fernando Earthquake (Boore, 1972), the 1985 Chile earthquake (Celebi, 1987), the 1987 Superstition Hills Earthquake (Celebi, 1991), and the 1989 Loma Prieta Earthquake (Hartzell et al., 1994; Sitar, 1990, and 1991).

One of the most recent examples of topographic amplification occurred in the coastal bluffs of the Pacific Palisades due to the January 17, 1994, Northridge Earthquake near Los Angeles, California. These coastal bluffs are located approximately 30 km south of the epicenter of the $M_W = 6.7$ earthquake. Strong motion records at the nearby Santa Monica Fire Station indicate a peak horizontal acceleration of 0.93g and a peak vertical acceleration of 0.25g. The failure of these bluffs closed the northbound lanes of the Pacific Coast Highway (State Route 1) for at least 4 days following the earthquake. Four large landslides were observed in this area, along with several smaller slides. One of the large slides carried a portion of a house down the slope, as shown in Figure 1. On properties adjacent to this house, shallow concrete piers and H-piles were observed hanging in mid-air at the crest of the slope, implying that they provided little benefit. The failures occurred in Quaternary age deposits of weakly cemented sand. The slopes on which the failures occurred were 40 to 60 m in height and moderately steep (between 45 and 60 degrees). The

Figure 1: Slope failure in marine terrace deposits in Pacific Palisades as a result of the January 17, 1994, Northridge earthquake.

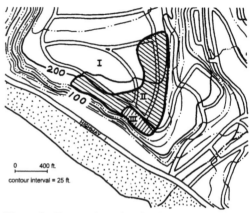

Figure 2: Zones of varying levels of damage on top of the bluff in the Pacific Palisades. Zone III indicates heavy damage, where severe structural damage was observed. Most buildings in this zone were "red-tagged." Zone II indicates a moderate level of damage consisting primarily of fallen chimneys, broken windows, and fallen masonry walls. Zone I indicates slight damage.

failure masses appeared to be only a few yards thick, subparallel to the slope, and had widths on the order of 100 m.

A detailed inventory of the damage sustained in the residential area on the top of the bluff was carried out within days of the earthquake. An overview of the damage inventory is presented in Figure 2. This figure shows three zones of damage. The heavily damaged zone (I) includes houses that sustained severe structural damage. While a portion of one of the homes in this zone slid down the face of the bluff (Figure 1), all the others sustained damaged from the strong shaking. A striking phenomenon in this zone were numerous toppled masonry garden walls generally less than 2 m high. The damage in the moderate damage zone (II) consisted primarily of fallen chimneys, broken windows, and fallen masonry walls. In some cases, the fallen chimneys resulted in damage to the roof. In the slight damage zone (I), essentially no damage was observed.

Figure 2 indicates that the most severe damage occurred within approximately 50 meters of the crest of the slope, a dimension approximately equal to the height of the bluffs. Most of the remaining damage occurred within approximately 100 m of the slope crest. It should also be noted that the most severe damage occurred nearest the steepest slope, and that all of the damage was concentrated in the southern corner of the terrace, essentially due south of the epicenter, indicating some directivity effects in the topographic amplification.

Qualitative understanding of the effects of topography that has developed following observations such as these suggests that earthquake-related damage tends to be more extensive on ridge tops (e.g. Celebi, 1987) and the crest of cliffs (e.g. Ashford and Sitar, 1994). Unfortunately, quantitative data on the effect of topography is generally limited to small strain response from blasting or earthquake aftershocks. In an attempt to improve our understanding of the effect of topography on site response, a considerable amount of work has been done in an attempt to model, quantify, and predict these effects. However, the corresponding attempts to model these effects have generally shown qualitative, but not quantitative, agreement with observed behavior.

Idriss and Seed (1967) conducted one of the first studies to specifically consider the seismic response of soil slopes, prompted by the extensive landslides generated during the 1964 Alaskan Earthquake. They conducted parametric studies of the response of 27-degree clay slopes, and later of 45-degree slopes (Idriss, 1968), using the finite element method. The authors found that magnitude of the peak surface acceleration was in all cases greater at the crest of the slope than at points lower on the slope. However, when comparing the peak surface acceleration at the crest to that at some distance behind the crest, they found that while in some cases the acceleration at the crest was much greater, in other cases that there was little difference between the response at the crest and that at some distance behind the crest. Their results suggest that the natural period of the soil column behind the crest of a slope, was responsible for much more amplification of the input motion than the slope geometry itself. The dominance of the soil amplification for these slopes was later confirmed by Kovacs et al. (1971) through laboratory shake table testing.

Later, Sitar and Clough (1983) used a finite element model to analyze the seismic response of steep slopes in weakly cemented sands and found that accelerations tended to be amplified in the vicinity of the slope face, to 70 percent as compared to the free field behind the crest. However, in contrast to Geli *et al.* (1988), they noted that these topographic effects tended to be small relative to the amplification that occurs in the free field due to the site period.

Ashford *et al* (1997) conducted a study of topographic amplification motivated by an interest in evaluating the response of steep coastal bluffs, such as are found along much of the California coast. Their frequency domain parametric study applied the generalized consistent transmitting boundary method (Deng, 1991) to evaluate the significance of topographic effects on the seismic response of steep slopes (30 to 90 degrees) Both vertically propagating SH- and SV-waves were considered; in addition, wave splitting due to vertical incidence on an inclined slope was incorporated into the analysis. In an attempt to address the relative effect of soil amplification, the responses of a stepped half-space and a stepped layer over a half-space were compared to determine the effect of the fundamental frequency of the material behind the slope crest.

Ashford *et al.* (1997) concluded that topographic effect of a steep slope on the seismic response of that slope can be normalized as a function of the ratio of the slope height (H) and the wavelength of the motion (λ). The relationship between slope height and wavelength was also noted by May (1980) for horizontally propagating SH-waves incident on a vertical scarp, and similar relationships were observed between structure dimension and wavelength by others (e.g. Boore, 1972; Geli *et al.*, 1988; Dakoulas, 1993). In addition, for both SH- and SV-waves, the peak topographic effect occurs at H/λ = 0.2. This amplification is on the order of 25% for SH-waves, and 50% for SV-waves. This first

peak, at H/λ = 0.2, approximately corresponds to the first mode of vibration of a soil column of thickness H (H/λ = 0.25), which is the frequency at which Boore (1972) and Geli *et al.* (1988) observed the peak response in their studies of ridges. Considering the effect of slope angle on response, the topographic effect is most apparent for slopes steeper than 60 degrees, and tends to decrease with slope angle. The initial peak horizontal response occurs at the topographic frequency, and tends to decrease with increasing slope angle (Figure 3).

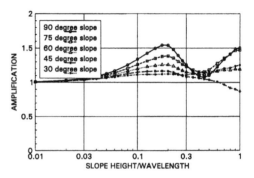

Figure 3. Amplification of horizontal motion at the crest as compared to the free filed behind the crest for a vertically incident SV-wave on an inclined slope in a stepped half space (from Ashford and Sitar, 1997)

For a stepped layer over a half-space, Ashford *et al.* concluded the natural frequency of the site behind the crest dominates the response, which agrees with observations by Sitar and Clough (1983). If the natural frequency of the site is approximately equal to the topographic frequency, i.e. $f_n = f_t$, then that response is amplified. In no case was the topographic effect found to be greater than the response at the natural frequency. Based on these results, it appears that the effect of topography can be handled separately from amplification due to the natural frequency of the layer behind the crest of the slope. This concept of separating the amplification caused by topography from that caused by the natural frequency is advantageous to the development of a simplified method to estimate topographic effects (Ashford and Sitar, 1994). Most importantly, however, these results show that more detailed studies of slope geometries and site stratigraphy, coupled with strong motion recordings, are needed in order to develop a fuller understanding of the role of topographic site amplification during strong shaking.

3 STRUCTURAL FILLS

Previous studies of the seismic performance of compacted earth structures have primarily focused on earth dams, which can pose a significant life safety hazard in the event of failure. Several recent earthquakes, however, have shed light on the seismic performance of hillside fills, a type of earth structure whose performance seldom threatens life safety, but for which even modest deformations often constitute unacceptable performance due to the sensitive nature of the residential construction for which hillside fills are often constructed. In contrast to dams, the study of hillside fills is a relatively recent endeavor brought on by the enormous economic ramifications of movements triggered by the 1994 Northridge Earthquake.

Relatively few studies have focused specifically on the seismic behavior of hillside fills or attempted to document their performance on a broad scale, though several researchers have made note of earthquake-induced deformations of unsaturated fills. Lawson (1908), in summarizing observations of ground cracking in hillside areas from the 1906 San Francisco Earthquake, noted "roadways and artificial embankments were particularly susceptible to ... cracks". In summarizing observations from the 1952 Kern County, 1960 Chilean, and 1957 Hebgen Lake earthquakes, Seed (1967) noted "the effect of earthquakes on banks of well-compacted fill constructed on firm foundations in which no significant increases in pore water pressure develop during the earthquake is characteristically a slumping of the fill varying from a fraction of an inch to several feet".

In the 1971 San Fernando Earthquake, a 12 m thick fill at the Jensen Filtration Plant composed of unsaturated clayey sands and underlain by bedrock underwent 10 to 15 cm settlements which significantly damaged a building constructed on spread footings (Pyke et al., 1975). A systematic survey of distress to single family dwellings from the San Fernando Earthquake (McClure, 1973) noted the strong impact of fills on structural damage patterns, particularly when residences were constructed over cut/fill contacts. Specifically, this study found that "...ground failure occurred on a higher percentage of sites that were on fill or cut and fill than on those sites which were on cut or natural grade" and "dwellings on cut and fill or fill had more relative damage than dwellings on cut or natural grade". In a separate report documenting earthquake effects in residential areas, Slosson (1975) noted that fills constructed after the implementation of relatively stringent grading codes in 1963 performed markedly better than pre-1963 fills. Incidents of hillside fill movements during the 1989 Loma Prieta Earthquake have been reported by several consultants; however, this information has not been compiled and little published information is available.

Most recently, however, the 1995 Kobe Earthquake caused extensive deformations of sidehill road embankments, deep structural fills, and sidehill fill pads for single family structures (Sassa et al., 1996; Sitar, 1995). These fills were largely composed of compacted decomposed granite soil. The damage ranged from differential settlement, to downhill extension and slumping depending on the fill configuration and slope geometry. In many instances the fill deformation was large enough to cause damage to underground utilities, sewers and water supply pipes in particular.

Figure 4. Deep extension cracks in a sidehill fill in Rokko Mountains above Kobe.

In order to analyze this phenomenon, Stewart et al. (1995) investigated over 250 sites affected by the 1994 Northridge Earthquake. Fill deformations were found to generally involve modest local displacements (i.e. < 10 cm), but such movements resulted in significant cumulative economic losses due to the large number of affected sites (over 2000) and the high cost of damages at these sites (often $50,000 to $100,000 per site). A number of characteristic surface deformation patterns were observed as illustrated in Fig. 5 and listed below:

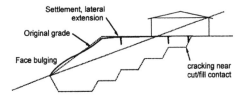

Figure 5. Schematic showing typical damage to fill slope (from Stewart *et al.*, 1995).

Cracks near cut/fill contacts: This was the most common location of ground cracking in building pads, with most cracks typically having less than 8 cm of lateral extension and 3 cm of localized differential settlement of the fill relative to the cut. Damage to str ⌐ these features was often significant (e.g. Fig. 6).

Figure 6. Cracked floor slab across cut/fill contact. Displacements are 1.9 cm (V) and 5 cm (H) (from Stewart *et al.*, 1995).

Figure 7. Evidence of extensional ground movement at back of a house. Top of slope is to the left (from Stewart *et al.*, 1995).

Lateral extension in fill pad: Tensile cracking was commonly observed in fill pads parallel to the top of slope (e.g. Fig. 7). While these crack widths and separations were typically about 2.5-10 cm, significantly larger cracks (up to 30 cm wide) were observed at some sites. The setback of tensile cracking from the top of slope tended to increase with fill depth, and houses constructed at a setback of one-third of the slope height were generally not damaged by this cracking.

Differential settlement: Fill pad settlements increased with fill depth, resulting in differential settlements across the surface of fills. These settlements often caused 1.0 to 1.5% slopes in house floors, significantly exceeding settlement standards for houses set forth by Los Angeles County, which allow for 1.25 cm settlement in 9 m (0.14% slope) (Pearson, 1995).

Face bulging: Slope face bulging was evident from movements of concrete surface drains running cross-slope (terrace drains) and down-slope (downdrains). Terrace drains were observed to have cracks oriented perpendicular to the slope contours which widened in the downslope direction, providing evidence for face bulging of the center of the fill, while downdrains had been uplifted at approximately one-third the slope height, indicating lower slope face compression or bulging.

A number ground failure mechanisms may have caused these surface deformations. Some fill movements were traced to permanent deformations of soils or rock underlying the fill (i.e. landsliding or liquefaction in these materials). However, most fill movements appeared to result from modest deformations within the fill mass itself, and can generally be attributed to the accumulation of plastic deviatoric and/or volumetric strains in the fill under seismic loading. The viability of these deformation mechanisms has been investigated using two-dimensional, finite element seismic response analyses employing an equivalent-linear characterization of dynamic soil properties (i.e. QUAD4M by Hudson, et al. 1994). Results of these analyses showed that the amplitudes of dynamic shear strains in fill are strongly dependent on the fill/base material impedance contrast. Further, for typical fill/bedrock shear wave velocities, it was found that the dynamic shear strains in typical slope geometries (subjected to Northridge strong motion recordings) were of a sufficiently high amplitude that significant elasto-plastic deviatoric and volumetric deformations in the fill soils were likely to have occurred (Stewart *et al.*, 1995). Hence, the manifestation of these strains was likely to have caused the damaging ground surface deformations.

While design and construction standards for hillside fills have evolved significantly over the last 50 years, these advances have primarily been directed towards addressing "static" stability concerns (e.g. settlement and landsliding), with little emphasis given to seismic considerations. Based on the poor performance of many hillside fills during recent earthquakes, including a number of fills constructed to relatively modern standards, further advances in the standards of practice appear to be warranted.

4 REINFORCED SOIL SLOPES AND WALLS

Reinforced soil slopes and walls are relatively new types of structures and, given the relatively slow recurrence of earthquakes, there is a need to develop a data base of observed seismic performance. Such data base can then be used to further develop and validate design procedures. Interestingly, many reinforced soil walls have not been specifically designed to withstand seismic loading, although they have generally performed well during earthquakes. A compilation of published data on the seismic performance of reinforced soil slopes and walls is presented in Table 1. This information has been compiled as a part of a larger study of seismic response of reinforced soil slopes and walls currently in progress and only the highlights of the observed performance are presented herein.

1976 Gemona Earthquake, Italy

An earthquake with Richter magnitude 6.4 shook the town of Gemona in northern Italy in May of 1976. Three 4 to 6 meter Reinforced Earth™ walls were located 25 to 40 kilometers from the epicenter. The walls were designed for only static conditions with a minimum factor of safety for pullout. No movements were reported for any of the walls.

1983 Honshu Earthquake, Japan

In May of 1983 an earthquake with Richter magnitude of 7.7 struck northern Honshu. The epicenter was located in the Sea of Japan west of the island. Maximum accelerations of 0.1g to 0.3g were recorded at a distance of 140 kilometers west of the epicenter. Liquefaction damage was also reported. Forty-nine Reinforced Earth™ structures of varying heights were located 80 to 275 kilometers from the epicenter. The walls showed no signs of damage except for one wall located closest to the epicenter.

Table 1. Seismic field performance of reinforced soil slopes and walls.

Earthquake/ Country	Year	Magn. (M$_L$)	Dist. to Epicenter (km)	Approx. Horiz. Acc. (g)	# of Walls	Wall Type	Wall Height (m)	Observations
Gemona Italy	1976	6.4	25-40		3	RE™	4-6	no signs of damage
Leige, Belgium	1983	5	0.8	0.15-0.2	2	RE™	4.5-6	no signs of damage
Honshu Japan	1983	7.7	80-275	0.1-0.3 *at 140 km*	49	RE™		no signs of damage* *one wall had few cms. of settlmt
Edgecumbe New Zealand	1987	6.3	30		1	bridge abutment	6	(unsecured) deck "danced" during EQ no signs of damage
Loma Prieta CA	1989	7.1	11-100	0.1-0.55	20	RE™	5-10	no signs of damage
			11-130	0.1-0.4	sev	geogrid	3-24	no signs of damage
			65	0.1	1	geogrid	21	La Honda slope: 2%H movement at top
Northridge CA	1994	6.7	2.5-84	0.1-0.9	20	RE™	4-17	panel spalling and minor cracking
			61	0.1	1	RE™	16	bulged at center (3%H)
			8-113	~0.2	sev	geogrid	3-15	no signs of damage
			19	0.35	1	MSE	12	continuous cracking, 2.5cm diff settlmt
Hyogoken-Nanbu Japan	1995	6.9 (M$_w$)	16-40	up to 0.8	3	fiber grid	3-8	no signs of damage
			16	up to 0.8	1	fiber grid	6	wall deformed about 30cm minor panel spalling and cracking

Although this wall settled a few centimeters relative to nearby pile-supported structures, it remained serviceable.

1983 Liege Earthquake, Belgium

In November of 1983, a shallow earthquake with Richter magnitude 5 occurred in Liege, Belgium. The earthquake, which initiated at a depth of roughly 4 kilometers, caused extensive damage to local structures. Two Reinforced Earth™ walls with heights of about 4.5 to 6 meters were located half a mile from the epicenter. Accelerations ranging from 0.15g to 0.2g were recorded near the walls, and caused 3 centimeters of lateral movement to the roof of a garage nearby. The walls showed no signs of damage.

1987 Edgecumbe Earthquake, New Zealand

The town of Edgecumbe was struck by a 6.3 Richter magnitude earthquake on March of 1987. The Maniatutu bridge, located 30 kilometers from the epicenter, was under construction at the time. Reinforced Earth™ walls about 6 meters tall were used for the bridge abutments (*Figure 20; RE Co., 1995*). The deck of the bridge had been placed but was not yet secured, and the backfilling of the walls had not been completed, leaving them with little support. During the earthquake, workers reported that the deck "danced" on its supports. However, no noticeable deformations were observed, and the bridge was completed.

1989 Loma Prieta Earthquake, California

On October 17, 1989, part of the San Andreas fault slipped 16 kilometers northeast of Santa Cruz, California. The earthquake measured 7.1 on the Richter scale and caused high levels of ground accelerations ranging from 0.5g to 0.6g. Twenty Reinforced Earth™ walls (RE Co., 1990) with heights of 5 to 10 meters were affected and were located 11 to 100 kilometers from the epicenter. Only two walls with heights of 5.5 and 6 meters were designed to withstand lateral loads of 0.2g and 0.1g, respectively. Both of the walls were located about 100 kilometers from the epicenter. There was no evidence of damage to any of the walls.

Likewise, embankments reinforced with geogrids with heights of 3 to 24 meters were affected by the earthquake (Collin et al., 1992). They were located 11 to 130 kilometers from the epicenter and

designed for about 0.15g. The embankments experienced maximum horizontal accelerations of 0.1g to 0.4g but showed no signs of distress. The La Honda slope, located 65 kilometers from the epicenter, did experience some movement. This 21-meter slope was reinforced with geogrids. Figure 8 shows a typical section of the La Honda reinforced soil slope.

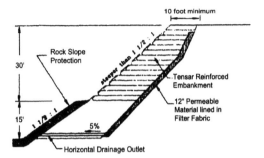

Figure 8. Typical Section through the La Honda Reinforced Soil Slope (after Collin et al., 1992).

In 1982-83, the slope had experienced landslides due to a series of heavy storms. The wall was repaired and showed no sign of distress until after the Loma Prieta earthquake. Lateral loads were estimated to be about 0.1g. Inclinometers of the California Department of Transportation showed movements less than 2 cm (2% H) in the upper wall, as shown in Figure 9 (Collin *et al.*, 1992).

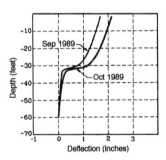

Figure 9. Measured Deflections of the La Honda Reinforced Soil Slope (after Collin *et al.*, 1992)

1994 Northridge Earthquake, California

The Northridge Earthquake had a Richter magnitude of 6.7 and occurred on January 17, 1994, in Southern California. There were 21 Reinforced Earth™

347

structures affected (RE Co., 1994). The walls were 4 to 17 meters tall and located 2.5 to 84 kilometers from the epicenter. Estimated peak horizontal and vertical ground accelerations ranged from 0.1g to 0.9g and 0.03g to 0.62g, respectively. The walls performed well with only minor problems such as panel spalling and minor cracking. Only one 16-meter wall, which was located 61 kilometers from the epicenter, bulged 46 centimeters (3%H) at the wall face due to excessive settlement.

The performance of geogrid-reinforced soil structures, reinforced soil earth walls, and soil-nailed excavations was also good (Stewart *et al.*, 1994). These walls were located 8 to 113 kilometers from the epicenter, and are estimated to have experienced a peak ground acceleration of about 0.2g (Sandri, 1994). Wall heights ranged from 3 to 15 meters. One 12-meter MSE wall, located about 19 kilometers from the epicenter, experienced continuous cracking, with about 2.5 centimeters of differential settlement (Stewart *et al.*, 1994). The cracking was parallel to the wall and appeared at distances of 2 and 7.5 meters behind the wall face. The crack farthest from the wall face occurred at the transition point between the retained backfill and the natural ground. Nearby recording stations indicated a maximum horizontal acceleration of 0.35g.

1995 Hyogoken-Nanbu Earthquake, Japan

An earthquake with a moment magnitude of 6.9 and located 20 kilometers southwest of downtown Kobe occurred on January 17, 1995. The performance of four 3 to 8 m high geogrid reinforced walls was described by Tatsuoka *et al.* (1996). The walls were pseudo-statically designed using a horizontal inertial acceleration of 0.2g. Three of the walls suffered no visual damage. However, the fourth wall did incur damage. This wall, known as the Tanata wall, was built using fiber-grid reinforcements and a cohesionless backfill (Tatsuoka *et al.*, 1996). After experiencing horizontal ground accelerations of up to 0.8g, the top and base of the wall deformed 30 cm horizontally. Fortunately, the wall was able to avoid catastrophic failure. It is speculated that the movements were probably due to the relatively short ties at the top of the wall (0.4H). The wall also suffered minor face panel spalling and cracking.

Overall, as can be seen from the accumulated data, reinforced soil walls perform very well during seismic loading. This good performance may be partly due to the inherently conservative design of reinforced soil structures. In particular, catastrophic failures have not occurred and the deformed structures generally remained functional.

Figure 10. Lateral displacement of the Tanata Wall.

CONCLUSIONS

Though recent research and observations show that the effect of topography on site response can be significant, it is becoming clear that these effects can be overshadowed by soil amplification. It is vital that strong motion data be obtained from actual earthquake to verify these trends, and that separate contributions from soil and topographic amplification be quantified for future as well as past earthquake records. Only with careful planning and forethought on proper instrument location can the effect of topography alone to quantified.

The damage caused by the deformations of structural fill during recent earthquakes points to the significance of this latent hazard in heavily developed urban areas. At this point, there is a need to develop simplified analytical techniques for estimating fill movements that are calibrated against field case histories. On an even more fundamental level, however, there is a need for an improved understanding of the behavior of compacted, unsaturated soil under cyclic loading.

Finally, the performance of reinforced soil slopes and walls in recent earthquakes suggest that these types of structures are well suited for seismically active regions. However, further studies are needed to develop procedures for evaluation of the degree of conservatism apparently inherent in their design and to develop criteria for estimating seismically induced deformations.

REFERENCES

Ashford, S.A. and Sitar, N. (1994). *Seismic Response of Steep Natural Slopes*, Report No. UCB/EERC 94-05, Earthquake Engineering Center, College of Engineering, University of California, Berkeley, CA.

Ashford, S.A., Sitar, N. Lysmer, J., and Deng, N. (1997). "Topographic Effects on the Seismic Response of Steep Slopes," Bulletin of the Seismological Society of America, 87(3).

Ashford, S.A. and Sitar, N. (1997). "Analysis of Topographic Amplification of Inclined Shear Waves in a Steep Coastal Bluff," Bulletin of the Seismological Society of America, 87(3).

Barlow, N. (1933). *Charles Darwin's Diary of the Voyage of H.M.S. Beagle*; Nora Barlow, editor, Cambridge University Press.

Boore, D. M. (1972). "A Note on the Effect of Simple Topography on Seismic SH Waves," *Bull. Seis. Soc. Am.*, 62(1) 275-284.

Celebi, M. (1987). "Topographic and Geological Amplification Determined from Strong-Motion and Aftershock Records of the 3 March 1985 Chile Earthquake," *Bull. Seis. Soc. Am.*, 77(4) 1147-1167.

Celebi, M. (1991). "Topographic and Geological Amplification: Case Studies and Engineering Implications," *Structural Safety*, 10 (1991) 199-217.

Collin, J.G., Chouery-Curtis, V.E., and Berg, R.R., "Field Observations of Reinforced Soil Structures under Seismic Loading," *Earth Reinforcement Practice*, Balkema, Rotterdam, ISBM 90 5410 093 1, January 1992.

Dakoulas, P. (1993). "Earth Dam-Canyon Interaction Effects for Obliquely Incident SH Waves." *J. Geotech. Engrg.*, ASCE, 119(11) 1696-1716.

Deng, N. (1991). "Two-Dimensional Site Response Analyses", thesis presented to University of California at Berkeley, Berkeley, California in partial satisfaction of the requirements for the degree of Doctor of Philosophy.

Geli, L., Bard, P-Y., and Jullien, B. (1988). "The Effect of Topography on Earthquake Ground Motion: A Review and New Results," *Bull. Seis. Soc. Amer.*, 78(1) 42-63.

Hartzell, S. H., Carver, D. L., and King, K. W. (1994). "Initial Investigation of Site and Topographic Effects at Robinwood Ridge, California," *Bull. Seis. Soc. Amer.*, 84(5) 1336-1349.

Hudson, M., Idriss, I.M., Beikae, M. (1994). "QUAD4M: A computer program to evaluate the seismic response of soil structures using finite element procedures and incorporating a compliant base," Ctr. for Geotech. Modeling, Univ. of California, Davis.

Idriss, I. M. and Seed, H. B. (1967). "Response of Earthbanks During Earthquakes," *J. Soil Mech. and Found. Div.*, ASCE, 93(SM3) 61-82.

Idriss, I.M. (1968). "Finite Element Analysis for the Seismic Response of Earth Banks," *J. Soil Mechanics and Foundations Division*, ASCE, 94(SM3), 617-636.

Kovacs, W.D., Seed, H.B. and Idriss, I.M. (1971). "Studies of Seismic Response of Clay Banks," *J. Soil Mechanics and Foundations Division*, ASCE, 97(SM2), 441-445.

Lawson, A.C., ed. (1908). "Minor geologic effects of the earthquake," *in* California Earthquake of April 18, 1906, *Carnegie Institution of Washington, D.C.*, Publication 87, Vol. 1, Part 2, 384-409.

McClure, F.E. (1973). "Performance of Single Family Dwellings in the San Fernando Earthquake of February 9, 1971," U.S. Dept. of Commerce, NOAA, May.

Pearson, D. (1995). "L.A. County 'Section 309' Statement," *Geogram*, California Geotech. Engineers Assoc.

Pyke, R., Seed, H.B., Chan, C.K. (1975). "Settlement of Sands Under Multidirectional Shaking," *J. Geotech. Engrg. Div.*, ASCE, 101 (4), 379-398.

Reinforced Earth Company, "An investigation of Reinforced Earth Structures Impacted by the Loma Prieta Earthquake," February 1990.

Reinforced Earth Company, "Performance of the Reinforced Earth Structures near the Epicenter of the Northridge Earthquake: January 17, 1994," April 1994.

Sandri, D. (1994). "Retaining Walls Stand Up to the Northridge Earthquake," *Geotechnical Fabrics Report*, June/July.

Sassa, K., Fukuoka, H., Scarascia-Mugnozza, G., and Evans, S. (1996). "Earthquake-Induced-Land-slides: Distribution, Motion, Mechanisms," in Special Issue on Geotechnical Aspects of the January 17, 1995, Hyogoken- Nambu Earthquake, *Soils and Foundations*, Japanese Geotechnical Society, Jan., pp. 53-64.

Seed, H.B. (1967). "Soil Stability Problems Caused by Earthquakes," Soil Mechanics and Bituminous Materials Research Laboratory, Univ. of California, Berkeley.

349

Sitar, N. and Clough, G.W. (1983). "Seismic Response of Steep Slopes in Cemented Soils," J. Geotech. Engrg. Division, ASCE, 109(GT2), 210-227.

Sitar, N. (1990). "Seismic Response of Steep Slopes in Weakly Cemented Sands and Gravels," *Proc., H. Bolton Seed Memorial Symposium*, BiTech Publishers, Vancouver, B.C., Vol. II, 67-82.

Sitar, N. (1991). "Earthquake-Induced Landslides in Coastal Bluffs and Marine Terrace Deposits," in *Special Publication No. 1, Loma Prieta Earthquake*, Assoc. of Eng. Geologists, 75-82.

Sitar, N., editor (1995). "Geotechnical Reconnaissance of the Effects of the January 17, 1995, Hyogoken-Nanbu Earthquake, Japan," *A Report for the National Science Foundation*, Report No. UCB/EERC-95/01.

Slosson, J.E. (1975). "Chapter 19, Effects of the Earthquake on Residential Areas," *San Fernando, California, Earthquake of 9 February 1971*, CA Div. Mines and Geology, Bulletin 196.

Stewart, J.P., Bray, J.D., Seed, R.B., and Sitar, N. (1994). "Preliminary Report on the Principal Geotechnical Aspects of the January 17, 1994 Northridge Earthquake," *Earthquake Engineering Research Center*, Report No. UCB/EERC-94/08, June.

Stewart, J.P., Bray, J.D., McMahon, D.J., and Kropp, A.L. (1995). "Seismic performance of hillside fills," in Landslides under Static and Dynamic Conditions - Analysis, Monitoring, and Mitigation, *Geotech. Special Publication No. 52*, ASCE, C.L. Ho and D.K. Keefer (editors), October, 76-95.

Tatsuoka, F., Tateyama, M., and Koseki, J., (1996). "Performance of Soil Retaining Walls for Railway Embankments," in Special Issue on Geotechnical Aspects of the January 17, 1995, Hyogoken- Nambu Earthquake, *Soils and Foundations*, Japanese Geotechnical Society, Jan., p. 311-324.

Seismic Behaviour of Ground and Geotechnical Structures, Sêco e Pinto (ed.) © 1997 Balkema, Rotterdam, ISBN 90 5410 887 8

Seismically induced displacement of retaining walls

R.S. Steedman
GIBB Ltd, Reading, UK

ABSTRACT: Field evidence of the behaviour of retaining walls in recent earthquakes has continued to show a wide range of performance, implying that despite widespread adoption of force-based (threshold) design approaches, techniques for the prediction of seismically induced wall displacement are less well developed. The background and limitations of force-based and displacement approaches are discussed in the light of the Japanese experience following the Hyogoken-Nambu (Kobe) earthquake in 1995, where a large proportion of caisson walls throughout the port of Kobe were severely damaged. The nature of seismically induced wall displacement is described and techniques are outlined for use in design.

1 INTRODUCTION

The adequacy of design provisions for retaining walls under earthquake loading was thrown into sharp focus again by the extensive damage to quay walls in the port of Kobe during the Hyogoken-Nambu earthquake on January 17, 1995. Most of the caisson walls throughout the port showed large permanent movements seaward, combined with settlement and rotation. Movements were reported as being up to 5m maximum, 3m on average, with settlements behind of the same order, Inagaki et al. (1996). This sort of damage, to quay walls in particular but also to other large retaining walls, has been seen at many other previous earthquake sites.

Standard practice in the earthquake design of retaining walls is to adopt the well-known pseudo-static Mononobe-Okabe calculation, following Coulomb, and to solve for the critical wedge angle and hence the maximum "dynamic" force on the wall. Most calculations follow a force based approach, with the implication that displacements under earthquake loading should be small.

In practice retaining walls, and particularly quay walls, commonly show significant movement after earthquakes. The evidence suggests that in practice, their behaviour is dominated by displacements rather than by stresses and it is therefore clear that simple reliance on force based approaches is not a reliable indicator of the ultimate performance of a retaining wall; displacement based approaches give a better indication of their ultimate capability (ie. the onset of failure). Although this has been recognised for some time standard design procedures are not yet widely available (with the exception of the Newmark approach for walls in sliding).

However, although these methods are inherently attractive because of their simplicity, in practice few retaining walls meet the assumptions which are implicit in their use. In the following sections, the relevance of simple pseudo-static approaches will be considered in relation to problems frequently encountered in design, such as amplification up the height of large walls, softening of the backfill and outward rotational and translational movements.

2 LIMITATIONS OF THE MONONOBE-OKABE APPROACH

There are three key aspects in the design of retaining walls for earthquake loading over which there has been considerable debate in the literature :

(a) the nature of the dynamic earth pressures on the retaining wall - during base shaking the forces acting to destabilise the wall will vary with elevation and time as functions of the relative stiffness, inertia and strength of the wall, backfill and foundation;

(b) the nature of the hydrodynamic pressures on the wall - which may exist on one or both sides of the wall in backfill or in free water;

(c) the evaluation of the likely recoverable and irrecoverable (ie residual) displacements.

The first studies of earthquake-induced earth pressures on a retaining structure were reported in Japan by Okabe (1924) and Mononobe and Matsuo (1929) who developed a pseudo-static calculation for the earth pressure on a retaining wall following the limiting equilibrium approach of Coulomb. (The Mononobe-Okabe equations are summarised in many references, see for example Seed and Whitman (1970), Krinitzsky et al. (1993), Chapter 15, or Ebeling and Morrison (1992), and therefore are not presented here.) This calculation has been the subject of many model experiments, but although most of the test data agree with the Mononobe-Okabe solution for the total seismic force, the acting point of the dynamic force has been found to vary considerably, eg. Seed and Whitman (1970), and this is discussed below in the context of analysis of large walls.

The hydrodynamic pressures on a quay wall have also been investigated. Steedman and Zeng (1990a) showed that the hydrodynamic pressure on the seaward side can be estimated using Westergaard's (1933) solution if the stiffness of the soil-wall system is high. If the wall is flexible, however, then significantly larger forces can be expected. The calculation of the hydrodynamic pressure on the backfill side is much more complicated. Here movement of the fluid is no longer free but is restricted by soil particles, and the hydrodynamic pressure therefore depends not only on the interaction of the fluid and the structure but also on the movement of fluid through the voids in the soil. The magnitude of the hydrodynamic pressure is likely to be related to the permeability of the soil, and Matsuzawa et al (1985) present an approach to determine whether the water in the backfill can be considered as 'free' or 'restrained'. Several approaches can be found in the literature for the assessment of the magnitude of the hydrodynamic force (see for example Amano et al (1956), Matsuo and Ohara (1960), Steedman and Zeng (1990a)).

In practice, designers may opt to compute the peak dynamic lateral force from the fill using both free and restrained water solutions and compare the two, which are unlikely to differ greatly. In the free water case, the hydrodynamic force acting through the fill should be calculated based on Westergaard but factored by the porosity (see Steedman and Zeng, 1990a), and added to the dynamic soil force which (in this case) is a function of the inertia of the soil skeleton. In the restrained water case, there is no hydrodynamic component (no 'free' water) but the dynamic soil force is proportional to the saturated or total unit weight and this gives a significantly larger 'seismic angle' (the ratio of the horizontal to the vertical inertia forces) and consequently a larger lateral earth pressure coefficient. This can be seen by comparing the seismic angles for the free and restrained water cases, θ_f and $_r\theta$ respectively, Equation (1), where γ_t , γ_d and γ_b are the total, dry and bouyant unit weights of the soil and k_h, k_v are the horizontal and vertical acceleration coefficients. The difference in the resulting lateral earth pressure coefficients increases with k_h, with the

restrained case being typically around 40% larger than the free case for $k_h = 0.25$, which broadly compensates for the effects of summing the lateral force due to the soil skeleton and the hydrodynamic force from the pore fluid (in the free case).

$$\theta_r = \tan^{-1}\left[\frac{\gamma_t \ k_h}{\gamma_b \ (1-k_v)}\right]$$

$$\theta_f = \tan^{-1}\left[\frac{\gamma_d \ k_h}{\gamma_b \ (1-k_v)}\right] \tag{1}$$

Data from model tests, such as Whitman and Ting (1994) confirm the earlier studies by Steedman and Zeng (1990a) that this approach to the hydrodynamic pore water force and the computation of the lateral force is reasonable, provided consideration is also made of the effects of amplification, discussed below.

For most retaining walls only a limited amount of permanent displacement is acceptable. It is helpful to distinguish between two mechanisms for the development of such displacement under seismic loading. First, the dynamic earth pressure on the wall during base shaking causes both elastic and plastic deformation in the backfill, leading to recoverable and irrecoverable (residual) displacements of the retaining wall, even in the absence of wall 'yield'. The magnitude of this residual displacement depends largely on the stiffness of the retaining wall and the intensity of base shaking. Such residual displacement is likely to be small. Large permanent displacements require a second stage, involving yielding of the wall or its foundation, to take place. This may occur through sliding, toppling, bearing failure or internal overstressing, depending on the nature of the wall, and a number of examples are considered below together with procedures for assessing the magnitude of such displacements.

The Mononobe-Okabe approach is valuable in providing a good assessment of the magnitude of the peak dynamic force acting on a retaining wall. However the method is based on three fundamental assumptions:

(i) The wall has already deformed outwards sufficiently to generate the minimum (active) earth pressure.

(ii) A soil wedge, with a planar sliding surface running through the base of the wall, is on the point of failure with a maximum shear strength mobilised along the length of the surface.

(iii) The soil behind the wall behaves as a rigid body so that accelerations can be assumed to be uniform throughout the backfill at the instant of failure.

With the possible exception of massive embedded structures, condition (i) is generally satisfied. Although values quoted in texts vary, a detailed study by Bolton (1991) shows clearly that 90% of active conditions are reached by outward movements as small as 0.1% of the wall height in dense sands, somewhat more in loose sands. Planar sliding surfaces have been observed in model tests reported by Murphy (1960) and Bolton and Steedman (1985) for walls which translate outwards. However, the Mononobe-Okabe approach may also provide useful guidance for the limiting forces on a wall constrained to rotate and not to slide. Similar values to the Mononobe-Okabe solution for the dynamic lateral earth pressure coefficient may be calculated by considering the equilibrium of stresses, without appeal to the kinematics of sliding wedges, provided that the wall is rough and the soil is in limiting equilibrium (Bolton and Steedman, 1982), and this has been confirmed by model tests, (Steedman, 1984).

An assumption of uniform acceleration thoughout the backfill is an assumption about not only the magnitude but also the phase of the acceleration. It would only hold if the soil were rigid and the shear wave velocity infinite. In practice, the finite shear modulus in the backfill, reducing towards the ground surface in a cohesionless backfill, will cause a phase change and an amplification of motion between the motion at the base of the wall and the motion at the ground surface. Clearly this will have an impact on the dynamic lateral pressure.

The effect of amplification on the Mononobe-Okabe solution was analysed by Steedman and Zeng (1990b) where it was shown that common levels of amplification in the backfill have a qualitatively similar effect to an increase in the base input lateral acceleration coefficient. Based on this finding, and comparisons with extensive model test data, it is considered that the mean of the ground surface and base input lateral acceleration coefficients can provide an appropriate and conservative estimate of the equivalent 'uniform' lateral acceleration for use with the Mononobe-Okabe solution. For large walls, accelerations may be calculated at several elevations over the full height of the wall and a mean acceleration may then be determined from this profile. The effect of phase changes and amplification on the distribution of the dynamic lateral earth pressure is developed further below.

3 FIELD PERFORMANCE: QUAY WALL BEHAVIOUR DURING THE 1995 HYOGOKEN-NAMBU EARTHQUAKE

The extensive damage to the caisson type quay walls in the port of Kobe during the January 17, 1995 Hyogoken-Nambu earthquake has been widely reported, see for example Japan Ministry of Transport (1995), Earthquake Engineering Research Center (1995), EQE International (1995). Detailed numerical

and physical model studies have now been conducted to investigate the observed mechanisms of behaviour, particularly of the caisson walls on Port Island where there was a vertical seismic array (Inagaki et al. 1996). A typical example of the forward movement of a caisson wall is shown in Fig. 1.

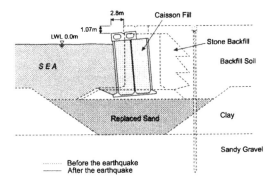

Figure 1. Sketch of outward movement of caisson wall at Port Island, Kobe (after Inagaki et al. 1996).

From the work by Inagaki et al. it is clear that degradation of the strength and stiffness (but not necessarily liquefaction) of the underlying foundation materials due to a progressive increase in the excess pore water pressures during the shaking is sufficient to explain the large permanent outward wall movements. The authors note that the extent of displacement (3m outward movement on average) appears to be closely correlated to the thickness of the underlying fill (termed 'replacement fill' as it is fill placed in a dredged cut), and conclude from close examination of the zones of shearing in the finite element and physical models that the behaviour is associated with deformation of the foundation material and not sliding between the base of the caisson and the foundation, as widely assumed in conventional pseudo-static models.

The assumption of a sliding mechanism with outward movement driven by liquefaction in the backfill is readily understandable and widely quoted as an explanation of the observed behaviour in most field reports of the earthquake damage. However, the modelling carried out by the Port and Harbour Research Institute shows clearly that this initial interpretation should be reconsidered, as it may have a misleading influence on design approaches.

In common with other standards, the Japanese Technical Standard for design of quay walls follows a simple force based approach and recommends a factor of safety of 1.0 in sliding and 1.1 in rotation under 'special conditions', which would cover earthquakes, Japan Ministry of Transport (1991) . The exact level of earthquake design of the caisson walls is uncertain (the design seismic coefficient being based partly on a coefficient of importance for the structure), but it is clear that the level of ground shaking which these walls experienced (0.54g horizontal and 0.45g vertical peak ground acceleration) was considerably in excess of any code requirement, and yet despite the large movements, it is reported that "no collapse or overturning of any wall section took place", Inagaki et al. (1996), or in other words, the damage was regarded as tolerable and not catastrophic.

Across the bay in Kobe, large dock walls retaining saturated fill were also subject to similar strong ground motions. Generally older structures, several of these dock walls predated any earthquake provision and relied generally on their massive construction to retain the soil behind. Although movement took place, and cracking led to leakage which was sufficient to fill at least one dock over a period of days, these walls did not suffer the forward movement or rotation exhibited by the caisson structures. Analysed by the Mononobe-Okabe method, it would be expected that both the dock walls and the caisson walls would have a threshold in sliding or rotation considerably less than the maximum lateral acceleration experienced during the earthquake. Both would be predicted to show substantial movements. However, where the caisson walls typically moved several metres, the dock walls moved a maximum of the order of 150 mm, with most damage confined to interfaces between major structural elements, such as the wall and entrance structures.

This example illustrates many of the difficulties in the application of pseudo-static methods to seismic retaining wall design. Whether designed for modest earthquake loading or not, all these walls were subject to forces considerably in excess of their design limit for lateral loading. A simple force-based design approach fails to distinguish adequately between the predicted performance of the different types of structure. It is clearly necessary to include, in the design approach, consideration of the potential gross displacement mechanisms, such as rocking, rotation and settlement (due to loss of bearing stiffness), internal overstressing or sliding on defined slip planes. It is also clear that some further guidance must be developed on how to accommodate excess pore pressures and ground softening.

4 DISPLACEMENT METHODS

As noted above, displacement of retaining walls under earthquake loading involves both recoverable and irrecoverable elements. Following the logic of the pseudo-static approach, the onset of permanent deformation should be a reflection of accelerations exceeding a yield or threshold value dependant on the initial parameters: geometry, strength, density etc.

For sliding, the approach proposed by Newmark (1965) is well established and widely used in design. In his Rankine lecture, Newmark noted that the slip displacement of a block on a plane subjected to lateral accelerations could be calculated by integrating the relative velocity of the block and the plane, having first defined a yield acceleration. Newmark had in mind modelling the behaviour of earth dams or embankments but this approach was subsequently extended to form the basis of a design approach for retaining walls by Richards and Elms (1979). The Richards and Elms approach followed the work of Franklin and Chang (1977), who had systematically integrated a large database of earthquake records using the Newmark approach as a function of the ratio of the threshold lateral acceleration to the peak value. Franklin and Chang scaled the earthquake records to a peak acceleration of 0.5g and peak velocity of 30 inches/sec, giving a 'standardised' relative displacement for each value of threshold acceleration and for each earthquake record.

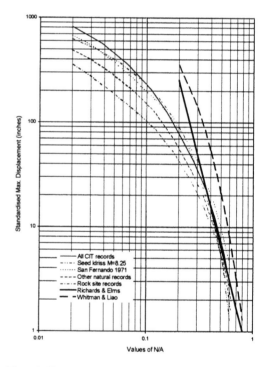

Figure 2. Upper bound envelopes to integrated displacements for natural and synthetic records, compared with Richards & Elms and Whitman & Liao design lines (after Franklin and Chang, 1977)

The new design approach developed by Richards and Elms was based on the view that it would be uneconomic to design walls which were so massive that no movement took place under earthquake loading, but that a realistic assessment could be made of the likely movement for walls with threshold accelerations slightly less than the peak by using an envelope relationship to the integrated earthquake records.

The expression proposed by Richards and Elms for the standardised maximum displacement, d, of a wall in sliding, as an approximation to Franklin and Chang's curves (Imperial units are used here as in the original paper) was given as:

$$d = 0.087 \ \frac{V^2}{A} \left(\frac{N}{A}\right)^{-4} \quad (\text{ inches }) \quad (2)$$

where V is the design earthquake peak velocity in inches/sec, N is the threshold acceleration for sliding (inches/sec^2) and A is the peak lateral ground acceleration (inches/sec^2). At high values of N/A, this relationship provides a reasonable upper bound to the actual records, as shown in Fig. 2. It should be noted that at low values of N/A, Equation (2) overpredicts the likely sliding displacement significantly.

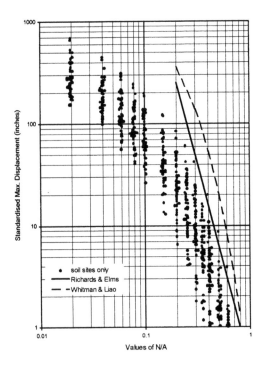

Figure 3. Range of predictions of permanent displacement for San Fernando, 1971 earthquake (after Franklin and Chang, 1977)

However, it is clear that such a calculation is sensitive to a number of key parameters in the prediction of the threshold acceleration and in the nature of the earthquake record. Indeed the scatter of results from which the upper bounds in Fig. 2 were drawn is large, as is shown by the different data points

obtained by Franklin and Chang from the integration of time histories of acceleration from just one single earthquake, the San Fernando earthquake of 1971, Fig. 3. Whitman and Liao (1985) developed this approach further, building on the work carried out at MIT, by analysing a selection of fourteen earthquake records (twelve from earthquakes with magnitudes between 6.3 and 6.7 and two larger) taking into account the effects of orientation, vertical motion and the nature of the site where possible. A simple expression was first derived for the mean displacement, $\bar{d}$, as a function of N/A, analogous to Equation (2), to match the results of Franklin and Chang (input data as for Equation 2):

$$\bar{d} = \frac{37 \ V^2}{A} \ e^{\ -9.4 \ \frac{N}{A}} \quad (\text{ inches }) \quad (3)$$

This expression then formed the basis for a statistical analysis of the likely sources of error, leading to an alternative recommended design line based on a safety factor of 4, equivalent to a probability of non-exceedance in excess of 95% :

$$\frac{N}{A} = 0.66 - \frac{1}{9.4} \ \ln \frac{d \ A}{V^2} \quad (4)$$

suitable for most (larger) levels of ground acceleration. Consistent with the approach used to derive this expression, N is used here as the expected or average threshold acceleration based on average values for the various friction angles, and is denoted as $\bar{N}$ in the original reference. The units are inches for d, inches/sec for V and inches/sec^2 for A, as for Equation (2). N/A is dimensionless.

For the designer these expressions for predicting the likely permanent displacement of a retaining wall simply as a function of a threshold acceleration have considerable merit. However, some cautionary notes are necessary.

Firstly, no account is taken of the character of the actual earthquake expected at the site in question. Integrating a database of world-wide records has the advantage of including some particularly 'damaging' earthquakes (long duration, many large cycles etc), but this may be grossly over-conservative for less damaging events (short duration, peak acceleration defined by one single spike) during which there is not the opportunity for sliding displacement to build up. Where site-specific seismic hazard studies have been undertaken, it is recommended that site specific 'design lines' are developed which will provide better insight into how the design is likely to perform in its specific location.

Secondly, the prediction of sliding displacement is critically dependent on the ratio of the threshold to the peak acceleration N/A. The selection of the peak acceleration should be treated with caution, as large high frequency 'spikes' in the record will reduce the ratio N/A and lead to an artificially high prediction of displacement. In practice, it is widely accepted that large structures such as retaining walls are not going to be affected by the higher frequency components. However there is no simple procedure in the literature which can provide guidance as to the nature of the filter which should be used to process an earthquake record prior to its use in design. In general, the designer will need to use judgement to ignore 'spikes' in the design input motion. Clearly the development of appropriate recommendations in this area would be valuable.

Thirdly, the sliding model used in the integration process assumes no degradation of soil properties and particularly no loss of bearing capacity, which is the most common cause of large wall movements. The analysis of the caisson wall performance at Kobe by the Japan Ministry of Transport Port and Harbour Research Institute (PHRI) shows clearly that simply trying to match the observed movements to some value of N/A based on sliding would be a gross misinterpretation of the actual displacement mechanism. The caisson walls moved several metres primarily through loss of stiffness in their foundations (this marking the onset of gross deformations even before bearing 'failure'); the dry dock walls moved a fraction of this through rocking, limited by three-dimensional effects along the wall and tensile strength of the mass concrete.

5 ROCKING DISPLACEMENT

The preceding section discussed the condition of walls which are prone to slide, rather than to rotate or to rock. In the work of Whitman and Liao an allowance for rotation of the wall, contributing to outward movement, was made based on earlier work by Nadim (1980) which suggested that permanent outward movement of walls which could tilt may be of the order of 50% greater than for walls which were constrained to slide only. However they acknowledged that this aspect was not well understood and recommended that further work be undertaken. In certain cases, depending on the wall geometry and characteristics, rotation of a gravity wall may be the critical mode of behaviour, 'triggered' before a sliding threshold is reached. Under these conditions, the use of Newmark's sliding mechanism is inappropriate to estimate the rotational displacement and an alternative model is necessary.

Steedman and Zeng (1996) provide guidance on the assessment of rocking displacement of walls using a parallel approach to Newmark. The model developed for predicting rocking movements defines a threshold

in rotation based on moment equilibrium which, once exceeded, leads to a relative angular velocity (and hence displacement) between the wall and the ground which can be readily calculated by integrating the equation of motion. Permanent rocking displacement is then accumulated until the rotational velocity is again zero, which occurs some time after the ground acceleration has fallen below the threshold in rocking (a solution analogous to the sliding model developed by Newmark). This approach showed good comparison with centrifuge model test data of the rocking of large concrete gravity walls under earthquake loading, as shown in Fig. 4 for a typical event, where 'ratcheting' of the outward permanent movement can be seen, (as predicted by the model) superimposed on an elastic (recoverable) displacement.

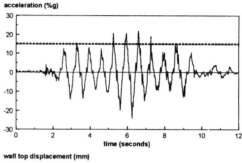

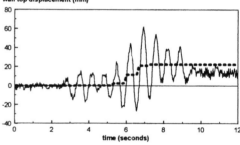

Figure 4. Comparison between rocking and actual performance of a large gravity wall in rocking, Steedman and Zeng (1996)

Fig. 5 shows a comparison between predicted displacement in sliding, following Richards and Elms and Whitman and Liao, and in rocking (at the wall top), following Steedman and Zeng, all based on a single earthquake similar to the class of earthquake shown in Fig. 4 with a peak acceleration of 15%g and a peak velocity of 0.11 m/s. The rocking model could have generated a wide variety of design lines depending on the character of the earthquake; the particular result shown in Fig. 5 is probably more damaging than most (because of the large number of large cycles in the record) but it is unlikely to be an upper bound. Nevertheless it demonstrates that the use of design 'lines' such as Whitman and Liao or particularly

Richards and Elms, which have been developed for sliding, may be unconservative in rocking.

6 ELASTIC DISPLACEMENT

It has already been noted that wall displacements comprise two elements: a recoverable and an irrecoverable element and these can both be seen in Fig. 4. Although the permanent displacement will generally be of most significance to designers, assessment of the elastic recoverable 'vibration' is often of interest. This can be calculated numerically using finite element models, and this may be essential for complex or high risk structures, but simple models based on the principles of applied mechanics can often also provide an order of magnitude estimate sufficient for most purposes.

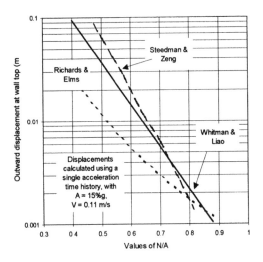

Figure 5. Comparison between rocking model, Whitman & Liao and Richards & Elms sliding models

For example, dynamic movements of a wall on the point of sliding or rotating will be dominated by the behaviour of the soil behind it and therefore it may be valuable as a first estimate for such walls to consider the behaviour of a shear beam representing the soil alone and to neglect the wall altogether. For a shear modulus G proportional to the square root of depth and a lateral acceleration coefficient k_h, the surface deflection, x, can be expressed as:

$$x = \frac{2}{3} \frac{k_h \gamma H^2}{G_b} \tag{5}$$

where G_b is the shear modulus at the base of the soil

column of depth H, unit weight γ. This dynamic or cyclical displacement element is superimposed on the permanent movements. For the example shown in Fig. 4, with a wall height of around 18 m, Equation (5) predicts an amplitude of surface deflection x = +/- 30 mm, for k_h = 0.2 and values of shear modulus consistent with the density and strain levels within the sand backfill, which compares favourably with the order of magnitude of the observed cyclical displacements.

7 MOVEMENT AND FLEXIBILITY

It has been noted above from the work of Bolton (1991) that only very small movements are necessary to reduce the soil pressures behind a wall to near active conditions. Even the elastic vibration of the large wall in the example above implies a deflection of +/- 0.17% of the wall height. However, some walls are clearly very massive, or stiff, or both and with compacted fill behind there is an implication that some higher level of lateral earth pressure may be imposed on them during earthquakes.

There is no validated design approach in this area but elastic analyses have produced estimates of lateral earth pressure which are significantly in excess of the Mononobe-Okabe solution. Wood's simplified solution, for example, described in Ebeling and Morrison (1992), dates from 1973 and essentially represents a static elastic finite element analysis of a wall bonded to an elastic medium. More recent studies have attempted to provide further guidance in this area, such as the work of Veletsos and Younan (1997). But it remains unclear in practice how the large lateral stresses which are predicted on the back of the wall could be sustained by conventional backfill without deterioration in the soil. Historically, this has led to a cycle of seismic wall design in which 'stiff' retaining walls generate high dynamic loads and are then further stiffened, which apparently attracts more load and so on.

A simple solution to this design problem is to redefine the acceptance criteria for the wall such that at the point of yield, some movement is allowable; provided this movement is sufficient to reduce earth pressures to near active, then seismic design can proceed on the basis of the approaches described herein, which should lead to a considerably more economic solution.

However, it may not be clear whether a wall will behave as a stiff or a flexible system prior to wall 'yield'. Clearly it will depend not only on the soil but also on the flexibility of the wall and the inertial loading on it due to its own self-weight. Simple elastic analyses of the type proposed in Equation (5) can be used to explore the relationship between stiffness and inertia in soil and wall. For a retaining wall of uniform thickness D, height H, unit weight γ_c and Young's Modulus E_w, cantilevered from a fixed base, Steedman (1984) showed that a soil-structure interaction parameter, C_i, defined non-dimensionally as:

$$C_i = \left(\frac{8\,G_b\,H^3}{5\,E_w\,D^3 + 8\,G_b\,H^3} \right) \left[\frac{15\,\gamma_c\,D}{4\,\gamma\,D} - \frac{5\,E_w\,D^3}{3\,G_b\,H^3} \right] \quad (6)$$

provided valuable insight into the transition between flexible and rigid response for a wall in bending, as shown in Fig. 6 for a typical set of input parameters. The value of C_i then gives an indication of the nature of the stress increment imposed between the soil and the wall under a pseudo-static lateral acceleration field. Flexible walls are defined as those where the wall moves in phase with the soil with either a reduction of stress on the interface (simply defined as $\Delta\sigma = C_i\,k_h\,\gamma\,z$), or a small increase compared with the static condition. In Fig. 6, a zone of behavior defined as 'flexible' is shown, with values of C_i which are positive (ie. a stress decrement) or have a small negative value. For rigid or near rigid walls, however, a large stress increase is predicted on the interface during seismic loading, and the value of C_i asymptotes at around -2.7 for this example. This approach could readily be extended to other modes of wall deformation such as rotation about the base or shear (for example in a reinforced earth structure).

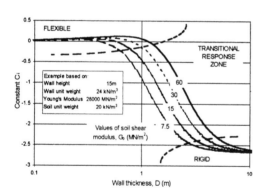

Figure 6. Transition between 'flexible' and 'rigid' wall behaviour defined by a soil-structure interaction parameter, C_i, for a uniform cantilever in bending.

Thus from Fig. 6 it can be seen that for a given direction of lateral acceleration (inwards or outwards from the wall), the increment of stress on the interface may be either positive or negative and hence the cyclic force exerted by the soil on the wall may be either in phase or out of phase with the base shaking, as noted also by Whitman and Ting (1994). Such analyses illustrate also how degradation of soil stiffness, for example through excess pore pressure generation, may change the nature of the soil-wall behaviour, both in phase and amplitude of force on the wall.

8 DEGRADATION OF SOIL STIFFNESS AND STRENGTH

The key difference between the behaviour of the caisson walls and the dock walls in Kobe was the performance of the foundation. Softening and liquefaction was clearly observed behind both types of wall, but only the caisson walls exhibited large movements. Therefore liquefaction alone behind a wall is not necessarily catastrophic, and must be combined with other factors to cause heavy damage.

In tie-back walls, such as sheet pile structures, loss of strength and stiffness below the ground surface may lead to degradation of the anchor system and thus to an overloading of the free part of the wall, which is essentially transformed from a propped cantilever (propped by the anchor) to an unpropped cantilever (no benefit from the anchor but still embedded over its lower section). This mechanism was well illustrated by Steedman and Zeng (1993) where model tests showed the consequences of loss of stiffness in the backfill on the soil-structure system.

Thus far, it has been recommended that pseudo-static approaches can be used successfully for retaining wall design for earthquake loading. However, an implicit assumption noted above is that as the wall and backfill are rigid elements, the stiffness of the soil-structure system is high and remains so throughout the earthquake. However, as excess pore pressures rise under earthquake loading, the stiffness of the soil reduces and this can lead to a dramatic reduction in the natural frequency of the soil-structure system, perhaps analogous to progressive failure in structural systems. In a tie-back wall, the difference in the natural frequency of the anchored system to a cantilever is large, and moreover as the natural frequency of the system falls, amplification of motion is inevitable, combined with phase changes as the system approaches resonance. All these effects have been observed in model tests, Steedman and Zeng (1993).

For the designer, the consequences of 'losing control' of the retaining wall are severe. As with structural design, avoidance of resonance is preferable to attempting to design for unpredictable dynamic effects. In geotechnical design, this may mean using drains to ensure that excess pore pressures are limited to a percentage of the overburden, or ensuring that the backfill is sufficiently dense or of high enough permeability that excess pore pressures cannot build up to unmanageable levels.

9 DISTRIBUTION OF DYNAMIC EARTH PRESSURE

A final comment is appropriate concerning the distribution of dynamic earth pressure with depth. Many authors have commented favourably on the magnitude of the Mononobe-Okabe lateral earth force but have disagreed on the point of application, which indicates uncertainty over the distribution of earth pressure with depth. There is no justification in the Coulomb approach adopted by Mononobe and Okabe for assuming any distribution other than triangular with depth. However, if the backfill is assumed to have a finite as opposed to an infinite modulus, then it is possible to solve for the Mononobe-Okabe force by integrating over the height of the wall to arrive at a pseudo-dynamic solution for the lateral earth pressure, Steedman and Zeng (1990b). This approach also provides an explanation as to why different authors have reported different points of application for the dynamic lateral force behind a wall.

In the Steedman and Zeng analysis a non-dimensional group H/TV_s is shown to be of significance, where H is the height of the wall, T the period of the vertically propagating shear wave and V_s the shear wave velocity in the fill. Clearly, for H/TV_s = 0 the solution should reduce to the standard Mononobe-Okabe solution as this implies infinite wave velocity (and hence modulus) in the backfill. As this group increases, a phase change is introduced between base and top of the wall, and the point of application of the dynamic force is predicted to rise slightly up the back of the wall. As the natural frequency falls, some amplification of motion through the soil is to be expected (which would tend to increase the total lateral force) but this is at least partially compensated by the phase lag which is inevitably introduced between top and base and which has the opposite effect (as not all of the wall is subject to the peak lateral acceleration simultaneously).

10 CONCLUSIONS

1. The problem of predicting wall displacements induced by earthquake loading remains complex. Solutions are available, and have been outlined above, which can provide a good basis for design, at least for standard structures. Design should be based on the premise that some movement is acceptable.

2. Simple reliance on a force-based approach using a sliding failure mechanism is not likely to be adequate, as field evidence and model tests have shown that rocking, bearing and dynamic effects are common modes of response which are not effectively addressed using the Mononobe-Okabe approach alone. Consideration of all the likely modes of deformation is

necessary to ensure that the design calculations have addressed the appropriate displacement mechanism.

3. Retaining wall movements under earthquake loading can be predicted, provided appropriate models are used. Sliding and rocking on rigid foundations can be assessed using the concept of thresholds and integration of relative rotational or translational velocity. Degradation of strength and stiffness through pore pressure rise, either in the foundation or in the backfill, requires a more complex physical or numerical approach.

ACKNOWLEDGEMENTS

Several of the key insights discussed in this paper developed out of the results of model tests and analyses carried out by a colleague, Dr X Zeng (University of Kentucky) as part of an extensive program of research into retaining wall behaviour at Cambridge University. The author gratefully acknowledges Dr Zeng's important contribution to this ongoing effort.

NOTATION

A	Peak acceleration
C_i	Soil-structure interaction constant
D	Wall thickness
E_w	Young's Modulus
d	displacement
$\bar{d}$	mean displacement
G	Shear Modulus
G_b	Shear Modulus at the base of a soil column
H	Height of wall or soil column
k_h	Lateral acceleration coefficient
k_v	Vertical acceleration coefficient
N	Threshold acceleration for permanent movement
$\bar{N}$	Threshold acceleration based on average parameter values
T	Period of a vertically propagating shear wave
V	Peak velocity
V_s	Velocity of shear wave
x	Horizontal displacement
γ	Unit weight
γ_b	Bouyant unit weight of soil
γ_c	Unit weight of wall
γ_d	Dry unit weight of soil
γ_t	Total unit weight of soil
θ_f	Seismic angle, free water case
θ_r	Seismic angle, restrained water case

REFERENCES

Amano R, Azuma H and Ishii Y (1956) Aseismic design of walls in Japan, Proc 1st World Conf Earthquake Eng, Paper 32, pp1-17.

Bolton M D (1991) Geotechnical stress analysis for bridge abutment design, Contractor report 270, Transport and Road Research Laboratory, Crowthorne, Berks.

Bolton M D and Steedman R S (1982) Centrifugal testing of microconcrete retaining walls subjected to base shaking, Proc Conf Soil Dynamics and Earthquake Eng, Southampton, pp 311-329, Balkema.

Bolton M D and Steedman R S (1985) Modelling the seismic resistance of retaining structures, Proc XI Int Conf Soil Mech Found Eng, San Francisco, Vol 4, pp1845-1848, Balkema.

Earthquake Engineering Research Center (1995) Geotechnical reconnaissance of the effects of the January 17 1995 Hyogoken-Nanbu earthquake, Japan, Report No. UCB/EERC-95/01 August, Berkeley.

Ebeling R M and Morrison E E (1992) The seismic design of waterfront retaining structures, Technical Report ITL-92-11, USAE, Waterways Experiment Station, Corps of Engineers, Vicksburg.

EQE International (1995) The January 17, 1995 Kobe earthquake: an EQE summary report, EQE International, San Francisco.

Franklin A G and Chang F K (1977) Earthquake resistance of earth and rockfill dams, Report 5: Permanent displacements of earth dams by Newmark analysis, Misc Paper S-71-17, USAE Waterways Experiment Station, November.

Japan Ministry of Transport (1991) Technical Standards for Port and Harbour Facilities in Japan, Ports and Harbours Bureau, English version by Overseas Coastal Area Development Institute of Japan, Tokyo.

Japan Ministry of Transport (1995) Report of a survey of the damage caused to port facilities etc by the Great Hanshin/Awaji Earthquake, Ministry of Transport Ports and Harbours Bureau, PHRI, No.3 Port Construction Bureau, May, (in Japanese).

Inagaki H, Iai S, Sugano T, Yamazaki H and Inatomi T (1996) Performance of caisson type walls at Kobe Port, Soils and Foundations, Special Issue, pp119-136, January.

Krinitzsky E L, Gould J P and Edinger P H (1993) Fundamentals of earthquake resistant construction, Wiley, New York.

Matsuo H and Ohara S (1960) Lateral earth pressure and stability of quay walls during earthquakes, Proc 2nd World Conf Earthquake Eng.

Matsuzawa H, Ishibashi I and Kawamura M (1985) Dynamic soil and water pressures of submerged soils, Proc ASCE, Journal Geotech Eng, Vol 111, No. 10.

Mononobe N and Matsuo H (1929) On the determination of earth pressure during earthquakes, Proc World Eng Congress, Vol. 9, pp177-185.

Murphy V A (1960) The effect of ground characteristics on the aseismic design of structures, Proc 2nd World Conf Earthquake Eng, pp231-247, Tokyo.

Nadim F (1980) Tilting and sliding of gravity retaining walls, SM Thesis, Dept Civil Eng, MIT.

Newmark N M (1965) Effect of earthquakes on dams and embankments, 5th Rankine Lecture, Geotechnique 15, No.2, pp139-160, June.

Okabe S (1924) General theory of earth pressure, Journal Japan Civil Eng Society, Vol 12 (1).

Richards R and Elms D G (1979) Seismic behaviour of gravity retaining walls, Proc ASCE, Journal Geotech Eng Div, Vol 105, No. GT4, pp449-464.

Seed H B and Whitman R V (1970) Design of earth retaining structures for dynamic loads, ASCE Spec Conf - Lateral Stresses in the Ground and Design of Earth Retaining Structures, Cornell, pp103-147.

Steedman R S (1984) Modelling the behaviour of retaining walls in earthquakes, PhD Thesis, Cambridge University.

Steedman R S and Zeng X (1990a) Hydrodynamic pressures on a flexible quay wall, Proc Eur Conf Structural Dynamics, Bochum, pp843-850, Balkema.

Steedman R S and Zeng X (1990b) The influence of phase on the calculation of pseudo-static earth pressure on a retaining wall, Geotechnique 40, No.1, pp103-112.

Steedman R S and Zeng X (1993) On the behaviour of quay walls in earthquakes, Geotechnique 43, No. 3, pp417-431.

Steedman R S and Zeng X (1996) Rotation of large gravity walls on rigid foundations under seismic loading, Proc ASCE Annual Convention, Washington, November.

Veletsos A S and Younan A H (1997) Dynamic response of cantilever retaining walls, Proc ASCE, Journal of Geotechnical and Geoenvironmental Engineering, pp161-172, February.

Westergaard H M (1933) Water pressure on dams during earthquakes, Trans Am Soc Civ Engrs, Vol 98, pp418-472.

Whitman R V and Liao S (1985) Seismic design of gravity retaining walls, Misc Paper GL-85-1, USAE Waterways Experiment Station, Vicksburg.

Whitman R V and Ting N-H (1994) Experimental results for tilting wall with saturated backfill, Proc Conf Verification of Numerical Procedures for the Analysis of Soil Liquefaction Problems, pp1515-1528, Arulanandan and Scott (eds), Balkema.

Seismic Behaviour of Ground and Geotechnical Structures, Sêco e Pinto (ed.)© 1997 Balkema, Rotterdam, ISBN 90 5410 887 8

Effects of liquefaction-induced ground displacement on pile damage in 1995 Hyogoken-Nanbu earthquake

Kohji Tokimatsu & Yoshiharu Asaka
Tokyo Institute of Technology, Japan

Akio Nakazawa & Shinsuke Nanba
Araigumi Co. Ltd, Japan

Hiroshi Oh-Oka
Building Research Institute, Japan

Yasuhiro Shamoto
Shimizu Corporation, Japan

ABSTRACT: Field investigation has been conducted to determine the failure and deformation modes of piles of three buildings that experienced liquefaction-induced permanent ground displacement in the 1995 Hyogoken-Nambu earthquake. Aerial photographic survey was also performed to estimate permanent displacements of the buildings and the surrounding ground surface. All the buildings tilted seaward by up to 1/39, while permanent ground displacements of 25 to 45 cm occurred in parallel with the shoreline throughout the site and seaward ground movements of 25 to 50 cm occurred on the sea side of the buildings. Despite the ground movements towards the sea on the south of the buildings, the pile heads did not move in the seaward direction and the piles on the land side did not show any significant deformation; however, the piles on the sea side failed at three depths. In the direction parallel to the shoreline, all the pile heads moved westward with the ground and failed near the pile head and the bottom of the liquefied layer. The different failure patterns between piles in the seaward direction are probably due to the difference in lateral ground movements in that direction. The similar failure patterns between piles in the direction parallel to the shoreline are due to the even permanent ground displacement in that direction. Thus, the variation of permanent ground displacements should be taken into account, in order to properly estimate the stresses in piles in liquefiable soils.

INTRODUCTION

Extensive soil liquefaction that occurred in the Hyogoken-Nambu earthquake (M=7.2) of January 17, 1995, damaged various structures and infrastructures in the reclaimed land areas between Kobe and Amagasaki cities. Particularly, many of the quay walls in these areas moved up to several meters towards the sea due to liquefaction of their foundation soils and/or back-fills. This induced large horizontal ground movements as well as differential ground settlements near the shoreline. As a result, not only buildings with spread foundations but also those supported on piles often settled and/or tilted without little damage to their superstructures. There was a serious concern therefore that their piles might have been damaged. Excavation surveys after the quake have shown this is the case for many pile heads. Integrity sonic testing (e.g., Oh-oka et al., 1996) further suggested that such failure might have occurred at some depths other than pile heads. In the area away from the shoreline, soil liquefaction often induced large permanent ground displacement as well as cyclic ground displacement, causing damage to piles. These findings indicate that the effects of ground displacements on piles should be properly taken into account in foundation design. However, little is known concerning the actual failure

and deformation patterns of piles subjected to permanent ground displacement. The object of this paper is to investigate both failure and deformation modes of piles of three buildings that experienced liquefaction-induced permanent ground displacement in the Hyogoken-Nambu earthquake and to discuss the effects of ground movements on pile damage.

OUTLINE OF BUILDING DAMAGE AND SOIL PROFILE

Building and Soil Conditions

The buildings, hereby called Buildings A, B, and, C, are situated near and parallel to the shoreline on the northeast corner of Ashiya-hama, Asahiya city (Fig. 1). Figs. 2 and 3 show the foundation plan and elevation of the buildings together with boring logs before construction. They are 11-story and in turn connected with each other using an expansion joint. Constructed in early 1980s, the buildings were supported on prestressed concrete piles 50 cm in diameter and 26 meters long. The piles were embedded 5 cm into their pile caps having a penetration depth of 1.5 m.

A reclaimed fill about 8 meters thick consisting of mainly gravelly sands is underlain by a layer of

Holocene sand having a thickness of 8 m and N-values of 2-40.   The top of Holocene sand is very loose and considered to be liquefiable.   Underlying below are a 2-m thick layer of Holocene clay with N-values of about 5 and alternate layers of Holocene sand and clay with a thickness of 6 m and N-values of 30-50.   A stiff layer of Pleistocene sand having N-values over 40 occurs at depths below 24 meters. The groundwater table is located at a depth of about 4 meters.

Performance of Buildings and Surrounding Soils during 1995 Event

Soil liquefaction occurred extensively around the site during the 1995 Hyogoken-Nambu earthquake. The ground surface of the parking lot on the north of the building settled by about 40 cm (Fig. 4), with many ground fissures running east and west. Many cracks 5 to 165 mm wide occurred on the top of the concrete revetment on the west, resulting in a sum of crack width over 50 cm. The ground surface between the

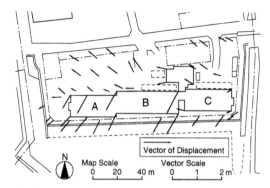

Fig. 1 Map showing location of buildings and vectors of permanent ground displacement

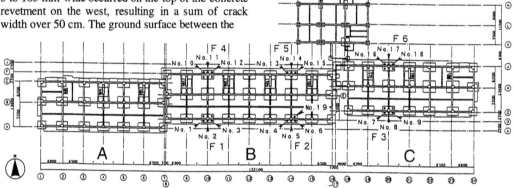

Fig. 2 Foundation plan of buildings (after Sotetsu, 1996)

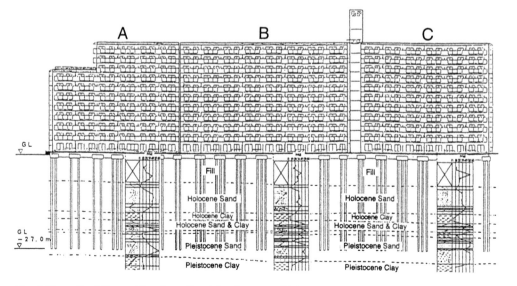

Fig. 3 Plan of buildings with soil profiles (after Sotetsu, 1996)

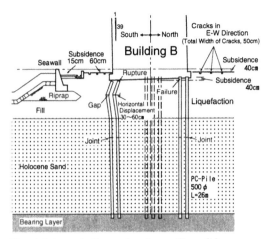

Fig. 4 Schematic figure showing outline of damage

building and the road on the south, and the road itself settled by 60 cm and 15 cm, respectively. The seawall on the south appeared to have moved seaward up to about 1 meter.

After the earthquake, Buildings A, B, and C tilted by 1/58, 1/39, and 1/1211, respectively, toward the sea. Due to the uneven tilt, a large horizontal gap of about 50 cm occurred on the 10th floor level between Buildings B and C as shown in Photo 1. In addition, the ground level of Buildings A and B are found to have displaced towards south by 3-4 cm with respect to Building C.

Although the inclination of the buildings in the east-west direction was negligibly small, Buildings A and B are found to have displaced westward by 12-15 cm, with respect to Building C. No significant damage to their superstructures was found, except for minor shear cracks on the north-south secondary walls of the first to third floor.

Permanent Displacement of Buildings and Ground Detected by Aerial Photographic Survey

The vector shown in Fig. 1 indicates the magnitude and direction of horizontal displacements of the ground surface as well as the roofs of buildings that have been determined from aerial photographs taken before and after the earthquake. The direction of the ground surface displacement varies around the buildings in such a way that westward to northwestward movement prevails on the north, while southwestward movement prevails on the south. It is noteworthy, however, that a westward ground movement of the order of 25 to 45 cm occurred throughout the site. In addition, the southward ground movements on the south of Buildings A and B (about 50 cm) are significantly greater than those on the south of Building C (about 25 cm).

Photo 1   A large gap between Buildings B and C due to seaward inclination of Building B

The aerial photographic survey further indicates that the roofs of the Buildings A, B, and C moved toward southwest, with southward movements of 60, 80, and 10 cm, and westward movements of 42, 43, and 28 cm, respectively. Most of the southward displacements of the roof correspond to those resulting from the inclination of the buildings, thereby suggesting that the southward displacements of the buildings on the ground level are considered to be small. The westward displacements of the buildings correspond to those of the ground surface detected by the aerial photographic survey as well as to the relative displacements between buildings that were observed after the earthquake. The good agreements between the results from different surveys suggest that the accuracy of the aerial photographic survey is reasonably high and that all the buildings did move mainly westward.

FAILURE AND DEFORMATION MODES OF PILES FROM TELEVISION CAMERA AND SLOPE-INDICATOR SURVEYS

Outlines of Field Investigation

Piles under 6 pile caps, labelled F1 to F6 shown in Figs. 2 and 5, were examined. The ground around the pile caps was excavated to a depth of about 3.5 m, and the failure and cracks of the exposed portion of the piles were directly observed as well as their slope angles. To investigate the damage below that level, a television camera and a slope-indicator (e.g., Yamagata, 1966; Shamoto et al., 1995) were in turn inserted into the hollow space of the pile through the opening made on the side of the pile, as shown in Photo 2.

The slope-indicator can provide separately the slope angles of two orthogonal components of a pile. Once the variation of the slope angles with depth is determined, the horizontal displacement and curvature of the pile can be estimated by either integrating or differentiating it with depth. The details of the test equipment and procedure are described elsewhere (Shamoto et al., 1995; Oh-oka et al., 1997).

Damage to Pile Heads and Their Slope Angles

Fig. 5 summarizes the slope angles of all piles investigated and Fig. 6 schematically shows major cracks of two piles. The piles on the sea side were cut off horizontally at the bottom of the pile cap, as shown in Photo 3, and the pile heads inclined extensively towards northwest. As a result, a vertical gap occurred on the northwest pile heads, while compression failure occurred on the southeast, with concrete being spalled. The longitudinal

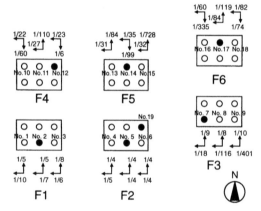

Fig. 5 Slope angles of pile heads

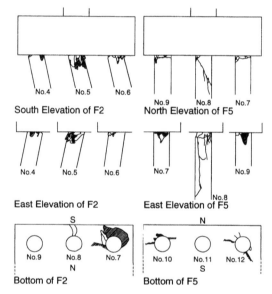

Fig. 6 Schematic figure showing damage to pile heads

Photo 2   A Snap taken during Television Observation

Photo 3   Damage to pile head on the sea side

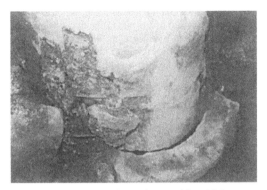

Photo 4   Failure of pile on the sea side with large horizontal gap

Photo 5   Shear crack near the pile head on the north side

reinforcements buckled and the transverse reinforcements fractured. The average slope angles of the pile head are 1/6 to the north and 1/8 to the west for F1, and 1/4 to the north and 1/4 to the west for F2. Similarly, the piles under F3 of Building C inclined, but with smaller slope angles of 1/9 to the north and less than 1/18 to the west.

Except for the damage near the pile cap, cracks or failures were invisible to a depth of 3.4 m. However, the piles were completely cut off at a depth of 3.4-4.2 meters, with failures of longitudinal reinforcements and extensive horizontal and/or diagonal cracks. Photo 4 shows such damage to a pile under F2 where the upper pile was displaced northwestward with respect to the lower one, causing a horizontal gap of 5 cm. Such damage was particularly extensive in the piles under F1 and F2.

On the north side of the buildings, the damage pattern near the pile heads was somewhat different. Large horizontal cracks occurred 20 cm below the pile cap of a pile under F4 with many longitudinal and diagonal cracks below, and concrete was spalled from the east face of the pile. Horizontal cracks were observed only on the north face in other piles of F4. Shear cracks occurred on the east faces of two piles of F5 that ran downward west, and horizontal cracks occurred near the other pile heads of F5, with compression failure on the east face. A large diagonal crack ran in a pile of F6 from the upward east to the downward west, with minor longitudinal cracks (Photo 5). Horizontal cracks occurred near the other pile heads of F6, with compression failure on the east face. The above damage patterns suggest that the piles on the north side of the buildings were deformed mainly to the west, which appears consistent with the slope angles of their pile heads shown in Fig. 5. The cause of the damage is considered to be either the horizontal force from the superstructure or ground displacement toward west or both.

Fig. 5 shows that all the pile heads of F4, F5, and F6, on the north side of the buildings, inclined

westward, while no definite trend is found in the north-south direction. The average slope angles of pile heads to the west are 1/24 and 1/33 for F4 and F5 of Building B, but it shows a smaller value of 1/75 for F6 of Building C.

Results from Television Observation

Fig. 7 shows the cracks and failures in piles detected by television observation for seven piles indicated with solid circles in Fig. 5. The maximum depths at which the television observation was made are indicated with solid triangles in Fig. 7. All the piles on the sea side of the buildings (No. 2, No. 5, No. 19, and No. 7) failed with a horizontal gap at a depth between 3.4 and 4.1 m (1.9 and 2.6 m below the pile cap), whereas the piles on the land side did not show any significant damage at the same depth. The damage patterns below a depth of 4m are very similar in such a way that large horizontal cracks occurred at a depth of 8.5 to 10.5 m (7 to 9 m). The SPT N-values below that depth become significantly greater than those of the above. Probably the sandy deposits above that depth did liquefied during the

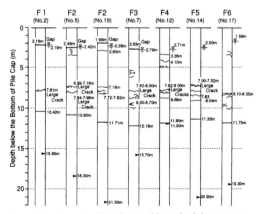

Fig. 7 Damage to piles detected by television camera

365

1995 event. Several horizontal cracks occurred above that depth, but no major crack was observed below.

Results from Slope-Indicator Survey

Fig. 8 shows the distribution of the two orthogonal slope angles with depth, measured for the piles of F1 to F6 using the slope-indicator. Fig. 9 shows the relative displacement of the pile in the two orthogonal directions, obtained by integrating the observed slope angles with depth. Fig. 10 shows the vector displacement relative to the pile tip, on the horizontal plane.

Figs. 9 and 10 indicate that all pile heads moved

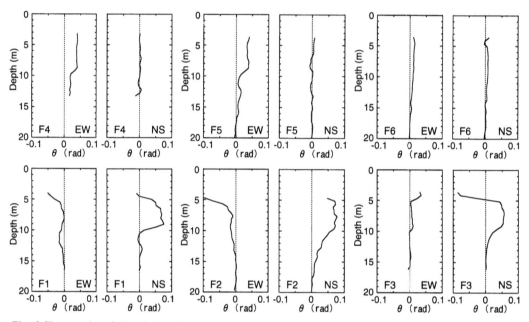

Fig. 8 Slope angles of piles detected by slope-indicator

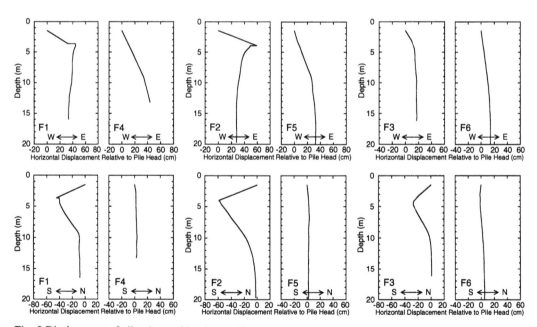

Fig. 9 Displacement of piles detected by slope-indicator

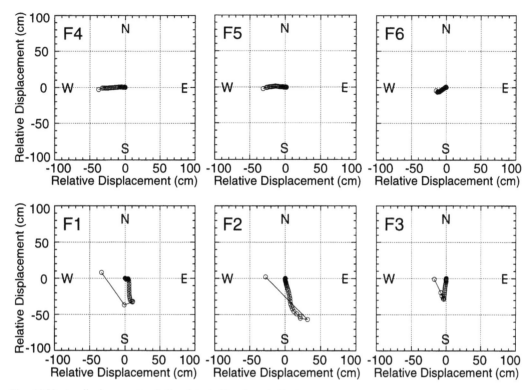

Fig. 10 Vector displacements of piles detected by slope-indicator

westward but did not move in the N-S direction. The westward displacements are 30-40 cm in Building B (F1, F2, F4, and F5) and about 20 cm in Building C (F3 and F6). It is noteworthy that the pile heads of the same buildings show the same vector displacements. Besides, these vector displacements are consistent with those of the buildings on the ground level estimated from the aerial photographic survey. The results from the slope-indicator survey therefore appear to be reasonable .

In spite of the same vector displacements of the pile heads under the same building, their deformation patterns with depth are extremely different between the sea (south) side and land (north) side. Namely, in the N-S direction, the piles on the north side, F4, F5, and F6, did not show significant deformation, while the piles on the south side, F1, F2, and F3, inclined toward sea below 5 m depth and against sea above 5 m depth. The reversals of the inclination indicate that extensive failures occurred at that depth, being consistent with the results from television observation shown in Fig. 7. Similar damage pattern was also observed at buildings located near the shoreline (e.g., Tokimatsu et al., 1996).

Similarly, in the E-W direction, the piles on the north side of Building B, F4 and F5, simply inclined westward, while the piles on the south side, F1 and F2, inclined eastward below 5 m depth and westward

above that depth. The depths where the reversals of the inclination occur also correspond to the depths where extensive failure occurred.

In spite of the difference in the deformation patterns, all the piles deformed only at depths of above 8.5 to 10.5 m where soil liquefaction is estimated to have occurred. The major directions of the displacement of piles in the lower part of this liquefied layer are south on the sea side and west on the north side. The directions of the displacements are consistent with those of the permanent ground displacement nearby, estimated from aerial photographic survey. In particular, the magnitudes of the pile displacements toward west are also consistent with those of the ground displacements from the aerial photographic survey. Thus, the ground displacements might have resulted from the residual deformation of the liquefied sandy layer, and the piles might have been deformed in the same direction.

EFFECTS OF PERMANENT GROUND DISPLACEMENT ON PILE DAMAGE

It is conceivable that the failure and deformation modes of the piles is due partly to the variation of permanent ground displacements around the buildings, though the horizontal forces from the

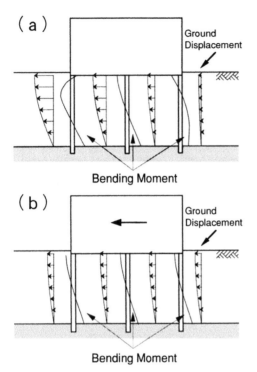

(a)

Ground
Displacement

Bending Moment

(b)

Ground
Displacement

Bending Moment

Fig. 11 Schematic figures showing effects of permanent ground displacement on pile damage

superstructures and the cyclic ground displacements during shaking are other major factors. Particularly, the difference in failure modes of piles in the N-S direction is considered to be due mainly to the permanent ground displacement. Fig. 11. schematically shows an a model to explain the effects of such ground displacements on the bending moments in the piles in E-W and N-S directions of the buildings. In the model, a group of piles connected with the same foundation beam is subjected to horizontal ground movements.

In the N-S direction, the permanent ground displacements on the south of the building were significantly larger than those on the north side. The soil on the south therefore pushes the pile seaward, while the soil on the north resists the pile to deform seaward. Thus, the bending moment gets large at three portions in the pile on the south, i.e., near the pile head, and the middle and the bottom of the liquefied layer; but only two parts on the north, i.e., near the pile head and the bottom of the liquefied layer. Considering that the failure can occur where large bending moment occurs, the above model is consistent with the actual damage patterns, such as shown in Figs. 7 and 9. The difference in the amount of ground displacements on the south of Buildings B and C might have affected the difference in the damage extent of piles between these buildings.

In the E-W direction, the permanent ground displacements were almost the same along the building. The soil therefore pushes the pile throughout the building. All the piles deform evenly westward, experiencing large bending moments and failures at two portions, i.e., near the pile head and the bottom of the liquefied layer. These failure and deformation patterns are very consistent with those of the north side of Building B, as shown in Figs. 7 and 9. The failure and deformation patterns in the E-W direction on the south side of the building might been complicated due to the fact that the piles suffered dominant southward ground movements in addition to westward ground movements.

Thus, the distribution of permanent ground displacement could have had significant effects on pile damage in the liquefied areas during the Hyogoken-Nambu earthquake. Such effects therefore should be properly taken into account in the foundation design in such areas.

CONCLUSIONS

Field investigation has been conducted to determine the failure and deformation modes of piles of the three buildings that experienced liquefaction-induced permanent ground displacement in the 1995 Hyogoken-Nambu earthquake. Aerial photographic survey was also performed to estimate permanent displacements of the buildings and the surrounding ground surface. Based on the comparison of the results, the following conclusions may be made:

(1) The results from the slope-indicator survey are consistent with those from the televiewer and aerial photographic surveys, indicating that the slope-indicator used herein can provide deformation patterns of piles with a reasonable degree of accuracy.

(2) All the buildings tilted seaward by up to 1/39, while permanent ground displacements of 25 to 45 cm occurred in parallel with the shoreline throughout the site and seaward ground movements of 25 to 50 cm occurred on the sea side of the buildings.

(3) Despite the ground movements towards the sea on the south of the buildings, the pile heads did not move in the seaward direction and the piles on the land side did not show any significant deformation; however, the piles on the sea side failed at three depths. In the direction parallel to the shoreline, all the pile heads moved westward with the ground and failed at two depths, i.e., near the pile head and the bottom of the liquefied layer.

(4) The different failure patterns between piles in the seaward direction are probably due to the difference in lateral ground movements in that direction. The similar failure patterns between piles in the direction parallel to the shoreline are due to the even permanent ground displacement in that direction. Thus, in addition to the horizontal forces from superstructures and cyclic ground displacements during shaking, the variation of permanent ground displacements should be taken into account, in order to properly estimate the stresses in piles in liquefiable soils.

ACKNOWLEDGMENTS

The investigation described herein was made possible through the consents given by the resident union of the buildings. The authors express their sincere thanks.

REFERENCES

Oh-oka, H., Iiba, M., Abe, A., and Tokimatsu, K. (1996): Investigation of earthquake-induced damage to pile foundation using televiewer observation and integrity sonic tests, Tsuchi-to-kiso, JSG, Vol. 44, No. 3, pp. 28-30 (in Japanese).

Oh-oka, H., Onishi, K., Nanba, S., Mori, T., Ishikawa, K., Koyama, S., and Shimazu, S. (1997): Liquefaction-induced failure of piles in 1995 Kobe earthquake, Proceedings, 3rd KANSAI International Geotechnical Forum, pp. 265-274.

Shamoto, Y., Sato, M., Futaki, M., and Shimazu, S. (1996): A site investigation of post-liquefaction lateral displacement of pile foundation in reclaimed land, Tsuchi-to-Kiso, JSG, Vol. 44, No. 3, pp. 25-27 (in Japanese).

Sotetsu, A. (1996): Damage investigation of foundation structure on the Hyogoken-Nambu earthquake, Arai Technical Research Report, Vol. 9, pp. 91-100, (in Japanese).

Tokimatsu, K., Mizuno, H., and Kakurai, M. (1996): Building damage associated with geotechnical problems, Soils and Foundations, Special Issue, pp. 219-234.

Yamagata, K. (1966): Measurement of inclination angle of PC piles with newly developed inclinometer, Tsuchi-to-Kiso, JGS, Vol. 14, No. 5, pp. 11-18 (in Japanese).

Seismic Behaviour of Ground and Geotechnical Structures, Sêco e Pinto (ed.) © 1997 Balkema, Rotterdam, ISBN 90 5410 887 8

Boundary simulation for dynamic soil-structure interaction problems

M. N. Viladkar, P. N. Godbole & S. K. Garg
Department of Civil Engineering, University of Roorkee, India

ABSTRACT: In the problem of dynamic soil-structure interaction, it is essential to appropriately model the boundary of soil domain for elimination of the influence of the reflected waves from the boundary. In this paper, some of the existing prominent models have been critically reviewed and an attempt has been made to model the soil domain using only the finite elements by placing the truncated (artificial) boundary in the soil mass at appropriate distances from the structure. An example of a rigid vibrating disc resting on the semi-infinite soil surface has been considered. The minimum essential distance of the boundary from the structure has been found out for various values of damping so as to cause negligible change in the response of the structure. The proposed technique has also been tested for nonlinear transient dynamic behaviour.

1 INTRODUCTION

Modelling of infinite soil media in soil-structure interaction plays a very vital role. In case of underground and semi-buried structures, the importance of soil-structure interaction is further enhanced. Dynamic soil-structure interaction analysis poses a special type of problem viz. the boundary simulation. Soil media being semi-infinite in extent, have to be terminated at some finite distance so as to enable its finite element modelling along with the structure. In this case, it becomes very important that the effect of soil mass beyond the assumed boundary, which has not been included in the discretisation, should be considered in some or other way so that the results are within the acceptable error range. Major problem faced at the abruptly terminated boundary is due to outgoing waves which are reflected from the boundary back towards the structure. This alters the response of the structure.

2 EXISTING MODELS

Several techniques have been suggested in the literature to overcome the above mentioned problems. This study is aimed at presenting a brief survey of the various techniques available for the purpose. At the end, an improved version of one of the existing techniques (elementary boundary) has been presented which has been found to provide satisfactory results and can be used very easily.

A critical review of literature indicates that the various available techniques can be grouped into three major categories :

1. Lumped (Discrete) Models
2. Special Techniques
3. Boundary Integral-Equation Method

These models have been presented in the following discussion :

2.1 Lumped Models

In lumped models, the properties of the soil domain extending to infinity, which is not being considered in the model, are lumped at nodes lying at the truncated boundary. These models, though simple to understand and use, are obviously approximate and observe a lot of constraints when put to use. Various models available in this category are :

1. elementary boundaries
2. local (viscous) boundaries
3. consistent boundaries

2.1.1 Elementary boundaries

Elementary boundaries can be formed by abruptly truncating the soil mass to form an artificial boundary. At the boundary, either zero displacements are specified by fixing the nodes lying at the boundary or zero surface tractions are enforced at the boundary by providing the rollers, moving tangential to the boundary. Both types of boundaries have been shown in Fig. 1. Both the representations provide a perfect reflection of the waves, impinging at the boundary, altering the response of the structure in turn. It is the simplest technique available for the boundary simulation and works well if damping is present and the boundary is chosen at a sufficient distance from the structure. The major question is that how far should the boundary be located from the structure so that it is unable to alter the response of the structure significantly.Ghosh and Wilson (1969), have suggested that if the boundary is kept at a distance of 3r to 4r in horizontal direction and 2r to 3r in vertical direction then satisfactory results can be obtained, where r is the radius of the foundation. A better model has been presented at the end of the paper which is based on the maximum wave length expected in the response. It is also to be noted that for zero or nearly zero damping ratios, this boundary should not be used as the distance required from the structure to the boundary becomes very large.

These consist of a single dashpot (damper) corresponding to each degree of freedom at each node lying at the boundary, as suggested by Lysmer and Kuhlemeyer (1969). The model has been shown in Fig. 2(a). The dashpot coefficients in normal and tangential directions, denoted by C_n and C_t respectively, are easy to calculate and are given by Eqn. (1) -

$$C_n = A . \rho . c_l \qquad (1a)$$
$$C_t = A . \rho . c_s \qquad (1b)$$

where, A is the area associated with the node, ρ, the mass density of the soil at the boundary and c_l and c_s are longitudinal and shear wave velocities respectively.

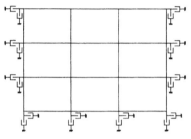

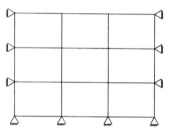

a) Zero displacements

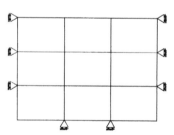

b) Zero surface tractions

Fig. 1. Elementary boundaries

a) Lysmer's (viscous) boundary

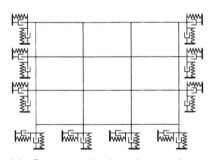

b) Spring-dashpot system

Fig. 2. Local (viscous) boundries

2.1.2 *Local (Viscous) boundaries*

The name has probably originated from the fact that the degrees of freedom lying at the boundary are not coupled to each other and are thus local in nature.

Though this simple boundary is able to fully transmit the waves, impinging normally to the boundary, it reflects the part of the obliquely incident waves. It also poorly performs for surface (Rayleigh) waves. It is also to be noted that viscous dampers lead to permanent deformations nearly in the entire mesh in the elastic systems, (Wolf, 1988) because the restoring forces provided by the dashpots are zero in the absence of the velocity.

A better local boundary model employs frequency dependent spring along with the dashpot as shown in Fig. 2(b). This model improves upon the drawbacks of the simple viscous damper model, but does not fully remove them. This model is suitable only for frequency domain analysis as spring stiffness used is frequency dependent.

2.1.3 Consistent boundaries

Consistent boundaries are able to transmit both types of waves, i.e. body and surface waves, for any angle of incidence resulting in no reflection. These consist of spring-dashpot pairs corresponding to each degree of freedom of each node, as shown in Fig. 3. All the degrees of freedom, lying at the boundary are fully coupled. Both the spring and dashpot have frequency dependent properties. As these can fully transmit all type of waves occurring in the soil mass, these can be placed right at the soil-structure interface. The only restraint imposed is that these observe frequency dependent properties, rendering them unfit for the time-domain dynamic analysis. To compute the properties of spring-dashpot system , discretised boundary-integral-equation procedure is used, (Wolf, 1988).

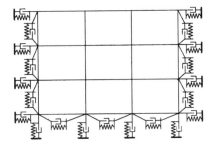

Fig. 3. Consistent boundary

2.2 Special Techniques

Some special techniques have also been developed to simulate the non-reflecting boundaries. These have a special advantage that these can be employed both for time domain as well as for frequency domain analysis. These are simple to understand and provide very good results. Only their actual implementation part is a little involved in nature. Superposition boundary and extrapolation algorithm are two such special techniques which provide satisfactory results.

2.2.1 Superposition boundary

It is a well known fact that the amplitude of a reflected wave from a hard surface is the same as that of a wave reflected by a free surface but later suffers a phase change of l/2. Hence, if the two solutions are superimposed, it will result in zero reflected wave amplitude. The same principle is used in superposition boundary. In superposition boundary, the boundary is split into two parts as suggested by Cundal et al. (1974) and Kunar and Marti (1981). The system is shown in Fig. 4. The system is solved twice once with free boundary and then with fixed boundary. Averaging of the solutions is carried out at every three to four time steps. Frequent averaging is necessary to avoid multiple reflections. The formulation is known to provide satisfactory results. Like the viscous boundary, superposition boundary also leads to permanent deformations nearly at all the nodes of the mesh even in the elastic systems, (Wolf, 1988).

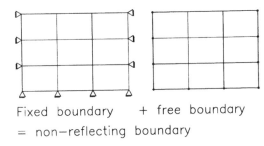

Fixed boundary   + free boundary
= non-reflecting boundary

(a) Conceptual model for 2-D wave propagation

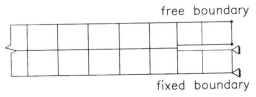

(b) Boundary for 1-dimensional wave propagation in a bar

Fig. 4. Superposition boundary

2.2.2 Extrapolation algorithms

These algorithms are based on the fact that boundary conditions at the artificial boundary for a particular time station can be extrapolated from the values obtained at the previous time stations, Liao and Wong (1984). The procedure is simple and reported to produce very good results. It can take care of both the body and surface waves. The procedure can be understood from the Fig. 5. At node i, a normal in is drawn. If c_a is apparent approach velocity of the wave and Δt is the time step, then displacement at the node i at N^{th} time station will be approximately the same as that at point j at the $(N-1)^{th}$ time station. Apparent approach velocity, c_a is taken as the product of actual wave velocity, c and cosine of incidence angle, ψ. In practice, the incidence angle, ψ may very from one time station to another, hence c_a too will very with time. Therefore, a weighted sum of $c.\cos\psi$ is used in practice. It has been reported that $c_a = c_s$ works well, Lieo and Wong (1984), c_s being the shear wave velocity.

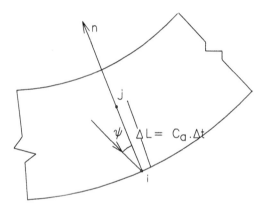

Fig. 5. Extrapolation point used in extrapolation algorithm

2.3 Boundary Integral-Equation Method

Wave propagation problem can more easily be handled in equation form where it is much easier to exclude the incoming (reflected) wave, resulting in perfect radiation condition. These equations more commonly known as boundary integral-equations can be written for any surface and and solution can be obtained. Using these basic equations, Green's functions can be constructed which give the amplitudes of displacements and surface tractions for the given amplitude of applied loads, Wolf (1985). As the boundary integral-equations form the analytical solution, these can be used in discretised form as in boundary element method, in the finite element model. Boundary integral-equations are also employed to calculate the properties of spring-dashpot system used in consistent boundaries.

Boundary elements can easily be developed from boundary integral equations, either using weighted-residual technique or reciprocity theorem. Solutions of boundary integral equations form a natural choice for shape functions. Green's functions are used to calculate surface tractions and displacements due to applied loads. When Green's functions for surface tractions are also chosen as weight functions, dynamic stiffness matrix for indirect boundary element is obtained. Indirect boundary element method is far superior from accuracy point of view in comparison to other boundary elements developed, (Wolf and Darbre, 1984a,1984b). Moreover, symmetry is guaranteed in case of indirect boundary elements, making it computationally more attractive.

Both boundary integral-equation and boundary element methods suffer from a major draw back i.e. their solutions are frequency dependent and these cannot be used directly for time-domain analysis. Some techniques have been reported which permit the transient dynamic analysis using boundary element methods. Banaugh and Goldsmith (1963) and Niwa et al. (1976) have suggested that using the steady state harmonic solution in frequency domain and then carrying out Fourier synthesis, the transient solution in time domain can be obtained. Cruse and Rizo (1968), Cruse (1968) and Monalis and Beskose (1981,1983) have obtained the solution first in Laplace transform domain using boundary element method and then applied inverse transformation to obtain the solution in time domain. Direct formulation in time domain with step by step integration either using implicit or explicit or combined implicit-explicit time integration schemes is also possible.

Time domain boundary element methods are computationally very much uneconomical. For the same mesh consisting of 10 elements, number of operations required to be performed in various methods is presented in Table 1., (Wolf, 1988). From computational point of view, it can be observed from this table that only direct stiffness formulation in time domain is closer to a frequency domain analysis to some extent.

Table 1. Number of operations for various methods

No. of time steps	Indirect BEM ×10^{11}	Direct BEM ×10^{10}	Weighted Residual BEM ×10^9	Time Domain Stiffness Formulation ×10^8	Frequency Domain Stiffness Formulation ×10^6
1024	1.15	5.82	1.11	9.45	4.67
2048	4.61	23.2	4.11	37.8	9.34

3 A SUMMERY OF AVAILABLE MODELS FOR TIME DOMAIN ANALYSIS

From the discussion presented above, it is clear that out of the many techniques discussed, only few can be used for time domain analysis as techniques which are frequency dependent can not be used. The techniques available for simulation of boundaries in time domain analysis are therefore -

1. Elementary boundaries
2. Viscous boundaries
3. Superposition boundaries
4. Extrapolation algorithm
5. Boundary element methods

Out of these, viscous and superposition boundaries lead to permanent displacements at the end of the analysis in the entire mesh even for the elastic system. Boundary element methods are computationally highly uneconomical as is clear from Table 1. Therefore, elementary boundaries and extrapolation algorithm become the natural choice for simulation of boundaries in time domain analysis.

In this paper, an attempt has been made to correlate the possible error in the response with the distance from the structure to the elementary boundary. Using the proposed non-dimensional curves, elementary boundaries can be used with much higher degree of confidence.

4 PROPOSED IMPROVEMENT IN IMPLEMENTATION OF ELEMENTARY BOUNDARIES

In the proposed model, the semi-infinite soil mass has been modelled as a finite soil-mass with fixity at the boundary (Fig. 6.) and an attempt has been made to find the distance of the soil boundary from the foundation at which the reflected wave will have a negligible amplitude. One may argue that it will provide a large model resulting in computational inefficiency but this may not be true due to the fact that -

1. infinite/boundary elements, which provide alternative approach are also not computationally economical and

2. these infinite elements can be placed only beyond the zone of non-linearity in soil mass.

In case of heavily loaded structures, this zone spreads to a large extent. This zone of non-linearity further increases when dynamic loads are applied resulting in a large soil model requirement.

To find the proper location of the boundary, a parametric study has been carried out in time domain by varying :

1. the distance of the boundary from the foundation and

2. material damping of the soil mass.

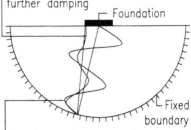

Fig. 6. Proposed soil model for dynamic analysis and its working principle

4.1 *Parametric Data*

Various parameters considered for the study of an axi-symmetric soil-foundation system, as shown in Fig. 7., are presented in Table 2.

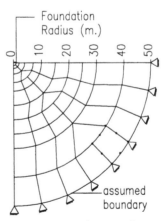

49 elements (8 noded) and 180 nodes

Fig. 7. Finite element model of disk on half space (axisymmetric)

Table 2. Parametric values adopted in the present study

S. N.	Parameter	Symbol	Units	Parametric Values
1.	Modulus of elasticity of disc,	E_d	N/m^2	2.5×10^{10}
2.	Modulus of elasticity of soil,	E_s	N/m^2	1.0×10^8
3.	Poisson's ratio of disc,	v_d		0.15
4.	Poisson's ratio of soil,	v_s		0.35
5.	Density of disc material,	ρ_d	g/m^3	2.2×10^3
6.	Density of soil material,	ρ_s	g/m^3	1.8×10^3
7.	Radius of foundation,	r		1.0
8.	Distance to soil boundary from center of foundation,	R		10,15,20,25, 30,35,40,45, 50
9.	Damping as % of critical damping,	ζ		2,5,10,20
10.	Longitudinal wave velocity in soil mass,	V_l	m/s	235.7
11.	Fundamental frequency of soil-foundation system (vertical mode),	f ω	Hz rad/s	22.0 138.2
12.	Maximum wave length,	λ_{max}		10.7
13.	Loading (Impulsive) - magnitude - duration		N s	90.0 10.0
14.	Time step		s	2.0
15.	Elements			8 noded iso-parametric elements

Impulse loading has been applied at the center of the foundation. To study the effect of location of boundary, the analysis has been carried out for various boundary locations, mentioned in table 2. To find the error in the vertical response of the disc, percentage error ε has been defined as -

$$e = \sqrt{\frac{\sum_i (d_{ref,i} - d_{R,i})^2}{\sum_i d_{ref,i}^2}} \qquad (2)$$

where, $d_{ref,i}$ = amplitude of the ith peak at reference boundary which was placed at a distance of 50 m for ζ=2 % and 4 % and at 40 m for ζ=10 % and 20 % and $d_{R,i}$ = amplitude of the i peak when boundary is at any distance, R. The denominator in eq. 2 has been calculated for only those peaks for which $d_{ref,i} \neq d_{R,i}$.

4.2 Discussion of Results

The response of circular foundation in terms of displacement versus time has been plotted in Fig. 8 for values of damping ratio, ζ = 5 % and for different values of the location of the assumed boundary, R = 10 m, 20 m, 30 m and 40 m. Similar plots for R = 15 m and 25 m have not been presented here due to limitation of space. Figure 9 shows the response of circular foundation for the assumed boundary location, R = 40 m and for different values of the damping ratio namely ζ = 2 %, 10 % and 20 %. Similar plots can also be made for other values of R.

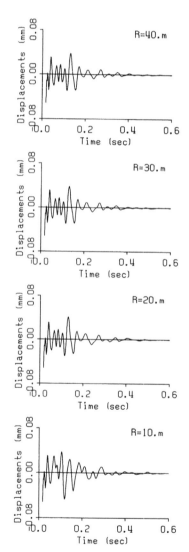

Fig. 8. Response of the circular foundation for ζ = 5.0%

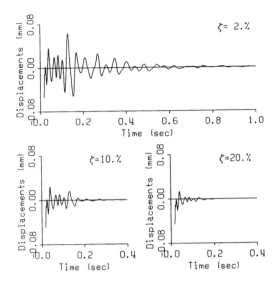

Fig. 9. Response of the circular foundation for R = 40. m

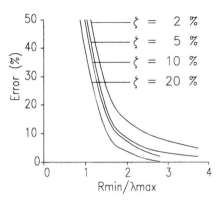

Fig. 10. Modelled soil domain V/s error curve

The errors calculated from the above study for various boundary locations and various damping ratios are presented in Table 3.

Table 3. Error in response at different boundary locations

ζ	Distance of boundary, R(m)					
(%)	10	15	20	25	30	40
2.0	67.2	23.7	15.2	11.0	7.8	5.0
5.0	59.4	15.4	9.4	5.4	3.2	1.9
10.0	54.8	14.3	8.2	3.9	2.1	-
20.0	42.9	6.6	5.4	1.1	0.5	-

Further, to present the results in a more generalised way, distance to the boundary has been non-dimensionalised with respect to the maximum wave length contained in the response spectrum, λ_{max}. Figure 10 shows the plot of percentage error in the response, ε versus the ratio, R_{min}/λ_{max} where R_{min} is the distance of the boundary up to which the soil should be modelled in order to allow a certain percentage error level, ε. This plot can be used to decide the location of the boundary for a specified permissible error in the response. As a typical example, for an error ≤ 5.0 % and $\zeta = 5.0$ %, the value of R_{min}/λ_{max} for Fig. 10 is 2.4 i.e. for a typical damping ratio, $\zeta = 5.0$ %, the boundary should be placed at a distance of 2.4 times the maximum wave length in order to have a maximum error of 5.0 % in the response.

From Fig. 8, it can also be observed that the effect of artificial boundary is to introduce two types of errors in the response, viz. i) change in the amplitudes of the resulting peaks and ii) frequency modulation of the response. From Fig. 10, it can be observed that the percentage error approaches exponentially to zero with increase in radial distance of the boundary and the damping ratio.

Singhal and Chandrasekeran (1988) have carried out numerical experiments on an oil tank having radius, r = 20. m and with $\zeta = 5.0$ %. Wave velocity, V and maximum wave length, λ_{max} in vertical mode were 250 m/s and 42 m respectively. Horizontal distance giving satisfactory results was found out to be as R = 4.r. Therefore, for $R/\lambda_{max} = 2.0$, Fig. 10. predicts an error of 9.% which can not be estimated using their results. Thus, the proposed model is more reliable.

As the present parametric study has been carried out for impulse loading only, the results are likely to be applicable to similar conditions e.g. impact loading, blast loading and free vibration after any type of loading. However, it is expected that error will be less than that predicted on basis of Fig. 10 for long duration cyclic loading and random excitations.

4.3 Mathematical Analogy

The ratio, R_{min} / λ_{max}, used in Fig. 10, can be interpreted as the minimum number of cycles which the wave will pass through before reaching the boundary. Equal number of cycles will be required for the wave to comeback to the foundation after reflection. This interpretation directly provides the analogy with the law of logarithmic decrement in which, amplitude d_2 after n cycles from the occurrence of amplitude d_1 is given by -

$$d_2 = d_1 / e^{2\pi\zeta n} \qquad (3)$$

377

but, d_2 is also the probable error in the response as it will be the amplitude of the reflected wave if n cycles are required to reach the wave up to the boundary. Therefore :

$$\varepsilon \ = \ (d_2 \ / \ d_1) \ 100. \ \% \ = \ 1 \ / \ e^{2\pi\zeta n} \qquad (4a)$$

$$\text{or}, \varepsilon \ \propto \ (d_2 \ / \ d_1) \ \text{and} \ e^{-n} \qquad (4b)$$

i.e. error is inversely proportional to e^n . This can also be observed from Fig. 10. which provides firm support to the proposed model and its reliability.

5 ELASTO-PLASTIC ANALYSIS USING PROPOSED METHOD

An elasto-plastic analysis of the same soil-foundation system (Fig. 6) has also been carried out to check the validity of observations made in the linear analysis. Various parameters used in the non-linear analysis have been presented in Table 4. The analysis has been carried out for a steady state cyclic loading because for the impulse loading, used in linear analysis, significant yielding of soil mass was not observed.

Table 4. Parametric values adopted for the elasto-plastic analysis

S. N.	Parameter	Symbol	nits	Parametric Values
1.	Cohesion of the soil,	c	/m²	1.0×10^4
2.	Angle of internal friction of soil,	φ	eg.	37.0
3.	Yield criterion,			Mohr-Coulomb
4.	Loading,			Cyclic (Steady state)
	-circular frequency,	ω	ad/sec	125.0
	-load amplitude,	n	N	90.0
5.	Distance to assumed boundary,	R		15 and 40
6.	Damping ratio	ζ		5.0

5.1 Discussion of Results

The response of the circular foundation-soil system obtained on the basis of elasto-plastic analysis has been plotted in Fig. 11 for the damping ratio, $\zeta = 5$ % and for values of location of the boundary, R = 15 m and 40 m. The value of percentage error, ε has been calculated for R = 15 m by taking the reference boundary location, R = 40 m and has been found to be 9.4 % as against the corresponding value of 15 %

predicted on basis of Fig. 10 for the linear analysis. This confirms the earlier statement that the error for cyclic loading will be less than that predicted on the basis of Fig. 10.

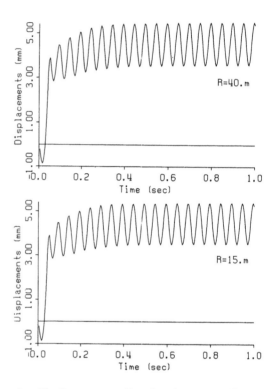

ig. 11. Response of the circular foundation for ζ = 5.0% for cyclic loading and non-linear soil properties.

6 CONCLUSIONS

The paper presents an improved version of elementary boundary model for semi-infinite soil media and the parametric study for location of the soil boundary. The finite element analysis has been carried out in time domain considering linear elastic and elasto-plastic characteristics of soil mass. It can be inferred from the study that -

1. The percentage error in response of the circular foundation approaches exponentially to zero with increase in radial distance of the assumed boundary and the damping ratio.
2. Figure 5 can be used with full reliability for deciding the location of boundary of the semi-infinite soil media for any type of loading and for both linear as well as non-linear soil behaviour.

REFERENCES

Banaugh R.P. & Goldsmith W. 1963. Diffraction of Steady Elastic Waves by Surface of arbitrary Shape. *Jr. Appl. Mech. ASME.* 30:589-597.

Cruse T.A. and Rizzo F.J. 1968. A Direct Formulation and Numerical Solution of the General Transient Elastodynamic Problem I. *Jr. Math. Anal. Appl.* 22:244-259.

Cruse T.A. 1968. A Direct Formulation and Numerical Solution of the General Transient Elastodynamic Problem II. *Jr. Math. Anal. Appl.* 22:341-355.

Cundal P.A., Kunar R.R., Carpenter P.C. and Marti J. 1974. Solutions of Infinite Dynamic Problems by Finite Modelling in Dime Domain. *Proceedings of the 2nd International Conference on Applied Numerical Modelling.* 339-351. Madrid. London:Pentech Press.

Ghosh S. and Wilson E.L. 1969. Analysis of Axisymmetric Structures Under Arbitrary Loading. *EERC Report No. 69-10,* Univ. of California. Berkeley:U.S.A.

Kunar R.R. and Marti J. 1981. A Non-reflecting Boundary for Explicit Calculations. *Computational Methods for Infinite Domain Media-Structure Interaction. Appl. Mech. Div. ASME.* 46:183-204.

Liao Z.P. and Wong H.L. 1984. A Transmitting Boundary for the Numerical Simulation of Elastic Wave Propagation. *Soil Dynamics and Earthquake Engineering.* 3:174-183.

Lysmer J. and Kuhlemeyer R.L. 1969. Finite Dynamic model for Infinite Media. *Jr. Engg. Mech. Div. ASCE.* 95:859-877.

Monalis G.D. and Beskose D.E. 1981. Dynamic Stress Concentration studies by Boundary Integrals and Laplace Transform. *Int. Jr. Numer. Methods Engg.* 17:573-599.

Monalis G.D. and Beskose D.E. 1983. Dynamic Response of Lined Tunnels by an Isoparametric Boundary Method. *Comp. Methods Appl. Mech. Engg.* 36:291-307.

Niwa Y., Kobayashi S. and Fukui T. 1976. Applications of Integral Equation Method to Some Geomechanical Problems. *Numerical Methods in Geomechanics.* (Ed. C.S. Desai). 120-131. New York:ASCE.

Singhal N.C. and Chandrasekeran A.R. 1988. Vibration Characteristics of a Soil-Structure System- An Appraisal of Foundation Soil Extent and its Inertia. *Indian Geotech. Conf.* 1:271-275. Allahabad.

Wolf J.P. 1985. *Dynamic Soil-Structure-Interaction.* New Jersey:Prentice-Hall.

Wolf J.P. 1988. *Soil-Structure-Interaction Analysis in Time Domain.* New Jersey:Prentice-Hall.

Wolf J.P. and Darbre G.R. 1984. Dynamic-Stiffness Matrix of Soil by the Boundary Element Method: Conceptual Aspects. *Earthquake Engineering and structural Dynamics.* 12:285-400.

Wolf J.P. and Darbre G.R. 1984. Dynamic-Stiffness Matrix of Soil by the Boundary-Element Method: Embedded Foundation. *Earthquake Engineering and structural Dynamics.* 12:401-416.

Seismic Behaviour of Ground and Geotechnical Structures, Sêco e Pinto (ed.) © 1997 Balkema, Rotterdam, ISBN 90 5410 887 8

Validation of analytical procedure on Daikai Subway Station damaged during the 1995 Hyogoken-Nanbu earthquake

Nozomu Yoshida, Susumu Nakamura, Iwao Suetomi & Jun-ichi Esaki
Sato Kogyo Co., Ltd, Tokyo, Japan

ABSTRACT: Presented are an analytical procedure to explain the mechanism of the collapse of the Daikai subway station during the 1995 Hyogoken-nanbu earthquake by considering nonlinear behaviors of both the subsoil and the structure. Then the validity of the method is examined by two approaches. In the first approach, computed displacement is compared with the observed permanent displacement of the small underground structure near the station. In the second approach, the same procedure is applied to the neighboring station which is similar structural situation, but not completely collapsed. Both approaches indicate that the method is effective. It is concluded that the station was collapsed by the lateral load caused by the nonlinear behavior of the ground.

1 INTRODUCTION

Underground subway structures were severely damaged during the 1995 Hyogoken-nanbu earthquake. Among them, complete collapse of the Daikai Subway Station was one of the amazing event because it has been believed that underground structures is safe against the earthquake. Because of its importance, many analyses have been done to explain the mechanism of the failure, and concluded that it is reasonable for the station to collapse under the earthquake, or that their method is valid or effective to explain the damage. Since the structure will eventually collapse under the very strong earthquake, however, even if the analysis indicates complete failure of the structure, it does not imply that the method is valid or effective; all the approach will give the same conclusion, i.e., collapse of the structure, under very severe external load. The validity of the method should not be evaluated by the collapse of the structure but should be evaluated by the explanation of the whole behavior.

The authors belongs to the company that constructed and restored the station, and have conducted the analysis to make clear the mechanism of the failure of the station and to ensure the accuracy of the analysis (Yamato *et al.*, 1996; Nakamura *et al*, 1996; Nakamura *et al.*, 1997; Yoshida and Nakamura, 1996; Yoshida, N. *et al.*, 1996). This includes in-situ measurement of the permanent displacement nearby the station and the application of the method to the neighboring station that was not completely collapsed. Those approaches are summarized in this paper.

2 ANALYSIS AT DAIKAI SUBWAY STATION

Figure 1 shows cross-section of the completely collapsed portion of the Daikai Subway Station. As reported before (Iida *et al.*, 1995), walls in the

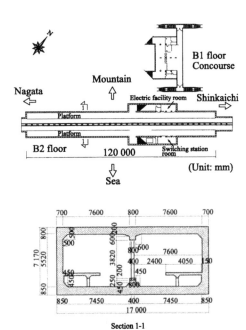

Fig.1 Plan and section of the Daikai Station

transverse direction seemed to play an important role to resist horizontal load, discussion is limited only on this portion where contribution of the transverse wall against horizontal load is hardly expected.

The analysis is conducted in two steps. In the first step, the whole soil-structural system is analyzed by the dynamic response analysis considering the nonlinear behavior of the ground. The structure is modeled into a frame with elastic behavior. In the second analysis, nonlinear behavior of the structure is considered in the static analysis, in which the external load computed by the dynamic response analysis is applied incrementally as the external load.

1.1 Soil-structure interaction analysis

Figure 2 shows the soil profile in the longitudinal direction obtained from the borehole test

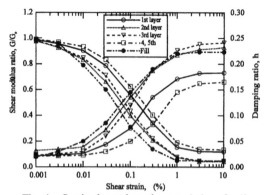

Fig. 4   Strain dependent characteristics of soil

surrounding the station. Since borehole data at D-1 and B-3 that are obtained at the mountain and sea sides of the station are nearly the same, the ground is modeled into a level ground in the transverse direction. The finite element mesh of the soil-structure system is shown in Fig. 3, where the station is modeled into a box-type frame with a column at the center and rigid region at the beam-to column intersection. Fill surrounding the station is also considered.

Strain dependent shear modulus and damping characteristics used in the analysis is shown in Fig. 4, where result of the dynamic deformation test is used for the fill material and the empirical equation proposed by Yasuda and Yamaguchi (1985) is employed to evaluate the nonlinear characteristics of the soil other than the fill from soil classification. Elastic property and unit weight is shown in Fig. 3 where γ denotes unit weight, V_s denotes shear wave velocity and ν denotes Poisson's ratio. Shear modulus of the frame is set 1.27×10^6 tf/m^2, Poisson's ratio is 0.2, and unit weight is 2.4 tf/m^3.

Earthquake wave recorded at the Kobe University is treated as outcrop wave at the base because the site is located on the rock. Both horizontal in the transverse direction and up-down component are applied simultaneously.   Equivalent

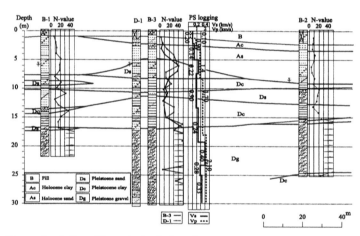

Figure 2   Soil profile at the Daikai station

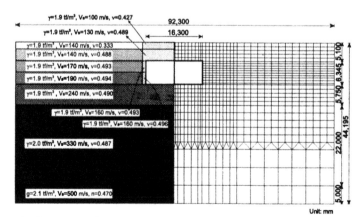

Fig. 3 FEM mesh and material property of soil for the Daikai station.

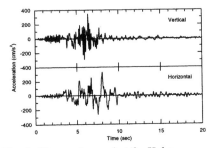

Fig. 5 Observed record at the Kobe University

linear method is employed in considering the nonlinear behavior of the soil.

The analysis is conducted in two steps in order to input both horizontal and vertical earthquake simultaneously because the program requires shear modulus and Poisson's ratio for two elastic constants and the latter is kept constant. Shear modulus in the equivalent linear analysis depends on the shear strain. When Poisson's ratio is kept constant, however, the bulk modulus also changes depending on the shear strain, which is against the experimental fact. In the first analysis, therefore, only the horizontal earthquake is applied. New Poisson's ratios are computed from reduced shear modulus keeping the bulk modulus constant. The second analysis is conducted without iteration using the new pairs of the shear modulus and bulk modulus.

Peak accelerations and displacements at the representative points are shown in Fig. 6. Peak response of the generalized stress of the frame is shown in Fig. 7. Moment-axial force interactions at three critical points of the structure are shown in Fig. 8, where chained lines denote the state where external fiber concrete at the compressive side reaches failure strain. Here, it is noted that, since the elastic behavior is assumed for the frame, the

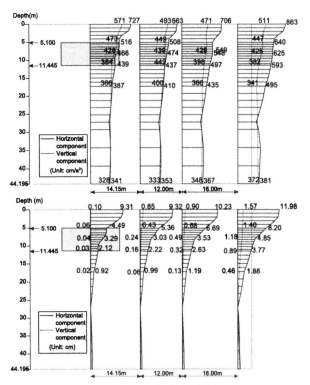

Fig. 6 Peak response under Kobe University wave

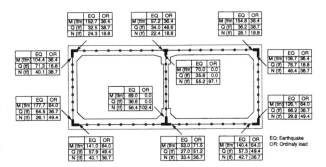

Fig. 7 Peak response of the structure

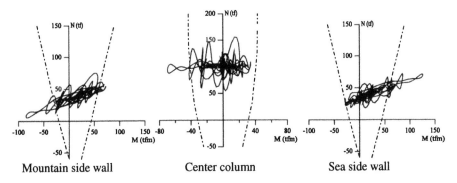

Mountain side wall   Center column   Sea side wall

Fig. 8 Moment-axial force interaction

383

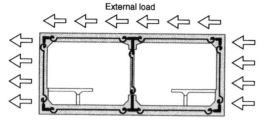

External load

Fig. 9 Frame model for nonlinear analysis

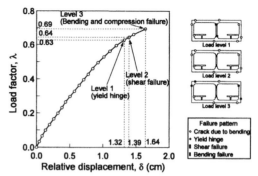

Fig. 11 Load-displacement relationship

stress point may exceed the failure surface. It is also seen that the vertical component of the earthquake input hardly affect the failure of the station because change in the moment carrying capacity under the change in the axial load is not much. It is noted that the inclination of the stress paths in the walls in Fig. 8 does not come from the vertical earthquake motion but comes from the equilibrium condition with the horizontal force.

1.2 Nonlinear analysis of the structure

Nonlinear behavior of the structure is considered in the static analysis. Figure 9 shows the frame model used in the analysis. Each structural members are modeled into rigid bar with nonlinear rotational spring at both ends. The moment-curvature relationship of each structural members are computed by considering the elastic-perfectly plastic behavior for the reinforcing bar and e-function for the concrete. The yield stress of the reinforcing bar

is 3.12 tf/m^2, in which effect of strain hardening as well as scattering of the yield stress are considered. Compressive strength of the concrete is set 3.8 tf/m^2, which is an average value obtained by the Schmidt hammer test. Then the curvature is integrated along the longitudinal axis by assuming the skew symmetrical moment distribution along the axis in order to obtain the rotational angle of the spring at the end. Finally, the moment-curvature relationship is modeled into tri-linear model, which is shown in Fig. 10.

Force acting on the frame when bending moment of the center column reaches its peak value is back analyzed from the bending moment distribution of the dynamic response analysis. It is applied incrementally and proportionally to the model shown in Fig. 9. Since elastic behavior is assumed in the dynamic response analysis, it is

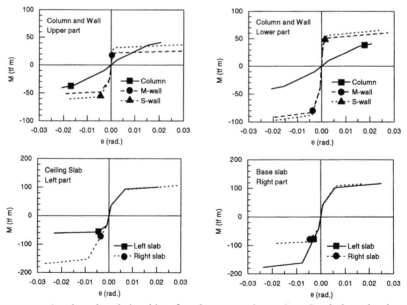

Fig. 10 Moment-rotational angle relationship of each structural member. Symbol marks denote the point when center column collapsed by shear.

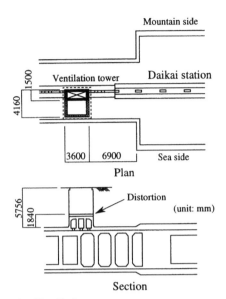

Fig. 12   Ventilation tower

Photo 1 Distortion of the upper part of the
ventilation tower

obvious that full application of back analyzed load may be greater than load carrying capacity of the frame, hence resulting in infinite displacement. Therefore, discussion is made by focusing on the displacement.

Figure 11 shows load-relative displacement relationship. Here, load factor λ is defined to be the ratio of the applied load to the load acting on the structure when center column reaches its peak value in the dynamic response analysis, and δ is the relative displacement between the ceiling and base slabs. At stage 1 where $\lambda = 0.63$, reinforcing bar at the center column yields. Then center column fails by shear at stage 2 where $\lambda=0.64$. Relative displacement δ between upper and lower slab is 1.39 cm at this stage. If center column does not fail at

this stage, then the concrete in the center column reaches compressive failure at stage 3 where relative displacement is 1.64cm. Reinforcing bar at the exterior wall already yields at this stage.

As seen in Fig. 10, when center column is supposed to collapse at stage 3, moment at exterior wall and slab just reaches yielding of reinforcing bar or less. Considering that ductility factor of wall is greater than 3.7, relative displacement when exterior wall collapse must be greater than $1.64 \times 3.7 \approx 6.1$cm. On the other hand, as seen in Fig. 6, relative displacement of the ground between the depths of both slabs is 2.7cm. Therefore, it is concluded that, under the Hyogoken-nanbu earthquake, the center column collapsed but exterior wall did not collapse although reinforcing bar yielded. This is consistent with actual damage.

3 DISPLACEMENT OF VENTILATION TOWER

There was a ventilation tower at about 70 m southeast from the Daikai station. Figure 12 shows the location and details of the tower. As shown in Photo 1, there found distortion of about 4 cm at about 4 m from the ground surface.

This indicates that the relative displacement of the ground surface from GL-4m was larger than 4 cm. Relative displacement between the ground surface and GL-4m obtained by the dynamic response analysis is, however, about 1.3 cm, which is much smaller than observed displacement. This indicates that, although analysis in the previous section seemed to explain the damage mechanism well, it was not sufficient to conclude that the method is effective.

Smaller evaluation of the displacement may be true by considering the difference of the location between the Kobe University and the Daikai station. The former is about double distant from the epicenter compared with the latter. Therefore, earthquake wave recorded at much closer site should be used in the analysis. In the next section, therefore, we repeat the same analysis using the earthquake wave at GL-83m of the Port Island. Since the wave observed at the site is affected by the reflecting wave from upward, incident wave at GL-83m is computed from one-dimensional nonlinear dynamic response analysis and is multiplied by a factor of 2 to obtain outcrop wave, which is shown in Fig. 13.

4 ANALYSIS BY PORT ISLAND WAVE

The same procedure with chapter 2 is conducted under the Port Island wave. Figure 14 shows peak response values obtained by the dynamic response analysis. The peak response becomes nearly double.

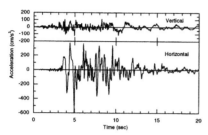

Fig. 13 Outcrop waveform at GL-83m at Port Island

The peak acceleration at the ground surface in free field, for example, increases from 863 to 1487 cm/s^2 by changing the input motion. In the same manner, the peak displacement increases from 12.0 to 22.5.

The maximum relative displacement between the ground surface and GL-4m at free field is about 3.9 cm, which is a little smaller than the permanent displacement observed at the ventilation tower, 4 cm, but nearly the same order. Therefore, the intensity of the earthquake input at the Daikai station may be only a little larger than Port Island wave, but not much. In the followings, therefore, we treat the Port Island wave as actual input wave.

Figure 15 shows horizontal load Q versus relative displacement δ relationships under Port

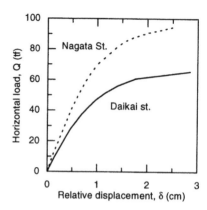

Fig. 15 Horizontal load versus relative displacement relationship

Island wave. The curve stopped when center column fails by the compressive failure under the action of the horizontal load. Figures 11 and 15 is the same types of the figure, but there is a significant difference between the failure displacement. The confinement effect of core concrete by hoop and increase of displacement due to pull out of reinforcing bar from the base (Railway Research Institute, 1992) are taken into account in this analysis.

Relative displacement of the ground nearby the station is 3.64 cm, which is much larger than the displacement corresponding to the failure, i.e. 2.85 cm. Moreover, relative displacement of the ground is still much smaller than failure displacement of the exterior walls. Therefore, mechanism explained in chapter 2 still holed; the center columns collapsed but exterior walls and slabs did not completely collapsed.

5 ANALYSIS AT NAGATA STATION

The Nagata station is located one station west to the Daikai station. It was also damaged during the Hyogoken-nanbu earthquake, but not so severe like the Daikai station; about 1/3 center columns were damaged. This station is also analyzed in the same procedure shown in the preceding. Figure 16 shows finite element mesh and elastic properties. Compared to the ground near the Daikai station, shear wave velocity beneath the station is a little smaller and that beside the station is a little larger in the Nagata station compared with the Daikai station.

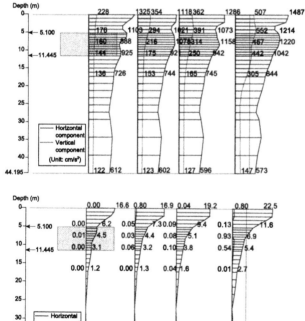

Fig. 14 Peak response of Daikai station under Port Island wave

Strain dependent characteristics are also computed based on the same procedure with above.

Figure 17 shows peak acceleration and displacement in the horizontal direction, in which those at the Daikai station are also shown. Peak acceleration is generally larger at the Daikai station than at the Nagata station, because, at the Nagata station, it decreases below the station because of the existence of softer layer. On the other hand, displacement is larger at the Nagata station than at the Daikai station.

Figure 18 shows time histories of the relative displacement between ceiling and base slabs at both station. The relative displacement of the ground at the Nagata station is 2.7 cm, which is smaller than that at Daikai station (3.64 cm). This is because the subsoil beside the Nagata station is stiffer than that that at the Daikai station and because the acceleration at the Nagata station decreases because of the above mentioned reason.

Horizontal load versus relative displacement relationship is shown in Fig. 14. By comparing with the Daikai station, load carrying capacity is larger at the Nagata station and relative displacement at the failure is smaller compared with those at the Daikai station. This seems to indicate that Nagata station is easier to collapse than the Daikai station. However, actual phenomena was inverse. This can be explained be comparing the ground displacement near the station. At the Daikai station, the relative displacement at failure by the static analysis is 2.85 cm whereas that by the dynamic analysis is 3.64 cm, which is much larger than the failure displacement. On the other hand, relative displacement at failure is 2.55 cm at the Nagata station and that by the dynamic analysis is 2.7 cm. They are nearly the same order although the displacement by the dynamic analysis is a little larger. This error may be caused since contribution of transverse wall is not taken into account, and because of the error in evaluating the material property of soil.

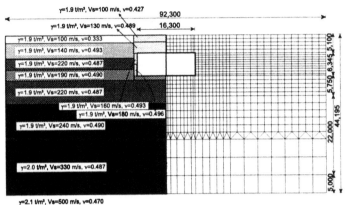

Fig. 16   Finite element mesh and material property used in the analysis at Nagata Station.

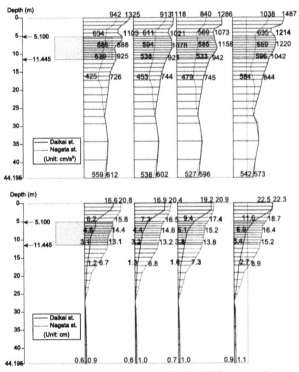

Fig. 17   Peak response value at Daikai and Nagata stations under Port Island wave

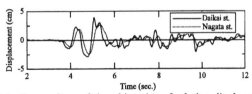

Fig. 18   Comparison of time histories of relative displacement at Daikai and Nagata station

387

6 CONCLUDING REMARKS

Mechanism of the failure of the Daikai subway station is explained by two step nonlinear analysis: the dynamic response analysis considering the nonlinear behavior of the ground and the static analysis considering the nonlinear behavior of the structure in which external load obtained by the dynamic response analysis is applied. The accuracy or effectiveness of the method is examined by two approaches in addition with the damage investigation.

Through the analysis, it is confirmed that center column collapsed by the combination of bending and shear, but lateral wall and slabs does not completely fail although tensile reinforcing bar yields.

Judging from the permanent displacement at the ventilation tower near the Daikai station, earthquake input to the station should be much larger than the record at the Kobe university and close or a little larger than the record at Port Island.

Although relative displacement at the failure between the ceiling and base slabs are smaller and the location is closer to the epicenter at the Nagata station than at the Daikai station, complete failure did not occur at the Nagata station whereas it occurred at the Daikai station. This can be explained by comparing the failure displacement with the relative displacement during the earthquake; they are the same order in the Nagata station whereas the former is much smaller in the Daikai station.

REFERENCES

Iida, H. et al. (1996): Damage to Daikai Subway Station, Special Issue of Soils and Foundation, JGS, pp.283-300

Nakamura, S., Suetomi, I. and Yoshida, N. (1996): Estimation of aseismic ground displacement around Daikai Subway Station based on earthquake damage, Proc., 31th Symposium of Geotechnical Engineering, pp.1275-1276 (in Japanese)

Nakamura, S., Esaki, J. and Suetomi, I. (1997): Analysis of Damage Factor of Subway Focusing on Difference of Damage, Proc., 2nd Symposium on the Great Hanshin-Awaji Earthquake Disaster, JSCE, pp. 171-178 (in Japanese)

Railway Research Institute (1992): Design Specification of Railway Structure, Concrete Structure, Maruzen, pp.425-428

Yamato, T. et al., Damage to Daikai Subway Station, Kobe Rapid Transit System and Estimation of Its Reason during Hyogoken-nanbu Earthquake, 1995, Proc. of Japan Society of Civil Engineer, No. 537/I-35, pp.303-320 (in Japanese)

Yasuda, S. and Yamaguchi, I. (1985): Dynamic Shear Modulus Obtained in the Laboratory and In-situ, Proc., Symposium on Evaluation of Deformation and Strength of Sandy Gravels, JSSMFE, pp.115-118 (in Japanese)

Yoshida, N. and Nakamura, S. (1996): Damage to Daikai Subway Station during the 1995 Hyogoken-nanbu Earthquake and Its Investigation, Proc., Eleventh World Conference on Earthquake Engineering, Acapulco, Mexico, Paper No. 2151

Yoshida, N. et al. (1996): Dynamic Analysis of Daikai Subway Station, Kobe Rapid Transit System, Proc. Seminar on Dynamic Analysis on the Great Hanshin-Awaji Earthquake Disaster, JGS, pp.38-53 (in Japanese)

3 Miscellaneous

Seismic Behaviour of Ground and Geotechnical Structures, Sêco e Pinto (ed.)© 1997 Balkema, Rotterdam, ISBN 90 5410 887 8

Damping ratio of soils from laboratory and in situ tests

Diego C. F. Lo Presti & Oronzo Pallara
Politecnico di Torino, Italy

Antonio Cavallaro
University of Catania, Italy

ABSTRACT: This paper considers the influence of N and $\dot{\gamma}$ on the damping ratio (D) of two undisturbed Italian clays and of three reconstituted granular soils. The experimental results were obtained in the laboratory by means of a Resonant Column/Torsional Shear apparatus.
The damping ratio of a reconstituted silica sand was also determined, at small strains, from seismic tests performed in a large Calibration Chamber. The spectral ratio and the spectral slope methods were used for this purpose. The capability and limits of these methods are discussed in the light of the strong influence of test conditions on damping ratio even at very small strains.
Finally the experimentally determined values of D are compared against those predicted by means of simple non-linear models that incorporate the Masing rules.

1. INTRODUCTION

Grade 3 methods of zonation for ground motion, that are reported in the Manual for Zonation (TC4 1993) suggest the use of linear equivalent analysis to assess the response of soil deposits to seismic motion.

The most important input parameters required for such an analysis are shear modulus (G) and damping ratio (D). In particular it is fundamental to know the dependence of these parameters on the shear strain level which is considered the most important factor of influence. In reality, both G and D are affected by several factors, such as the frequency or strain rate (f, $\dot{\gamma}$), the drainage conditions, the number of loading cycles (N), the cyclic prestraining at larger strains and so on. The damping ratio is in particular more sensitive than stiffness to the above mentioned factors as shown in recent works by Tatsuoka and Kohata 1995, Toki et al. 1995, Shibuya et al. 1995, Stokoe et al. 1995, Tatsuoka et al. 1995 and Ashmawy et al. 1995.

This paper tries to evaluate the influence of N and f (or $\dot{\gamma}$) on the damping ratio of two natural clays and three different reconstituted sands from cyclic loading torsional shear tests (CLTST) and Resonant Column Tests (RCT) performed with a Torsional Shear/Resonant Column apparatus (Lo Presti et al. 1993).

The damping ratio values obtained from seismic tests performed in a large Calibration Chamber (CC) on HK-CLK sand (Fioravante et al. 1997) are also presented in this paper.

Seismic measurements were performed with arrays of three miniature geophones, using the geophone located at one end of the arrays as a source and the other two as receivers. The damping ratio values were determined by analysing the traces of shear waves propagating along the horizontal direction with particle motion in the vertical direction (S_{hv}) and those propagating in the vertical direction with particle motion in the horizontal direction (S_{vh}). These shear waves practically correspond to those generated during Cross hole (CH) and Down Hole (DH) tests, respectively. It is reasonable to assume that the seismic tests in CC simulate CH and DH tests performed in ideal conditions: i) ideal coupling between soil and geophone, ii) homogeneous soil, iii) well known boundary conditions, soil density and imposed stresses.

Thanks to the use of a couple of receivers, it was possible to analyse the data by means of the so called "spectral ratio" and "spectral slope" methods (Redpath et al. 1982, Mok 1987, Stewart and Campanella 1991, Fuhriman 1993, Mancuso 1994). Both methods consider that the vibration amplitude A at a given distance from the source can be analytically expressed in the following way:

$$A = S \cdot I \cdot e^{-\alpha R} / R \qquad (1)$$

where S is the amplitude of the source impulse, R is the distance of the receiver from the source, α is the material attenuation and I is the soil-receiver interaction function.

The expression of eq. (1) implicitly assumes that near field effects can be neglected as only the far field term 1/R accounts for geometric spreading.

When a seismic test is performed using a pair of receivers located at different distances from the source, it is possible to define a spectral ratio of two body wave amplitudes in the following way:

$$SR(f) = \frac{A_1(f)}{A_2(f)} = \frac{R_1 I_2}{R_2 I_1} e^{-\alpha(R_2 - R_1)} \qquad (2)$$

From the spectral ratio, it is possible to determine attenuation and hence damping, if one makes some assumptions. A common assumption for both methods is that $I_2 = I_1$. Moreover the spectral ratio method assumes that attenuation is frequency independent ($\alpha = const.$), while the spectral slope method assumes frequency dependent attenuation ($\alpha = k \cdot f$).

The assumption of the same interaction function for both receivers (i.e. $I_2 = I_1$) seems, to the authors, to be unrealistic when applied to in situ seismic tests such as CH or DH. Infact in this case, the receiver is not directly in contact with the soil and there is no guarantee that the grouting between the soil and PVC casing is perfectly homogeneous. However, in the case of the tests performed in CC it is possible to consider that the geophones are ideally installed and that the specimen is homogeneous, at least in the horizontal plane, so that the assumption that $I_2 = I_1$ seems reasonable.

When the propagation of body waves takes place in heterogeneous soils (for instance: soils with wave propagation velocities that increase with depth) the wave amplitude is attenuated because of the occurrence of reflection and refraction phenomena at the interface of layers with different velocities. According to Stewart and Campanella (1991) the measured wave amplitude should be corrected to take these phenomena into account by expressing the far field geometric damping by means of the term $1/R^n$ (n>1). In principle, the S_{vh} waves in CC travel through a heterogeneous medium. Such a heterogeneity is a consequence of the σ'_{vc} increase with depth due to the own weight of the sand. In practice, the degree of heterogeneity in CC is relevant only at very low consolidation stresses. The wave amplitude correction above indicated is necessary only when the Spectral Ratio method is used to interpret the data. Such a correction has not been applied to the data of this research.

Seismic tests in a CC on reconstituted mortar sand specimens have also been performed by Mok (1987) and Khwaja (1993) with conditions similar to those above described.

2. EXPERIMENTAL SET-UP

The Torsional Shear/Resonant Column apparatus can house solid and hollow specimens and is equipped with an electric motor (SBEL, Arizona) which is capable of a maximum torque of about 1 Nm and can operate under stress control conditions.

An accelerometer is used to determine resonance during RCT. The overall rotational shear strain ($\gamma = \frac{2}{3} \cdot \frac{\theta}{h} \cdot \frac{R_o^3 - R_i^3}{R_o^2 - R_i^2}$; with: θ=rotation, h=specimen height, R_o=outer radius and R_i=inner radius) during CLTST is measured from platen-to-platen by means of a pair of proximity transducers with a resolution of 0.1 μm. This resolution corresponds to a shear strain of about 0.0001%.

The applied torque is computed by means of a torque-voltage calibration curve (Lo Presti et al. 1993). Acquisition of the applied voltage and of the output of the two proximity transducers used for the rotation measurement was carried out by means of three simultaneously triggered digital multimeters. Thus, no delay exists between the torque and rotation measurements. According to Toki et al. (1995) a delay between the torque and rotation acquisition could introduce a certain error in damping determination.

In the case of RCT, the damping ratio was determined during forced vibrations by means of the so called half-power method.

In the case of cyclic torsional shear tests, it was decided the acquire a minimum of 30 data points per cycle. This limit was established by Pallara (1995) in order to accurately define the area enclosed by the loop and hence to obtain stable and accurate damping measurements. The indication given by Toki et al. (1995) who suggest more than forty data points per cycle, is more severe.

Seismic tests were performed in the ISMES CC which houses a cylindrical specimen, 1.2 m in diameter and 1.5 m in height.

The specimens were reconstituted by pluviation through air. During formation of the specimens, the pluviation process was interrupted at a pre-established elevation in order to place several miniature cylindrical geophones on the specimen surface. The specimen was saturated with deaired water after formation.

The source geophone was excited by means of a

50 V peak-to-peak sine wave with frequencies of 2000 Hz. Typical lengths of the travel path between two receivers ranged from 45 to 60 cm. The above conditions minimise the so called "near field" effects.

The CC specimens were subjected to stepwise loading with stress increments for σ_1 of 25-50 kPa with σ_1 corresponding to either σ_v or σ_h depending on the chosen path, where: σ_1 is the major principal effective stress, σ_v is the vertical effective stress and σ_h is the horizontal effective stress. The seismic tests were performed, at the end of each consolidation step, by generating the groups of body waves and measuring their velocities. Only a limited number of the acquired wave traces have been analysed in this paper. Typical wave traces obtained in the case of HK-CLK sand are shown in Fig. 1. These waveforms have been pre-processed by means of a cosine bell filter before their use for damping determination.

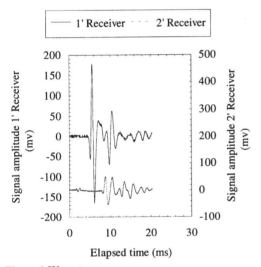

Figure 1 Wave traces

3. SOILS AND TEST PROGRAMME

Laboratory tests were performed on Pisa and Augusta clay specimens. The Pisa clay samples were retrieved from depths of between 12 and 17 meters at the site of the Leaning Tower by means of a Laval sampler (La Rochelle et al. 1991). The tested samples belong to the deposit that is locally called "Upper Pancone clay" which is a soft, lightly overconsolidated (OCR = 1.5 to 2.0), quaternary, marine clay with low to medium PI. Detailed information on the Pisa clay deposit is given by Costanzo et al. (1994). The Augusta site is located

on the east coast of Sicily. The Augusta clay is a medium stiff, overconsolidated (OCR = 2.0 to 6.0), quaternary marine clay with low to medium PI. Samples were retrieved with a Shelby sampler of 86 mm diameter. Detailed information on the Augusta clay deposit is given by Maugeri et al. (1994) and Cavallaro (1997).

The clay specimens were reconsolidated to $\sigma'_{vc} = \sigma'_{vo}$ (i.e. the best estimate of the in situ vertical stress). The majority of the tests were isotropically consolidated. Only a few specimens were reconsolidated with $K_c = \sigma'_{hc} / \sigma'_{vc} = K_o$. The K_o coefficient of earth pressure at rest, for the considered Pisa clay specimens, is about 0.65. In the case of Augusta clay $K_o \cong 1$. After reconsolidation the specimens were subjected to three different kinds of tests:
- monotonic loading torsional shear tests at constant stress rate (MLTST)
- cyclic loading torsional shear tests (CLTST) performed under stress control with decreasing frequencies in order to keep the equivalent cyclic strain rate constant ($\dot{\gamma} = 240 \cdot f \cdot \gamma [\% / \min]$; with γ = single amplitude shear strain). Values of $\dot{\gamma} = 0.04$ and $0.4 [\% / \min]$ were used in the experiments.
- resonant column tests (RCT). In this case, the equivalent strain rates imposed on the specimen increases with an increase of the strain level. The tested specimens experienced equivalent strain rates of 10 to 2500 %/min.

Table 1 Test Conditions for Clay Specimens

Site Test No.	σ'_{vc} [kPa]	K_c	e	PI	MLTST	CLTST	RCT	Specimen
Pisa 1	112	0.65	1.606	55	D		U	H
Pisa 2	112	1	1.572	47	D		U	H
Pisa 3	138	1	1.023	21		U	U	S
Pisa 4	114	1	0.901	22	U			S
Pisa 5	104	1	0.896	22		U		S
Augusta 6	182	1	0.684	29		U	U	S
Augusta 7	377	1	0.834	38	U		U	H
Augusta 8	398	1	0.768	38		U	U	H

where: D= Drained U=Undrained; H= Hollow cylindrical specimen ($R_o = 25$ mm $R_i = 15$ mm h=100 mm); S= Solid cylindrical specimen (R = 25 mm h=100 mm)

In some cases, the same specimen was first subjected to monotonic loading, then to cyclic static loading, after a rest period of 24 hrs with opened drainage, and eventually, after another 24 hrs of rest with opened drainage, to a RC test. A list of the tests performed on Pisa and Augusta clays is reported in Table 1. The size and shape of the specimens are also indicated in the same Table.

Table 2 Main Characteristics of the Tested Granular Soils

Soil	FC	G_s	D_{50}	U_c	e_{max}	e_{min}	Mineral
	[%]	[-]	[mm]	[-]	[-]	[-]	
Toyoura	-	2.650	0.22	1.35	0.985	0.611	quarz
Quiou	2	2.716	0.75	4.40	1.281	0.831	carbonate
Catania	-	2.683	0.24	2.20	0.850	0.592	silica
HK CLK	1.6	2.652	1.37	4.97	0.682	0.411	silica

where: FC=Fine Content; G_s =specific gravity; D_{50}=mean grain size; U_c=coefficiente of uniformity e_{max} e_{min} = maximum and minimum void ratio

A total of 52 tests were performed on dry, reconstituted hollow cylindrical specimens of three different kinds of sands [Toyoura (TS), Quiou (QS) and Catania (CS)] with an external diameter of 70 mm, an internal diameter of 50 mm and a height of 140 mm. Table 2 summarises the main characteristics of the tested sands.

MLTST, CLTST and RCT were performed on these sands. MLTST and CLTST were performed on virgin specimens, whilst RCT were performed on the same specimens previously subjected to CLTST or MLTST. CLTST were performed under stress control with frequencies equal to f=0.01, 0.1 and 1 Hz.

Seismic tests in CC have been performed on 7 different granular soils. Only the results obtained in the case of HK-CLK sand are presented in this paper. The characteristics of this sand are listed in Table 2.

4. EXPERIMENTAL RESULTS.

Laboratory tests
The damping ratio is hardly influenced by N at very small strains. The influence of N on D of dry sands becomes more and more relevant as the shear strain level increases. The decrease of D with an increase of N is shown in Fig. 2, in the case of a virgin specimen of QS (i.e. a specimen not subjected to any prior stress-strain history). Results of a stepwise test on QS are also shown in the same Figure. In both cases an initial single amplitude shear strain of about 0.04 % was imposed on the specimens. In the case of the virgin specimen, γ decreased from 0.044 % to about 0.028 % after 30 cycles, whilst in the case of the stepwise test the shear strain remained more or less constant and equal to 0.041 %. It is

possible to see that the damping ratio is strongly influenced by N in both cases, even though the effects of N on D are more evident in the case of the virgin specimen, especially during the first cycles. Moreover, even after hundreds of cycles D still decreases. The D values, obtained for the same shear strain level from the RCT performed after the CLTST, are also plotted in Fig. 2. It is possible to see that this values matches the trend of the CLTST results.

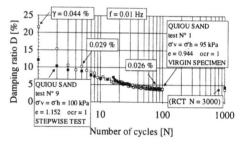

Figure 2 Damping ratio vs. Number of loading cycles (Quiou sand)

The damping values obtained from RCT and CLTST are compared in Fig. 3 for TS. As far as CLTST are concerned, the damping ratio values plotted in Fig. 3 are those obtained after 20 or 30 cycles when the variation of D with N becomes relatively important. It is possible to see that, at very small strains, the damping ratio obtained from CLTST is almost zero, whilst that from RCT is about 2 %. Moreover for a given strain level, the damping ratio from RCT is greater than that from CLTST in the small strain range, whilst, at larger strains, the opposite is true. Similar results were obtained for QS and for CS (see also Cascone 1996). The data shown by Papa et al. (1988) and by Stokoe et al. (1995) are in good agreement with those of Fig. 3. At large strains, the effect of N on D can explain the differences between the results of CLTST and those of RCT. According to Stokoe et al. (1995) the differences between the values of D from CLTST and those from RCT, in the small strain interval, could be a consequence of the equipment-generated damping in RCT. Infact, according to Stokoe et al. (1995) the D values obtained with a RC apparatus should be corrected by subtracting the equipment-generated damping (D_{app}) which is a frequency dependent parameter. The D_{app}-f calibration curve is, of course, different for each apparatus (Stokoe et al. 1995). However, considering that the apparatus available at the authors' laboratory is quite similar to that used by Stokoe et al. (1995), it is acceptable to use their

D_{app}-f calibration curve to correct the experimental measurements of D. This calibration curve indicates that, for the CLTST with f<1 [Hz], $D_{app} \cong 0$ %, while, in the case of the RCT (f≅80-100 [Hz]), $D_{app} \cong 0.5$ %. Such corrections are not capable of explaining the differences observed between CLTST and RCT results.

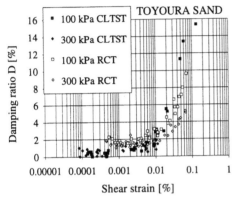

Figure 3 Damping ratio of Toyoura sand vs. Shear strain from CLTST and RCT

The rate effect, due to the viscosity of the pore fluid (not necessarily water, Kokusho 1987), could be considered as a possible explanation of the fact that, at small strains, the D values from RCT are larger than those obtained from CLTST. Fig. 4 shows the increase of D of QS with an increase of the loading frequency for three different strain levels no greater than 0.006%.

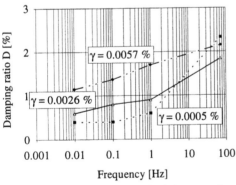

Figure 4 Damping ratio of Quiou sand vs. Frequency from CLTST and RCT

A comparison between the damping ratio values obtained from RCT and those obtained from CLTST is shown in Fig. 5 for Augusta clay. It is possible to see that the damping ratio from CLTST, at very small strains, is equal to about 2 %. Greater

values of D are obtained from RCT for the whole strain interval investigated. The correction for equipment-generated damping (D_{app}) is more important in this case than for sand specimens. However, also in this case, such a correction is not sufficient to explain the differences between CLTST and RCT results.

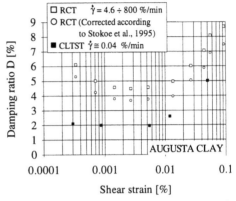

Figure 5 Damping ratio of Augusta clay from CLTST and RCT tests

A comprehensive comparison between the results of CLTST and RCT, performed on Pisa and Augusta clays, is shown in Fig. 6. This comparison confirms what can be observed in Figs. 3 and 5.

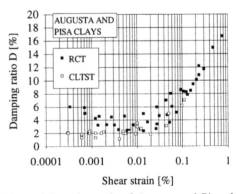

Figure 6 Damping ratio of Augusta and Pisa clays vs. Shear strain

Considering that the influence of N on D has been found to be negligible in the case of clayey soils for strain levels of less than 0.1 %, it has been supposed that RCT provide larger values of D than CLTST because of the rate (frequency) effect, in agreement with data shown by Shibuya et al. (1995) and Tatsuoka et al. (1995). According to these researchers the nature of soil damping in soils can be linked to the following phenomena:

- Non-linearity which governs the so called hysteretic damping controlled by the current shear strain level. This kind of material damping is absent or negligible at very small strains.
- Viscosity of the soil skeleton (creep) which is relevant at very small strain rates.
- Viscosity of the pore fluid which is relevant at very high frequencies.

Soil damping at very small strains is mainly due to the viscosity of the soil skeleton or of the pore fluid, depending on the strain rates or frequencies. Moreover, according to Tatsuoka and Kohata (1995) and Tatsuoka et al. (1995) a partial drainage condition can provide very high values of the damping ratio. Shibuya et al. (1995) indicate that, for a given strain level, the damping ratio of cohesive soils increases when the loading frequency is smaller than 0.1 Hz (because of the creep effects), is more or less constant for loading frequency between 0.1 and 10 Hz (non linearity is dominant) and increases for frequencies greater than 10 Hz (because of pore fluid viscosity).

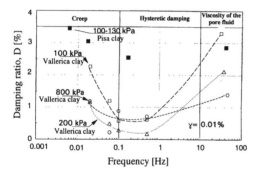

Figure 7 Damping ratio vs. Frequency (adapted from d'Onofrio, 1996)

Fig. 7 shows the damping ratio of Vallericca clay (Italy) vs. the frequency, for a strain level of 0.01 % and consolidation pressure of between 100 and 800 kPa. This data was obtained by d'Onofrio 1996. The considered soil is a stiff, highly overconsolidated, Pliopleistocene, marine clay with a PI of about 26 %. The data of the Pisa clay, obtained in this research, has been reported in the same Figure. The trend of the whole data is in good agreement with the findings of Shibuya et al. (1995).

Similar results have been obtained by Santucci (1996) from triaxial tests performed on reconstituted silty and clayey sand form Metramo (Italy) with a 5 % of bentonite and by Olivares (1996) from triaxial tests on the shale clay from Bisaccia (Italy).

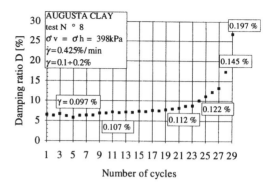

Figure 8a Damping ratio vs. Number of cycles

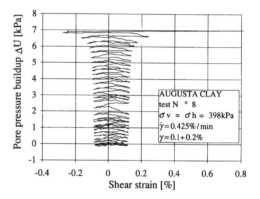

Figure 8b Pore pressure buildup

Fig. 8a and 8b show, respectively, the increase of D and pore pressure with an increase of N because of cyclic material degradation as obtained from a CLTST on Augusta clay. In this case, D from CLTST can become higher than that obtained from RCT. It is not easy, from an experimental point of view, to observe the degradation phenomena during RCT. However, if it is considered acceptable to assume the accumulated pore pressure increase as an index of material degradation, the data shown in Fig. 9 clearly indicates that such phenomena have more relevance in CLTST than in RCT. In particular the occurrence of material degradation seems to depend on the shear strain level and on the loading rate. In the case of Augusta clay with PI=38 %, the threshold strain beyond which material degradation occurs during CLTST is in between 0.05 and 0.1 % which is in agreement with other available data (Lo Presti 1989).

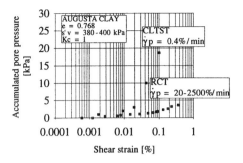

Figure 9 Accumulated pore pressure vs. Shear strain

Calibration Chamber results
Examples of the damping assessment from the spectral slope and spectral ratio methods, in the case of HK-CLK sand, are shown in figs. 10a and 10b respectively. The values of D obtained for S_{hv} waves by means of the Spectral Ratio and Spectral Slope methods are plotted in Fig. 11 vs. the effective consolidation stresses acting along the directions of wave propagation and particle motion..

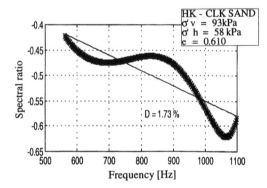

Figure 10a Damping ratio determination with the spectral slope method

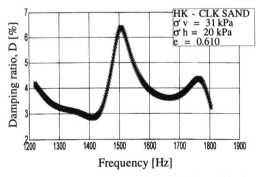

Figure 10b Damping ratio determination with the spectral ratio method

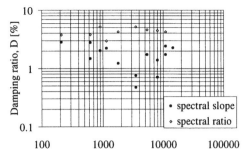

Figure 11 Damping ratio of S_{hv} waves from the spectral ratio and spectral slope methods

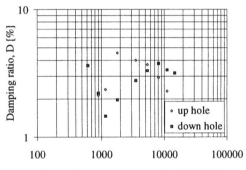

Figure 12 Damping ratio of S_{vh} waves from the spectral slope method

It is possible to see that D decreases with an increase of the consolidation stresses, as expected, but the obtained values are greater than those usually determined in the laboratory which are of less than 2 % (Tatsuoka et al. 1995, Jamiolkowski et al. 1995, Stewart and Campanella 1991). In particular the largest values have been obtained with the Spectral Ratio method. In Fig. 12 the D values obtained in the case of S_{vh} waves by means of the Spectral Slope method are shown. In this case, the decrease of D with consolidation stresses is not very clear for the S_{vh} waves propagated from top to bottom of the CC (down hole). The obtained D values are similar to those of S_{hv} waves.

The high values of damping ratio obtained from seismic tests in CC could be a consequence of the fact that the waves are propagated at very high frequencies (1000-2000 Hz) Therefore, the observed very high dissipation could be due to the viscosity of the pore fluid (water in this case).

It was attempted to estimate the maximum shear strain level occurring during wave propagation by means of the following eq:

$$\gamma = \frac{\dot{u}^s_{peak}}{V_s} \qquad (3)$$

where: $\dot{u}^s_{peak}$=0-peak velocity of particles measured by geophones during propagation of S_{hv} and S_{vh} waves; V_s=propagation velocity of S_{hv} and S_{vh} waves.

According to eq. (1) the shear strain resulted to be equal to $1-2\cdot10^{-4}$ %. Considering that the waveforms have frequencies of about 1000-2000 Hz, the equivalent strain rates for seismic tests is of about 20-40%/min. These values are similar to those employed in RCT at small strains.

Model for Damping
In the case of one-dimensional loading conditions, it is common practice to use simple non linear models which generate the unloading and reloading branches of the stress-strain relationship by the knowledge of the parameters defining the backbone (or skeleton) curve and by means of second the Masing rule (Masing 1926). In particular this rule generates a loop by enlarging the backbone curve by a factor of two in stresses and strains. Therefore, the second Masing rule automatically defines damping once the parameter of the backbone curve are known. In particular, when using the Ramberg Osgood (1943) relationship for the backbone curve, the damping ratio results to be function of the R exponent of the Ramberg Osgood equation according to the following expression:

$$D = \frac{2}{\pi} \cdot \frac{R-1}{R+1}\left(1 - \frac{G}{G_o}\right) \qquad (4)$$

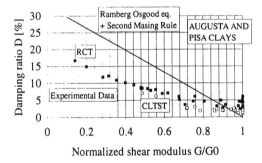

Figure 13 Experimentally determined and predicted damping ratio of Pisa and Augusta clays

The Ramberg Osgood parameters have been computed for Augusta clay using the data of RCT. In particular the exponent R resulted to be equal to 3.16. Eq. (4), assuming R=3.16, is plotted in Fig. 13. In the same Figure the experimentally determined damping values are also plotted. It is possible to see that Ramberg Osgood eq. and the second Masing rule generate too large damping values, at large strains, while at small strains, the considered model assume zero damping which is also unrealistic. Moreover, the supposed linearity of the relationship D-G/Go is not perfectly verified by the experimental data of Figure 13.

5. CONCLUSIONS

On the basis of the experimental results obtained in this research for dry reconstituted sands and natural saturated clays it is possible to draw the following conclusions:

- damping ratio at small strains is not equal to zero for both sands and clays;
- larger values of the damping ratio are observed for clays than for sands, at small strains;
- the damping ratio of sands decreases with an increase of the number of loading cycles. This influence is negligible at small strains and becomes more and more important with increasing γ;
- the damping ratio of clays is affected in a negligible way by N until material degradation does not occur;
- degradation phenomena have been observed for shear strain levels of between 0.05 to 0.1 %; the material degradation also depend on the strain rate as it is more relevant at lower strain rates;
- the experimental results have shown the dependence of the damping ratio of sands at small strains on the frequency; the increase of D with f is probably due to the viscosity of the pore fluid (air);
- the damping ratio of clays resulted to be a complex parameter that is influenced in a measurable way by creep, soil non linearity and the viscosity of the pore fluid in accordance with other data available in literature;
- the damping ratio of HK-CLK sand, obtained from seismic tests resulted to be greater than those usually obtained in the laboratory for similar soils;
- the Ramberg Osgood eq. and the second Masing rule generate very high values of D in comparison to those experimentally determined.

Due to the high variability of D, especially with frequency, it is extremely important to choose the most appropriate value of damping in relation to the examined problem when computing the seismic response of soil deposits by means of linear equivalent methods.

REFERENCES

Ashmawy A.K., Salgado R. Guha S. and Drnevich V.P., 1995 Soil damping and its use in dynamic analyses. *3rd Int. Conf. on Recent Advances in Geotechnical Earthquake Engineering and Soil Dynamics*. Volume 1, pp.35-42.

Bellotti R., Jamiolkowski M, Lo Presti D.C.F. and D.A. Neill. 1996 Anisotropy of small strain stiffness in Ticino sand *Geotechnique* 46, No.1, pp. 115-131.

Cascone E. 1997 Analisi sperimentale e modellazione del comportamento dinamico dei muri di sostegno. *Ph. D. Thesis*, University of Catania.

Cavallaro A. 1997, Influenza della velocità di deformazione sul modulo di taglio e sullo smorzamento delle argille. *Ph. D. Thesis*, University of Catania

Costanzo D., Jamiolkowski M., Lancellotta R. and Pepe M.C. 1994, Leaning Tower of Pisa - Description of the Behaviour. Invited Lecture - *Proceedings Settlement '94*, Texas A&M University

d'Onofrio A. 1996 Comportamento Meccanico dell'argilla di Vallericca in condizioni lontane dalla rottura, *Ph. D. Thesis*, University of Naples, Department of Geotechnical Engineering

Fioravante V., Jamiolkowski M., Lo Presti D.C.F., Manfredini G. and Pedroni S. 1997 "Coefficient of the earth pressure at rest on cohesionless soils from in situ tests" *submitted to Geotechnique* Symposium in Print

Fuhriman, M.D., 1993, Cross-Hole Seismic Tests at two Northern California Sites Affected by the 1989 Loma Prieta Earthquake, *M. Sc. Thesis*, The University of Texas at Austin.

Jamiolkowski M., Lo Presti D.C.F. and Pallara O. 1995 Role of In-Situ Testing in Geotechnical earthquake Engineering, *3rd International Conference on Recent Advances in Geotechnical Earthquake Engineering and Soil Dynamic*, State of the Art 7, Vol. 3, pp. 1523-1546.

Khawaja, A.S., 1993, Damping Ratios from Compression and Shear Wave Measurements in the Large Scale Triaxial Chamber, *M. Sc Thesis*, The University of Texas at Austin.

Kokusho, T. (1987) In-situ Dynamic Soil Properties and Their Evaluation. Proc. 8th Asian regional Conference on SMFE, Vol. 2, pp. 215-340.

La Rochelle, P., Sarrailh, J., Tavenas, F. and Leroueil, S. 1981, Causes of Sampling Disturbance and Design of a New Sampler for Sensitive Soils. *Canadian Geotechnical Journal*, Vol. 18, pp. 52-66.

Lo Presti D.C.F. 1989 Proprietà Dinamiche dei Terreni," *XIV Conferenza Geotecnica di Torino, Dipartimento di Ingegneria Strutturale*, Poltecnico di Torino, 62 pp.

Lo Presti, D.C.F., Pallara, O., Lancellotta, R., Armandi, M. and Maniscalco, R. 1993 Monotonic and Cyclic loading Behaviour of Two Sands at Small Strains. *Geotechnical Testing Journal* 16, No. 4, 409-424.

Mancuso, C., 1994, Damping of Soil by Cross-Hole Method, *Proc. of XIII ICSMFE*, New Delhi, India, Vol. 3, pp. 1337-1340.

Masing G. 1926, Eigenspannungen und verfestigung beim messing. *Proc. 2nd International Congress of Applied Mechanics*, Zurich, Swisse. (in German)

Maugeri M., Castelli F. and Motta E. 1994, Pile Foundation Performance of an Earthquake Damaged Building, Proc. of the Italian-French Symposium on Strengthening and Repair of Structures in Seismic Areas, Nice, France

Mok, Y.J., (1987), "Analytical and Experimental Studies of Borehole Seismic Methods," Ph.D. Thesis, Univ. of Texas at Austin, Austin, TX.

Olivares L. 1996, Caratterizzazione dell'argilla di Bisaccia in condizioni monotone, cicliche e dinamiche e riflessi sul comportamento del Colle a seguito del terremoto del 1980. Ph. D. Thesis, University of Naples, Department of Geotechnical Engineering).

Pallara, O. (1995) Comportamento Sforzi-Deformazioni di Due Sabbie Soggette a Sollecitazioni Monotone e Cicliche. Ph. D. Thesis, Department of Structural Engineering, Politecnico di Torino, (in Italian)

Papa V., Silvestri F. and Vinale F. 1988 Analisi delle Proprietà di un Tipico Terreno Piroclastico mediante Prove di taglio Semplice. *Proc. Gruppo Nazionale di Coordinamento per gli Studi di Ingegneria Geotecnica*. Monselice, Italy, Vol. 1, pp. 265-286.

Ramberg W. and Osgood W.T. 1943. Description of Stress-Strain Curves by Three Parameters. *Tech Note 902*, National Advisory Committee for Aeronautics.

Redpath, B.B., Edwards, R.B., Hale, R.J. and Kintzer, F.C., 1982, Development of Field Techniques to Measure Damping Values for Near Surface Rocks and Soils, *Prepared for NSF* Grant No. PFR-7900192.

Santucci de Magistris F., 1996, Comportamento di un limo sabbioso ed argilloso costipato ed addizionato con bentonite, *Ph. D Thesis*, University of Naples, department of Geotechnical Engineering.

Shibuya S., Mitachi T., Fukuda F. and Degoshi T. 1995 Strain Rate Effect on Shear Modulus and Damping of Normally Consolidated Clay. *Geotechnical Testing Journal* 18:3, 365-375.

Stewart, W.P. and Campanella, R.G., 1991, In Situ Measurement of Damping of Soils, *Proc. of Second International Conference on Recent Advances in Geotechnical Earthquake Engineering and Soil Dynamics*, St. Louis, Missouri, Vol. I, pp. 83-92.

Stokoe K.H. II, Hwang S.K. Lee J.N.-K and Andrus R. 1995 Effects of Various Parameters on the Stiffness and Damping of Soils at Small to Medium Strains. *Keynote Lecture 2, IS Hokkaido 1994* 2, 785-816.

Tatsuoka, F. and Shibuya, S. 1992 Deformation Characteristics of Soil and Rocks from Field and Laboratory Tests. *Keynote Lecture, Proc. IX Asian Conference on SMFE*, Bangkok, 1991 2, 101-190.

Tatsuoka F. and Kohata Y. 1995 Stiffness of Hard Soils and Soft Rocks in Engineering Applications. Keynote Lecture 8, *IS Hokkaido 1994* 2, 947-1066.

Tatsuoka F., Lo Presti D.C.F. and Kohata Y. 1995 Deformation Characteristics of Soils and Soft Rocks Under Monotonic and Cyclic Loads and Their Relations. *3rd International Conference on Recent Advances in Geotechnical Earthquake Engineering and Soil Dynamic*, State of the Art 1, 2, 851-879.

TC4 (Technical Committee for Earthquake Geotechnical Engineering, ISSMFE) 1993 Manual for Zonation on Seismic Geotechnical Hazards, *The Japanese Society for Soil Mechanics and Foundation Engineering,* Tokyo, Japan.

Toki, S., Shibuya S. and Yamashita S. 1995 Standardization of Laboratory Test Methods to Determine the Cyclic Deformation properties of Geomaterials in Japan. *Keynote Lecture 1, IS Hokkaido 1994* 2, 741-784.

Author index